Fundamentals of Engineering Geology

Fundamentals of Engineering Geology

F G Bell

Butterworths
London · Boston · Durban · Singapore · Sydney · Toronto · Wellington

First published 1983

© Butterworth & Co. (Publishers) Ltd 1983

British Library Cataloguing in Publication Data
Bell, F. G.
  Fundamentals of engineering geology.
  1. Geology
  I. Title
  550'.246231   QE26.2

  ISBN 0-408-01169-6

Filmset by Northumberland Press Ltd, Gateshead, Tyne and Wear
Printed in England by Butler & Tanner Ltd, Frome and London

Contents

Foreword

Few of today's technical problems can be solved merely by the application of the disciplines of a single field of knowledge. In engineering construction, in particular, the interface between structural loadings imposed and the capacity of the geological conditions to support these loadings must be more carefully studied than was necessary only a few decades ago. Sites which have been previously unloaded are hard to find. The relationship between the nature of the geological support and its concomitant loading becomes of increasing significance.

This volume considers the wider view by describing—in welcome detail—the changes which may take place in various geological sites. Dr Bell emphasises this aspect by interleaving geomorphological processes between studies of general geological conditions and descriptions of tests of soils and rocks. Such an attack underlines the need for a study of the indicators presented to the Engineer by the geological history of the site. Aerial photographic site investigation can be used to depict not only the surface of the ground, but also sub-surface features with evidences of past and possible future movements and erosions. Such surveys give confidence that the link between loading and the reaction of the site can be predicted. For example, warning of possible movement in periglacial material, left at the edge of the last melting ice sheet of 10 000 years ago, may be given.

This volume adds to the series of excellent treatises, which Dr Bell has either written or edited, on the characteristics of the support given by soils and rocks to engineering construction. He is to be congratulated on his latest thorough study and on the masterly linkages he has achieved between geology, geotechnics, rock and soil mechanics and foundation design.

W. Fisher Cassie
Emeritus professor of civil engineering
University of Newcastle upon Tyne

Preface

This text concerns itself with the basic principles or fundamentals of engineering geology. It therefore deliberately avoids those applied aspects of the subject such as site investigation, geophysical exploration, etc. One of the reasons for this is that the text has been written as a companion volume to the author's *Engineering Geology and Geotechnics* (Butterworth, 1980) which deals with the applied aspects of engineering geology. A more obvious reason for writing this book is that there is no other which covers the subject matter in such depth. Yet another reason for this emphasis is the pace at which geotechnical theory has developed. Indeed, most of the current journals in engineering geology and geotechnical subjects have made their appearance within the last twenty years. This pace of change brings with it the need for a textbook which surveys the relevant advances and in so doing allows the student or professional engineer to avoid being overcome by too much information. In this context particular attention has been given to that work done by the International Association of Engineering Geology, the International Society of Rock Mechanics and the Engineering Group of the Geological Society of London which have attempted to systematise and standardise engineering geology.

The book is aimed at those who will be or who are in some way or other involved with ground conditions. In other words it is primarily written for engineering geologists, civil engineers and mining engineers. It is hoped that the text will be bought by the student and that it will continue to be of value to him throughout his subsequent professional career. This does not mean to infer that the book was written solely or mainly with the student in mind. Indeed from the point of view of the student in civil or mining engineering it could be argued that this text contains more detail than he will require in his initial studies. If this is true, then no apology is made since the text also has to consider the needs of students of engineering geology. In addition, if the book is to be of service to the practising engineer, then it must contain more detail than general texts of engineering geology written for students. With this in mind, numerous references have been provided at the end of each chapter.

Nevertheless obtaining a happy balance when writing a text for both the student and professional who are being or have been educated in three different subjects is not the simplest of tasks. It is a continuous compromise and one is acutely aware that every academic and professional engineer in this field has his own opinions as to what should or should not be emphasised, etc. The author is, of course, entitled to his own opinions and it is hoped that at least some of these will coincide with those of others.

Engineering geology can be simply defined as the application of geology to engineering practice. As such it draws on several geological disciplines such as petrology, sedimentology, stratigraphy, structural geology and geomorphology, as well as hydrology. Accordingly, the applied or engineering aspects of these disciplines constitute the bulk of the text. In addition, some of the pioneers of engineering geology, notably Terzaghi, regarded soil and rock mechanics as integral parts of engineering geology. Although soil and rock mechanics have now developed into independent disciplines, they still form part of engineering geology in that anyone concerned with ground conditions must appreciate how these materials behave. It is for this reason that a sizeable part of the text is devoted to these two disciplines. Admittedly civil engineers and mining engineers need more soil mechanics and rock mechanics respectively than contained herein, but generally speaking these particular chapters should satisfy the requirements of engineering geologists. Hopefully the chapters concerned will also provide a summary for those civil and mining engineers who may require a quick reminder of such subject matter.

In writing the text the author assumed that many student readers would have little basic knowledge of geology. He is aware that geology, like all other sciences, has its own language and that this sometimes presents engineering students with difficulties. Although geological terminology inevitably occurs in the first few chapters in particular, taking the book as a whole, it is kept at a low level. It certainly should not be beyond the wit of any undergraduate, ultimately worthy of a degree, to cope in this respect.

Many concerns have supplied material, especially illustrations, and due acknowledgement is given in the text. The author wishes to offer his thanks to all those involved. All extracts from BS 5930:1981 are reproduced by permission of the British Standards Institution, 2 Park Street, London W1A 2BS, from whom complete copies of the standard may be obtained. Thanks are also due to Dr A. C. Waltham of Trent Polytechnic who kindly supplied the cover photograph. If any person or concern has inadvertently not been afforded due acknowledgement, then apologies are given. In particular, the author would like to record his grateful appreciation for the help given by Professor W. F. Cassie, CBE, who meticulously ploughed his way through the manuscript and subsequently provided much useful advice. Lastly, but by no means least, one must thank Pauline Marchant of Butterworths whose endeavour during sub-editing was commendable.

F.G.B.
1983

Chapter 1

Igneous rocks

1.1 Introduction

A rock is formed of one or more minerals. The essential minerals, frequently not more than three in number, account for 95% or more of the volume of a rock, the accessory minerals comprising the rest. Obviously the character of its essential minerals influences the properties of a rock. The texture, that is, the way in which minerals are arranged in a rock, also influences its properties.

Rocks are divided according to their origin, into three groups, namely, the igneous, metamorphic and sedimentary rocks. The origin of a rock determines its composition and texture. As pointed out above, these two factors influence its mechanical behaviour.

Igneous rocks are formed when hot molten rock material called *magma* solidifies. Magmas are developed either within or beneath the Earth's crust, that is, in the mantle. They comprise hot solutions of several liquid phases, the most conspicuous of which is invariably a complex silicate phase. Thus the igneous rocks, which according to Clarke and Washington (1924) form 95% of the Earth's crust (a figure which obviously includes the metamorphic rocks), are principally composed of silicate minerals. Indeed the nonsilicates total less than 1% of these rocks. Furthermore, of the silicate minerals, six families, the olivines, the pyroxenes (e.g. augite), the amphiboles (e.g. hornblende), the micas (e.g. biotite or muscovite), the feldspars (e.g. orthoclase, albite or anorthite) and the silica minerals (e.g. quartz), are quantitatively by far the most important constituents. Figure 1.1 shows the approximate distribution of these minerals in the commonest igneous rocks. (See Appendix 1 for composition and physical properties of the above minerals.)

Water and volatiles are held in solution in magmas by high pressure and they influence the behaviour of magmas, notably their crystallisation. Magmas are not necessarily completely molten, for they may contain a certain amount of growing crystals. Nevertheless, if a magma is to be intruded along fissures or erupted at the surface it must not become too choked with solid matter.

The magmas which are generated when melting occurs in the mantle or crust are named *primary magmas*. They represent the parent material from which

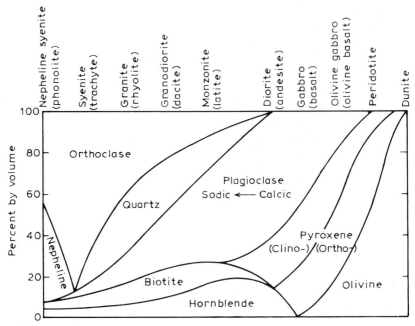

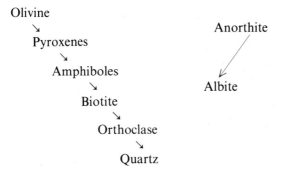

Figure 1.1 Approximate mineralogical composition of the commoner types of igneous rocks (plutonic types without brackets; volcanic equivalents in brackets)

secondary or *derived magmas* may arise due to differentiation or contamination.

Differentiation is brought about due to the fact that different minerals crystallise at different temperatures. Thus the following order of crystallisation was distinguished by Bowen (1928).

Olivine
  ↘
  Pyroxenes   Anorthite
    ↘   ╱
    Amphiboles   ╱
      ↘   ↙
      Biotite   Albite
        ↘
        Orthoclase
          ↘
          Quartz

The order indicates that when those minerals which crystallise at high temperatures have formed, the composition of the remaining magma is changed. This process, known as fractional crystallisation, can produce different types of rock from the original magma.

A magma becomes contaminated when country rock is incorporated into it. This can alter its composition. Evidence of contamination is exhibited, for example, by the presence of xenoliths (Figure 1.2). These are fragments of the country rock which have not been completely assimilated by the host magma.

Figure 1.2 Xenolith in block of Shap adamellite, Shap, Cumbria

It would appear that most granite igneous rocks are developed by other processes, namely, by granitisation and anatexis (see Marmo (1971)). Read (1948) defined grantisation as the process by which solid rocks are converted to rocks of granitic character without passing through a magmatic stage. Anatectic processes, which lead to the remelting of rocks, were not included within granitisation. Rocks formed from remelted material frequently have a mixed or hybrid appearance and were termed migmatites by Sederholm (1967); (see also Mehnert (1968)). From his experiments on argillaceous rocks and greywackes, Winkler (1967) concluded that their anatexis would lead to the formation of granitic and granodioritic magmas.

Igneous rocks may be divided into intrusive and extrusive types according to their mode of occurrence. In the former type, the magma crystallises within the Earth's crust, whereas in the latter it solidifies at the surface, having been erupted as lavas and/or pyroclasts from a volcano. The intrusions may be further subdivided on a basis of their size, into major and minor categories; the former are developed in a plutonic, the latter in a hypabyssal environment. About 95% of the plutonic intrusions have a granite–granodiorite composition and basaltic rocks account for approximately 98% of the extrusives (see Barth (1962)).

1.2 Igneous intrusions

The form which intrusions adopt may be influenced by the structure of the country rocks. This applies particularly to intrusions of small size.

1.2.1 Minor intrusions

Dykes are discordant igneous intrusions, that is, they traverse their host rocks at an angle and are steeply dipping (Figure 1.3). As a consequence their surface outcrop is little affected by topography and, in fact, they usually strike in a straight course. Dykes range in width up to several tens of metres but their average width is measured in a few metres. The length of their surface outcrop also varies, for example, the Cleveland dyke in the north of England can be traced over some 200 km. Conversely, some of the dykes on the Isle of Arran are only a metre or so wide and can be traced for a few hundred metres. Dykelets may extend from and run parallel to large dykes, irregular offshoots may also branch away from large dykes. Dykes do not usually have an upward termination, although they may act as feeders for lava flows and sills. However, on occasions they fail to penetrate massive rock horizons.

Figure 1.3 Basalt dykes (courtesy the Institute of Geological Sciences)

Dykes often occur along faults, which provide a natural path of escape for the injected magma. Nevertheless, some dykes have to force open fractures in the Earth's crust, hence they are most common in regions which have suffered crustal tension. Since their length of surface outcrop is great in comparison with their width it can be inferred that dykes are injected rapidly and that their parent magmas are mobile.

Most dykes are of basaltic composition (see Frankel (1967)). However, dykes may be multiple or composite. Multiple dykes are formed by two or more injections of the same material which occur at different times so that the different phases are distinctly discernible. A composite dyke involves two or more injections of magma of different composition.

An ideal ring dyke when exposed by erosion would exhibit a ring-shaped outcrop (Figure 1.4), but oval or arcuate intrusions which are localised in extent and do not form a complete ring are also regarded as ring dykes (see Richey (1932)). All ring dykes dip steeply outward from their centres. They are usually of uniform composition, being frequently formed of acidic rocks.

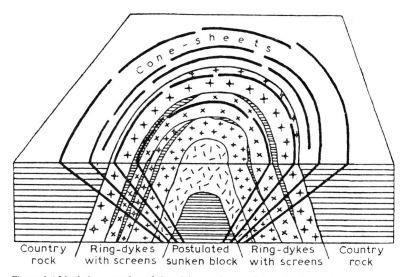

Figure 1.4 Ideal ring complex of ring-dykes and cone-sheets, in plan and section

Cone sheets, like ring dykes, are arcuate in outcrop and surround a central area free of intrusions (Figure 1.4). However, they are inclined inwards towards a central focus and are often formed of basic igneous material. Like ring dykes, they frequently occur in series. Nearer the centre of a series the cone sheets dip more steeply, for example, angles up to 70° have been recorded, whilst the outer sheets may dip at about 30°. Cone sheets are not as common as ring dykes, in fact they are quite rare.

Sills, like dykes, are comparatively thin, parallel-sided igneous intrusions which occur over relatively extensive areas. Their thickness varies up to several hundred metres. However, unlike dykes, they are injected in an approximately horizontal direction although their attitude may be subsequently altered by folding. When sills form in a series of sedimentary rocks the magma is injected along bedding planes (Figure 1.5). Nevertheless, an individual sill may transgress upwards from one horizon to another. Because sills are intruded along bedding planes they are said to be concordant and their outcrop is similar to that of the country rocks. Sills may be fed from

Figure 1.5 Composite sill (basalt and quartz porphyry) at Drumadoon Point, Isle of Arran

dykes and small dykes may arise from sills. Such arrangements depend largely upon the path of easiest escape for the parent magma. Most sills are composed of basic igneous material and the magma must have been hot and mobile when intruded. Sills may be multiple or composite in character.

Laccoliths are mushroom-shaped intrusions (Figure 1.6) which up-dome the stratified rocks into which they are injected (see Gilbert (1877)). However, the base of a laccolith and the conduit by which it is fed, are very rarely exposed. Accordingly their existence is largely based on inference. Laccoliths have an approximately circular plan which may reach a diameter of about 8 km. In thickness they range from a few to several hundred metres.

Phaccoliths are concordant intrusions found in folded terrains where they occupy the crests of anticlines or the troughs of synclines.

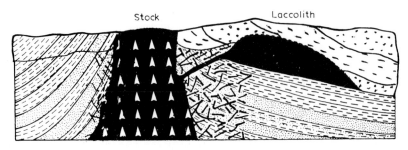

Figure 1.6 Section through a laccolith showing its relationship to an adjacent stock

1.2.2 Major intrusions

Lopoliths are large concordant intrusions, generally gabbroic in composition, which are convex downwards and may be concave upwards, that is, basin-like in shape. They may measure hundreds of kilometres in diameter and their thickness runs into thousands of metres. In fact, the thickness is usually between one-tenth and one-twentieth the diameter. Lopoliths form some of the largest basic intrusions in the world. The type example is the Duluth intrusion (see Grout (1918)). This feature has a diameter of 240 km and extends over approximately 38 000 km². Its estimated thickness is in the order of 15 000 m and its volume is believed to total some 200 000 km³. The basin shape has been attributed to sagging within the Earth's crust.

The major intrusions include batholiths, stocks and bosses. Batholiths are very large in size, and are generally of granitic or granodioritic composition. Indeed many batholiths have an immense surface exposure. For instance, the Coast Range batholith of Alaska and British Columbia is over 1000 km in length by approximately 130 to 190 km in width. Batholiths are associated with orogenic regions. They appear to have no visible base and their contacts are well defined and dip steeply outwards (see Daly (1933)). However, some granitic batholiths appear to be made up of composite irregular sheets. They are more or less stratified and are not bottomless. Ranguin (1965) termed such features stratiform granitic massifs. He regarded them as large concordant sheets which occur along major structures of the regional tectonic plan. Bosses are distinguished from stocks in that they have a circular outcrop. Their surface exposures are of limited size, frequently less than 100 km². They may represent upward extensions from deep-seated batholiths. Like batholiths, they are generally composed of granitic–granodioritic rocks.

Subvolcanic massifs are small features, usually less than 24 km in diameter, which are found in volcanic areas. These massifs are often circular or oval in outcrop, and they may be of composite character. For example, the northern granite of the Isle of Arran consists of an outer ring of coarse-grained granite which encloses a central core of fine-grained granite. Indications of vertical displacement have frequently been observed about these subvolcanic massifs, witness the effects of sinking which have been noted in the Scottish complexes. Clough *et al.* (1909) explained the Loch Etive granite complex (Devonian) as having formed as a result of sinking *en masse* of a cylindrical block of the crust, the void so produced being filled with granitic magma. A repetition of the sinking process caused further descent of the massif relative to the adjacent rocks and successive intrusion of granitic material, which adopted a concentric arrangement.

1.2.3 Structures associated with plutonic intrusions

In the 1920s Hans Cloos pioneered the study of megastructures in plutonic intrusions (see Balk (1938)). These structures are best developed around the margins of granite massifs. It was assumed that variation in the viscosity of a magma consequent upon changes in stress, temperature and frictional resistance caused differences in the speed of its movement and led to the development of stream-lines. As a result particles of elongate habit are aligned with their long axes parallel to the direction of stream flow. This

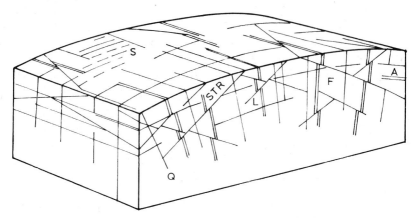

Figure 1.7 Block diagram showing the types structures in a batholith. Q = cross joints; S = longitudinal joints; L = flat-lying joints; STR = planes of stretching; F = linear flow structures; A = aplite dykes (after Balk (1938))

gives rise to linear flow structures (Figure 1.7). Where particles of platey habit or schlieren, that is, layers which possess the same minerals as the rock itself but in different proportions, are orientated parallel to one another, this is described as platey flow structure. The latter structure is developed parallel to the contacts since they exert frictional effects upon the moving magma. A rock may be referred to as lineated or foliated if it exhibits linear flow or platey flow structures respectively.

Cloos (1923) suggested that most joints and minor faults in batholiths were caused by the continuing activity of the stresses which were responsible for emplacement and that these features possessed a definite relationship with the shape of the intrusion. Fractures are first developed in the solidified margins of plutonic masses and may be filled with material from the still liquid interior. Cloos distinguished four types of joints (Figure 1.7). Cross joints or Q joints lie at right angles to the flow lines. They tend to radiate from the centre of the dome or arch (the term *dome* is used to describe a massif if the flow structures extend over the entire massif, whereas if they are absent from its centre the arrangement is termed an *arch*). Joints which strike parallel to the flow lines and are steeply dipping are known as longitudinal or S joints. Pegmatites or aplites may be injected along both types of joints mentioned. Diagonal joints are orientated at 45° to the direction of the flow lines. Flat-lying joints may be developed during or after emplacement of the intrusion and they may be distinguished as primary and secondary, respectively. Normal faults and thrusts occur in the marginal zones of large intrusions and the adjacent country rocks. The thrusts are arranged *en echelon* and the displacement along them is usually measured in a few centimetres. They are numerous. The angle at which a marginal thrust dips is dependent upon its position in relation to the intrusion, for example, against steep contacts the dip is low whilst it is higher in the roof zone. Minor sets of pinnate shear joints may be found along marginal thrusts. Flat-lying normal faults form as a result of tension developed parallel to the flow lines. They are generally restricted to the upper parts of a massif.

1.3 Volcanicity

Like earthquake zones, volcanic zones are also associated with the boundaries of the crustal plates (Figure 1.8). The type of plate boundary offers some indication of the type of volcano which is likely to develop. Plates can be largely continental, oceanic, or both oceanic and continental. Oceanic crust is composed of basaltic material whereas continental crust varies from granitic to basaltic in composition. At destructive plate margins oceanic plates are overriden by continental plates. The descent of the oceanic plate, together with any associated sediments, into zones of higher temperature leads to melting and the formation of magmas. Such magmas vary in composition but some may be richer in silica. The latter magmas are often responsible for violent eruptions. By contrast at constructive plate margins, where plates are diverging, the associated volcanic activity is a consequence of magma formation in the upper mantle. The magma is of basaltic composition which is less viscous than andesitic or rhyolitic magma. Hence there is relatively little explosive activity and associated lava flows are more mobile. However, certain volcanoes, for example, those of the Hawaiian Islands, are located in the centres of plates. Obviously these volcanoes are totally unrelated to plate boundaries. They owe their origins to hot spots in the Earth's internal structure which have 'burned' holes through the overlying plates.

Volcanic activity is a surface manifestation of a disordered state within the Earth's interior which has led to the melting of rock material and the consequent formation of a magma. This magma travels to the surface where it is extravasated either from a fissure or a central vent. In some cases instead of flowing from the volcano as a lava the magma is exploded into the air by the rapid escape of the gases from within it. The fragments produced by explosive activity are collectively known as pyroclasts.

Eruptions from volcanoes are spasmodic rather than continuous. Between eruptions, activity may still be witnessed in the form of steam and vapours issuing from small vents named fumaroles or solfataras. But in some volcanoes even this form of surface manifestation ceases and such a dormant state may continue for centuries. To all intents and purposes these volcanoes appear extinct. For example, this was believed about Vesuvius prior to the fateful eruption of AD 79. In old age the activity of a volcano becomes limited to emissions of gases from fumaroles and hot water from geysers and hot springs.

Most volcanic material is of basaltic composition. As mentioned above, it is believed that basaltic magmas are generated within the upper mantle where temperatures approach the fusion point of the material from which they are derived. Their development appears to be associated with huge rifts and thrusts which fissure the crust and extend into the mantle. However, the small, local magma chambers which supply individual volcanoes need not be located at such depths. Indeed it has been estimated that the chamber beneath Monte Somma is situated at not more than 3 km below the surface. By contrast, Eaton and Murato (1960) noted continuous tremors starting at a depth of 60 km a few months before an eruption of Kilauea. It was inferred that the tremors were caused by magma streaming into a conduit leading to a chamber immediately beneath the crater.

According to Rittmann (1962) basalt magmas generated in the mantle,

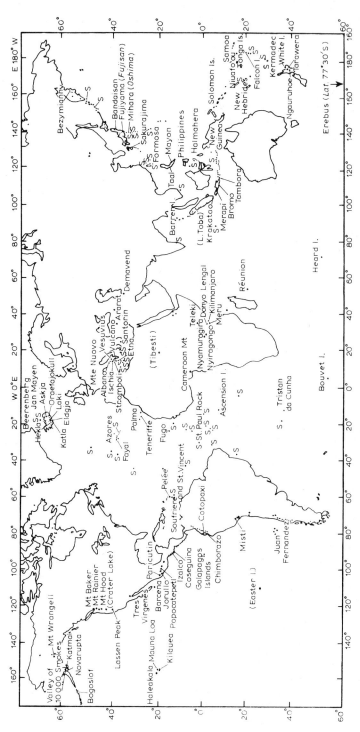

Figure 1.8 Distribution of the active volcanoes of the world (S, submarine eruptions)

presumably in the low-velocity layer, cannot reach the Earth's surface under their own power but do so via open fissures which act as channels of escape. These fissures are generated by the convection currents in the mantle. The hydrostatic pressure of the magma which penetrates a fissure helps widen it and escaping gases from the magma help clear the channel by blast action. Furthermore the reduction of pressure on the magma as it ascends causes a lowering of its viscosity which means that it can penetrate more easily along the fissure. The potential eruptive energy of the magma is governed by the quantity of volatile constituents it contains. Indeed basalt magmas formed in the mantle could not reach the continental surface if they lacked volatiles since their density would be higher than that of the upper regions of the crust. They would consequently obtain a state of hydrostatic equilibrium some kilometres below the surface. The release of original gaseous constituents in the magma lowers its density and thereby aids its ascent.

1.4 Volatiles

When a magma is erupted it separates at low pressures into incandescent lava and a gaseous phase. If the magma is viscous (the viscosity is to a large extent governed by the silica content) then separation is accompanied by explosive activity. On the other hand volatiles escape quietly from very fluid magmas.

Steam may account for 90% of the gases emitted during a volcanic eruption. For instance, it was estimated that water vapour constitutes 99% of the exhalations from fumaroles in the Valley of Ten Thousand Smokes (see Fenner (1923)). Other gases present include carbon dioxide, carbon monoxide, sulphur dioxide, sulphur trioxide, hydrogen sulphide, hydrogen chloride and hydrogen fluoride. Small quantities of methane, ammonia, nitrogen, hydrogen thiocyanate, carbonyl sulphide, silicon tetrafluoride, ferric chloride, aluminium chloride, ammonium chloride and argon have also been noted in volcanic gases. It has often been found that hydrogen chloride is, next to steam, the major gas evolved during an eruption but that in the later stages the sulphurous gases take over this role.

Water tends to move towards the top of a body of magma where the temperatures and pressures are lower. For example, in a chamber extending from 7–14 km depth the saturation water content would occur at the top whereas at the bottom it would be 2%. This relationship offers an explanation of a common sequence in volcanic eruptions, that is, a highly explosive opening phase with a resultant vigorous gas column carrying pumice to a great height, followed by the formation of ash flows, and lastly by the comparatively quiet effusion of lava flows.

At high pressures gas is held in solution but as they fall gas is released by the magma. The amount and rate at which it escapes determines the explosiveness of the eruption. An explosive eruption occurs when, because of its high viscosity, magma cannot readily allow the escape of gas. It is only secondarily related to the amount of gas a magma holds. Nevertheless because obsidians and basaltic glasses contain approximately 0.1% original water, it can be inferred therefrom that explosive eruption is impossible when a magma possesses this amount of dissolved water or less.

Lavas frequently contain bubble-like holes or vesicles which have been left behind by gas action. Even the escape of small quantities of gas is sufficient to cause frothing of the surface of a lava since pumice can be produced by the exsolution of less than 0.1% of dissolved water.

Solfataras and fumaroles occur in groups and are referred to as solfatara or fumarole fields. They are commonly found in the craters of dormant volcanoes or are associated with volcanoes which have reached the age of senility in their life cycle. Superheated steam may issue continuously from the fissures of larger solfataras and at irregular intervals from the smaller ones. The steam has a temperature of between 100 and 300°C and commonly contains carbon dioxide and hydrogen sulphide. Atmospheric oxygen reacts with the hydrogen sulphide to form water and free sulphur, the latter being deposited around the steam holes. Sulphuric acid is also formed by oxidation of hydrogen sulphide and this, together with superheated steam, frequently causes complete decomposition of the rocks in the immediate neighbourhood, leaching out their bases and replacing them with sulphates. Boiling mud pits may be formed on the floor of a crater where steam bubbles through fine dust and ash.

The composition of fumarolic gases depends not only upon their initial composition in the magma but also upon their temperature; the length of time since they began to form, the more insoluble gases are more abundant in the early emanations; the place where the gases are emitted, whether from the eruptive vent or from a lava flow; the extent of mixing with air or meteoric water; and upon reactions with air, water and country rock. Cool fumaroles, the gas temperatures of which only exceed the boiling point of water by a few degrees, are more frequent in their occurrence than are solfataras. The water vapour emitted generally contains some carbon dioxide but no hydrogen sulphide. In cooler fumaroles the gases are often only a few degrees warmer than air temperature, and have either been cooled during their ascent or arise from evaporating thermal water. By contrast the temperature of steam may reach 900°C in very hot fumaroles which occur on active volcanoes. These emissions always contain hydrogen chloride and volatile chlorides, particularly sodium chloride and ferric chloride, together with the usual constituents of solfataric gases. In addition small quantities of carbon dioxide, carbon monoxide, carbonyl sulphide, hydrogen fluoride, hydrogen thiocyanate and silicon tetrafluoride have been noted.

Hot springs are found in all volcanic districts, even some of those where the volcanoes are extinct. They originate from hot steam and gases given off by masses of intruded magma which are in the last stages of crystallisation. On their passage to the surface the steam and gases often encounter groundwater which is thereby heated and forms part of the hot springs. Many hot springs contain carbon dioxide and hydrogen sulphide together with dissolved salts. Indeed some very hot springs found in active volcanic districts may contain dissolved silica, which on cooling is deposited to form sinter terraces. If the hot springs contain dissolved calcium carbonate this is precipitated to form travertine terraces. Geysers are hot springs from which a column of hot water is explosively discharged at intervals, the water spout in some cases rising over 100 m (Figure 1.9). The periodicity of their ejections varies from a matter of minutes to many days and changes with time. Geysers are generally short lived.

Figure 1.9 Old Faithful geyser, Yellowstone National Park (courtesy United States Department of the Interior). Drillholes through rhyolitic lavas encountered vast quantities of superheated steam under high pressure. In one case, the temperature was 205°C at 75 m depth

1.5 Pyroclasts

The term *pyroclast* is collectively applied to material which has been fragmented by explosive volcanic action. Tephra is a synonym for the phrase
'pyroclastic material'. Pyroclasts may consist of fragments of lava exploded
on eruption, of fragments of pre-existing solidified lava or pyroclasts, or of
fragments of country rock, which, in both latter instances, have been blown
from the neck of the volcano. These three types have been distinguished
as essential, accessory and accidental ejectamenta respectively.

The size of pyroclasts varies enormously. It is dependent upon the viscosity of the magma, the violence of the explosive activity, the amount of
gas coming out of solution during the flight of the pyroclast, and the
height to which it is thrown. The largest blocks thrown into the air may
weigh over 100 tonnes whereas the smallest consist of very fine ash which
may take years to fall back to the Earth's surface. The largest pyroclasts are
referred to as volcanic bombs. These consist of clots of lava or of fragments of
wall rock.

The term lapilli is applied to pyroclastic material which has a diameter
varying from approximately 10–50 mm. Cinder or scoria is irregular shaped
material of lapilli size. It is usually glassy and fairly to highly vesicular and
represents the ejected froth of a magma.

The finest pyroclastic material is called ash. Much more ash is produced on
eruption of acidic than basaltic magma. This is because acidic material is
more viscous and so gas cannot escape as readily from it as it can from
basaltic lava. For example, ash forms less than 1% of those parts of the
Hawaiian shield volcanoes exposed above sea level. These are formed of
basalt lava flows. By contrast Monte Nuovo, near Naples, is mostly formed
of ash and cinders of trachytic composition.

Beds of ash commonly show lateral variation as well as vertical. In other
words with increasing distance from the vent the ash becomes finer and, in the
second case, because the heavier material falls first, ashes frequently exhibit
graded bedding, coarser material occurring at the base of a bed—it becoming
finer towards the top. Reverse grading may occur as a consequence of an
increase in the violence of eruption or changes in wind velocity. The spatial
distribution of ash is very much influenced by wind direction and deposits
on the leeside of a volcano may be much more extensive than on the windward; indeed they may be virtually absent from the latter side.

Rocks which consist of fragments of volcanic ejectamenta set in a fine-
grained groundmass are referred to as agglomerate or volcanic breccia,
depending on whether the fragments are rounded or angular respectively
(Figure 1.10).

After pyroclastic material has fallen back to the surface it eventually
becomes indurated. It is then described as tuff. According to the material
of which tuff is composed, distinction can be drawn between ash tuff,
pumiceous tuff and tuff breccia. Tuffs are usually well bedded and the
deposits of individual eruptions may be separated by thin bands of fossil
soil or old erosion surfaces. Chaotic tuffs are formed from the deposits of
glowing clouds and mud streams. Glowing clouds give rise to chaotic tuffs
in which blocks of all dimensions are present along with very fine ash. Lenses
of breccia, pumice and volcanic sand are found in chaotic tuffs which

Figure 1.10 Volcanic breccia, Isle of Arran

are formed by mud flows and they indicate that some amount of incomplete sorting has occurred during flow. Pyroclastic deposits which accumulate beneath the sea are often mixed with a varying amount of sediment and are referred to as tuffites. They are generally well sorted and well bedded.

When clouds or showers of intensely heated, incandescent lava spray fall to the ground, they weld together. Because the particles become intimately fused with each other they attain a largely pseudo-viscous state, especially in the deeper parts of the deposit. The resultant massive rock frequently exhibits columnar jointing. The term ignimbrite is used to describe these rocks (see Cook (1966)). If ignimbrites are deposited on a steep slope then they begin to flow, hence they resemble lava flows. The considerable mobility of some pyroclastic flows, which allows them to move over distances which may be measured in tens of kilometres, has been explained by the process of fluidisation. Fluidisation involves the rapid escape of gas in which pyroclastic material becomes suspended. Ignimbrites are associated with nuées ardentes.

1.6 Lava flows

Lavas are emitted from volcanoes at temperatures only slightly above their freezing points. During the course of their flow the temperature falls from within outwards until solidification takes place somewhere between 600 and 900°C, depending upon their chemical composition and gas content. Basic lavas solidify at a higher temperature than do acidic ones.

Generally, flow within a laval stream is laminar. The rate of flow of a lava is determined by the gradient of the slope down which it moves and by its viscosity which, in turn, is governed by its composition, temperature and volatile content. It has long been realised that the greater the silica content of a lava, the greater is its viscosity. Thus basic lavas flow much faster and further than do acid lavas. Indeed the former type have been known to travel at speeds of up to 80 km/h.

Many lava flows consist of several flow units which represent separate sheets of liquid that were poured over one another during a single eruption. According to MacDonald (1967) the basaltic lava flows on the slopes of Hawaiian volcanoes range in thickness from a few hundred millimetres to about 20 m. Richey (1961) noted that the plateau lavas of the Inner Hebrides were commonly 10 m thick, although many were thinner and some reached over 30 m in thickness.

1.6.1 Types of lava flows

The upper surface of a recently solidified lava flow develops a hummocky, ropy (*pahoehoe*); rough, fragmental, clinkery, spiny (*aa*); or blocky structure. The reasons for the formation of these different structures are not fully understood but certainly the physical properties of the lava and the amount of disturbance it has to undergo must play an important part. The *pahoehoe* is the most fundamental type, however, some way downslope from the vent it may give way to *aa* or block lava. In other cases *aa* or block lava may be traceable into the vent. It would appear that the change from *pahoehoe* to *aa* takes place as a result of increasing viscosity or stirring of the lava. Increasing viscosity occurs due to loss of volatiles, cooling and progressive crystallisation, whilst a lava flow may be stirred by an increase in gradient of the slope down which it is travelling. Moreover if strong fountaining occurs in the lava whilst it is in the vent this increases stirring and it may either issue as an *aa* flow or the likelihood of it changing to *aa* is accordingly increased. Melts that give rise to block lavas are more viscous than those which form *aa*, for example, they are typically andesitic although many are basaltic.

Pahoehoe is typified by a smooth, billowy or rolling, and locally ropy surface (Figure 1.11a). Such surfaces are developed by dragging, folding and twisting of the still plastic crust of the flow, due to the movement of the liquid lava beneath. Although the hummocky surface may give the appearance of smoothness it is usually interrupted by small sharp projections which mark the places where bubbles have burst into the air. Lava which has oozed through the crust may be drawn into threads orientated in the direction of flow movement. The ridges of ropy lavas are commonly curved, the convex sides pointing in the downstream direction of flow.

The skin of an active *pahoehoe* flow is very tough and flexible and impedes the escape of gas from the lava. As a consequence it is sometimes found that newly consolidated flows have a skin a millimetre or so in thickness which overlies a vesicular layer, at times the vesicles become so abundant that they merge. The skin tends to flake off, revealing the vesicular surface beneath. Sometimes the crust of the lava may be broken into a series of slabs by the movement of the still liquid material beneath.

With the exception of flows of flood eruptions all large *pahoehoe* flows consist of several units. Large flows are fed by a complex of internal streams beneath the crust, each stream being surrounded by less mobile lava. When the supply of lava is exhausted the stream of liquid may drain out of the tunnel through which it has been flowing. Cross sections of open tunnels or tubes are circular or ovoid. Tunnels up to 15 m in diameter have been observed in Hawaii and in vertical cross section the larger may extend across several flow units. Tunnels repeatedly branch and rejoin to form an anastomosing internal feeding system.

The edge of a large, slow-moving *pahoehoe* flow does not generally advance as a single unit but rather by the extension of one toe of lava after another. These toes may reach a metre across and rarely advance more than two metres before they become immobile. When this happens the crust of the flow front cracks and another toe emerges. Some toes are hollow due to them being inflated with gas.

Aa lava flows are characterised by very rough fragmental surfaces (Figure 1.11b). The fragments are commonly referred to as clinker and have numerous sharp, jagged spines. Most fragments are less than 150 mm across but some may be twice this dimension. Clinker is not readily compacted. This is illustrated by highly permeable clinker horizons in Hawaiian lavas several million years old and buried at depths of 500–1000 m. The spinose nature of the clinker is partly due to the still plastic surface being pulled apart when lava breaks into fragments. However, the crusts of some spinose *aa* flows are unbroken and spinose protruberances have been observed growing on clinker-free *aa* surfaces. The massive part of *aa* flows is usually much less vesicular than *pahoehoe*, the vesicularity frequently being less than 30%.

Aa flows are fed by streams that lie approximately in the centre of each flow. These streams are usually a few metres in width. Levees, a metre or so in height, may be constructed along the sides of the streams as a result of numerous overflows congealing. These streams are only rarely roofed over by solidified lava and if they are, it is only for a short distance. The stream retains a close connection with the interior of the flow beneath the clinker. After an eruption much of the lava drains out of the stream and leaves behind a channel which may be several metres deep.

A steep bank of clinker several metres high develops at the margin of a slowly moving *aa* flow. Movement takes place by a part of the front slowly bulging forward until it becomes unstable and separates from the parent body. The process is repeated over and over again and so a talus of clinkery deposits builds up along the foot of the lava front. These deposits are eventually buried under the flow. At the margin of a more active flow the process is just the same except that bulges grow more quickly and they form almost in a continual line along the front. Clinker on the surface of the flow is carried forward and deposited over the front. Much of this is broken in the process. The actual motion of the flow is simply brought about by its interior—still a viscous liquid—spreading. Because this pasty material overrides the talus deposits of clinker, a congealed flow has a tripartite structure, that is, a massive inner layer sandwiched between two layers of clinker.

Although the term block lava refers to lavas with fragmented surfaces it is usually restricted to those flows in which the fragments are much more regular in shape than are those of *aa* flows. Most of the flows of orogenic regions are of block lava. The individual fragments of block lava are polyhedral in shape and they have smooth surfaces although some may develop spines.

Block lava flows have very uneven surfaces. What is more they generally exhibit a series of fairly regularly arranged ridges at right angles to the direction of flow. The ridges may be several metres high and are also covered with blocky rubble. Although a central massive layer is usually present the fragmental material at times may form the whole of the flow.

It appears that the movement of block lava is similar to that of *aa* flows. However, individual parts of the lava front probably move forwards more quickly and there is a tendency for discrete layers to push forwards, which sometimes results in several layers shearing over one another. Such action is responsible for the production of considerable quantities of crushed fragmental material. Since this material is still plastic it may weld together. Block lavas may also move by sliding over the underlying surface.

(a)

(b)

(c)

Figure 1.11 (a) *Pahoehoe* or ropy lava, Devil's Sewer, Idaho (courtesy J. M. Coulthard). (b) *Aa* lava flow, Craters of the Moon, Idaho (courtesy J. M. Coulthard). (c) Pillow lavas, Isle of Arran

The vesicles in block lava are not as common as in *aa* lava. It has been found that both the blocks and a massive portion of the flow generally contain more glassy material than do the corresponding parts of *aa* flows.

Pillow lavas consist of a mass of pillow-like bodies which may be in contact with each other (Figure 1.11c). Sometimes, however, they are partially or wholly separated by detrital material which is either derived from the lavas themselves or is of sedimentary origin. The pillows usually have their shortest axis at right angles to the bedding. Pillow lavas are usually spilitic or basaltic in composition and the pillows range in size from several centimetres to a few metres in diameter. Pillows often exhibit radial jointing, vesicles and amygdales may be elongated in a radial direction, and even phenocrysts sometimes have a radial arrangement. Such a radial pattern may be due to expansion which occurred whilst the rock was still in a fluid condition. On occasions these features may be accompanied by slight cracking of the glassy crust of the pillow in which they are found.

1.6.2 Structures associated with lava flows

The surface of lava solidifies before the main body of the flow beneath. If this surface crust cracks before the lava has completely solidified then the fluid lava below may ooze up through the crack to form a squeeze-up. Pressure ridges are built on the surface of lava flows where the solidified crustal zone is pushed into a linear fold (Figure 1.12). Some ridges look like asymmetrical or overturned anticlines—the overturning being in the direction of flow—but most are simply irregular blocks of crustal material which exhibit a general fold-like form. These ridges often occur near the edge of the lava flow where the advance of the crust is retarded, thus the continual nearly horizontal thrust of the flow moving behind pushes the crust into pressure ridges. The region beneath the ridge may be unoccupied or it may have filled with lava and the axial zone may be cracked. Pressure ridges may extend over 500 m, but most are less than 100 m in length, being up to 40 m wide and 15 m in height.

Figure 1.12 Squeeze-ups and pressure ridges, Craters of the Moon, Idaho (courtesy J. M. Coulthard)

Tumuli are upheavals of dome-like shape whose formation may be aided by a localised increase in hydrostatic pressure in the fluid lava beneath the crust. They are almost always cracked and lava may have squeezed through the crack and have dribbled down the dome.

Pipes, vesicle trains or spiracles may be developed in the lava depending on the amount of gas given off, the resistance offered by the lava and the speed at which the flow is travelling. Pipes are tubes which project upwards from the base and are usually several centimetres in length and a centimetre or less in diameter. Vesicle trains form when gas action has not been strong enough to produce pipes. Nevertheless they represent zones in which vesicles are notably more abundant than in the neighbouring rock. They may either arise from the base of a flow or represent a continuation of a pipe. Spiracles are openings formed by explosive disruption of the still-fluid lava by gas generated beneath it. Generally they are roughly cylindrical in shape, with a height of up to 5 m and a diameter of less than 1 m. Some are irregular in shape.

Thin lava flows are broken by joints which may either run at right angles or parallel to the direction of flow. Joints do occur with other orientations but they are much less common. Those joints which are normal to the surface usually display a polygonal arrangement but only rarely do they give rise to columnar jointing. The joints develop as the lava cools. First primary joints form, from which secondary joints arise and so it continues.

Typical columnar jointing is developed in thick flows of basalt (Figure 1.13). Columnar jointed flows may exhibit a two- or three-tiered arrangement. Tomkeieff (1940) called the lower zone, which was characterised by well developed polygonal jointing, the colonnade. Overlying the colonnade was the entablature which he divided into a lower curvi-columnar zone and an upper pseudo-columnar zone. In the former zone, the columns are irregularly arranged in various patterns, they are often thick and may be waved. The pseudo-columnar zone is composed of short, thick-joint blocks which sometimes resemble real columns, and may be slaggy and vesicular. The relative thickness of the different parts of the flow vary appreciably. For example, in the Giant's Causeway the colonnade ranges from less than 3 to about 20 m in thickness, the total thickness of the flows varying from 20–33 m. Columnar jointing is the result of shrinkage which takes place on cooling.

The columns in columnar jointing are interrupted by cross joints which may be either flat or saucer-shaped. The latter may be convex up or down. These are not to be confused with platy joints which are developed in lavas as they become more viscous on cooling, so that slight shearing occurs along flow planes.

1.7 Volcanic form and structure

The form and structure which a volcano adopts depends upon the type of magma feeder channel, the character of the material emitted and the number of eruptions which occur. As far as the feeder channel is concerned this may be either a central vent or a fissure, which gives rise to radically different forms. The composition and viscosity of a magma influence its eruption. For instance, acid magmas are more viscous than basic and so gas cannot escape as readily from them. As a consequence the more acid, viscous magmas are generally associated with explosive activity and the volcanoes they give rise to may be built mainly of pyroclasts. Alternatively, fluid basalt magmas construct volcanoes which consist of piles of lavas with very little pyroclastic material. Volcanoes built largely of pyroclasts grow in height much more rapidly than do those formed of lavas. As an illustration, in 1538 Monte Nuovo, on the edge of the Bay of Naples, grew to a height of 134 m in a single day. It would take anything up to a million years for lava volcanoes like those of Hawaii to grow to a similar height. The number of eruptions which take place from the same vent allows the recognition of monogenetic (single eruption) and polygenetic (multiple eruption) volcanoes. Monogenetic central vent volcanoes are always small and have a simple structure. Polygenetic central vent volcanoes are much larger and more complicated. The influence of the original topography upon their form is obscured. Displacements of the vents frequently occur in polygenetic types. Fissure volcanoes are always monogenetic.

Some initial volcanic perforations only emit gas but the explosive force of the escaping gas may be sufficient to produce explosion vents. These are

(a)

(b)

Figure 1.13 (a) Giant's Causeway, Co. Antrim, Northern Ireland. View of the Grand Causeway, showing vertical columns (courtesy of the Northern Ireland Tourist Board). (b) Fingal's Cave, Island of Staffa, west of Mull, Argyllshire (courtesy of Popperfoto)

usually small in size and are surrounded by angular pyroclastic material formed from the country rock. If the explosive activity is weaker, then the country rocks are broken and pulverised in place rather than thrown from the vent. Accordingly breccia-filled vents are formed.

Pyroclastic volcanoes are formed when viscous magma is explosively erupted. They are often monogenetic and are generally found in groups, their deposits interdigitating with one another. They are rather small when compared with some shield and composite volcanoes. Some of the earlier formed cones in this category are often destroyed or buried by later outbreaks. In pumice cones, banks of ash often alternate with layers of pumice. Cinder cones are not as common as those of pumice and ash. They are formed by explosive eruptions of the Strombolian or Vulcanian type. Some of these cones are symmetrical with an almost circular ground plan, the diameter of which may measure several kilometres. They may reach several hundred metres in height. Parasitic cones may arise from the sides of these volcanoes.

Fissure pyroclastic volcanoes arising from basic and intermediate magmas are not common. By contrast, sheets of ignimbrite of rhyolitic composition have often been erupted from fissures. When enormous quantities of material are ejected from fissures associated with large volcanoes, because their magma chambers are emptied, they may collapse to form huge volcano-tectonic sinks.

Mixed or composite volcanoes have an explosive index in excess of 10 (according to Rittmann (1962) the explosive index (E) is the percentage of fragmentary material in the total material erupted). They are the most common type of volcano.

Strato-volcanoes are polygenetic and consist of alternating layers of lava flows and pyroclasts. The simplest form of strato-volcano is cone shaped with concave slopes and a crater at the summit from which eruptions take place. However, as it grows in height the pressure exerted by the magma against the conduit walls increases and eventually the sides are ruptured by radial fissures from which new eruptions take place. Cinder cones form around the uppermost centres whilst lava wells from the lower. The shape of strato-volcanoes may be changed by a number of factors, for example, migration of their vents is not uncommon so that they may exhibit two or more summit craters. Significant changes in form are also brought about by violent explosions which may blow part of the volcano or even the uppermost portion of the magma chamber away. When the latter occurs the central part of the volcanic structure collapses because of loss of support. In this way huge summit craters are formed in which new volcanoes may subsequently develop.

Most calderas measure several kilometres across and are thought to be formed by collapse of the superstructure of a volcano into the magma chamber below, since this accounts for the small proportion of pyroclastic deposits surrounding the crater. If they had been formed as a result of tremendous explosions then fragmentary material would be commonplace.

Rittmann (1962) used the term lava volcano to include those volcanic structures which had an explosive index of less than 10; in fact it is usually between two and three. Monogenetic central lava volcanoes rarely occur as independent structures and are generally developed as parasitic types on the flanks of larger lava volcanoes. The former type of volcano invariably consists of a small cone surrounding the vent from which a lava stream has issued. Polygenetic lava volcanoes are represented by shield volcanoes such as those of the Hawaiian islands. These volcanoes are built by successive outpourings of basaltic lava from their lava-lakes. The latter occur at the summit of the volcanoes in steep-sided craters. When emitted, the lavas spread in all directions and because of their fluidity cover large areas. As a consequence the slopes of these volcanoes are very shallow, usually between 4 and 6°.

Endogenous lava domes are usually steep sided, for example, the Puy de Sarcoui in Auvergne reaches a height of 150 m with a base only 400 m across. They are formed as a result of viscous lava tending to block the vent. Consequently rising lava accumulates beneath the obstruction and exerts an increasing pressure on the sides of the volcano until they are eventually fissured. The lava is then forcibly intruded through the cracks and emerges to form streams which flow down the flanks of the dome.

The term flood basalt was introduced by Tyrrell (1937) to describe large areas, usually at least 130 000 km², which are covered by basaltic lava flows. These vast outpourings have tended to build up plateaux, for instance, the Deccan plateau extends over some 640 000 km² and at Bombay reaches a thickness of approximately 3000 m.

The individual lava flows which comprise these plateau basalt areas are relatively thin, varying between 5 and 13 m in thickness. In fact, some are less than 1 m thick. They form the vast majority of the sequence, pyroclastic material being of very minor importance. That the lavas were erupted intermittently is shown by their upper parts which are stained red by weathering. Where weathering has proceeded further red earth or bole has been developed.

It is likely that flood basalts were erupted by both fissures and central vent volcanoes. For instance, Anderson and Dunham (1966) wrote that the distribution of lava types on Skye suggests that they were extruded by several fissures which were related to a central volcano. The several groups into which the lavas of Skye have been divided all thin away from centres which have been regarded as the sites of their feeders. Consequently the flood basalts are believed to have been built up by flows from several fissures operating at different times and in different areas, the lavas from which met and overlapped to form a succession which does not exceed 1200 m.

The most notable flood basalt eruption which occurred in historic times took place at Laki in Iceland in 1783 (see Thorarinsson (1970)). Torrents of lava were emitted from a fissure approximately 32 km long and overwhelmed 560 km². As the volcanic energy declined the fissure was choked but eventually pent-up gases broke through to the surface at innumerable points and small cones, ranging up to 30 m or so high, were constructed.

1.8 Types of central eruption

The Hawaiian type of central eruption is characterised in the earliest stages by the effusion of mobile lava flows from the places where rifts intersect, the volcano growing with each emission (Figure 1.14). When eruptive action was taking place under the sea, steam-blast activity caused a much higher proportion of pyroclasts to be formed. Above sea level Hawaiian volcanoes

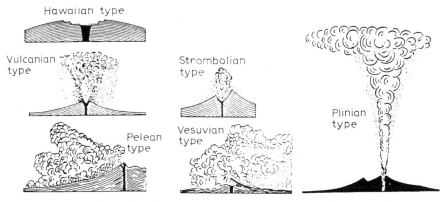

Figure 1.14 Types of central eruption

are typified by quiet emissions of lavas. This is due to their low viscosity which permits the ready escape of gas. Low viscosity also accounts for the fact that certain flows at outbreak have been observed to travel at speeds of up to 55 km/h. Lava flows are emitted both from a summit crater and from rift zones on the flanks of the volcano. Eruptive activity may follow a pattern similar to that associated with Mauna Loa (see Bullard (1976)). There, after the initial earthquakes, bulging of the volcano is caused by up-welling magma. Fissures open across the floor of the summit crater from which lavas, under initial release gas pressures, shoot into the air to form a curtain of 'fire' (Figure 1.15). After a day or so the fissures are sealed with freshly congealed lava and emission becomes localised in the largest vents. A major lava flow is erupted from Mauna Loa on average once in every three or four years.

If the lava emitted by a volcano is somewhat more viscous than the basaltic types which flow from Hawaiian volcanoes then, since gas cannot escape as readily, moderate explosions ensue. Clots of lava are thrown into the air to form cinders and bombs and so pyroclasts accompany the extrusion of lavas. Stromboli typifies this kind of central eruption, hence the designation Strombolian type (Figure 1.14). This volcano has erupted with amazing consistency throughout historic times. After the ejection of lavas some of the molten material congeals in the vent. At the next eruption, gas breaks through this thin crust and throws blocks of it, together with fragments of lava, into the air. Most of this material falls back into the crater. The steam and gases which are emitted form clouds.

Vulcano gives its name to the Vulcanian type of central eruption (Figure 1.14). In this type the lava is somewhat more viscous than that of the Strombolian type and so after an eruption, the lava remaining in the conduit of the volcano quickly solidifies. This means that gases accumulate beneath the obstruction until they have gathered enough strength to blow it into the air. The fragmented crust together with exploded lava and gas give rise to large clouds. The lavas which are emitted tend to fill the crater and spill over the sides. After a time they congeal and the process begins again.

The Vesuvian type of central eruption is yet more violent and may be preceded by eruptions of the Strombolian and Vulcanian type (Figure 1.14).

Figure 1.15 Incandescent fountain above the fissure from which the lava is flowing into the lava lake occupying the pit crater in front. (Kilauea Iki eruption of 1959, photographed at 6 am, 19 November) (courtesy US Geological Survey, photo: J. P. Eaton)

Once again the conduit of the volcano becomes sealed with a plug of solidified lava so that great quantities of gas become pent up in the magma below. This may cause lava to escape through fissures or parasitic vents on the slopes of the volcano and so the conduit may be emptied down to a considerable depth. The release of pressure so created allows gas, held in solution in the remaining magma, to escape. It does so with great force thereby clearing the conduit of its obstruction, throwing its fragmented remains and exploded lava far into the air, so forming great clouds of ash. Vesuvian eruptions are followed by periods of long quiescence; for example, since AD 79 Vesuvius itself has only broken into violent eruption ten times.

The Plinian type of central eruption is the most violent form of the Vesuvian type (Figure 1.14). Such an eruption from Vesuvius was responsible for the destruction of Pompeii and Herculaneum in AD 79. This tremendous eruption was recorded by the Roman historian, Pliny, hence the name. The Plinian type of eruption is usually preceded by the Vesuvian and is characterised by an extremely violent upsurge of gas shooting from the neck of the volcano to heights of 12 km or more, where it levels out to form a canopy. The amount of ash involved in such eruptions is relatively small and consists chiefly of material eroded from the conduit wall by the tremendous gas activity. The last Plinian eruption given off by Vesuvius occurred in 1944. It lasted for a few hours but had been preceded by nine days of Vesuvian paroxysms. These increased in gas activity and periodically fountains of lava were thrown into the air, sometimes reaching heights of 1000 m above the summit. Indeed the preceding Vesuvian eruptions almost exhausted the lava supply. The eruption of 1944 ended with a few days of pseudo-Vulcanian eruptions.

In the Pelean type of central eruption the magma concerned is extremely viscous, which leads to catastrophic explosions (Figure 1.14). Prior to the eruption of Mt Pelee in 1902 this type of volcanic activity had been unknown. Once again the neck of the volcano is blocked with congealed lava so that

Figure 1.16 Nuées ardentes associated with the eruption of Mt Pelee in 1902 which obliterated St Pierre, Martinique

the magma beneath becomes very highly charged with gas. The pressures which are thereby developed reach a point at which they have sufficient strength to fissure the weakest zone in the flanks of the volcano. The lava which is extruded through the fissure is violently exploded by the escape of gas which is held in solution. In this way a cloud of very hot gas and ash is formed. This rushes with tremendous speed down the side of the volcano destroying everything in its path. Such glowing clouds are referred to as nuées ardentes (Figure 1.16). In the case of the Mt Pelee eruption of 1902, Vulcanian activity began in the early spring but the crater filled up and was sealed by the growth of a dome within the conduit. Hot mud continued to pour down the Rivière Blanche from a notch in the rim and it was from this point of weakness that all the nuées ardentes were later to arise. On May 8 the first, and easily the most violent nuée, rushed down the slope to envelop the port of St Pierre within seconds, killing all but two of its inhabitants. Because of its high density the cloud hugged the ground and because of its intense heat, when it reached the sea, the sea boiled. Nuée activity continued and the dome blocking the conduit was forced slowly upwards until it reached about 300 m above the summit. However, it was soon reduced to a large heap of rubble by weathering.

1.9 Prediction of volcanic activity

The obvious first step in any attempt at predicting volcanic eruption is to determine whether the volcano concerned is active or extinct. If a volcano has erupted during historic times it should be regarded as active. However, many volcanoes which have long periods of repose will not have erupted in historic times. Nonetheless they are likely to erupt at some future date. It has been estimated that the active life span of most volcanoes is probably between one and two million years (see Baker (1979)).

Determination of the expected recurrence interval of particular types of eruption, the distribution of their deposits, the magnitude of events and the recognition of short-term cycles or patterns of volcanic activity are all of value as far as prediction is concerned. It must be admitted, however, that no volcano has revealed a recognisable cycle of activity which could be used to predict the time of an eruption within a decade. Furthermore volcanoes thought to be extinct may erupt—witness the violent eruptions of Mount Lamington, Papua, 1951, and Bezymianny, USSR, 1956. Consequently, individual volcanoes require mapping in order that their evolution can be reconstructed. The geological data so gathered are used in conjunction with historical records, where available, to postulate future events. If it is assumed that future volcanic activity will be similar to previously observed activity, then it is possible to make certain general predictions about hazard.

Tazieff (1979) pointed out that no volcanic catastrophies have occurred at the very start of an eruption. Consequently this affords a certain length of time to take protective measures. Even so because less than one out of several hundred eruptions proves dangerous for a neighbouring population, evacuation, presuming that an accurate prediction could be made, would not take place before an eruption became alarming. Nevertheless it is still important to predict whether or not a developing eruption will culminate in a dangerous climax and, if it does, when and how. According to Tazieff it is more difficult to predict the evolution of a developing eruption than to predict the initial outbreak.

1.9.1 Assessment of risk and violence hazard

In any assessment of risk due to volcanic activity the number of lives at stake, the capital value of the property and the productive capacity of the area concerned have to be taken into account. Evacuation from danger areas is possible if enough time is available. However, the vulnerability of property is frequently close to 100% in the case of most violent volcanic eruptions. Hazard must also be taken into account in such an assessment. It is a complex function of the probability of eruptions of various intensities at a given volcano and of the location of the site in question with respect to the volcano. Hazard is the most difficult of factors to estimate, mainly because violent eruptions are rare events about which there are insufficient observational data for effective analysis. For example, in the case of many volcanoes, large eruptions occur at intervals of hundreds or thousands of years. Thorough stratigraphic study and dating of the deposits will help to provide the evidence needed to calculate the risk factors of such volcanoes. Because in the foreseeable future man is unlikely to influence the degree of hazard, then the reduction of risk can only be achieved by reducing the exposure of life and property to volcanic hazards. This can be assessed by balancing the loss of income resulting from non-exploitation of a particular area against the risk of loss in the event of an eruption.

Most dangerous volcanic phenomena happen very quickly. For instance, the time interval between the beginning of an eruption and the appearance of the first nuées ardentes may be only a matter of hours. Fortunately such events are usually preceded by visible signs of eruption.

Booth (1979) divided volcanic hazards into six categories, namely, premonitory earthquakes, pyroclast falls, pyroclast flows and surges, lava flows, structural collapse, and associated hazards. Each type represents a specific phase of activity during a major eruptive cycle of a polygenetic volcano and may occur singly or in combination with other types. Damage resulting from volcano-seismic activity is rare. However, intensities on the Mercalli scale varying from 6–9 have been recorded over limited areas.

There are, on average, about 60 pyroclast or tephra falls per century which are of social importance. In violent eruptions intense falls of ash interrupt human activities and cause serious damage. They can affect areas up to several tens of kilometres from a volcano within a few hours from the commencement of an eruption.

Approximately 20 pyroclastic flows and surges occur every 100 years. Because of their high mobility (up to 160 km/h on the steeper slopes of volcanoes) they constitute a great potential danger to many populated areas. What is more they are hot enough to kill anything in their path instantly. Fortunately such flows and surges tend to affect limited areas. They are generally associated with Plinian eruptions. Pyroclast flows may be classified according to the ratio of gas to water they contain on the one hand and their temperature on the other (Table 1.1).

If violent Strombolian eruptions occur near a centre of population then

Table 1.1. Classification of volcanoclastic flows for risk assessment (after Booth (1979))

Gas concentration in flow (air, volcanic gas, or water vapour)					Water concentration in flow	
High gas/solid ratio Hot Pyroclast surges		Low gas/solid ratio Hot Pyroclast flows		Cold	High solid/ water ratio	Low solid/ water ratio
Cold	Hot	Non-vesiculated	Pumice or scoria			
Base surges	Nuées ardentes	Lava debris flows	Pumice slurry flows Ash flows Scoria flows	Rock avalanches	Mudflows Contain as little as 20% water	Torrents Contain as much as 80 to 90% water
Base surge deposit	Nuée ardente/ Surge deposit		Ignimbrite (Welded tuffs)		Stratified sedimentary deposits (Commonly with cross bedding)	

the rain of incandescent pyroclasts causes fires, and buildings with flat roofs can collapse under the weight of ash they collect. Permanent damage to vegetation can be brought about by toxic gases and they make evacuation more difficult. Plinian eruptions are several orders of magnitude greater than those of Strombolian type, for instance, Booth (1973) recorded that Plinian eruptions on Tenerife, Canary Islands, have covered 4200 km² with 1 m or more of ash. Welded air-fall ash can occur during a phase of high emission during a Plinian eruption. Although this action generally is concentrated near the parent crater such deposits have occasionally been found up to 6 or 7 km distant.

Lahars occur when an eruption has led to large quantities of loose material accumulating on the upper slopes of a volcano. Torrential rain storms, which often accompany eruptions, rapidly turn such loose ash into highly destructive mudflows which may travel long distances in a matter of minutes. Indeed lahars may prove as destructive as pyroclastic flows over limited areas. Destructive lahars average 50 per century.

Lava effusions of social consequence average 60 per century. Fortunately because their rate of flow is usually sufficiently slow and along courses which are predetermined by topography, they rarely pose a serious threat to life. Damage to property, however, may be complete. Lava flood eruptions are the most serious and may cover large areas with immense volumes of lava, for example, the Laki eruption, Iceland, of 1783 produced 11.67 km³ (see Thorarinsson (1970)).

The likelihood of a given location being inundated with lava at a given time can be estimated from information relating to the periodicity of eruptions in time and space, the distribution of rift zones on the flanks of a volcano, the topographic constraints on the directions of flow of lavas, and the rate of covering of the volcano by lava. The length of a lava flow is dependent upon the rate of eruption, the viscosity of the lava and the topography of the area involved. Given the rate of eruption it may be

possible to estimate the length of flow. Each new eruption of lava alters the topography of the slopes of a volcano to a certain extent and therefore flow paths may change. What is more, prolonged eruptions of lava may eventually surmount obstacles which lie in their path and which act as temporary dams. This may then mean that the lava invades areas which were formerly considered safe.

The collapse of lava domes and coulees on the steep slopes of volcanoes can give rise to lava debris flows or pyroclast flows, especially if vesiculation occurs in the freshly exposed hot lava interior.

The formation of calderas and landslip scars due to the structural collapse of large volcanoes are rare events (0.5–1 per century). They are caused by magma reservoirs being evacuated during violent Plinian eruptions. Because calderas develop near the summits of volcanoes and subside progressively as evacuation of their magma chambers takes place, caldera collapses do not offer such a threat to life and property as does sector collapse. Sector collapse landslips take place over comparatively short periods of time.

Hazards associated with volcanic activity include destructive floods and mudflows caused by sudden melting of snow and ice which cap high volcanoes, or by heavy downfalls of rain (vast quantities of steam may be given off during an eruption), or the rapid collapse of a crater lake. Far more dangerous are the tsunamis (huge sea waves, see Chapter 4) generated by violent explosive eruptions and sector collapse. Tsunamis may decimate coastal areas. Dense poisonous gases offer a greater threat to livestock than to man. Air blasts, shock waves and counter blasts are relatively minor hazards, although they can break windows several tens of kilometres away from major eruptions.

According to Booth (1979), four categories of hazard have been distinguished in Italy.

(1) Very high frequency events with mean recurrence intervals (MRI) of less than two years. The area affected by such events is usually less than 1 km².
(2) High frequency events with MRI values of 2–200 years. In this category damage may extend up to 10 km².
(3) Low frequency events with MRI values of 200–2000 years. Areal damage may cover 1000 km².
(4) Very low frequency events are associated with the most destructive eruptions and have MRI values in excess of 2000 years. The area affected may be greater than 10 000 km².

Hazard zoning involves mapping deposits which have formed during particular phases of volcanic activity and their extrapolation to identify areas which would be likely to suffer a similar fate at some future time (Figure 1.17a). Volcanic risk maps indicating the specified maximum extents of particular hazards are needed by local government and civil defence authorities (Figure 1.17b). It has been suggested by Fournier d'Albe (1979) that events with MRI of less than 5000 years should be taken into account in the production of maps of volcanic hazard zoning. He further suggested that data on any events which have taken place in the last 50 000 years are probably significant. He proposed that two types of map would be useful for economic and social planning. One type would indicate areas liable to suffer total destruction by lava flows, nuées ardentes and lahars. The other would show areas likely to be affected temporarily by damaging, but not destructive phenomena, such as heavy falls of ash, toxic emissions, pollution of surface or underground waters, etc.

1.9.2 Methods of prediction

Geophysical observations, principally tiltmetry, seismography and thermometry, provide the basis for forecasting volcanic eruptions. A volcanic eruption involves the transfer towards the surface of millions of tonnes of magma. This leads to the volcano concerned undergoing uplift. For instance, the summit of Kilauea, Hawaii, has been known to rise by almost a metre in the months preceding an eruption. The uplift is usually measured by a network of geodimeters and tiltmeters set up around a volcano. Gravity meters can be used to detect any vertical swelling.

Such uplift also means that rocks are fractured so that volcanic eruptions are generally preceded by seismic activity. However, this is not always the case. For instance, this did not happen at Heimaey, Iceland, in 1973. Conversely earthquake swarms need not be followed by eruptions—those tremors which were felt on Guadeloupe in 1976 were not followed by an eruption. A network of seismographs is set up to monitor the tremors and from the data obtained, the position and depth of origin of the tremors can be ascertained. The number of tremors increases as the time of eruption approaches. For example, tremors average six per day on Kilauea but at the beginning of 1955 these increased markedly, 600 being detected on February 26. Two days later an eruption occurred.

Infrared techniques have been used in the prediction of volcanic eruptions since, due to the rising magma, the volcano area usually becomes hotter than its surroundings. Thermal maps of volcanoes can be produced quickly by ground-based surveys using infrared telescopes (see Francis (1979)). How-

---

Figure 1.17 (a) Volcanic hazard maps for Sao Miguel, Azores. (i) Hazard from trachytic eruptions. Light shading-areas liable to burial by over 0.25 m of air fall pumice ($0.25 \ t/m^2$ damp ash); cross-shaded areas liable to burial by over 1.0 m of air-fall pumice ($1 \ t/m^2$ damp ash); dot-shaded area represents the probable route of hot lahars and/or pyroclast flows. Some principal towns: PD = Ponta Delgade; RG = Ribeira Grande; VF = Villa Franca. Other villages indicated by black squares. Calderas; SC = Sete Cidades with its 6 internal craters and 2 crater lakes; AP = Agua de Pau with its four internal craters; C = Congro diatreme; F = Furnas with its four internal craters, one of which is occupied by Furnas village; P = Povoacao volcano, believed extinct. (ii) Hazard from basaltic eruptions. Shaded area is that in danger of burial by Strombolian ashes or lava extrusions. S1–S5 are the locations of ancient phreato-magmatic (Surtseyan) eruptions. (b) Volcanic risk map for Guadalajara City and surrounding villages, western Mexico. The isopleths show the percentage probability that the area which they enclose will be buried by over 1 m of pumice ash (over $1 \ t/m^2$ damp ash) during the next major eruption. The dotted line encloses the area at risk from 0.1 m lithic missiles. Pyroclast flow hazard is not shown, but encompasses the whole map area and equates with pyroclast fall probabilities in so far as Guadalajara City is concerned (after Booth (1979))

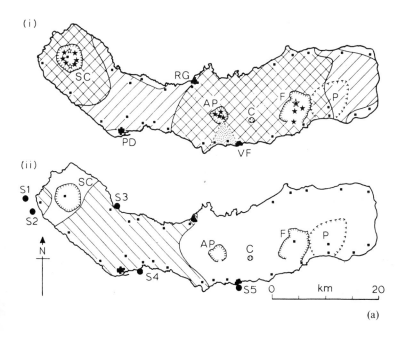

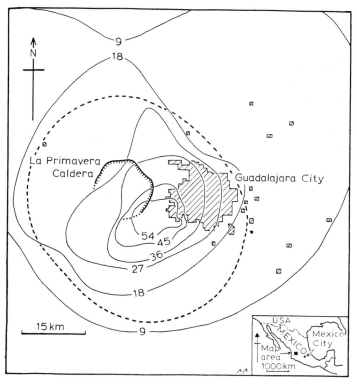

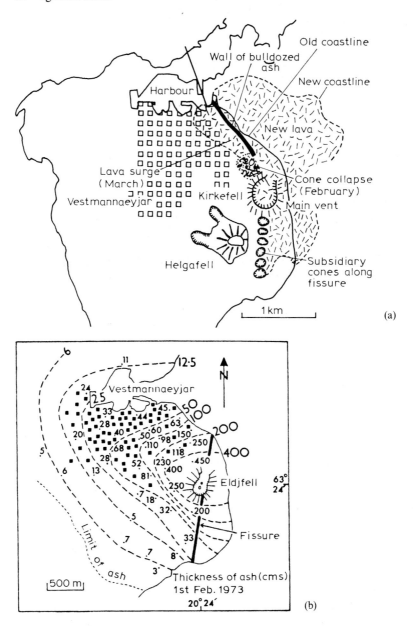

Old coastline

Wall of bulldozed ash

New coastline

Harbour

New lava

Lava surge (March)

Cone collapse (February)

Vestmannaeyjar

Kirkefell

Main vent

Helgafell

Subsidiary cones along fissure

1 km

(a)

Vestmannaeyjar

N

Eldjfell

63° 24'

Fissure

Limit of ash

500 m

Thickness of ash (cms) 1st Feb. 1973

20° 24'

(b)

ever, consistent monitoring is necessary in order to distinguish between real and apparent thermal anomalies. Aerial surveys provide better data but are too expensive to be used for routine monitoring. Satellite infrared systems, although unsatisfactory at present, may offer great potential for monitoring volcanoes in the future.

According to Baker (1979) basaltic magmas of low viscosity prove an exception if their ascent is more rapid than the rate at which heat is conducted from them. Indeed he suggested thermal anomalies are more likely

(c)

Figure 1.18 (a) The island of Heimaey, showing the town of
Vestmannaeyjar threatened and encroached upon by the 1973 eruption of
the Kirkefell volcano. (b) Isopach map of the Strombolian scoria fall
deposit on Heimaey, Iceland, after only 8 days' activity; later, much of the
eastern part of the settlement was totally buried by ash and then destroyed
by a lava flow (after Booth (1979)). (c) The Heimaey volcanic eruption, 26
January, 1973 with ash entombing houses (courtesy The Icelandic
Embassy)

to be produced by rising andesitic or rhyolitic magmas. A more or less
contrary view has been advanced by Tazieff (1979). He maintained that
thermal techniques have not proved satisfactory when monitoring explosive
andesitic and dacitic volcanoes, since the ascent of highly viscous acid
magmas presumably is too slow to give rise to easily detectable temperature
changes. Perhaps both workers are correct in that the rise of magma can
be too rapid or too slow to produce thermal anomalies which can be readily
monitored.

Be that as it may, detectable anomalies are more likely to develop when
heat is transferred by circulating groundwater rather than by conduction
from a magma. Hot groundwater also gives rise to the appearance of
new fumaroles or to an increase in the temperatures at the existing fumaroles.
Again caution must be exercised since fumarole temperatures vary due to
other factors such as the amount of rain which has fallen. Steam gauges
are used to monitor changes in gas temperatures, pH values and amounts
of suspended mineral matter. The eruption of Taal, in the Phillippines, in
1965, was predicted because of the rise in temperature of the water in the
crater lake. This allowed evacuation to take place.

An increase in temperature also leads to demagnetisation of rock, as the
magnetic minerals are heated above their Curie points. This can be monitored
by magnetic surveying.

Geophysical methods, however, cannot detect the climax of an eruption.
This is because the magma is already very close to the surface and therefore
rock fracturing, volcano inflation and increased heat transfer are not signifi-

cant enough to record. As pointed out above, it is more important to forecast the climax, and its character, rather than the outbreak of an eruption.

The evolution of an eruption involves changes in matter and energy. The most significant variations take place in the gas phase and therefore it seems appropriate to gather as much information relating to this phase as possible, especially when this phase is the active agent of the eruptive phenomenon. Gas sensors can be stationed on the ground or carried in aircraft and record the changes in the composition of gases. Unfortunately reliable data are extremely scarce and the interpretation of what is available is in its infancy. A point of interest is that the fumarole gases from some Japanese volcanoes have been found to contain an increased amount of chlorine and sulphur dioxide just before eruption.

1.9.3 Dealing with volcanic activity

The threat of lava flows can be dealt with, with varying degrees of success, by diverting, disrupting or stopping them. For example, during the eruption from Kirkefell on the island of Heimaey, Iceland, in 1973, ash from the volcano was bulldozed to form a wall in order to divert lava flows away from the town of Vestmannaeyjar (Figure 1.18). Masonry and earth walls have been used in Japan and Hawaii for the same purpose. They should be constructed at an angle to the direction of flow so that it can be diverted rather than dammed. In the latter case a lava flow would eventually spill over the dam. Topographic expression is therefore important: it should permit diversion so that no serious damage results. Lavas frequently flow along pre-existing channels across which such diversionary barriers can be built. Another diversionary technique is to dam the summit crater of a volcano at the usual exit and to breach it somewhere else so that the lavas then will flow in a different direction, to where they do little or no harm.

Bombing was used successfully in 1935 and 1942 to disrupt lava flowing from Mauna Loa, Hawaii, away from the town of Hilo. The technique might also be used to breach a summit wall in order to release lava in a harmless direction.

Water has been sprayed on to advancing lavas to cool them and thereby cause them to solidify and stop. This technique eventually proved successful on Heimaey in 1973.

1.10 Texture of igneous rocks

The degree of crystallinity is one of the most important items of texture. An igneous rock may be composed of an aggregate of crystals, of natural glass, or of crystals and glass in varying proportions. This depends on the rate of cooling and composition of the magma on the one hand and the environment under which the rock developed on the other. If a rock is completely composed of crystalline mineral material then it is described as holocrystalline. Most rocks are holocrystalline. Conversely, rocks which consist entirely of glassy material are referred to as holohyaline. The terms hypo-, hemi- or merocrystalline are given to rocks which are made up of intermediate proportions of crystalline and glassy material.

When referring to the size of individual crystals they are described as cryptocrystalline if they can just be seen under the highest resolution of the microscope or as microcrystalline if they can be seen at a lower magnification. These two types, together with glassy rocks, are collectively described as aphanitic, which means that the individual minerals cannot be distinguished with the naked eye. When the minerals of which a rock is composed are mega- or macroscopic, that is, they can be recognised with the unaided eye, it is described as phanerocrystalline. Three grades of megascopic texture are usually distinguished, fine grained, medium grained and coarse grained, the limits being under 1 mm diameter, between 1 and 5 mm diameter, and over 5 mm diameter, respectively.

The shape of individual minerals within a rock varies. Crystals are described as equidimensional if they are of equal size and development in all directions. Those which are tablet-shaped, that is, possess two, well-developed longer dimensions are termed tabular, whilst prismatic refers to crystals which possess one notably long axis. Minerals with no particular shape, which occur between their better-developed neighbours, are simply described as irregular. Such minerals, which have not formed crystal faces, are also termed anhedral. When all the crystal faces are well developed the mineral is said to be euhedral, whereas if only some of the faces are developed it is called subhedral. A rock in which most of the minerals are euhedral in shape possesses a panidiomorphic texture; if most of the minerals are anhedral then it has an allotriomorphic texture; whilst if the majority are subhedral then it has a hypidiomorphic texture.

A granular texture is one in which there is no glassy material and the individual crystals have a grain-like appearance. If the minerals are approximately the same size then the texture is described as equigranular whereas if this is not the case it is referred to as inequigranular. Equigranular textures are more typically found in plutonic igneous rocks.

Many volcanic and hypabyssal rocks display inequigranular textures, the two most important types being porphyritic and poikilitic. In the former case large crystals or phenocrysts are set in a fine-grained groundmass. A porphyritic texture may be distinguished as macro- or microporphyritic according to whether or not it may be observed with the unaided eye respectively. Plutonic rocks may occasionally exhibit porphyritic textures, a notable example being provided by Shap adamellite with its large phenocrysts of orthoclase. The poikilitic texture is characterised by the presence of small crystals enclosed within larger ones. An ophitic texture is a type of poikilitic texture in which plagioclase laths are enclosed within crystals of augite. It is typically displayed by dolerites.

The most important rock-forming minerals are often referred to as felsic and mafic depending upon whether they are light or dark coloured respectively. Felsic minerals include quartz, feldspars and feldspathoids, whilst olivines, pyroxenes, amphiboles and biotite are mafic minerals. The colour index of a rock is an expression of the percentage of mafic minerals which it contains. Four categories have been distinguished

(1) leucocratic rocks contain less than 30% dark minerals,
(2) mesocratic rocks contain between 30 and 60% dark minerals,

(3) melanocratic rocks contain between 60 and 90% dark minerals, and

(4) hypermelanic rocks contain over 90% dark minerals.

Usually acidic rocks are leucocratic whilst basic and ultrabasic rocks are melanocratic and hypermelanic respectively.

Flow structures develop in igneous rocks as a result of the alignment of crystals and xenoliths during intrusion or extrusion (Figure 1.19). Mineral alignment is most clearly demonstrated by crystals of prismatic or tabular habit, the feldspars in some Cornish granites providing good examples.

Figure 1.19 Flow banding in 'Schoolhouse' pitchstone, Brodick, Isle of Arran

1.11 Classification of igneous rocks

A classification must attempt to systematise the subject matter with which it is dealing, to catalogue it and to present it in summary form (Figure 1.20). Whenever possible scientific classification should show the genetic relationships between the individual groups of which it is made. Rocks must be classified in order that they may be compared with others of similar appearance and composition which have been previously described. Unfortunately there is no standard classification of igneous rocks. This is partly because of the enormity and complexity of the subject matter involved and partly because there are very few definite natural boundaries which divide one group of rocks from another.

Since silica (SiO_2) is the most important constituent of igneous rocks it has been frequently used in their classification. Four groups have been defined in terms of silica percentage, the usual division being as follows

A field classification of the igneous rocks (rotated table):

	Acid	Intermediate	Basic	Ultra basic
Specific gravity	2.65 ————————————————————→ 3.0			
Colour index	Leucocratic —— Mesocratic —— Melanocratic —— Hypermelanic			
Silica Percentage / Silica Saturation	Oversaturated ←	→ Saturated ←	→ Under saturated →	
Mineral Composition — Feldspar	Orthoclase Feldspar dominant	Plagioclase Feldspar Dominant — Oligoclase Andesine	Andesine Labradorite etc.	No Feldspar
Ferro-magnesian minerals	Biotite Mainly ←	→ Hornblende mainly ←	→ Augite →	Augite and/or hornblende → Olivine
	← No olivine	← No olivine →	Olivine →	No olivine → Olivine
	Quartz >10%	Quartz <10%	Quartz very rare to absent ——————→	

Grain Size

Degree of crystallinity	Granite	Syenite	Grano-diorite	Diorite	Gabbro	Olivine Gabbro	Pyroxenite Hornblendite	Peridotite Dunite	Mode of occurrence
Phanerocratic — Holocrystalline Equigranular	Granite	Syenite	Granodiorite	Diorite	Gabbro	Olivine Gabbro	Pyroxenite / Hornblendite	Peridotite / Dunite	Plutonic Batholiths Stocks Bosses (Intrusive)
Phanerocratic — Porphyritic	Quartz porphyry	Syenite porphyry	Quartz porphyry	Diorite porphyry	Dolerite	Olivine-dolerite	Rare	Rare	Hypabyssal dykes Sills (Intrusive)
Aphanitic — Crystalline To glassy Some Porphyritic	Rhyolite	Trachyte	Dacite	Andesite	Basalt	Olivine Basalt	Rare	Rare	Volcanic lava flows (Extrusive)
Aphanitic — Glassy	Obsidian / Pitchstone	←	←	→ Tachylite →			Rare		

Figure 1.20 A field classification of the igneous rocks

(1) acid igneous rocks, over 65%,
(2) intermediate igneous rocks, 55–65%,
(3) basic igneous rocks, 45–55%,
(4) ultrabasic igneous rocks, under 45%.

However, it is now realised that the boundaries of these four groups cannot be assigned fixed limits since this cannot be reconciled with mineralogical composition. Nevertheless the terms acid and basic have become deeply entrenched in the geological literature and through their constant use have acquired a much wider significance than their original purely chemical relationship. All four terms are retained but used in a qualitative sense; acid, intermediate, basic and ultrabasic being immediately associated with granitic, rhyolitic; dioritic, andesitic; gabbroic, basaltic; and peridotitic rocks respectively.

The saturation principle has also been used in geological classifications of igneous rocks. It depends upon the fact that certain minerals can, whilst others cannot, exist in the presence of excess silica. These are referred to as saturated and unsaturated minerals respectively. This allowed Shand (1913) to distinguish three groups of igneous rocks, namely, oversaturated, saturated and undersaturated. The oversaturated rocks contain free silica of primary origin. There are no unsaturated minerals or free silica in the saturated rocks (in practice this group contains rocks which may possess up to 10% free quartz). The undersaturated rocks are composed either in part or wholly of unsaturated minerals.

Shand (1927) also advanced a two-fold division of igneous rocks based on grain size, namely, the eucrystalline and dyscrystalline rocks. In the former the essential minerals can be seen with the unaided eye whereas in the latter they cannot. The terms more or less correspond with the terms phanerocrystalline and aphanitic respectively. In the case of a porphyritic rock if it contains glassy material in the groundmass then it was placed in the dyscrystalline division.

1.11.1. Oversaturated igneous rocks, eucrystalline types

Granites and granodiorites are by far the commonest rocks of the plutonic association. They are characterised by a coarse grained, holocrystalline, hypidiomorphic granular texture. Although the term *granite* lacks precision, Johannsen (1939) defined a normal granite as a rock in which quartz forms more than 5% and less than 50% of the quarfeloids (quartz, feldspar, feldspathoid content), potash feldspar constitutes 50–95% of the total feldspar content, the plagioclase is sodi-calcic, and the mafites form more than 5% and less than 50% of the total constituents (Figure 1.21).

Adamellites are coarse grained rocks which contain similar amounts of alkali and plagioclase feldspars. They are intermediate between granites and granodiorites. The intrusion at Shap Fell, Cumbria, provides a notable example of adamellite and a typical mineralogical analysis of it would contain about 24% quartz, 36% orthoclase, 34% oligoclase and 6% biotite.

In granodiorite the plagioclase is oligoclase or andesine and is at least double the amount of the alkali feldspar present, the latter forming 8–20% of the rock. The plagioclases are nearly always euhedral, as may be biotite

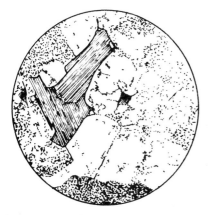

Figure 1.21 Granite from Isle of Arran containing feldspar, quartz and biotite (× 20)

Figure 1.22 Partially kaolinised pegmatite vein in Shap adamellite, Shap, Cumbria

and hornblende. These minerals are set in a quartz–potash feldspar matrix.

The term pegmatite refers to coarse or very coarse grained rocks which are formed during the last stages of differentiation of a magma. Pegmatitic facies, although commonly associated with granitic rocks, are found in association with all types of plutonites. Pegmatites occur as dykes, sills, veins, lenses or irregular pockets in the host rocks, with which they rarely have sharp contacts (Figure 1.22).

Aplites occur as veins, usually several tens of millimetres thick, in granites, although like pegmatites they are found in association with other plutonites. They possess a fine grained, hypidiomorphic, equigranular texture. There is no important chemical difference between aplite and pegmatite and it is assumed that they both have crystallised from residual magmatic solutions. Indeed they may be associated with each other in composite intrusions.

1.11.2 Oversaturated igneous rocks, dyscrystalline types

Rhyolites, toscanites and dacites are acidic extrusive rocks which are commonly associated with andesites. They are generally regarded as representing the volcanic equivalents of granite, adamellite and granodiorite, respectively. They are usually leucocratic and sometimes exhibit flow banding. They may be holocrystalline but very often they contain an appreciable amount of glass. Rhyolites, toscanites and dacites are frequently porphyritic, the phenocrysts varying in size and abundance. They occur in a glassy, cryptocrystalline or microcrystalline groundmass. Vesicles are usually found in these rocks.

Because acid lavas are highly viscous, rhyolites, toscanites and dacites do not form extensive spreads as do basic lavas. Moreover the high viscosity and gas content of acid lavas frequently leads to them being exploded from their parent volcanoes. Thus rhyolites, toscanites and dacites are associated with pyroclastic deposits. Indeed many rock masses which were formerly regarded as rhyolitic lavas are now interpreted as ignimbrites.

Because of the high viscosity of acidic lavas, crystal growth in the rocks which develop from them is often impeded with the result that they contain a greater proportion of glass than other rock types. Indeed some extrusive acid rocks are largely composed of glass. For example, obsidian is a volcanic acid rock composed almost entirely of natural glass. Pitchstone is another acid natural glass which contains more crystallites than does obsidian.

Acidic rocks of hypabyssal occurrence are often porphyritic, quartz porphyry being the commonest example. Quartz porphyry is similar in composition to rhyolite but it occurs in sills and dykes. Quartz porphyrite is a rock with a hypabyssal mode of occurrence and a composition similar to dacite.

1.11.3 Saturated igneous rocks, eucrystalline types

The three major families within the saturated eucrystalline igneous rocks are the syenites, the monzonites and the diorites. They are distinguished from each other by the type and proportions of feldspar which they contain. Syenites are plutonic rocks which have a hypidiomorphic granular texture and consist of potash feldspar, a subordinate amount of acid plagioclase and some mafic minerals, usually biotite or hornblende. The essential feature of monzonite is that it contains similar proportions of alkali and plagioclase feldspar. Usually about two-thirds of monzonite consists of feldspar, the plagioclase of which falls within the oligoclase–andesine range. Diorite has been defined as an intermediate plutonic, granular rock composed of plagioclase and hornblende, although at times the latter may be partially or completely replaced by biotite and/or pyroxene. Plagioclase, in the form of oligoclase and andesine, is the dominant feldspar. If orthoclase is present it acts only as an accessory mineral.

Syenites, monzonites and diorites are holocrystalline, usually equigranular rocks, in which the constituent crystals are subhedral and anhedral. Porphyritic types are not common but when they do occur the phenocrysts are usually feldspar. The feldspars are commonly lath shaped and in syenite they may display a trachytoid texture. In some monzonites orthoclase has a poikilitic relationship with plagioclase. The ferromagnesian minerals are frequently anhedral and in diorite symplectic intergrowths of andesine and hornblende occasionally may be observed. Miarolitic cavities occur in diorites and are lined with secondary minerals such as calcite and chlorite.

1.11.4 Saturated igneous rocks: dyscrystalline types

Trachytes, latites and andesites are the fine-grained equivalents of syenites, monzonites and diorites respectively. Andesite is probably the commonest of the three types, followed by trachyte. Latite is not particularly frequent in its occurrence.

Trachytes are extrusive rocks, which are often porphyritic, in which alkali feldspars are dominant. Most phenocrysts are composed of alkali feldspar, and to a lesser extent of alkali-lime feldspar. More rarely biotite, hornblende and/or augite may form phenocrysts. The groundmass is usually a holocrystalline aggregate of sanidine laths which typically have a trachytic texture.

Latites are commonly porphyritic and their matrices vary from holocrystalline to holohyaline. The phenocrysts are chiefly composed of plagioclase (andesine or oligoclase). Sanidine, together with plagioclase, occurs in the groundmass, it rarely forms phenocrysts. The mafic minerals may also be represented as phenocrysts. Biotite may occur alone or with hornblende and/or pyroxene.

Andesites are commonly porphyritic with a holocrystalline groundmass. Plagioclase (oligoclase–andesine), which is the dominant feldspar, forms most of the phenocrysts. The plagioclases of the groundmass are more sodic than those of the phenocrysts. Sanidine and anorthoclase rarely form phenocrysts but the former mineral does occur in the groundmass and may encircle some of the plagioclase phenocrysts. Hornblende is the commonest of the ferromagnesian minerals and may occur as phenocrysts or in the groundmass, as may biotite and pyroxene.

As can be inferred from above, trachytes, latites and andesites are commonly porphyritic with a holocrystalline groundmass. The characteristic texture of trachytes is trachytic in which lath-shaped feldspars, with little or no interstitial material between, have a roughly parallel alignment due to flow movement. If on the other hand the feldspars are mainly euhedral a fine grained idiomorphic texture results which is called orthophyric. Vitrophyric textures, that is, where the phenocrysts are set in a glassy matrix, together with spherulites and vesicles are rare in trachytes. A vitreous matrix is also rare in latites. They usually have a trachytic or pilotaxitic groundmass. The latter texture consists of a groundmass in which lath-shaped microlites are closely packed with no glass between. Latites may also have a microgranular texture. Andesites occasionally may possess a vitrophyric groundmass but it is more often trachytic or pilotaxitic. In the more silicic types glass may occur in a felted mass of microlites and such a texture is described as hyalopilitic.

1.11.5 Undersaturated igneous rocks, eucrystalline types

Gabbros and norites are plutonic igneous rocks with hypidiomorphic or allotriomorphic granular textures. They are dark in colour. Plagioclase, commonly labradorite (but bytownite also occurs) is usually the dominant mineral in gabbros and norites. The pyroxenes found in gabbros are typically augite, diopsidic augite and diallage. They are usually subhedral or anhedral. Norites, unlike gabbros, contain orthopyroxenes instead of clinopyroxenes, hypersthene being the principal pyroxene.

Gabbros which occur in basic complexes generally exhibit a successive pattern of banding. The individual bands vary in thickness from a few millimetres to several metres and may continue in length over distances measured in kilometres. One of the most notable features of such banding is its constancy. The banding results from a concentration of plagioclase and mafic minerals in alternating layers. One of the most striking examples of a layered basic complex is the Skaergaard intrusion in Greenland (see Wager and Deer (1939); Wager and Brown (1968)).

1.11.6 Undersaturated igneous rocks, dyscrystalline types

Basalts are the extrusive equivalents of gabbros and norites and are principally composed of basic plagioclase and pyroxene in roughly equal amounts, or there may be an excess of plagioclase. It is by far the most important type of extrusive rock. Basalts also occur in dykes, cone sheets, sills and volcanic plugs.

Tholeiite consists essentially of pyroxene and basic plagioclase. The pyroxenes occur both as phenocrysts and in the groundmass of these rocks, the former are represented by augite, subcalcic augite and occasionally hypersthene, whilst augite, subcalcic augite and pigeonite are found in the matrix. Tholeiite may contain up to 10% of quartz, which usually occurs as an interstitial mineral in the groundmass, where it is occasionally associated with orthoclase. Tridymite may be found in the matrix.

Basalts exhibit a great variety of textures and may be holocrystalline or merocrystalline, equigranular or macro- or microporphyritic (Figure 1.23). Vitrophyric basalts are those which possess a dominantly vitreous groundmass with a comparatively small number of phenocrysts. The spherulitic texture characteristic of acidic lavas is represented in basalts by a variolitic texture which usually consists of radially arranged plagioclase microlites set in a glassy or cryptocrystalline matrix. On the other hand if the matrix is

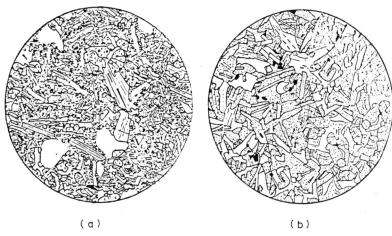

(a)   (b)

Figure 1.23 (a) Basalt, Yellowstone Park. A holocrystalline lava with laths
of labradorite embedded in fine-grained augite (stippled), and a few grains
of olivine and fine-grained magnetite. Irregular clear areas are vesicles (× 24).
(b) Basalt, Deccan, India, made up of plates of augite enclosing some of the
labradorite laths, in a matrix of altered glass (light stipple), some
magnetite (× 24)

dominantly composed of plagioclase microlites which display a semi-parallel
orientation then the texture is referred to as pilotaxitic. If wedges of glass
occur within the plagioclase microlites then the term hyalopilitic is used to
describe the texture. A distinctive feature of the holocrystalline basalts is
the lath-like shape of the plagioclase crystals of the groundmass. They
frequently possess a preferred alignment imposed upon them by flow movement. Crystals of pyroxene adopt an interstitial relationship with the plagioclase laths which they may partially or even completely enclose. Such an
arrangement is referred to as an ophitic texture. The phenocrysts consist of
olivine, when present, pyroxene and plagioclase. The olivines are often
euhedral and may show a varying degree of serpentinisation.

Spilites are basic, fine-grained rocks with a silica content ranging between
45 and 55%. Petrographically they are distinguished by their feldspar. This
is albite although they may contain minor amounts of oligoclase. Chlorite
is usually the most important ferromagnesian mineral present. Clinopyroxene
is represented by augite. Olivine rarely occurs and when it does it is usually
highly serpentinised. Many spilite lavas exhibit a pillow structure but this
is not an essential feature nor are all pillow lavas spilitic in composition.
The larger pillows often show a concentric zoning due to a radial variation
in their composition and texture. The inner zone is richer in lime and somewhat poorer in soda than the outer and is also coarser grained.

Dolerites are commonly found in minor intrusions. They consist primarily
of plagioclase, usually labradorite, and pyroxene, usually augite. The plagioclase occurs both as phenocrysts and in the groundmass. Dolerites are fine
to medium grained and are typified by an ophitic texture. They usually are
equigranular but as they grade towards basalts they tend to become porphyritic, this is particularly so in the case of plagioclase. Nevertheless the
phenocrysts generally constitute less than 10% of the rock. The groundmass

consists of plagioclase laths, small anhedral pyroxenes and minor amounts of ores.

Occasionally dolerites may show a pegmatitic development. Such pegmatites are found in sills where they occur as coarse grained schlieren, irregular patches, or irregular trending veins. These features usually show sharp contacts with the host rock.

1.11.7   The eucrystalline ultramafic igneous rocks

The ultramafic rocks are principally composed of ferromagnesian minerals; feldspars and feldspathoids, if present, only play a very minor role. Some ultramafic rocks are predominantly monomineralic and they almost invariably form a small part of a large intrusion. Peridotite is a feldspar-free rock in which olivine is the dominant mineral, constituting more than 30% of the rock. Olivine may be accompanied by pyroxene, hornblende and/or mica. Dunite is composed essentially of olivine. Similarly pyroxenite is a monomineralic rock. The fine-grained ultramafic rocks are very rare.

References

ANDERSON, F. W. & DUNHAM, C. K. (1966). 'The geology of northern Skye', *Mem. Geol. Soc. GB*, HMSO, London.

BAKER, P.E. (1979). 'Geological aspects of volcano prediction', *J. Geol. Soc.*, **136**, 341–46.

BALK, R. (1938). 'Structural behaviour of igneous rocks', *Geol. Soc. Am. Mem.*, **6**.

BARTH, T. F. W. (1962). *Theoretical Petrology*, Wiley.

BOOTH, B. (1973). 'The Granadilla pumice deposit southern Tenerife, Canary Islands', *Proc. Geol. Ass.*, **84**, 353–70.

BOOTH, B. (1979). 'Assessing volcanic risk', *J. Geol. Soc.*, **136**, 331–40.

BOWEN, N. L. (1928). *Igneous Rocks*, Dover, New York.

BULLARD, F. M. (1976). *Volcanoes of the Earth*, University of Texas Press, Austin.

CLARK, F. W. & WASHINGTON, H. S. (1924). 'The composition of the Earth's crust', *US Geol. Surv.*, Prof. Paper 127.

CLOOS, H. (1923). Das Batholithenproblem, *Fortschr. Geologie und Palaeontologie*, HT, **1**.

CLOUGH, C. T., MAUFE, H. B. & BAILEY, E. B. (1909). 'The cauldron subsidence of Glencoe', *Q. J. Geol. Soc.*, **65**, 611–78.

COOK, E. F. (1966). *Tufflavas and Ignimbrites*, Elsevier, New York.

DALY, R. A. (1933). *Igneous Rocks and the Depth of the Earth*, McGraw-Hill, New York.

EATON, J. P. & MURATO, K. J. (1960). 'How volcanoes grow', *Science*, **132**, 925–38.

FENNER, C. N. (1923). 'The origin and mode of emplacement of the great tuff deposit of the Valley of Ten Thousand Smokes', *Nat. Geog. Soc. Am. Tech. Paper Katmai Ser.*, No. 1, 74 pp.

FOURNIER D'ALBE, E. M. (1979). 'Objectives of volcanic monitoring and prediction', *J. Geol. Soc.*, **136**, 321–6.

FRANCIS, P. W. (1979). 'Infrared techniques for volcano monitoring and prediction—a review', *J. Geol. Soc.*, **136**, 355–60.

FRANKEL, J. J. (1967). 'Forms and structures of intrusive basaltic rocks' (in *Basalts*), Hess, H. H. & Poldervaart, A. (eds), Interscience, New York.

GILBERT, W. S. (1877). 'The geology of the Henry Mountains', *Mem. US Geol. Surv.*, Washington.

GROUT, F. F. (1918). 'The lopolith—an igneous form exemplified by the Duluth gabbro', *Am. J. Sci.*, **46**, 516–22.

JOHANNSEN, A. (1939). *A Descriptive Petrography of the Igneous Rocks*, 4 vols, Chicago University Press, Chicago.

MACDONALD, G. A. (1967). 'Forms and structures of extrusive basaltic rocks' (in *Basalts*), Hess, H. H. & Poldervaart, A. (eds), Interscience, New York.

MARMO, V. (1971). *Granite Petrology*, Elsevier, Amsterdam.

MEHNERT, K. R. (1968). *Migmamites*, Elsevier, Amsterdam.

RANGUIN, E. (1965). *Geology of Granite*, Kranck, E. H. & Eakins, P. R. (translated by), Interscience, New York.

READ, H. H. (1948). *The Granite Controversy*, Murby, London.

RICHEY, J. E. (1932). 'Tertiary ring structures in Britain', *Trans. Geol. Soc. Glasgow*, **19**, 41–53.

RICHEY, J. E. (1961). *Scotland: The Tertiary Volcanic Districts*, McGregor, A. G. & Anderson, F. W. (revised by), British Regional Geol., HMSO, London.

RITTMANN, A. (1962). *Volcanoes and their Activity*, Vincent, E. A. (translated by), Interscience, London.

SEDERHOLM, J. J. (1967). *Selected Works: Granite and Migmatites*, Oliver & Boyd, Edinburgh.

SHAND, S. J. (1913). 'The principle of saturation in petrology', *Geol. Mag.*, Decade 6, **1**, 485–93.

SHAND, S. J. (1927). *Erupture Rocks*, Murby, London.

TAZIEFF, H. (1979). 'What is to be forecast: outbreak of eruption of possible paroxysm? The example of the Guadeloupe Soufrière', *J. Geol. Soc.*, **136**, 327–30.

THORARINSSON, S. (1970). 'The Lakigigar eruptions of 1783', *Bull. Volcanologique Ser. 2*, **33**, 910–27.

TOMKEIEFF, S. I. (1940). 'The basalt lavas of the Giant's Causeway district of Northern Ireland', *Bull. Volcanologique Ser. 2*, **3**, 89–143.

TYRRELL, G. W. (1937). 'Flood basalts and fissure eruptions', *Bull. Volcanologique Ser. 2*, **1**, 89–111.

WAGER, L. R. & BROWN, G. M. (1968). *Layered Igneous Rocks*, Oliver & Boyd, Edinburgh.

WAGER, L. R. & DEER, W. A. (1939). 'The petrology of the Skaergaard Intrusion, Kangerdlugssueq, East Greenland', *Meddeleser om Gronland*, **105**, No. 4.

WINKLER, H. G. F. (1967). *Petrogenesis of Igneous and Metamorphic Rocks*, Springer-Verlag, Berlin.

Chapter 2

Metamorphism and metamorphic rocks

2.1 Introduction

Metamorphic rocks are derived from pre-existing rock types and have undergone mineralogical, textural and structural changes. The latter have been brought about by changes which have taken place in the physical and chemical environments in which the rocks existed. The processes responsible for change give rise to progressive transformation which takes place in the solid state. The changing conditions of temperature and/or pressure are the primary agents causing metamorphic reactions in rocks. Individual minerals are stable over limited temperature–pressure conditions which means that when these limits are exceeded mineralogical adjustment has to be made to establish equilibrium with the new environment.

When metamorphism occurs there is usually little alteration in the bulk chemical composition of the rocks involved, that is, with the exception of water and volatile constituents such as carbon dioxide, little material is lost or gained (see Vernon (1976)). This type of alteration is described as an isochemical change. By contrast allochemical changes are brought about by metasomatic processes which introduce material into or remove it from the rocks they affect. Metasomatic changes are brought about by hot gases or solutions permeating through rocks (see Section 2.7).

Metamorphic reactions are influenced by the presence of fluids or gases in the pores of the rocks concerned. For instance, due to the low conductivity of rocks pore fluids may act as a medium of heat transfer. Except at low temperatures the fluid phase in metamorphism is represented by gas with a high density which has many of the properties of a liquid.

Not only does water act as an agent of transfer in metamorphism, but it also acts as a catalyst in many chemical reactions. It is a constituent in many minerals in metamorphic rocks of low and medium grade. Grade refers to the range of temperature under which metamorphism occurred. Certainly the presence of H_2O has been shown in experimental work to accelerate many metamorphic reactions, indeed some might never be completed without its aid (see Yoder (1955)). Above about 410°C, water turns to water vapour no matter what the pressure and as this temperature is approached the effectiveness of water as a solvent seems to be reduced. However, water vapour still attacks minerals, dissolving silica above 700°C. The quantity of water held in a rock depends upon its permeability and porosity. These two factors depend in turn upon the type of rock and the depth at which it is buried, the deeper the burial the lower the amount of pore fluid. Nonetheless, the H_2O contained within a mineral structure can be liberated by a rise in temperature, for example, this occurs when, because of rising temperatures, muscovite gives place to orthoclase. Thus hydration and dehydration are principally determined by changes in temperature conditions. Minerals such as chlorite, epidote, serpentine, and talc are formed below a certain critical temperature when the pressure is high enough to ensure that water vapour is not a separate phase. According to Winkler (1967) if carbonates and water or OH bearing minerals take part in a metamorphic reaction, then carbon dioxide and water are liberated. The higher the temperature at which the reaction occurs then the smaller the amounts of these two components which are combined in the new minerals.

The phyllosilicates are common minerals in metamorphic rocks of medium and low grade, especially muscovite, biotite and chlorite, and to a lesser extent talc and serpentine. Of the inosilicates the amphiboles are more typically found in metamorphic rocks of low and medium grade, pyroxenes being developed at higher temperatures. Some of the minerals of this family are almost restricted to the metamorphic rocks, notable examples being actinolite, anthophyllite, cummingtonite and glaucophane amongst the amphiboles, and the pyroxene, jadeite. The phyllosilicates and the inosilicates are amongst the most important minerals in the metamorphic rocks as their structures allow appreciable atomic substitution, so they can adjust to changing conditions. Furthermore, being minerals of fairly high density, their formation is aided by increases in pressure. Nesosilicates like epidote, garnet, staurolite, kyanite, sillimanite and andalusite are typical of metamorphic rocks. However, many of the tectosilicates are unstable under metamorphic conditions, although quartz occurs throughout almost the whole range of metamorphism. Albite is also found over a wide range of conditions and the calcium content of plagioclase increases directly with the grade of metamorphism. Anorthite, however, is rarely found in metamorphic rocks. As far as potash feldspar is concerned microcline is usually of more common occurrence than orthoclase.

Two major types of metamorphism may be distinguished on the basis of geological setting. One type is of local extent whereas the other extends over a large area. The first type includes thermal or contact metamorphism (see Section 2.3) and the latter refers to regional metamorphism (see Section 2.5).

2.2 Metamorphic textures and structures

When referring to the textures displayed by metamorphic rocks 'blasto' is used to prefix a textural term if the feature being described is partly of a residual nature whilst the suffix 'blastic' indicates that the feature is totally metamorphic in origin (see Mason (1978)). Thus the general term crystalloblastic describes the crystalline texture developed by a metamorphic rock. Granoblastic refers to a granular texture, xenoblastic to one in which the crystals are not bounded by crystal faces whereas idioblastic describes minerals which are outlined by crystal faces which have grown *in situ*. Large metamorphic minerals set in a fine grained groundmass are termed porphyroblasts. A sieve-like texture, which is characteristic of minerals carrying inclusions such as cordierite and staurolite, is called poikiloblastic. If minerals are intergrown then the term diablastic is used to describe such a texture.

The preservation of residual or palimpest textures and structures in metamorphic rocks depends principally upon the grade and length of time the particular conditions persisted. The higher the grade and the greater its duration, then the more likely the rock is to have been totally reconstituted. Reconstitution of a rock mass also depends upon its mineralogical composition and texture, coarse textured rocks being less readily altered than fine types.

2.2.1 Preferred orientation

Most deformed metamorphic rocks possess some kind of preferred orientation of which two types may be distinguished, namely, dimensional preferred orientation and lattice preferred orientation. In the first type inequidimensional units such as acicular, platy or prismatic crystals tend towards parallel alignment of their long axes. Secondly, lattice preferred orientation refers to crystal structures and therefore optic directions and various crystallographic features possessing related orientations.

Preferred orientations in metamorphic rocks are commonly exhibited as mesoscopic linear or planar structures which allow the rocks to split more easily in one direction than others. One of the most familiar examples is cleavage in a slate, a similar type of structure in metamorphic rocks of higher grade is schistosity. At this point it may be mentioned that the term foliation is now frequently used to include cleavage and schistosity. For example, Turner and Weiss (1963) use it in this way, regarding foliation as essentially consisting of penetrative surfaces of discontinuity in deformed rocks which have been formed by metamorphic processes. However, according to Harker (1939), foliation comprised a segregation of particular minerals into inconstant bands or contiguous lenticles. These bands exhibit a common parallel orientation. This phenomenon Turner and Weiss referred to as layering or lamination. In this text foliation is used in the sense defined by Harker.

Preferred orientation represents one aspect of the strain suffered by a metamorphic rock. Most of these fabrics must be attributed either directly or indirectly to permanent deformation, although Flinn (1965) suggested that a preferred orientation might arise as a consequence of preferred nucleation or due to a preferred growth mechanism during elastic strain. Certainly the relation with deformation can be demonstrated by the general lack of preferred orientations in undeformed rocks, by the fact that preferred orientations can be produced experimentally by deformation and more simply by the occurrence of preferred orientations in rocks which show obvious signs of having suffered strain.

Spry (1969) stated that preferred orientations of various kinds could be produced during three different stages in the metamorphism of a rock. First, pre-metamorphic minerals may be mechanically rotated in order to adopt related orientations without their identity being lost or their internal structure and boundaries being appreciably altered. Secondly, minerals may be initially developed in a preferred position during crystallisation of a randomly orientated aggregate which is being strained. Thirdly, minerals may be formed in preferred positions because of some previously existing orientation. As can be deduced from the foregoing, no one explanation of preferred orientation is universally applicable; sometimes it is a slip phenomenon, sometimes a product of compression, then again it may be the consequence of post-tectonic growth of mica flakes parallel to pre-existing planar structures which themselves may be of diverse origin. The style of preferred orientation like the style of the folding of the rocks in which it occurs, reflects their rheological state at the time of their deformation. This is a function of the physical conditions prevailing, such as temperature and pressure, as well as the phases (minerals and intergranular fluid) of which the tectonite was composed. Accordingly it is not surprising to find that preferred orientations in slates and low-grade schists are frequently very different from those typical of granulites and amphibolites.

2.2.2 Cleavage

Slaty cleavage is probably the most familiar type of preferred orientation and occurs in rocks of low metamorphic grade (Figure 2.1a). Slaty cleavage is characteristic of slates and phyllites. It is independent of bedding, which it commonly intersects at high angles; and it reflects a highly developed preferred orientation of mineral boundaries, particularly of those belonging to the mica family. As slates and phyllites are the products of lower grade metamorphism, Turner and Weiss (1963) contended that appropriate values for their formation perhaps would be 300–350°C, with confining pressures in excess of three kilobars (300 MPa). Moreover metamorphism from shale to slate, phyllite and ultimately schist involves progressive dehydration. This means that water pressures equalling or exceeding load pressures are constantly maintained. Under these conditions, together with regional stress of long duration, the rheological condition of developing slate is perhaps similar to that of a viscous rather than a plastic body. Such material could then be deformed by laminar flow on planes of very low shear stress.

Fracture cleavage (see also Chapter 4) is a parting defined by closely spaced parallel fractures which are usually independent of any planar preferred orientation of mineral boundaries that may be present in a rock mass. It is closely related to strain-slip cleavage which is characterised by discrete surfaces of incipient transposition of a formerly existing cleavage. Unlike slaty cleavage, fracture cleavage is not restricted to one type of rock but like the former type it may take the form of axial plane cleavage (Figure 2.1b).

Strain-slip cleavage occurs in fine grained metamorphic rocks where it may maintain a regular, though not necessarily constant, orientation. This regularity suggests some simple relationship between the cleavage and movement under regionally homogeneous stress in the final phase of deformation. However, the development of strain-slip cleavage is not a simple process and it is not yet fully understood. It has been suggested that in its initial stages it is a simple slip structure which has a constant displacement in relation to the limbs of folds. With progressive strain the microlithons between adjacent

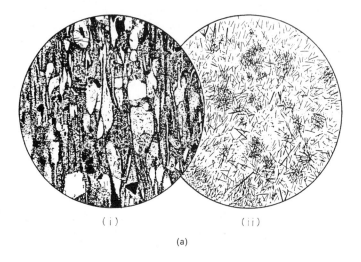

(i)   (ii)

(a)

(b)

Figure 2.1 (a) (i) Slate, Moel Tryfaen, Caernarvonshire. Cut transversely to the cleavage-planes and showing a parallel arrangement of the elements which is sufficient to impart a perfect cleavage. The constituents are sericitic mica, chlorite and haematite, with abundant clastic granules of quartz. (ii) Slate, Morben, near Machynlleth, Montgomeryshire. Cut parallel to the cleavage to show the swarm of minute needles of rutile. The tendency of clustering indicates vanished shreds of biotite (after Harker (1939)). (b) Slate quarry, Dinorwic, North Wales. Slates were worked along the cleavage planes, which are dipping steeply

slip surfaces are deformed by flattening, and this is illustrated by progressive folding in earlier surfaces. From experimental evidence it would appear that the slip surfaces are progressively rotated towards the normal to the principal compressive stress.

Turner and Weiss (1963) proposed that when a finely cleaved, and therefore mechanically anisotropic, rock such as a slate or phyllite was subjected to stress in such a way that the principal maximum axis of stress was inclined at a high angle to the initial cleavage, then failure by slippage took place along surfaces cutting across the cleavage. The slip surfaces represented laminar domains of intense strain which became the foci of syntectonic or post-tectonic recrystallisation of mica so that eventually the strain-slip structure develops into schistosity.

2.2.3 Schistosity and foliation

Harker (1939) maintained that schistosity was developed in a rock when it was subjected to increased temperatures and stress which involved its reconstitution, which was brought about by localised solution of mineral material and recrystallisation. In all types of metamorphism the growth of new crystals takes place in an attempt to minimise stress. When recrystallisation occurs under conditions which include shearing stress then a directional element is imparted to the newly formed rock. Minerals are arranged in parallel layers along the direction normal to the plane of shearing stress giving the rock its schistose character (Figure 2.2). The most important minerals responsible for the development of schistosity are those which possess an acicular, flaky or tabular habit, the micas being the principal family involved. The more abundant flaky and tabular minerals are in such rocks, the more pronounced is the schistosity. If a rock is crowded with acicular crystals then it may possess a linear or nematoblastic schistosity, as opposed to a planar or lepidoblastic schistosity characteristic of rocks containing crystals of flaky or tabular habit.

Foliation in a metamorphic rock, as defined by Harker (1939), is a most conspicuous feature consisting of parallel bands or tabular lenticles formed of contrasting mineral assemblages such as quartz–feldspar and mica–chlorite–amphibole. This parallel orientation agrees with the direction of schistosity, if any is present in nearby rocks. Foliation would therefore seem to be related to the same system of stress and strain responsible for the development of schistosity. However, at higher temperatures the influence of stress becomes less and so schistosity tends to disappear in rocks of high-grade metamorphism. By contrast foliation becomes a more significant feature. What is more, minerals of flaky habit are replaced in the higher grades of metamorphism by minerals, like garnet, kyanite, sillimanite, diopside and orthoclase. The segregation of different minerals into bands was thought by Harker to have been effected by local solution, diffusion and recrystallisation. The diffusion responsible for the transference of material was supposedly most effective in the plane perpendicular to maximum pressure. He went on to state that segregation implies a nucleus about which growth has taken place. If this was so then the heterogeneous nature of a foliated rock suggests some original heterogeneity which has been emphasised by the segregation process. In other words foliation is based upon

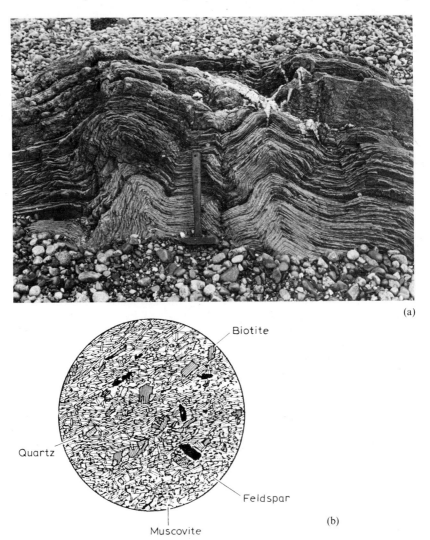

(a)

(b)

Figure 2.2 (a) Folded schist showing development of a quartz rod, Isle of Arran.
(b) Mica schist (× 35)

pre-existing structures in rocks. He illustrated his view by noting that the peculiar features of foliation seen in high grade schists seemed to be due to the process of segregation affecting rocks which had already acquired a strong schistosity. In such instances foliation developed along the planes of schistosity.

2.3 Thermal or contact metamorphism

Thermal metamorphism occurs around igneous intrusions so that the principal factor controlling these reactions is temperature, shearing stress being of negligible importance. Hydrostatic pressure also plays a minor part in thermal metamorphism, especially where the adjustments which occur involve little change in volume. However, if the chemical reactions involved in the metamorphic processes generate a gas phase then high confining pressures may inhibit its escape which, in turn, may indirectly affect the reactions.

Individual minerals possess a definite temperature range over which they are stable so that when these limits are exceeded the mineral must either change so that it may remain in equilibrium with the new conditions or remain in a metastable condition. The presence of metastable minerals suggests that the new conditions did not prevail long enough for stability to be established. The rate at which chemical reactions take place during thermal metamorphism is exceedingly slow and depends upon the rock types and temperatures involved. It has been estimated that the reaction rate is doubled by a rise of 10°C, whilst a rise of 100°C may give rise to an increase of a thousand-fold and 200°C of a million-fold. Equilibrium in metamorphic rocks is therefore attained more readily at a higher grade, than at lower grade because reaction proceeds more rapidly.

Thermal metamorphism commonly has taken place around igneous intrusions and the encircling zone of metamorphic rocks is referred to as the contact aureole (Figure 2.3). The size of an aureole depends upon the temperature and size of the intrusion, the quantity of hot gases and hydrothermal solutions which emanated from it, and the type of country rocks involved. Aureoles developed in argillaceous sediments are more impressive than those found in arenaceous or calcareous rocks. This is because clay minerals, which account for a large proportion of the composition of argillaceous rocks, are more susceptible to temperature changes than quartz or calcite. Aureoles formed in igneous or previously metamorphosed terrains are also less significant than those developed in argillaceous sediments. Nevertheless the capricious nature of thermal metamorphism must be emphasised for even within one formation of the same rock type the width of the aureole may vary. This need not be solely a reflection of fluctuations in the thermal gradient for the attitude of the hidden contact will probably vary.

The temperature in the aureole at any distance from the contact at a given time is governed by the size and temperature of the intruded body of magma; the thermal conductivity, specific heat, density and diffusivity of the rocks involved; the initial temperature and water content of the country rocks; the crystallisation temperature and latent heat of crystallisation of the magma; and the amount of heat generated by metamorphic reactions in the aureole. In fact the size of an intrusion is the most important factor determining the distance from the contact at which metamorphic temperatures can develop. From Jaeger's (1959) work it would appear that temperatures of 400°C or more could be maintained for tens of thousands of years at approximately 500 m from the contacts of granitic stocks one or two kilometres in width (the parent sediments were assumed dry and had an initial temperature of 100°C). However, if the parent sediments contain pore water then its vaporisation may lower the temperatures at the contact by about 100°C.

Within a contact aureole there is usually a sequence of mineralogical changes from the country rocks to the intrusion, which have been brought about by the effects of a decreasing thermal gradient whose source was in the

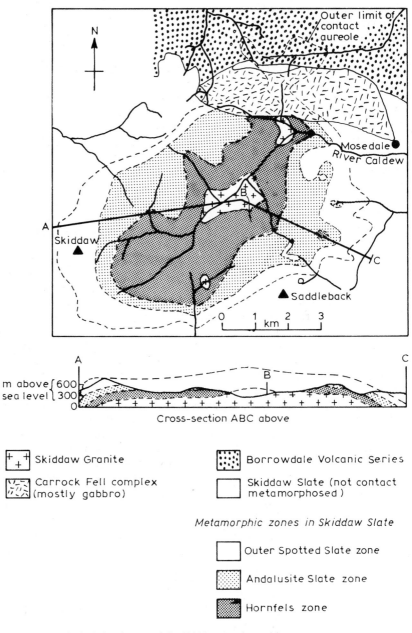

Figure 2.3 Geological sketch map of the Skiddaw granite and its contact aureole (from Eastwood *et al.* (1968))

hot magma. Indeed, aureoles in argillaceous (or pelitic) rocks may be concentrically zoned with respect to the intrusion. A frequently developed sequence varies inward from spotted slates to schists then hornfelses. It is characterised mineralogically by chlorite and muscovite in the outer zone, biotite with or without andalusite in the next zone, and biotite, cordierite and sillimanite in the zone nearest the contact.

Hornfelses are characteristic products of thermal metamorphism (Figure 2.4). They are dark coloured rocks with a fine grained decussate, that is, interlocking texture, containing andalusite, cordierite, quartz, biotite, muscovite, microcline or orthoclase, and sodic plagioclase. The higher temperature varieties may possess garnet, hypersthene, diopside and sillimanite, and, if they are silica deficient, corundum and spinel. Cordierite and andalusite (or chiastolite) usually form porphyroblasts, the former being usually irregular in shape and poikiloblastic. Garnets and micas may also be porphyroblastic.

Aureoles formed in calcareous rocks frequently exhibit greater mineralogical variation and less regularity. Zoning, except on a small and localised scale, is commonly obscure. The width of, and the mineral assemblage developed in the aureole appear to be related to the chemical composition and permeability of the parent calcareous beds. Marbles may be found in these aureoles, forming when limestones undergo metamorphism. If a limestone contains a significant amount of siliceous material then, with rising temperatures, this will combine with lime to form wollastonite marble. Brucite marbles are formed when dolostones suffer metamorphism and if they contain siliceous material forsterite marbles result. Forsterite is frequently hydrated to give serpentine. A pure marble usually has a mosaic texture due to the recrystallisation of calcite. Accessory minerals frequently occur as porphyroblasts. Calc-silicate hornfelses principally consist of calcium-bearing silicates arranged in a decussate fabric. Their mineralogical composition is generally variable with diopside and augite figuring largely and hedenbirgite occurring to a lesser extent. Wollastonite, andradite, grossularite and epidote minerals are other common minerals (these give place to calcic plagioclases at higher grade). Porphyroblasts are not conspicuous in these rocks.

The reactions which occur when arenaceous sediments are subjected to

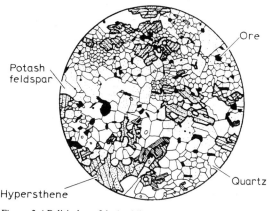

Figure 2.4 Pelitic hornfels (× 35)

thermal metamorphism are usually less complicated than those which take place in their argillaceous or calcareous counterparts. For example, the metamorphism of a quartz arenite leads to the recrystallisation of quartz to form a quartzite with a mosaic texture and the higher the grade, the coarser the fabric. It is the impurities in a sandstone which give rise to new minerals upon metamorphism. Common impurities found in the matrix of sandstones are clay minerals, sericite, chlorite, limonite and calcite. Biotite may form from sericite, chlorite and limonite; augite from chlorite, calcite and quartz; epidote from calcite and clay minerals; magnetite from limonite; cordierite from chlorite and clay minerals and andalusite from clay minerals. Garnets may appear in the higher grades whilst the highest grades may be marked by the presence of sillimanite. Also at these grades a foliation structure tends to develop and a gneissose rock is produced.

The acid and intermediate igneous rocks are resistant to thermal metamorphism, indeed they are usually only affected at very high grades. For instance, where a granite has been invaded by a later granite it usually remains unaltered texturally, any changes which take place doing so on an ionic scale. If mineralogical changes do occur then they are restricted to sericitisation or saussuritisation of feldspar exsolution of feldspar to produce perthites or anti-perthites, and chloritisation of biotite. However, when granites are intruded by basic igneous masses total recrystallisation may be brought about in the immediate neighbourhood of the contact to produce a gneissose rock. Recrystallisation takes place more easily in the fine-grained varieties. If glass is present in such rock types then, because it is very unstable, it crystallises to give a granoblastic aggregate of feldspar and quartz.

Basic igneous rocks undergo a number of changes when subjected to thermal metamorphism. They consist essentially of pyroxenes and plagioclase and the first changes take place in the ferromagnesian minerals, that is, in the outermost region of an aureole the plagioclases are unaffected thereby leaving the parental igneous texture intact. The pyroxenes may be altered to pseudomorphs of hornblende and if olivine was present it may have been replaced by serpentine or chlorite. As the intrusion is approached, the rocks become completely recrystallised and the mineral assemblage is mainly composed of actinolite and albite, hence albite–actinolite hornfelses make their appearance. At medium grade metamorphism hornblende hornfelses are common, which are characterised by the development of calcic plagioclase and frequently diopside, in addition to hornblende. Nearest the contact the high grade rocks are typically represented by pyroxene hornfelses. Under these conditions augite and olivine may be recrystallised or developed from hornblende and serpentine respectively.

2.4 Dynamic metamorphism*

Dynamic metamorphism, like contact metamorphism, is produced on a comparatively small scale and is usually highly localised, for example, its

---

*Some geologists do not regard this as a metamorphic process since it leads to deformation not transformation. This view is the more acceptable, bearing in mind the definition of metamorphism.

effects may be found in association with large faults or thrusts. On a larger scale it is associated with folding, however, in the latter case it is difficult to distinguish between the processes and effects of dynamic metamorphism and those of low grade regional metamorphism. What can be said is that at low temperatures recrystallisation is at a minimum and the texture of a rock is largely governed by the mechanical processes which have been operative. The processes of dynamic metamorphism include brecciation, cataclasis, granulation, mylonitisation, pressure solution, partial melting and slight recrystallisation.

Stress is the most important factor in dynamic metamorphism. When a body is subjected to stresses which exceed its limit of elasticity it is permanently strained or deformed. If the stresses are equal in all directions then the body simply undergoes a change in volume, whereas if they are directional its shape is changed. Stress also affects solubility; for instance, an increase in hydrostatic pressure raises the melting point of a solid mass. Also when a mineral is subjected to shearing stress, solution may occur at the point of maximum stress. The pore liquid about the mineral then becomes supersaturated with mineral material and secondary crystallisation takes place around the mineral at the region(s) of least stress. This process of secondary growth relieves the internal stress set up within the mineral and leads to the establishment of a phase which is stable under the prevailing conditions. What is more, shearing stress tends to increase the rate of chemical activity and unlike hydrostatic pressure it facilitates reactions which involve a gaseous or liquid phase since the directional character of shear aids the escape of these phases.

The textures of dynamically metamorphosed rocks are complex and depend upon the inter-relationship of temperature, stress, strain rate, the presence of solvents and the properties of the rocks concerned. Under conditions of relatively low shearing stress fracturing takes place within a rock and crushing leads to some minerals being broken. With increasing strain the minerals are more and more affected, micas are bent, garnets are fractured, quartzes develop undulose extinction, feldspars develop deformation twins and gliding takes place in calcite. Other mineralogical changes which occur under dynamic metamorphism include exsolution in feldspars to form perthites, sericitisation of orthoclase, saussuritisation of plagioclase, chloritisation of amphiboles, uralitisation of pyroxenes and serpentinisation of olivines. In other words, silicates become hydrated. As shearing stresses continue the rock becomes more and more broken until it is more or less powdered. Under extreme stress this material is fused.

Brecciation is the process by which a rock is fractured, the angular fragments produced being of varying size. It is commonly associated with faulting and thrusting. The fragments of a crush breccia may themselves be fractured and the mineral components may exhibit permanent strain phenomena. If during the process of fragmentation pieces are rotated, then they are eventually rounded and embedded in the worn-down powdered material. The resultant rock is referred to as a crush conglomerate. A classification of dynamically metamorphosed rocks is given in Table 2.1.

Mylonites are produced by the pulverisation of rocks, which not only involves extreme shearing stress but also considerable confining pressure. Mylonitisation is therefore associated with major faults, for example,

Table 2.1. Classification of dynamically metamorphosed rocks (after Spry (1969))

Nature of matrix		Properties of matrix			
Crushed		0–10%	10–50%	50–90%	90–100%
	Foliated	Crush breccia or con- glomerate	Protomylonite	Mylonite	Ultramylonite
	Massive		Protocataclasite	Cataclasite	Ultracataclasite
Recrystallised	Minor		Hartschiefer		
	Major		Blastomylonite		
Glassy			Pseudotachylite		

mylonites are found along the Highland Boundary Fault in Scotland. Mylonites are composed of strained porphyroclasts set in an abundant matrix of fine grained or cryptocrystalline material. Quartzes in the ground-mass are frequently elongated and may exhibit undulose extinction. Those mylonites which have suffered great stress lack porphyroclasts, having a laminated structure with a fine granular texture. The individual laminae are generally distinguishable because of their different colour or due to trains of magnetite. Protomylonite is transitional between micro-crush breccia and mylonite, whilst ultramylonite is a banded or structureless rock in which the material has been reduced to powder. Mylonites are distinguished from cataclasites by their banded or laminated appearance, a cataclasite having a structureless matrix.

A hartschiefer is a micro-brecciated rock in which some recrystallisation has occurred in the matrix. It is typically laminated. In a blastomylonite recrystallisation has taken place to a much greater extent, in fact it is not restricted to the groundmass being also observed in the porphyroblasts.

In the most extreme cases of dynamic metamorphism the resultant crushed material may be fused to produce a vitrified rock referred to as a pseudo-tachylite. It usually occurs as very small, discontinuous lenticular bodies or branching veins in granite, quartzite, amphibolite and gneiss. Quartz and feldspar fragments are usually found in a dark coloured glassy base. In some instances, because of the onset of devitrification, the glassy matrix is cloudy.

2.5 Regional metamorphism

Metamorphic rocks extending over hundreds or even thousands of square kilometres are found exposed in the Pre-Cambrian shields and the eroded roots of fold mountains (see Miyashiro (1973)). As a consequence the term *regional* has been applied to this type of metamorphism. Regional meta-morphism involves both the processes of changing temperature and stress. The principal factor is temperature of which the maximum figure concerned in regional metamorphism is probably in the neighbourhood of 800°C. Igneous intrusions are found within areas of regional metamorphism, but their influence is restricted. Although Dachille and Roy (1964) demonstrated experimentally that the stability field of metamorphic assemblages is not influenced by shearing stress, it does nevertheless increase the rate at which reaction occurs. Regional metamorphism may be regarded as taking place when the confining pressures are in excess of three kilobars, whilst below that figure, certainly below two kilobars, falls within the field of contact metamorphism. What is more, temperatures and pressures conducive to regional metamorphism must have been maintained over millions of years. That temperatures rose and fell is indicated by the evidence of repeated cycles of metamorphism. These are not only demonstrated by mineralogical evidence but also by that of structures. For example, cleavage and schistosity are the result of deformation which is approximately synchronous with metamorphism but many rocks show evidence of more than one cleavage or schistosity which implies repeated deformation and metamorphism. In spite of these changes in conditions it is usually possible to recognise a regular, progressive alteration which is the expression of a single thermal gradient. Where traces of partial reaction are observed in minerals they may be interpreted as a response to a falling temperature gradient which must have been maintained for a lengthy period of time.

2.5.1 Metamorphic zones

Regional metamorphism is a progressive process, that is, in any given terrain formed initially of rocks of the same composition, zones of increasing grade may be defined by different mineral assemblages. The pioneer of this concept of progressive zones of regional metamorphism was George Barrow (1912). At the turn of the twentieth century Barrow was working in the southeast of the Scottish Highlands. The metamorphic rocks of that region, which are mainly schists, are largely of pelitic origin and he was able to establish a sequence of zones of progressive metamorphism based upon mineralogical and, to a lesser extent, textural changes. Each zone was defined by a significant mineral and their mineralogical variation could be correlated with changing temperature–pressure conditions. The boundaries of each zone were therefore regarded as isograds, lines of equal metamorphic conditions. Barrow's zones included the zone of clastic mica, the zone of digested clastic mica, the biotite zone, the garnet zone, the staurolite zone, the kyanite zone and the sillimanite zone (Figure 2.5). Tilley (1925) substituted the term chlorite zone for Barrow's zones of clastic mica and digested clastic mica. He also showed that the Barrovian zones could be traced westward across the outcrop of Dalradian Schists into Ireland.

Similar zonal sequences to those found in the south-east Highlands have been observed in other areas of regional metamorphism. However, the zones need not appear in the exact same order as that which Barrow outlined. Furthermore Read (1936) established a series of zones in Banffshire based on the index minerals, biotite, andalusite-cordierite and sillimanite-cordierite. He argued that the temperature range experienced in this region must have been similar to that of the south-east Highland region, for the lowest and highest index minerals are the same in both. However, in Banffshire shearing stress was of little importance since metamorphism took place after the orogenic movements responsible for the formation of the Highlands had ceased. This illustrates the fact that while deformation is highly characteristic of regional metamorphism it is not an essential feature.

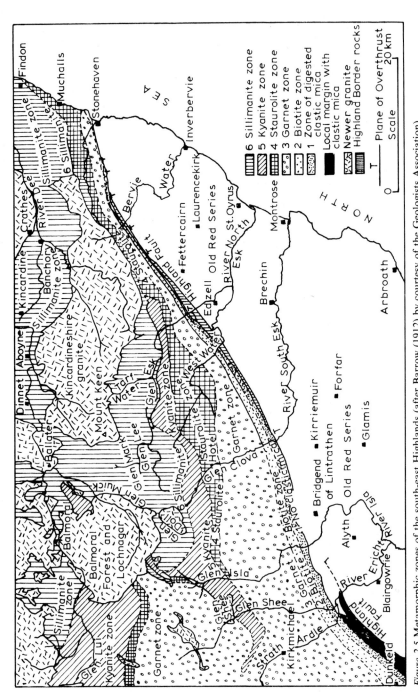

Figure 2.5 Metamorphic zones of the south-east Highlands (after Barrow (1912) by courtesy of the Geologists Association)

2.5.2 Regional metamorphism of sedimentary rocks

The above account of progressive metamorphism illustrates the changes which take place in argillaceous rocks when subjected to regional metamorphism. However, one change was barely mentioned and that was to slate. Slates are the products of low grade regional metamorphism of pelitic sediments. Fine grained, clayey material is structurally weak and tends to flow in a plastic manner when deformed. This leads to the formation of cleavage planes and recrystallisation, although the new minerals are very small. Flaky minerals are either rotated into parallel alignment or recrystallise in this preferred direction. Illite, sericite and chlorite are the principal flaky minerals, quartz, feldspar, hematite and some carbonaceous matter form the other components.

When sandstones are subjected to low grade regional metamorphism the quartz grains develop undulose extinction and are often fractured, whilst the material of the groundmass tends to recrystallise. Potash feldspar may be sericitised, plagioclase saussuritised, and biotite and hornblende chloritised. As metamorphism proceeds so quartz and then sodic and potassic feldspar begin to recrystallise. In this way a quartzite develops which has a granoblastic texture. A micaceous sandstone or one in which there is an appreciable amount of argillaceous material, on metamorphism yields a quartz–mica schist. Garnets may develop from the pelitic matrix at higher grades so that garnet–quartz–mica schists are formed. At even more advanced grades staurolite, kyanite or possibly sillimanite may make their appearance. Metamorphism of arkoses and feldspathic sandstones leads to the recrystallisation of feldspar and quartz so that granulites, with a granoblastic texture, are produced. Biotite is a more common accessory than muscovite and if it occurs in significant amounts it serves to distinguish biotite granulites.

Because confining pressure plays an important role in regional metamorphism, the escape of carbon dioxide from reactions involving a carbonate is impeded. In consequence the changes which take place in calcareous sediments under regional metamorphism differ somewhat from those which are associated with thermal metamorphism. Pure carbonate rocks when subjected

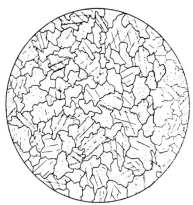

Figure 2.6 Calcite–marble, Carrara, Italy
(× 20)

to regional metamorphism simply recrystallise to form either a calcite or dolomite marble with a granoblastic texture (Figure 2.6).

Any silica present in a limestone tends to reform as quartz and even at the highest grades of metamorphism it does not react with calcite to give wollastonite. However, silica may react with dolomite to form talc. Muscovite and chlorite, when present in the matrix of a limestone, at lower grades recrystallise but as metamorphism intensifies, and if they are present in sufficient amount, they give rise to biotite, the excess alumina being used to produce zoisite. As metamorphism proceeds still further grossularite and idocrase may develop, together with the formation of diopside and microcline from biotite and muscovite respectively. Sodic feldspar is also produced at this stage whilst calcic plagioclase makes its appearance in the higher grades of metamorphism where it replaces epidote minerals. When clay is present in the groundmass of a limestone this gives rise to muscovite or phlogopite at low grade metamorphism, and with increasing grade these micas behave in the manner outlined above. The presence of micas in these rocks tends to give them a schistose appearance, schistose marbles or calc-schists being developed. Where mica is abundant it forms lenses or continuous layers giving the rock a foliated structure. If the supply of aluminous and siliceous material is plentiful abundant calcic plagioclase may be produced and the impure limestone converted into a rock consisting of plagioclase, pyroxene and/or amphibole with a granoblastic texture. When iron and magnesia are also abundant they give rise to amphibolites made up of hornblende, plagioclase and garnet. At high grade, metamorphic differentiation may take place, the rock separating into layers of pure marble and amphibolite.

2.5.3 Regional metamorphism of igneous rocks

In regionally metamorphosed rocks derived from acid igneous parents quartz and white mica are important components, muscovite–quartz schist being a typical product of the lower grades. Sericitisation of potash feldspar may occur at this stage if sufficient water is available. Conversely at high grades white mica is converted to potash feldspar. In the medium and high grades quartzo-feldspathic gneisses and granulites are common. Some of the gneisses

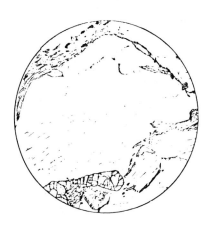

Figure 2.7 Augen-Gneiss, Glen Doll, Forfarshire in which the oligoclase crystals have the lenticular shape and mica exhibits a fluxional arrangement ($\times$ 25) (after Harker (1939))

are strongly foliated and may develop 'augen' (Figure 2.7). Biotite persists into the highest grades but there its dissociation may lead to the appearance of garnet, and the excess alumina which is liberated may then combine with silica to form kyanite or sillimanite.

If intermediate igneous rocks undergo regional metamorphism the first rocks which are formed are calc–albite–sericite–chlorite schists which are followed by albite–epidote–sericite–chlorite schists. With an advance in grade the place of chlorite and epidote is taken by biotite and hornblende, first in biotite–hornblende schists then in biotite–hornblende gneisses.

Basic rocks are converted into greenschists by low grade regional metamorphism, to amphibolites at medium grade and pyroxene granulites and eclogites at high grades.

2.6 The facies concept

The facies concept has been advanced as a method of classifying metamorphic rocks. Rocks belonging to a particular metamorphic facies have supposedly attained equilibrium under a given set of conditions. Their mineralogical composition was determined by the bulk chemical composition of their parent rocks and varies directly with changes in physical conditions (see Fyfe *et al.* (1958)).

Eskola (1915) defined a mineral facies as 'comprising all rocks which have originated under temperature–pressure conditions so similar that a definite chemical composition has resulted in the same set of minerals regardless of their mode of crystallisation'. In 1939 (Eskola *et al.* (1939)) wrote that a particular metamorphic facies consisted of rocks which possessed a characteristic correlation between chemical and mineralogical composition, that is, that rocks of a given chemical composition always had the same mineralogical composition and that differences in chemical composition from one individual rock type to another were reflected in the systematic differences in their mineralogical composition. A metamorphic facies involves more than one rock type and therefore comprises a group of mineral parageneses. The different parageneses are derived from rocks of different chemical composition but are presumed to be developed under specific metamorphic conditions, peculiar to the particular facies in question. Thus any facies is distinguished on the basis of a set of mutually associated rocks which together embrace a wide range of compositions. Nonetheless all rocks with the same chemical composition have the same mineralogical compositions if they fall within the same facies. One of the more recent definitions of metamorphic facies has been provided by Fyfe and Turner (1966) who have described it as 'a set of metamorphic mineral assemblages, repeatedly associated in space and time, such that there is a constant and therefore predictable relation between mineral composition and chemical composition'.

Certain minerals are stable over a wide range of temperature–pressure conditions and are consequently found in a number of metamorphic facies. Hence in order to define a metamorphic facies a critical mineral assemblage, stable over a limited field, is chosen, the facies itself being named after the diagnostic rock type or the diagnostic mineral suite. The number of metamorphic facies which will ultimately be recognised is limited only by the sensitivity of the common mineralogical assemblages to temperature–pressure changes. As a result, as knowledge of metamorphic paragenesis increases, the number of facies recognised is likely to increase.

In 1939 Eskola *et al.* distinguished eight facies which they correlated with temperature and pressure conditions (Figure 2.8). Subsequently other facies have been established and two of the most recent systems have been advanced by Winkler (1967) on the one hand and Turner (1968) on the other. A comparison of the two is given in Table 2.2.

Table 2.2. A comparison of the metamorphic facies of Winkler and Turner

After Winkler (1967)	*After Turner (1968)*
(1) Shallow contact metamorphism Very low fluid pressures of generally less than 1500 bar; sequence of facies in response to rising temperature	(1) Facies of low pressure
(a) Albite-epidote-hornfels	(a) Albite-epidote-hornfels
(b) Hornblende-hornfels	(b) Hornblende-hornfels
(c) K-feldspar-cordierite-hornfels	(c) Pyroxene-hornfels
(d) Sanidinite	(d) Sanidinite
(2) Regional metamorphism **A** Low and intermediate pressure of about 2000–6000 bar; sequence of facies in response to rising temperature	(2) Facies of medium to high pressure
(a) Greenschist	(a) Zeolite
(b) Cordierite-amphibolite	(b) Prehnite-pumpellyite-metagreywacke
B Higher and very high pressures, in excess of 6000 bar; sequence of facies in response to rising temperatures	(c) Greenschist
	(d) Amphibolite
(a) Greenschist (may be glaucophanitic)	(e) Granulite
(b) Almandine-amphibolite	
(c) Granulite	
(3) Burial metamorphism **A** Low temperatures, intermediate pressures; sequence of facies in response to slightly rising temperatures	(3) Facies of very high pressure
(a) Laumontite-prehnite-quartz	(a) Glaucophane-lawsonite
(b) Pumpellyite-prehnite-quartz	(b) Eclogite
B Low temperatures; high and very high pressures; sequence of facies in response to rising pressure	
(a) Lawsonite-albite	
(b) Lawsonite-glaucophane	

1 bar = 100 kPa

2.7 Metasomatism

Metasomatic activity involves the introduction of material into, as well as removal from, a rock mass by a gaseous or aqueous medium, the resultant chemical reactions lead to mineral replacement (see Ramberg (1952)). Thus two types of metasomatic action can be distinguished, pneumatolytic—that of gas, and hydrothermal—that of hot solutions. Replacement is brought about by atomic or molecular substitution so that there is usually little change

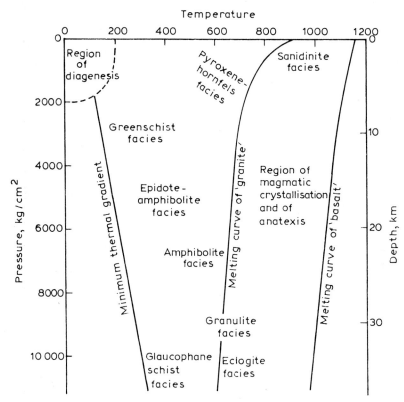

Figure 2.8 The principal metamorphic facies in relation to temperature and pressure (after Eskola (1939))

in rock texture and indeed fossils are often preserved intact. The composition of the transporting medium is continuously changing because of material being dissolved out of, and emplaced into the rocks which are affected. However, a rock developed through metasomatism is not solely governed by the active constituents in the transporting medium but it is also significantly determined by the rate of diffusion of material being lost and gained. The rate at which reaction and replacement occurs depends upon the character of the material involved, the concentration in the transporting medium, the rate of infiltration, the solubility and reactivity of the rocks, and the temperature–pressure conditions.

The gases and hot solutions involved usually emanate from an igneous source and the effects of metasomatism are often particularly notable about an intrusion of granitic character. Indeed there is a greater concentration of volatiles in acid than in basic magmas. On a smaller scale metamorphic activity liberates some carbon dioxide and water which may have some metasomatic effects.

Both gases and solutions make use of any structural weaknesses such as faults or joint planes, in the rocks they invade. Because these provide easier paths of escape, metasomatic activity is concentrated along them. They also travel through the pore spaces in rocks, the rate of infiltration being affected by the porosity, the shape of the pores and the temperature–pressure gradients.

A number of types of metasomatism have been recognised depending upon the principal constituent involved in replacement, for example, alkali metasomatism which gives rise to feldspathisation, boron metasomatism which is responsible for tourmalinisation and fluorine metasomatism which leads to the formation of white micas.

Metasomatic action, especially when it is concentrated along fissures and veins, may bring about severe alteration of certain minerals. For instance, feldspars in granite or gneiss may be highly kaolinised as a result of metasomatism; limestone may be reduced to a weakly bonded granular aggregate. Such alteration significantly lowers the engineering performance of rock concerned.

References

BARROW, G. (1912). 'On the geology of lower Deeside and the southern Highland border', *Proc. Geol. Ass.*, **23**, 268–84.

DACHILLE, F. & ROY, R. (1964). 'Effectiveness of shearing stresses in accelerating solid phase reactions at low temperatures and high pressures', *Geol. J.*, **72**, 243–7.

EASTWOOD, T., HOLLINGWORTH, S. E., ROSE, W. C. C. & TROTTER, F. M. (1968). 'Geology of the country around Cockermouth and Caldbeck', *Mem. Inst. Geol. Sci.*, HMSO, London.

ESKOLA, P. (1915). 'On the relations between chemical and mineralogical composition in the metamorphic rocks of the Orijarvi Region', *Bull. Comm. Geol.*, No. 44.

ESKOLA, P., BARTH, T. F. W. & CORRENS, C. W. (1939). *Die Enstehung der Gesteine*, Springer-Verlag, Berlin.

FLINN, D. (1965). 'Deformation in metamorphism' (in *Controls of Metamorphism*), Pitcher, W. S. & Flinn, G. W. (eds), Oliver & Boyd, Edinburgh.

FYFE, W. S. & TURNER, F. J. (1966). 'Reappraisal of the metamorphic facies concept', *Contrib. Mineral Petrol.*, **12**, 354–64.

FYFE, W. S., TURNER, F. J. & VERHOOGEN, J. (1958). 'Metamorphic reactions and metamorphic facies', *Geol. Soc. Am. Mem.*, **73**.

HARKER, A. (1939). *Metamorphism*, Chapman & Hall, London.

JAEGER, J. C. (1959). 'Temperature outside a cooling intrusive sheet', *Am. J. Sci.*, **257**, 44–54.

MASON, R. (1978). *Petrology of the Metamorphic Rocks*, Allen & Unwin, London.

MIYASHIRO, A. (1973). *Metamorphism and Metamorphic Belts*, Allen & Unwin, London.

RAMBERG, H. (1952). *The Origin of Metamorphic and Metasomatic Rocks*, University of Chicago Press, Chicago.

READ, H. H. (1936). 'Stratigraphical order of the Dalradian rocks of the Banffshire coast', *Geol. Mag.*, **73**, 468–82.

SPRY, A. (1969). *Textures of Metamorphic Rocks*, Pergamon Press, Oxford.

TILLEY, C. E. (1925). 'A preliminary survey of metamorphic zones in the southern Highlands of Scotland', *Q. J. Geol. Soc.*, **81**, 100–112.

TURNER, F. J. (1968). *Metamorphic Petrology*, McGraw-Hill, New York.

TURNER, F. J. & WEISS, L. E. (1963). *Structural Analysis of Metamorphic Tectonites*, McGraw-Hill, New York.

VERNON, R. H. (1976). *Metamorphic Processes*, Allen & Unwin, London.

WINKLER, H. G. F. (1967). *Petrogenesis of Igneous and Metamorphic Rocks*, Springer-Verlag, Berlin.

YODER, H. S. (1955). 'Role of water in metamorphism', *Geol. Soc. Am.*, Spec. Paper 62.

Chapter 3

Sedimentary rocks and stratigraphy

3.1 Introduction

The sedimentary rocks form an outer skin on the Earth's crust, covering three-quarters of the continental areas and most of the sea floor. They vary in thickness up to ten kilometres. Nevertheless they only comprise about 5% of the crust.

Most sedimentary rocks are of secondary origin in that they consist of detrital material derived by the breakdown of pre-existing rocks. Indeed it has been variously estimated that shales and sandstones, both of mechanical derivation, account for between 80 and 95% of all sedimentary rocks. Certain sedimentary rocks are the products of chemical or biochemical precipitation whilst others are of organic origin. Thus as long ago as 1904 Grabau suggested that the sedimentary rocks could be divided into two principal groups, namely, the clastic or exogenetic and the non-clastic or endogenetic types (see Grabau (1904)). The latter type was further subdivided into precipitated and organic subgroups. However, one factor which all sedimentary rocks have in common is that they are deposited and this gives rise to their most noteworthy characteristic, that is, they are bedded or stratified.

3.1.1 Formation of clastic sediments

As noted previously, most sedimentary rocks are formed from the breakdown products of pre-existing rocks. Accordingly the rate at which denudation takes place acts as a control on the rate of sedimentation, which in turn affects the character of a sediment. However, the rate of denudation is not only determined by the agents at work, that is, by weathering, or by river, marine, wind or ice action, but also by the nature of the surface. In other words upland areas are more rapidly worn away than are lowlands (see Krynine (1941)). Indeed denudation may be regarded as a cyclic process, in that it begins with or is furthered by the elevation of a land surface and as this is gradually worn down the rate of denudation slackens (Figure 3.1). Each cycle of erosion is accompanied by a cycle of sedimentation. In addition the harder the rock, the more able it is to resist denudation. Geological structure also influences the rate of breakdown. A further point to bear in mind

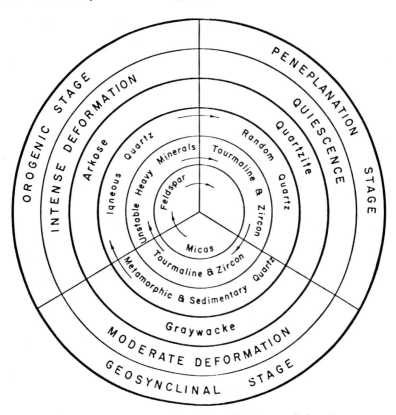

Figure 3.1 Krynine's tectonic cycle (from Distinguished Lecture Series, *Am. Ass. Petrol. Geologists* (1943))

regarding sedimentation is that the amount is affected by the amount of subsidence which occurs in a basin of deposition.

The particles of which most sedimentary rocks are composed have undergone varying amounts of transportation (the length of travel undergone by a sedimentary particle depends upon its size, shape and relative density as well as the density, viscosity, velocity and turbulence of the transporting medium). The amount of transport together with the agent responsible, be it water, wind or ice, plays an important role in determining the character of a sediment. For instance, transport over short distances usually means that the sediment is unsorted (the exception being beach sands), as does transportation by ice. With lengthier transport by water or wind not only does the material become sorted but it is further reduced in size. The character of a sedimentary rock also is influenced by the type of environment in which it has been deposited, witness the presence of ripple marks and cross bedding in sediments which accumulate in shallow water.

The composition of a sedimentary rock depends partly on the composition of the parent material and the stability of its component minerals (see Goldich (1938)) and partly on the type of action to which the parent rock was subjected and the length of time it had to suffer such action. The least stable minerals tend to be those which are developed in environments very different from those experienced at the Earth's surface. In fact quartz, and to a much lesser extent, mica, are the only common constituents of igneous and metamorphic rocks which are found in abundance in sediments. Most of the others ultimately giving rise to clay minerals. The more mature a sedimentary rock is, the more it approaches a stable end product and very mature sediments are likely to have experienced more than one cycle of sedimentation.

The type of climatic regime in which a deposit accumulates and the rate at which this takes place also affect the stability and maturity of the resultant sedimentary product. For example, chemical decay is inhibited in arid regions so that less stable minerals are more likely to survive than in humid regions. However, even in humid regions immature sediments may form when basins are rapidly filled with detritus derived from neighbouring mountains, the rapid burial affording protection against the attack of subaerial agencies.

As can be inferred from the above the maturity of a sediment can be assessed in terms of texture as well as composition. Several stages of textural maturity were recognised by Folk (1951) and are illustrated in Figure 3.2.

3.1.2 Lithification

In order to turn an unconsolidated sediment into a solid rock it must be lithified. Lithification involves two processes, consolidation and cementation. The amount of consolidation which takes place within a sediment depends, first, upon its composition and texture and, second, upon the pressures acting on it, notably that due to the weight of overburden. Consolidation of sediments deposited in water also involves dewatering, that is, the expulsion of connate water from the sediments. The porosity of a sediment is reduced as consolidation takes place and as the individual particles become more closely packed they may even be deformed. Pressures developed during consolidation may lead to the differential solution of minerals and the authigenic growth of new ones.

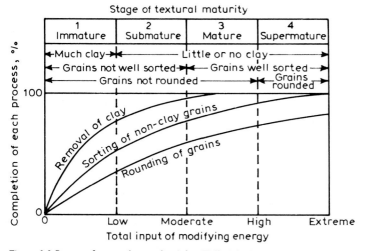

Figure 3.2 Stages of textural maturity (after Folk (1951))

Fine grained sediments possess a higher porosity than do coarser types and therefore undergo a greater amount of consolidation (see Jones (1944)). For instance, muds and clays may have original porosities ranging up to 80% compared with 45–50% in sands and silts. Hence, if muds and clays could be completely consolidated (they never are) they would occupy only 20–45% of their original volume. The amount of consolidation which takes place in sands and silts varies from 15 to 25%.

Cementation involves the bonding together of sedimentary particles by the precipitation of material in the pore spaces. This reduces the porosity. The cementing material may be derived by partial intrastratal solution of grains or may be introduced into the pore spaces from an extraneous source by circulating waters. Conversely cement may be removed from a sedimentary rock by leaching. The type of cement and, more importantly, the amount, affect the strength of a sedimentary rock. The type also influences its colour. For example, sandstones with a siliceous or calcium carbonate cement are usually whitish grey, those with a sideritic cement are buff coloured, whilst a red colour is indicative of a hematitic cement and brown of limonite. However, sedimentary rocks are frequently cemented by more than one material.

The matrix of a sedimentary rock refers to the fine material trapped within the pore spaces between the particles. It helps to bind the latter together.

3.2 The texture of sedimentary rocks

The texture of a sedimentary rock refers to the size, shape and arrangement of its constituent particles. The size and shape of sedimentary particles are initially controlled by the fracture pattern of the parent rock. This initial shape is often an influential factor and indeed it may have an important bearing upon the final shape of a particle. What is more the strength and durability of an individual fragment affects its further comminution, as does the length of transport and the nature of the transporting medium (contrast the likelihood of breakdown occurring to material entombed in ice with that of material carried by turbulent water).

3.2.1 Particle size distribution

Size is a property which is not easy to assess accurately, for the grains and pebbles of which clastic sediments are composed are irregular, three-dimensional objects. Direct measurement can only be applied to large individual fragments where the length of the three principal axes can be recorded. But even this rarely affords a true picture of size. Estimation of volume by displacement may provide a better measure. Because of their smallness, the size of grains of sands and silts has to be measured indirectly by sieving and sedimentation techniques respectively (see Chapter 9). If individual particles of clay have to be measured, then this is done with the aid of an electron microscope.

Several methods have been used in attempts to disaggregate fine-grained rocks like shale but none have proved completely satisfactory (see Krumbein

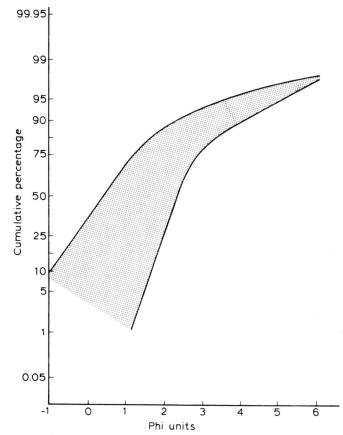

Figure 3.3 Diagram showing size range of sands as depicted by the two extreme cumulative curves

and Pettijohn (1938); Milner (1962)). As a consequence comparatively few size analyses of shales have been carried out.

If a rock is strongly indurated then its disaggregation is impossible without fracturing many of the grains. In such a case a thin section of the rock is made and size analysis is carried out with the aid of a petrological microscope, mechanical stage and micrometer. The Humphries micrometer and counter unit is one of the most useful in this respect.

The results of a size analysis may be represented graphically by a frequency curve or histogram. More frequently, however, they are used to draw a cumulative curve (see Chapter 9). The latter may be drawn on semilogarithmic (Figure 9.2) or probability paper (Figure 3.3). Probability paper is used when the curve is plotted against the phi scale rather than in millimetres. The phi scale was introduced by Krumbein (1938) and is based upon the Wentworth scale (see below). Krumbein recognised that statistical computation using the Wentworth scale was laborious and that it would be advantageous to convert the millimetre scale to a logarithmic scale in which the ratios of the Wentworth scale became equal arithmetic intervals. He therefore made use of the $\sqrt{2}$ ratio of the Wentworth scale as follows

$$d = 2^{-\phi}$$

where d is the diameter in millimetres, then

$$\log d = -\phi \log 2$$

The diameters of the grains in most sediments are less than one millimetre, consequently the use of the minus sign allows fractions of a millimetre to be expressed as positive integers on the phi scale.

Various statistical parameters such as median and mean size, deviation, skewness and kurtosis can be calculated from data derived from cumulative curves (see Griffiths (1967)). The median or mean size permits the determination of the grade of gravel, sand or silt, or their lithified equivalents. Deviation affords a measure of sorting. However, the latter can be quickly and simply estimated by visual examination of the curve in that the steeper it is, the better the sorting of the sediment. Trask (1930) introduced the term sorting coefficient (So) which was defined as

$$So = \sqrt{(Q_3/Q_1)}$$

where Q_3 and Q_1 are the third and first quartiles respectively. Although the concept has been much criticised, Friedman (1962) assigned new class limits to it, which appear to be more satisfactory:

very well sorted,	Less than 1.17
well sorted,	1.17–1.20
moderately well sorted,	1.20–1.35
moderately sorted,	1.35–1.87
poorly sorted,	1.87–2.75
very poorly sorted,	Over 2.75

It must be pointed out that the geologist and civil engineer place different meanings upon the term well sorted. To the former it means that most of the sediment falls within one size range, say medium grained sand, whilst the civil engineer refers to this as uniform sorting. By well sorted he means that the sediment is well distributed throughout its size grades. Unevenly sorted is frequently used to describe a sediment which has a bimodal frequency curve while poorly sorted refers to an irregular distribution throughout the size grades (see also Chapter 9).

Whether or not a curve is normally distributed can be determined from the skewness and kurtosis measures derived from a cumulative curve, or by visual examination of a frequency curve. However, if phi probability paper is used, then a Gaussian curve plots as a straight line. On the other hand if a curve on such paper breaks sharply to the right or left, then there is an excess of fine or coarse material respectively in that region of the curve (it is customary for the sedimentologist to represent the coarse fraction of a size analysis on the left of the abscissa and the fine on the right whilst in civil engineering the reverse is the case).

The size of the particles of a clastic sedimentary rock allows it to be placed in one of three groups which are termed rudaceous or psephitic, arenaceous or psammitic and argillaceous or pelitic. The size limits set out in Table 3.1 are those of the Wentworth scale which was introduced in 1922 by Wentworth (Wentworth (1922a)). This scale is the one most frequently used by

Table 3.1. Classification of size grade of clastic sedimentary rocks (after Wentworth (1922a))

Grade size (mm)	(phi units)	Examples	Descriptive category
256	−8	Boulder gravel, breccia, conglomerate	Rudaceous
64	−6	Cobble gravel, breccia, conglomerate	
4	−2	Pebble gravel, breccia, conglomerate	or
2	−1	Granule gravel, breccia, conglomerate	psephtic
1	0	Very coarse sand, sandstone, greywacke	Arenaceous
0.5	1	Coarse sand, sandstone, greywacke	
0.25	2	Medium sand, sandstone, greywacke	or
0.125	3	Fine sand, sandstone, greywacke	
0.0625	4	Very fine sand, sandstone, greywacke	psammitic
0.0312	5	Coarse silt, siltstone	Argillaceous
0.0156	6	Medium silt, siltstone	
0.0078	7	Fine silt, siltstone	
0.0039	8	Very fine silt, siltstone	or
0.00195	9	Coarse clay, shale, mudstone	
0.000975	10	Medium clay, shale, mudstone	
		Fine clay, shale, mudstone	pelitic

sedimentologists. Reference to the British Standard scale, which is used in soil mechanics, is made in Chapter 9 where a description of mixed aggregates is also provided.

3.2.2 Particle shape

Shape is probably the most fundamental property of any particle but unfortunately it is one of the most difficult to quantify. Nevertheless, measurement of shape may be made with reference to regular solids. Wentworth (1922b) was one of the first workers to develop a quantitative system of shape measurement, expressing the shape of pebbles by a roundness–flatness ratio. Subsequently Wadell (1932) differentiated between shape and roundness, stating that roundness was concerned with the sharpness of the edges and corners of a particle whereas shape expressed its form. He defined roundness as

$$\text{Roundness} = \frac{\sum r_1, r_2, r_3, \ldots, r_n}{R_i \times n}$$

where r_1, r_2, etc. are the radii of curvature of the corners of a particle, n is the number of corners, and R_i is the radius of the largest inscribed circle (Figure 3.4). Perfect roundness is expressed as 1, anything less as a fraction. Powers (1953) established the following grades of roundness

Description	Class intervals
Very angular	Under 0.17
Angular	0.17–0.25
Subangular	0.25–0.35
Subrounded	0.35–0.49
Rounded	0.49–0.70
Well rounded	Over 0.70

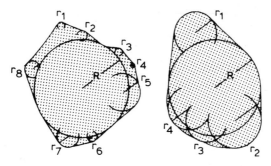

Figure 3.4 Assessment of roundness of grains (after
Wadell (1932))

As fragments are abraded they become smaller and, unless they split across
the middle, they are progressively rounded by wearing away of their corners.
Accordingly it is possible to visualise an optimum size at which the
maximum degree of roundness is attained. Beyond this optimum size further
abrasion leads to a reduction in roundness. Nevertheless the more mature a
sediment is, the more rounded its component particles are likely to be.

Sphericity is a more difficult quantity to define than roundness. Wadell
(1933) initially expressed it as follows

$$\text{Sphericity} = \frac{\text{Surface area of a sphere of the same volume as particle}}{\text{Actual surface area of particle}}$$

However, owing to the difficulty of determining the surface area and
volume of a small grain he alternatively suggested that the degree of sphericity
could be represented by

$$\text{Sphericity} = \frac{\text{Diameter of a circle equal in area to that of the particle}}{\text{Diameter of the largest inscribed circle}}$$

Even so, measuring the area of a particle is very time consuming. Consequently Riley (1941) introduced the concept of projection sphericity which
he defined as

$$\text{Projection sphericity} = \sqrt{\frac{\text{Diameter of largest inscribed circle}}{\text{Diameter of largest circumscribed circle}}}$$

Measurement of the roundness and projection sphericity of large fragments
can be done by projecting their areas onto paper and then drawing the
various circles involved. As far as assessment of the roundness and
projection sphericity of small particles is concerned, then a petrological
microscope can be used together with a special graticule, on which there
are circles of varying sizes, which is fitted into the ocular (see Robson
(1958)).

Alternatively, roundness and sphericity may be estimated visually by comparison with standard images (Figure 3.5). However, because the latter is
a subjective assessment, the values obtained suffer accordingly.

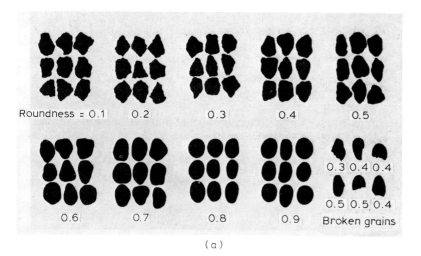

Roundness = 0.1   0.2   0.3   0.4   0.5

0.6   0.7   0.8   0.9   Broken grains

0.3 0.4 0.4
0.5 0.5 0.4

(a)

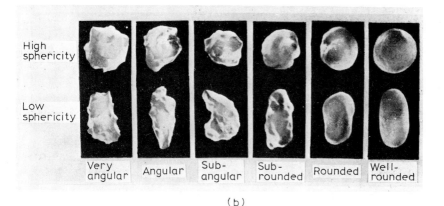

High
sphericity

Low
sphericity

Very angular   Angular   Sub-angular   Sub-rounded   Rounded   Well-rounded

(b)

Figure 3.5 (a) Images for estimating roundness values (after Krumbein, W. C. (1941), 'Measurement and geological significance of shape and roundness of sedimentary particles', *J. Sed. Pet.*, **11**, 64–72) (b) Models of roundness and sphericity (after Powers (1934))

3.2.3 Grain orientation

A sedimentary rock is an aggregate of particles and some of its characteristics depend upon the position of these particles in space. Ideally the position of a particle may be defined with reference to some standard axes. In practice, however, the measurement of grain orientation has been reduced from the ideal concept to an attempt to specify the degree of parallelism or preferred orientation of apparent long axes of grains, measured in thin sections cut with reference to the bedding plane. The degree of grain orientation varies between perfect preferred orientation, in which all the long axes run in the same direction, and perfect random orientation, where the long axes point in all directions. The latter is only infrequently found as most aggregates possess some degree of grain orientation. Four types of grain orientation have been distinguished

(1) sediments with a homogeneous high degree of preferred grain orientation,
(2) sediments with a high degree of preferred grain orientation in which the direction of orientation and its degree of perfection varies in different laminae,
(3) heterogeneous grain orientation in which the orientation is patchy,
(4) sediments in which grain orientation is lacking.

Beach and river deposits may exhibit pronounced preferred orientation and glacial deposits often show preferred orientation of their larger particles.

3.2.4 The concept of packing

The arrangement of particles in a sedimentary rock involves the concept of packing which refers to the spatial density of the particles in the aggregate. Taylor (1950) was the first to attempt a thin section measurement procedure of the arrangement of grains in sandstone. She considered two sources of variation, namely, the shape of the contact and the number of contacts per grain. The contacts were classified as tangential, long, concavo–convex, sutured and floating grains (Figure 3.6). Taylor argued that floating grains and tangential contacts were the result of original packing; long contacts were due to original packing, overburden pressure and precipitated cement; and concavo–convex and sutured contacts were generally brought about by high overburden pressure.

One of the most detailed analyses of packing was undertaken by Kahn (1956) who defined packing as the mutual spatial relationship among the grains. He concluded that the concept of packing could be resolved into two basic aspects which he termed the unit properties and the aggregate properties of packing respectively. The former included grain-to-grain contacts and the shape of the contact. In both cases the grains were considered indi-

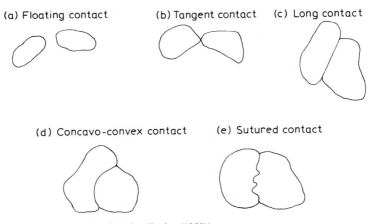

Figure 3.6 Types of grain (after Taylor (1950))

vidually. The latter involved the closeness or spread of particles and was thereby concerned with all the particles in the aggregate. In other words the aggregate property resolves the question of how much space in a given area is occupied by grains. The grain-to-grain contacts were expressed in Kahn's concept of packing proximity

$$\text{Packing proximity} = (q/n) \times 100$$

where q is the number of grain-to-grain contacts and n is the total number of contacts (as well as the total number of grains). Kahn did not consider the shape of grain contacts in his own analysis but he did introduce a further concept, namely, that of packing density which affords a means of expressing quantitatively the aggregate property of packing. If all the grains are considered as an aggregate, then the sum of the grain intercept values divided by the total length of the traverse across a rock section, corrected for magnification, would provide an expression for that amount of space in the traverse occupied by grains. Thus the packing density was defined as

$$\text{Packing density} = \frac{m\sum g_1, g_2, g_3, \ldots, g_n \times 100}{t}$$

where m is the correction for magnification, $\sum g_1, g_2, g_3, \ldots, g_n$ the sum of the grain intercept values in a given traverse, and t the total length of the traverse across the rock section.

Packing is an important property of sedimentary rocks for it is related to their degree of consolidation, density, porosity and strength.

3.3 Texture of non-clastic sedimentary rocks

Non-clastic sedimentary rocks are of chemical, biochemical or organic origin. As such they frequently exhibit a crystalline texture and the size of the crystals allows the rock to be designated coarse-, medium- or fine-grained. The crystalline texture may be of primary or secondary origin, an example of the latter occurring when limestone is dolomitised.

Oolitic and pisolitic textures are commonly found in limestone. Individual oolites and pisolites are approximately spherical in shape with a concentric layered structure. The difference between the two is simply one of size, oolites being less than 2 mm in diameter whilst pisolites exceed this dimension. Although oolites and pisolites are usually formed of calcite or dolomite, they more rarely may consist of silica, phosphate, hematite, limonite, siderite, laterite, bauxite or barite. These non-calcareous types may be original or secondary, in the latter case calcareous material is replaced by one of the other substances.

3.4 Bedding and sedimentary structures

Sedimentary rocks are characterised by their stratification, and bedding planes are frequently the dominant discontinuity in sedimentary rock masses. As such their spacing and character (are they irregular, waved or straight, tight or open, rough or smooth?) are of particular importance to the engineer. Several spacing classifications have been advanced, that given below being one of the most commonly accepted and systematic (see Anon (1970)), what is more it is related to the scale used in the mechanical analysis of soils

Description	Bedding plane spacing	Soil grading
Very thickly bedded	Over 2 m	
Thickly bedded	0.6–2 m	Boulders
Medium bedded	0.2–0.6 m	
Thinly bedded	60 mm–0.2 m	Cobbles
Very thinly bedded	20–60 mm	Coarse gravel
Laminated	6–20 mm	Medium gravel
Thinly laminated	Under 6 mm	Sand and fine gravel

An individual bed may be regarded as a thickness of sediment of the same composition which was deposited under the same conditions. However, lamination results from minor fluctuations in the velocity of the transporting medium or the supply of material, both of which produce alternating thin layers of slightly different grain size. Generally lamination is associated with the presence of thin layers of platy minerals, notably micas. These have a marked preferred orientation, usually parallel to the bedding planes, and are responsible for the fissility of the rock. The surfaces of these laminae are usually smooth and straight. Although lamination is most characteristic of shales it may also be present in siltstones and sandstones, and occasionally in some limestones.

Cross or current bedding is a depositional feature which occurs in sediments of fluvial, littoral, marine and aeolian origin and is most notably found in sandstones. In wind-blown sediments it is generally referred to as dune bedding. Cross bedding is confined within an individual sedimentation unit and consists of cross laminae inclined to the true bedding planes. The original dip of these cross laminae is frequently between 20 and 30°. The size of the sedimentation unit in which they occur varies enormously. For example, in microcross bedding (Figure 3.7a) it measures only a few millimetres, whilst in dune bedding the unit may exceed 100 m (Figure 3.7b). Several types of cross bedding have been recognised, the most common of which is planar or tabular cross bedding (Figure 3.8a). In this type the cross laminae are all inclined in the same direction. Herring-bone cross bedding differs from planar cross bedding in that the cross laminae in alternating sedimentation units dip in opposite directions. In festoon or trough cross bedding the cross laminae are contained within individual lobes which presumably are first scoured out of the sediment and then filled with the cross laminae (Figure 3.8b). First class reviews of such sedimentary structures have been provided by Allen (1968; 1970).

Although graded bedding occurs in several different types of sedimentary rock, it is characteristic of greywackes. As the name suggests the sedimentation unit exhibits a grading from coarser grain size at the bottom to finer at the top (Figure 3.9). Individual graded beds range in thickness from a few millimetres to several metres. Usually the thicker the bed, the coarser it is overall. Although greywackes are graded they are poorly sorted and their grains are notably angular. These features have been explained by

Figure 3.7 (a) Microcross bedding in fine-grained sandstone, Cementstone Group, Northumberland. (b) Dune bedding in Corrie Sandstone (Permian), Isle of Arran

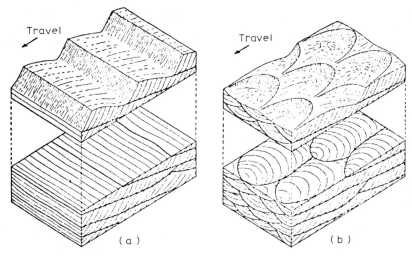

Figure 3.8 Two principal types of cross bedding (a) planar (b) festoon (or trough) (after Allen, J. R. L. (1970))

Figure 3.9 Graded bedding in greywacke, the lighter bands are the coarser material

Kuenen (1953), who proposed that greywackes were deposited by turbidity currents.

Ripple marks are ridge and furrow structures which are produced by currents of wind or water flowing at a critical velocity (between 0.96 and 2.78 km/h) over loose, granular sediments. They are very rarely seen in muddy or clayey sediments because the cohesive properties of these deposits inhibit ripple formation. Current and wind ripple marks are asymmetrical and they

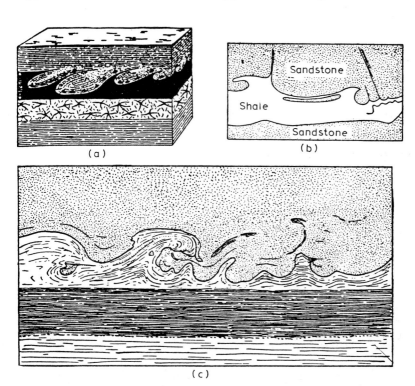

Figure 3.10 Flow-cast structures resting on coal, mudstone, and shale. All are ascribed to differential loading of a soft, viscous sediment by sand. The sand, being full of water at the beginning of the deformation, could flow readily. Later, upon losing some of the water, it became rigid and thereafter failed by shear.

(a) Sandstone flow cast resting on a coal bed 1.2 m thick. The first sand that was spread over the soft carbonaceous material sank into it, forming rolls in which some of the original stratification is still preserved. Later sand was deposited without disturbance.

(b) Basal sandstone deformations from Ordovician rocks of central Victoria, Australia. The sketch is one-fourth natural size. Prominent fissures rise from re-entrants in the base of the sandstone, and a thin sandstone lamina lies completely detached in the shale. This type of structure has been designated *basal sandstone deformations* ascribed to penecontemporaneous deformation connected with consolidation.

(c) Sketch of flow cast in a fine-grained ash layer that was deposited on a clay while the latter was still in a semi-liquid condition. About 25–35 mm of the clay was affected by the sand, which appears to have been swept in by a current moving from left to right. Wisps of the clay were caught in the flowing sand, and their present position indicates the extent of flowage in the sand (from Shrock (1948) used with permission of McGraw-Hill Book Company)

tend to form in roughly parallel sets (adjacent ripple marks which have the same orientation and occur on the same surface are referred to as sets). On the other hand, the oscillatory motion of water in wave action tends to produce symmetrical ripple marks, the crests of which are often sharp.

When soft, unconsolidated, saturated sediments are unequally loaded by later deposits they tend to flow, thereby producing folds and depressions on the surface of the loaded sediment, around which the overlying material is moulded. This occurs when muds are unevenly loaded with sands. The features, which are preserved in the sands rather than the muds because of the plastic nature of the latter, are termed flow casts (Figure 3.10).

The effects of intrastratal flowage may be seen in interbedded successions of sandstones, siltstones and shales. Such movement takes place before consolidation and the structures produced are usually confined to the sandstones and siltstones. For instance, they may be spirally wrapped to produce ball-like structures and after a period of long continual flowage a pseudoconglomerate may result. Sandstone dykes and sills may be found intruded, either from below or above, into adjacent shales. Saturated sands are more easily mobilised than muds and hence they are injected, under hydrostatic pressure, as slurries into the cracks or fissures in the muds.

Desiccation cracks develop in muddy sediments which are exposed to the air for appreciable periods of time. In this way the surface layers dry and the consequent shrinkage causes cracking (Figure 3.11). The cracks are polygonal in outline and the polygons vary in size. The cracks vary up to 100 mm or so in width. Due to differential drying mud crack polygons may turn up at their edges and the greatest amount of up-turning appears to occur in thinly bedded, fine grained muds.

Figure 3.11 Desiccation cracks in mudstone. The cracks have been filled with sandy material

Concretionary bodies are quite common in sedimentary rocks. They represent chemical precipitations of various shapes and are usually composed of one substance. Concretions may replace the host rock, thrust it aside as they develop or simply occupy cavities within it. Moreover they may form at any time during or after the sediment has accumulated, that is, they may be syngenetic or epigenetic respectively. Common examples are ironstone nodules in shales, cherts in limestones and flints in chalk. Septarian nodules, which are often found in clays and shales, are calcareous in composition and may exceed a metre in diameter.

Stylolites are post-consolidation, pressure solution phenomena, which in outline resemble irregular sutures. Usually stylolites run parallel to the bedding planes and they vary in length from a few millimetres to several metres. The widths of the largest seams are measured in tens of millimetres.

3.5 Classification of sedimentary rocks

Grabau (1904) proposed a systematic and comprehensive scheme for classifying sedimentary rocks. His system embodied a new nomenclature whereby the name of each rock provided an indication of its chemical composition and texture, as well as the agent which was largely responsible for its deposition. However, the introduction of new names has proved one of the main reasons why Grabau's system has not been widely adopted. Nevertheless, his division of sediments into two fundamental types, namely, *exogenetic* and *endogenetic* sediments forms the basis of most modern classifications. It must, however, be pointed out that even the common sediments are rarely exclusively exogenetic or endogenetic in origin. For example, in an exogenetic rock like sandstone, some amount of chemical activity goes on during diagenesis. Diagenesis refers to the reactions which take place between component minerals and pore fluids.

Because sedimentary rocks are of polygenetic origin a single workable system of classification for all sediments proves difficult, if not impossible. But partial classifications, which have a genetic basis, can be developed for certain groups of sediment.

3.5.1 Clastic sediments: rudaceous deposits

A gravel is an unconsolidated accumulation of rounded fragments, the lower size limit of which is 2 mm. The term *rubble* has been used to describe those deposits with angular fragments. The composition of a gravel deposit reflects not only the source rocks of the area from which it was derived but is also influenced by the agent(s) responsible for its formation and the climatic regime in which it was, or is being deposited. The latter two factors have a varying tendency to reduce the proportion of unstable material. Relief also influences the nature of a gravel deposit. For example, gravel production under low relief is small and the pebbles tend to be inert residues such as vein quartz, quartzite, chert and flint. Conversely high relief and accompanying rapid erosion yield coarse, immature gravels.

When a gravel becomes indurated it forms a conglomerate, when a rubble is indurated it is termed a breccia (Figure 3.12). Conglomerates and

Figure 3.12 Breccia, Isle of Arran

breccias fall into two categories. Terrigenous conglomerates are the most common type, being derived from the breakdown of pre-existing rocks outside the basin of deposition. On the other hand, intraformational conglomerates are formed entirely within the basin of deposition. Consequently the fragmentary material is of highly local origin and is only slightly worn, having undergone little or no transportation. Intraformational conglomerates are characterised by their thinness, flat-pebble form and restricted composition of the fragments.

Those conglomerates in which the fragments are in contact and so make up a framework are referred to as orthoconglomerates. By contrast those deposits in which the larger fragments are separated by matrix are referred to as paraconglomerates. The latter are poorly sorted and possess a polymodal particle size distribution, the principal mode being in the finer grades (they are in effect conglomeratic mudstones).

The fragments in orthoconglomerates may consist of a single rock type and these supermature varieties are described as oligomictic. Conversely the fragments in polymictic conglomerates are of mixed composition.

3.5.2 Clastic sediments: arenaceous deposits

Sands consist of a loose mixture of mineral grains and rock fragments. Generally they tend to be dominated by a few minerals, the chief of which is quartz. There is a presumed dearth of material in those grades transitional to gravel on the one hand and silt on the other.

Alluvial sands include those found in alluvial fans, in river channels, on flood plains and in deltaic deposits. Marine sands occur as beach deposits, as offshore bars and barriers, in tidal deltas and in tidal flats. They are mainly deposited on the continental shelf but some sand is carried over the continental edge by turbidity currents. Wind-blown sands accumulate as coastal dunes and as extensive dune fields in deserts. Sands are also produced as a result of glacial and fluvio-glacial action.

Sands tend to be close packed and usually the grains show some degree of orientation, presumably related to the direction of flow of the transporting medium. They vary greatly in maturity, the ultimate end product being a uniformly sorted quartz sand with rounded grains.

The process by which a sand is turned into a sandstone is partly mechanical, involving grain fracturing, bending and deformation. However, chemical activity is much more important. The latter includes decomposition and solution of grains, precipitation of material from pore fluids and intergranular reactions. Redistribution of material, as for instance, the solution of quartz at the points of grain contact and its precipitation as grain overgrowths and in the void space, cements grains together and reduces the porosity. In sands the porosity may average 30–35% whereas in a sandstone reduction in pore space may mean that the porosity is halved, or in extreme cases it may be reduced to more or less zero.

Solution transfer is characteristic of mature quartz sands, converting them to orthoquartzites (quartz arenites). Diagenetic activity in immature sands, especially lithic arenites with a high proportion of unstable rock fragments, leads to the breakdown of unstable material and the formation of greywacke.

Silica is the commonest cementing agent in sandstones, particularly older sandstones (see Bell (1982)). Various carbonate cements, especially calcite, are also common cementing materials. Ferruginous and gypsiferous cements are also found in sandstones. Cement, notably the carbonate types, may be removed in solution by percolating pore fluids. This brings about varying degrees of decementation.

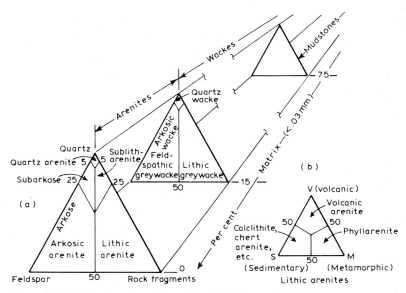

Figure 3.13 Classification of sandstones (from Pettijohn *et al.* (1972))

Most classifications of sandstone have been based upon composition. For example, quartz, feldspar and rock fragments are the principal detrital components of which sandstones are composed, and consequently they have been used to define the major classes of sandstone. Pettijohn *et al.* (1972) also used the type of matrix in their classification. In other words those sandstones with more than 15% matrix were termed wackes. The chief type of wacke is greywacke which can be subdivided into lithic and feldspathic varieties (Figure 3.13). Quartzwackes are a very uncommon group of wackes. Those sandstones with less than 15% matrix were divided into three families as follows

    (1) orthoquartzites or quartz arenites, in which 95% or more of the detrital material consists of quartz (Figure 3.14),

    (2) arkoses, in which 25% or more of the detrital material consists of feldspar (which exceeds the amount of rock particles),

    (3) lithic sandstones, in which 25% or more of the detrital material consists of rock fragments.

3.5.3 Clastic sediments: argillaceous deposits

Silts are clastic sediments derived from pre-existing rocks, chiefly by mechanical breakdown processes. They are mainly composed of fine quartz material. Silts may occur in residual soils but in such instances they are not important. However, silts are commonly found in alluvial, lacustrine, and marine deposits. As far as alluvial sediments are concerned silts typically are present in flood plain deposits and they may also occur on terraces which border such plains. These silts tend to interdigitate with deposits of sand and clay. Silts are also present with sands and clays in estuarine and deltaic

Figure 3.14 Photomicrograph of an orthoquartzite, showing grains of quartz, grain of microcline (cross-hatched) and a flake of muscovite. Occasional fragment of shaly material and matrix have fine-grained appearance

sediments. Lacustrine silts are often banded and may be associated with varved clays. Indeed varved clays contain significant proportions of particles of silt size. Marine silts also may be banded. Wind-blown silts are generally uniformly sorted.

Siltstones may be massive or laminated, the individual laminae being picked out by mica and/or carbonaceous material. Micro-cross bedding is frequently developed and in some siltstones the laminations may be convoluted. Siltstones have a high quartz content with a predominantly siliceous cement. Frequently siltstones are interbedded with shales or fine grained sandstones, the siltstones occurring as thin ribs.

Loess is a wind-blown deposit which is mainly of silt size and consists mostly of quartz particles, with lesser amounts of feldspar and clay minerals. It is commonly buff in colour, although locally it may be grey, yellow, brown or red. Loess is characterised by a lack of stratification and uniform sorting, and occurs as blanket deposits in western Europe, the United States, Russia and China. Deposits of loess are of Pleistocene age and because they show a close resemblance to fine-grained glacial debris their origin has customarily been assigned a glacial association. It is presumed that winds blowing from the arid interiors of the northern continents during glacial times picked up fine glacial outwash material and carried it for hundreds or thousands of kilometres before deposition took place. Deposition occurred over steppe lands and the grasses left behind fossil root-holes which typify loess. These

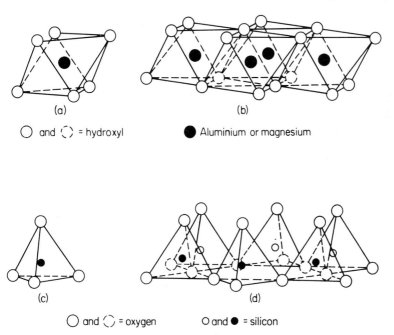

(a)

(b)

○ and ⊝ = hydroxyl   ● Aluminium or magnesium

(c)

(d)

○ and ⊝ = oxygen   ○ and ● = silicon

Figure 3.15 Fundamental units comprising the structure of clay minerals. (a) A single octahedral unit, in gibbite Al is surrounded by 6 oxygens whereas in brucite they surround Mg in six-fold coordination. (b) The sheet structure of the octahedral units. (c) The silica tetrahedron. (d) The sheet structure of silica tetrahedrons arranged in a hexagonal network (after Grim, R. E. (1960). *Clay Mineralogy*, McGraw Hill, New York.)

account for its crude columnar structure. The lengthy transport explains the uniform sorting of loess.

Deposits of clay are principally composed of fine quartz and clay minerals. The latter represent the commonest breakdown products of most of the chief rock forming silicate minerals. The clay minerals are phyllosilicates and their atomic structures can usually be regarded as consisting of two units. One of the units is composed of two sheets of closely packed oxygens or hydroxyls in which atoms of aluminium, magnesium or iron are arranged

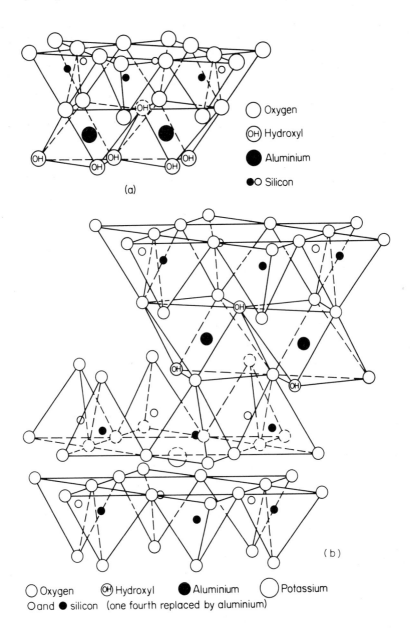

(a)

○ Oxygen
(OH) Hydroxyl
● Aluminium
●O Silicon

(b)

○ Oxygen   (OH) Hydroxyl   ● Aluminium   ○ Potassium
O and ● silicon (one fourth replaced by aluminium)

so that they have octahedral coordination (Figure 3.15a). If aluminium is present then only two-thirds of the possible positions are filled, the mineral is dioctahedral and is said to possess a gibbsite structure, $Al_2(OH)_6$ (Figure 3.15b). Conversely if magnesium or iron is present then, in order to balance the structure, all the positions are filled. The mineral is then described as trioctahedral and as possessing a brucite structure, $Mg_3(OH)_6$. The other units are formed of SiO_4 tetrahedra and these are arranged in layers, each of which have a hexagonal network, the composition of which is $Si_4O_6(OH)_4$ (Figures 3.15c and d).

The chemical composition of clay minerals varies according to the amount of aluminium which is substituted for silicon and also with the replacement of magnesium by other ions. The three major types of clay mineral are kaolinite, illite and montmorillonite, in which the fundamental units are arranged in the respective atomic lattices shown in Figures 3.16a, b and c. They are all hydrated aluminium silicates from which water can be driven on heating.

Kaolinite $(Al_4Si_4O_{10}(OH)_8)$ is principally formed by the alteration of feldspars, feldspathoids and other aluminium silicates due to hydrothermal action. Weathering under acidic conditions is also responsible for the kaolinisation of feldspars. Deposits of kaolin are associated with acid igneous rocks such as granites, granodiorites and tonalites, and with gneisses and granulites. Kaolinite, the chief clay mineral in most residual and transported clays, is important in shales, and is found in variable amounts in fire-

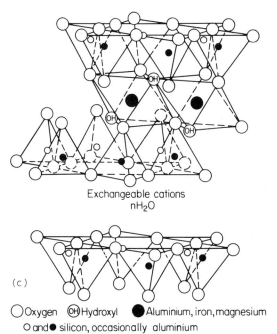

Exchangeable cations
nH_2O

(c)

○ Oxygen ⊙ Hydroxyl ● Aluminium, iron, magnesium
○ and ● silicon, occasionally aluminium

Figure 3.16 (a) Diagrammatic sketch of the kaolinite structure. (b) The structure is that of muscovite which is regarded as essentially the same as that of illite. (c) Diagrammatic sketch of the montmorillonite structure

clays, bauxites and soils. It is the most important mineral in china clays and ball clays.

Illite $(K_{2-3}Al_8(Al_{2-3}, Si_{13-14})O_{40}(OH)_8)$ is of common occurrence in clays and shales, and is found in variable amounts in tills and loess, but is less common in soils. It develops as an alteration product of feldspars, micas or ferromagnesian silicates upon weathering or may form from other clay minerals during diagenesis. Like kaolinite, illite also may be of hydro-thermal origin. For example, illite has been found in alteration zones surrounding hot springs and in the altered wall rocks around ore deposits of such origin. The development of illites, both under weathering and by hydro-thermal processes, is favoured by an alkaline environment.

Montmorillonite $(Al_4Si_8O_{20}(OH)_4 . nH_2O)$ develops when basic igneous rocks, in badly drained areas, are subjected to weathering. The presence of magnesium is necessary for this mineral to form, if the rocks were well drained then it would be carried away and kaolinite would develop. An alkaline environment favours the formation of montmorillonite. Mont-morillonites occur in soils and argillaceous sediments such as shales formed from basic rocks. It is the principal constituent of bentonic clays, which are formed by the weathering of basic volcanic ash, and of Fuller's earth, which is also formed when basic igneous rocks are weathered. In addition hydro-thermal action may lead to the development of montmorillonite; for instance, it is found in mineral veins and wall rocks, and around hot springs.

Residual clays develop in place and are the products of weathering. In humid regions residual clays tend to become enriched in hydroxides of ferric iron and aluminium, and impoverished in lime, magnesia and alkalies. Even silica is removed in hot humid regions, resulting in the formation of hydrated alumina or iron oxide, as in laterite (see Chapter 10). Both bauxitic and ferruginous laterites are typified by concretionary structures and piso-lites.

The composition of transported clays varies because these materials consist mainly of abrasion products (usually silty particles) and transported residual clay material.

Very high water contents are typical of newly deposited muds. Indeed Trask (1931) suggested that initial porosities in such muds may range between 70 and 80%. Consolidation due to the weight of overburden is one of the first post-depositional changes to occur in muds, the original material undergoing considerable consolidation and dewatering. Muds are principally river, lacustrine or marine deposits.

Shale is the commonest sedimentary rock and is characterised by its lamina-tion. Sedimentary rock of similar size range and composition, but which is not laminated is usually referred to as mudstone. In fact there is no sharp distinction between shale and mudstone, one grading into the other.

Lamination in shales is primarily due to the orientation of minerals of platy habit, chiefly micas, parallel to the bedding. However, in some shales, clay minerals exhibit a random orientation and in some instances this may be due to the authigenic growth of the minerals concerned. An increasing content of siliceous or calcareous material decreases the fissility of a shale whereas shales which have a high organic content are finely laminated (see Ingram (1953), Spears (1980)). Laminae range from 0.05 to 1.0 mm in thick-ness, with most in the range of 0.1–0.4 mm. Pettijohn (1975) recognised three kinds of lamination, namely, alternations of course and fine particles, such as silt and clay; alternations of light and dark layers distinguished only by organic content and alternations of calcium carbonate and silt.

Clay minerals and quartz are the principal constituents of mudstones and shales. Feldspars often occur in the siltier shales. Shales may also contain appreciable quantities of carbonate, particularly calcite, and gypsum. Indeed calcareous shales frequently grade into shaly limestones.

Most shales contain a large proportion of silt, which consists mainly of quartz. Indeed in a recent classification of shales Spears (1980) suggested that the boundary between mudrock and siltstone could be taken as 40% quartz content and he proposed the following classification of mudrocks

	Fissile	Non-fissile
Over 40% quartz	Flaggy* siltstone	Massive siltstone
30–40% quartz	Very coarse shale	Very coarse mudstone
20–30% quartz	Coarse shale	Coarse mudstone
10–20% quartz	Fine shale	Fine mudstone
Under 10% quartz	Very fine shale	Very fine mudstone

* Parting planes 10–50 mm apart.

Calcareous shales are very common and, as remarked, often grade into shaly limestones. The carbonate may be either calcite or dolomite and may be the dominant material. The clay mineral fraction generally consists of illite, and may constitute up to 40% of the rock. Quartz may account for up to 30%.

Siliceous shales possess an abnormally high content of silica. This may be as high as 85% as compared with 55–60% in average shale. Consequently these shales represent hard, durable rocks. The siliceous character is not attributable simply to detrital quartz but has been derived from amorphous silica such as opal or from volcanic ash or diatoms.

Carbonaceous black shales are usually rich in organic matter and are finely laminated. Indeed some carbonaceous shales have been referred to as 'paper' shales. Black shales accumulated slowly under anaerobic conditions, which accounts for their relatively high content of organic matter. It also accounts for the significant amount of iron sulphide, notably pyrite, which they contain. However, the amount of pyrite varies appreciably, over 30% having been recorded from some rare black shales. Carbonaceous shales may contain from 3–15% of carbon as compared with about 1% in average shales. Some of these shales contain unusual concentrations of certain trace elements such as vanadium, uranium, nickel and copper. Concretionary layers of carbonate or carbonate nodules are present in some black shales. Silt-size quartz may account for up to one-third of black shale, and clay minerals, mainly illite, for another third. Red shales and mudstones presumably are of terrestrial origin, being deposited in arid or semi-arid conditions. Accumulation in an oxidising environment explains their red colour.

The term *marl* has been assigned various meanings, although it has recently been defined by Pettijohn (1975) as a rock with 35–65% carbonate and a complementary content of clay. However, this definition cannot be applied to many of the rocks in Britain which are referred to as marls, for example, most of the marls of the Keuper series contain less than 20% carbonate material. Such rocks according to the classification of clay–lime carbonate

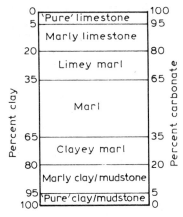

Figure 3.17 Classification of carbonate–clay mixtures (modified after Barth *et al.* (1939))

mixtures, after Barth *et al.* (1939), are marly clays or mudstones (Figure 3.17).

Marls are particularly well developed in the Old Red Sandstone, Permian and Trias systems in Britain. It is assumed that many of these marls were formed by the deposition of vast quantities of fine sediment in shallow inland lakes or seas in regions of high evaporation, others may represent aeolian deposits.

3.5.4 Carbonate rocks: limestones and dolostones

The term *limestone* is applied to those rocks in which the carbonate fraction exceeds 50%, over half of which is calcite or aragonite. If the carbonate material is made up chiefly of dolomite then the rock is named dolostone (this rock is generally referred to as dolomite but this term can be confused with that of the mineral of the same name). Limestones and dolostones constitute about 20–25% of the sedimentary rocks according to Pettijohn (1975). This figure is much higher than some of the estimates provided by previous authors.

Limestones are polygenetic. Some are of mechanical origin representing carbonate detritus which has been transported and deposited or that has accumulated *in situ*. Others represent chemical or biochemical precipitates which have formed in place. Allochthonous or transported limestone has a fabric similar to that of sandstone and may also display current structures such as cross bedding and ripple marks. By contrast carbonate rocks which have formed *in situ*, that is, autochthonous types, show no evidence of sorting or current action and at best possess a poorly developed stratification (see Bathurst (1975)). Exceptionally some autochthonous limestones show growth bedding, the most striking of which is stromatolitic bedding as seen in algal limestones.

Pettijohn (1975) distinguished five groups of modern carbonate deposits, namely, shallow water marine deposits, deep water marine carbonates, carbonates of evaporitic basins, carbonates of freshwater lakes and springs, and aeolian carbonates. Most deposits of the geological past belong to the first category, although deep-sea carbonates are the most widespread at the present day. Volume-wise the three latter types are not significant. The shallow water environment includes tidal and supratidal flats, shelf and bank areas, marginal reefs and back-reef lagoons. Shallow water limestones may be mechanically derived, accumulating as calcareous sands and muds. Deep-water carbonates are either turbidite (basinal) deposits or pelagic pteropod and globigerina oozes. The turbidite basins bordering carbonate platforms are more important environments than the abyssal environments in which pelagic deposits are laid down. Although the latter are widespread in the modern world, they leave little or no geological record. Caliche is the most widespread carbonate deposit formed in arid regions (see calcrete deposits, Chapter 10). Tufa is a porous carbonate deposit precipitated by springs and streams associated with limestone terrains. Travertine is a dense, banded deposit which is often present in caverns in limestone, again being precipitated from carbonate-rich waters. Both these freshwater deposits are generally of limited extent. Small deposits of wind-blown carbonate sand may form on beaches and as dunes on coral islands.

Lithification of carbonate sediments is often initiated as cementation at points of intergranular contact rather than as consolidation. Indeed, Bat-hurst (1975) maintained that carbonate muds consolidate very little because of this early cementation. The rigidity of the weakest carbonate rocks, such as chalk, may be attributed to mechanical interlocking of grains with little or no cement. Nevertheless, cementation may take place more or less at the same time as deposition, but cemented and uncemented assemblages may be found within short, horizontal distances. Indeed a recently cemented carbonate layer may overlie uncemented material. Because cementation occurs concurrently with, or soon after deposition, carbonate sediments can support high overburden pressures before consolidation takes place. Hence high values of porosity may be retained to considerable depths of burial. Eventually, however, the porosity is greatly reduced by post-depositional changes which bring about recrystallisation (see Choquette and Pray (1970)). Thus a crystalline limestone is formed. Terzaghi (1940) and Robertson (1965) advanced an alternative view, contending that carbonate muds which do not undergo early cementation do consolidate, which in turn reduces their porosity.

Limestone is perhaps more prone to pre- and post-consolidation changes than any other rock type. For example, after burial, limestones can be modi-fied to such an extent that their original characteristics are obscured or even obliterated. The most profound changes in composition and texture are those which lead to replacement of calcite by dolomite, silica, phosphate and so forth. But even in limestones many changes take place during transforma-tion from a soft and generally porous carbonate sediment to a dense, hard limestone with a low porosity. Furthermore carbonate rocks are susceptible to solution, which commonly removes shell fragments, or in rarer cases ooids are removed. Many limestones exhibit evidence of increases in grain size and crystallinity with increasing age of the deposit. The exact mechanisms by which this is brought about are not clearly understood (see Bathurst (1975)).

Limestones are composed of large complex grains or allochems; coarsely crystalline calcite called spar, which in many limestones constitutes the cement binding the allochems; and micrite, microcrystalline calcium car-bonate material which commonly acts as matrix. Folk (1959; 1962) recog-nised four types of allochems—intraclasts, pellets, oolites and fossil remains. Intraclasts are generally weakly cemented fragments that have undergone penecontemporaneous erosion and redeposition as clasts. They vary from fine sand size to slab-like pieces of intraformational limestone in conglomerates. Fine grained intraclasts are difficult to distinguish from pellets. The latter are small ovoid bodies or aggregates of microcrystalline calcite, devoid of internal structure. In any given rock pellets tend to be the same size and shape. They are difficult to see even with the binocular microscope. Oolites are small, roughly spherical, accretionary bodies, 0.25–2.0 mm in diameter. They have a concentric or radial structure and each oolite appears to have grown from a central nucleus (see Carozzi (1960)). If these bodies exceed 2.0 mm diameter then they are known as pisolites. Fossils are often very abundant in limestones.

Autochthonous limestones are generally fine grained and only infrequently contain oolites or intraclasts. However, they may contain abundant fossil material and pellets may be present in a fine grained matrix.

Folk (1973) distinguished two types of dolostone. First, he recognised an extremely fine grained crystalline dolomicrite (less than 20 microns grain diameter), and secondly, a more coarsely grained or saccharoidal dolostone in which there was plentiful evidence of replacement. He regarded the first type as of primary origin, and the second as being formed as a result of diagenetic replacement of calcite by dolomite in limestone. Primary dolostones tend to be thinly laminated and are generally unfossiliferous. They are commonly associated with evaporites and may contain either nodules or scattered crystals of gypsum or anhydrite. In those dolostones formed by dolomitisation the original textures and structures, as well as the fossil content, may be obscured or even have disappeared. Partial dolomitisation gives rise to a patchy distribution of dolomite within the rock mass whereas total recrystallisation produces a medium to coarsely crystalline mosaic.

Folk (1959) developed a nomenclature for his classification of carbonate rocks which was based on the type and proportions of allochems and the spar-matrix ratio (Table 3.2). The prefix in each term indicates the nature of the allochem, the stem indicates the nature of the cement or matrix, and the suffix indicates the texture or grain size. A later and simpler classification has been provided by Dunham (1962) and is based primarily on depositional texture (Table 3.3). As far as engineering purposes are con-

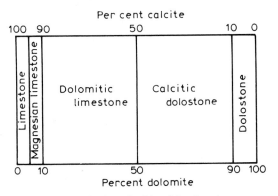

Figure 3.18 Classification of calcite–dolomite mixtures
(after Pettijohn (1975))

Table 3.2. Classification of carbonate rocks (after Folk (1959))

	Limestones, partially dolomitised limestones and primary dolostones					Non-'clastic' limestone Class IV	Replacement dolomites Class V	
	← Percent allochems ('clast'–matrix ratio) →						Allochem ghosts	No allochem ghosts
	Over 10% allochems ('clasts')		<10% allochems					
Allochem composition (varying kinds of 'clasts')	← Cement–matrix ratio →		1–10%	<1%	Microcrystalline rocks Class III			Coarse, medium, fine crystalline dolomites
	spar > matrix Class I	spar < matrix Class II						
>25% intraclasts	Intrasparrudite (intraformational conglomerate) — Intrasparite (lithic calcarenite)	Intramicrudite — Intramicrite			Micrite and dolomicrite (calcilutite)	Biohermite (klintite)	Intra-clastic dolostone	
<25% intraclasts, >25% oolites	Oosparrudite (pisolite) — Oosparite (oolitic calcarenite)	Oomicrudite — Oomicrite					Oolitic dolostone	
<25% oolites, Fossil–pellet ratio 3:1	Biosparrudite (coquina)	Biomicrudite (coquinoid limestone)					Biogenic dolostone	
3:1 to 1:3	Biosparite (biocalcarenite)	Biomicrite (fossiliferous calcilutite)						
1:3	Pelsparite (pellet calcarenite)	Pelmicrite (pelletiferous calcilutite)					Pellet dolostone	

Common terms, as used by Pettijohn (1975) given in brackets

Table 3.3. Classification of limestones according to depositional texture (after Dunham (1962))

Depositional texture recognisable					*Depositional texture not recognisable*
Original components not bound together during deposition				Original components bound together during deposition	Crystalline carbonate (subdivide according to physical or diagenetic texture)
Contains mud (fine silt and clay size particles)			Lacks mud		
Mud-supported		Grain-supported	Grain-supported		
Less than 10% grains	More than 10% grains				
Mudstone	Wackestone	Packstone	Grainstone	Boundstone	

cerned, a classification of carbonate sediments has been given by Fookes and Higginbottom (1975) (see Figure 13.1).

Rocks of intermediate composition between dolostone and limestone have been classified by Pettijohn (1975), as shown in Figure 3.18, but unfortunately no general agreement exists regarding calcite–dolomite mixtures. Such mixtures are in fact less common than the end members.

Carbonate rocks may contain various amounts of impurities, notably quartz and clay minerals. A classification of impure limestones and dolostones can be made with reference to a ternary diagram (Figure 3.19a), as can a classification of carbonate, sand (quartz) and clay mixtures (Figure 3.19b). Fine grained argillaceous limestones are frequently wavy bedded or nodular bedded, the beds being separated by thin shaly partings. The nodules range from one to several centimetres in thickness and may be 10 cm or more in length.

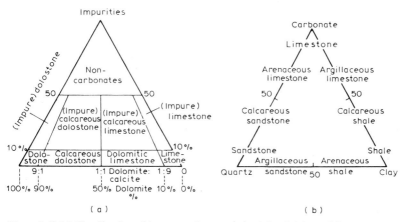

Figure 3.19 (a) Classification of impure carbonate rocks (after Leighton, M. W. & Pendexter, C. (1962). 'Carbonate rock types' (in *Classification of Carbonate Rocks*), Ham, W. E. (ed), *Mem. Am. Ass. Petrol. Geol.*, 62–85). (b) Classification of carbonate–sand–clay mixtures

3.5.5 Evaporitic rocks

Evaporitic deposits are quantitatively unimportant as sediments. They are formed by precipitation from saline waters, the high salt content being brought about by evaporation from inland seas or lakes in arid areas. Salts can also be deposited from subsurface brines, brought to the surface of a playa or sabkha flat by capillary action.

Sea water contains approximately 3.5%, by weight, of dissolved salts, about 80% of which is sodium chloride. Experimental work has shown that when the original volume of sea water is reduced by evaporation to about half, then a little iron oxide and some calcium carbonate are precipitated (see Hollingworth (1948)). Gypsum begins to form when the volume is reduced to about one-fifth of the original, rock salt begins to precipitate when about one-tenth of the volume remains, and finally when only 1.5% of the sea water is left potash and magnesium salts start to crystallise. This order agrees in a general way with the sequence found in some evaporitic deposits; however, many exceptions are known. Many complex replacement sequences occur amongst evaporitic rocks, for example, carbonate rocks may be replaced by anhydrite and sulphate rocks by halite (see Stewart (1963)).

Experimental work has also indicated that if all the salt in a column of sea water 305 m deep were precipitated it would produce a deposit 4.6 m thick. Of this 0.15 m would be calcium sulphate, 3.6 m would be rock salt and 0.8 m would be potash–magnesium salts.

Work carried out by Hardie (1967) on the gypsum–anhydrite equilibrium demonstrated that although anhydrite could be formed by dehydration at one atmosphere pressure and under geologically reasonable conditions of temperature and activity of water in a geologically reasonable time, primary precipitation of anhydrite could not be achieved. Considerably higher temperatures were required for the primary formation of anhydrite. This suggests that anhydrite is subsequently developed from gypsum, possibly by dehydration due to depth of burial.

Because of the high solubility of most evaporites they only outcrop at the surface in very arid regions. Moreover deposits of very soluble materials are rare. Gypsum and anhydrite are the exception, they occur at the surface and are the commonest evaporites. They frequently occur in a relatively pure state and beds may be of considerable thickness and extent.

Gypsum occurs in a number of ways dependent upon the process by which it was initially deposited and on post-depositional changes. Rock gypsum varies from coarsely crystalline to finely crystalline, the latter being the more common. In some cases gypsum is thinly laminated and it may be interlaminated with dolostone. It also occurs in some deposits as scattered nodules, generally in a carbonate matrix. Gypsum frequently forms veins in rock gypsum and associated beds.

Anhydrite may occur as thick extensive beds. These are usually finely granular, although fibrous and coarse grained varieties do occur.

3.5.6 Carbonaceous rocks

Organic residues which accumulate as sediments are of two major types, namely, peaty material which when buried gives rise to coal, and sapropelic residues. Sapropel is silt rich in, or composed wholly of, organic compounds which collect at the bottom of still bodies of water. Such deposits may give rise to cannel or boghead coals. Decomposition takes place in the presence of little oxygen and various hydrocarbons are formed. Sapropelic coals usually have a significant amount of inorganic matter as opposed to humic coals in which the inorganic content is low. The former are not generally extensive and are not underlain by seat earths. Sapropelites such as oil shales are simply shales rich in sapropelic residues from which oil can be distilled.

Peat deposits accumulate in poorly drained environments where the formation of humic acid gives rise to deoxygenated conditions (see Raistrick and Marshall (1938)). These inhibit the bacterial decay of organic matter. Peat accumulates wherever the deposition of plant debris exceeds the rate of its decomposition. A massive deposit of peat is required to produce a thick seam of coal, for example, a seam 1 m thick probably represents 15 m of peat. As many coal seams of Upper Carboniferous age are thicker than one metre

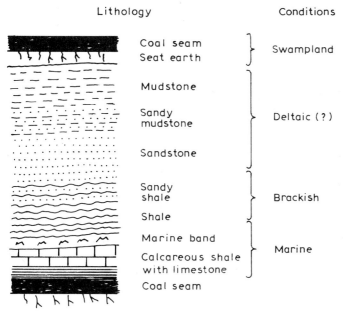

Lithology Conditions

Figure 3.20 Diagrammatic representation of a cyclotherm in the coal measures

and are extensively distributed, it has been assumed that they are the remains of vast forests which grew in low-lying swamplands. After a thick deposit of peaty material had been formed, it needed to be buried in order to be preserved. Burial usually corresponded with a period of more rapid subsidence which led to the deposition of muds and sands. The latter would then build up to form a new surface on which a prolific cover of vegetation could again exist. The pattern was repeated again and again so that the Coal Measures are characterised by cyclic sedimentation, each rhythmic unit being termed a cyclothem (Figure 3.20).

In order to convert peat to coal the carbon content must be increased, with a concomitant decrease in oxygen and a small reduction in hydrogen.

Rank of coal	Carbon (%)	Hydrogen (%)	Oxygen (%)
Peat	57	6.5	34.5
Lignite	70	5.5	23
Bituminous coal	86	5.5	6.8
Anthracite	94	3.5	1.5

The degree of alteration determines the rank of coal.

3.5.7 Other types of sedimentary rock

Chert and flint are the two most common siliceous sediments of chemical origin. Chert is a dense rock composed of one or more forms of silica such as opal, chalcedony or microcrystalline quartz. Sponge spicules and radiolarian remains may be found in some cherts and carbonate material may be scattered throughout impure varieties. Gradations occur from chert to sandstone with chert cement, although sandy cherts are not common. Cherts may suffer varying degrees of devitrification.

Chert may occur as thin beds or as nodules in carbonate host rocks. Both types are of polygenetic origin. In other words chert may be a replacement product, as for example, in siliceous limestones, or it may represent a biochemical accumulate formed in a basin below the calcium carbonate compensation depth. In yet other cases chert may be a product of an ephemeral silica-rich alkaline lake environment.

Cherts are found in cratonic, geosynclinal and evaporitic environments. Cratonic cherts are silicification products associated with shallow water limestones and orthoquartzites. Geosynclinal cherts are associated with siliceous black shales. They represent biochemical accumulates derived mainly from radiolarian and diatomaceous oozes. Evaporitic, or at least hypersaline deposits occur as nodules and layers interbedded with mudrocks.

Some sediments may have a high content of iron. The iron carbonate, siderite, often occurs interbedded with chert or mixed in varying proportions with clay, as in clay ironstones. Some iron-bearing formations are formed mainly of iron oxide, hematite being the most common mineral. Hematite-rich beds are generally oolitic. Limonite occurs in oolitic form in some ironstones. Bog iron ore is chiefly an earthy mixture of ferric hydroxides. Siliceous iron ores include chamositic ironstones, which are also typically oolitic. Glauconitic sandstones and limestones may contain 20% or more FeO and Fe_2O_3. On rare occasions bedded pyrite has been found in black shale.

Rocks which contain 50% or more of apatite have been termed phosphorites. These rocks may initially have had a high content of phosphate or subsequently they may have been phosphatised. For instance, some limestones have been altered by percolating phosphate-bearing solutions. Phosphate may occur in nodular form. Residual phosphate is a surface deposit of insoluble phosphate material left as a residuum from solution of limestone in which it was once dispersed. Bone phosphate and guano represent organic remains which are rich in phosphate.

3.6 Stratigraphy and stratification

Weller (1960) defined stratigraphy as 'the branch of geology that deals with the study and interpretation of stratified and sedimentary rocks and with the identification, description, sequence, both vertical and horizontal, mapping and correlation of stratigraphic rock units'. He went on to state that stratigraphy begins with the discrimination and description of stratigraphical units such as formations. This is necessary so that the complexities present in every stratigraphical section may be reduced, simplified and organised. Indeed without the recognition of stratigraphical units the stratigrapher would founder in a welter of detail.

Modern stratigraphy is based on three fundamental principles, namely, the *Law of Superposition*, the *Law of Faunal Succession* and the *Doctrine of Uniformitarianism*. The two former are explained below. The doctrine of uniformitarianism is one of the most important concepts in geology and was formulated towards the end of the eighteenth century by James Hutton. It postulates that geological processes which operated in the past are similar to those which are observed at the present day. Hence the present provides the key for the interpretation of the past. However, geological processes may not always have proceeded at the same rate as at the present. Moreover not all geological processes can be observed directly. For instance, the effects of those which occur at depth within the Earth's crust, such as the emplacement of granite or regional metamorphism of rocks, can only be seen after being exposed by millions of years of erosion. Despite these limitations the principle of uniformitarianism has established a basis from which geological thought can proceed.

Deposition involves the build-up of material on a given surface, either as a consequence of chemical or biological growth or, far more commonly, due to mechanically broken particles being laid down on such a surface. Hence this surface exerts an important influence on the attitude of the beds which are formed. Gravity may also exert an influence on the attitude of the bedding planes in mechanically derived sediments. At the start of deposition the layers of sediment more or less conform to the surface on which accumulation is occurring, provided this is not too irregular. With continued deposition any irregularities in the original surface are filled, the strata which then form tend to lie in a horizontal plane. However, it should be borne in mind that once a layer is formed, and before lithification occurs, it may be disturbed by subsequent deposition. Furthermore, differential consolidation of different materials, for example, sand and mud, or differential consolidation over buried hills may give rise to inclined bedding.

The changes which occur during deposition are responsible for stratification, that is, the layering which characterises sedimentary rocks. The simple cessation of deposition ordinarily does not produce stratification. The most obvious change which gives rise to stratification is that in the composition of the material being deposited. Even minor changes in the type of material may lead to distinct stratification, especially if they affect the colour of the rocks concerned. Changes in grain size also may cause notable layering and changes in other textural characteristics, such as roundness, may help distinguish one bed from another. Variations in the degree of consolidation or cementation, or variations in the fabric of sedimentary rocks may be significant in this context.

The extent and regularity of beds of sedimentary rocks vary within wide limits. This is because lateral persistence and regularity of stratification reflect the persistence and regularity of the agent responsible for deposition. For instance, sands may have been deposited in one area whilst in a neighbouring area muds were being deposited. Hence lateral changes in lithology reflect differences in the environments in which deposition took place. On the other hand a formation with a particular lithology, which is mappable as a stratigraphic unit, may not have been laid down at the same time wherever it occurs. The base of such a formation is described as diachronous. Diachronism is brought about when a basin of deposition is advancing or retreating, as for example, in a marine transgression or regression. In an expanding basin the lowest sediments to accumulate are not as extensive as those succeeding. The latter are said to overlap the lowermost deposits. Conversely, if the basin of deposition is shrinking the opposite situation arises in that succeeding beds are less extensive. This phenomenon is termed offlap. Examples of overlap and offlap generally are found in association with cyclic sedimentation due to oscillating crustal movements.

Agents which are confined to channels or deposited over small areas produce irregular strata that are not persistent. By contrast, strata that are very persistent are produced by agents which operate over wide areas. In addition folding and faulting of strata, along with subsequent erosion, give rise to discontinuous outcrops.

Since sediments are deposited it follows that the topmost layer in any succession of strata is the youngest. Also any particular stratum in a sequence can be dated by its position in the sequence relative to other strata. This is the *Law of Superposition* which was formulated as long ago as 1669 by Steno. This principle applies to all sedimentary rocks except, of

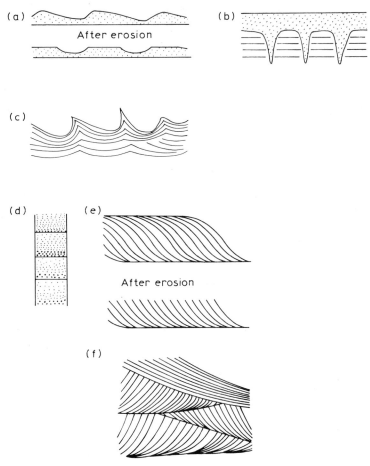

Figure 3.21 Some examples of 'way-up' criteria. (a) Asymmetrical ripple marks, before and after erosion. (b) Desiccation cracks in mudstone subsequently filled with another material. (c) Basal deformation in sandstone overlying shale. (d) Graded bedding, typical of rocks like greywacke. (e) Planar cross bedding before and after erosion. (f) Dune bedding in aeolian sandstone

course, those which have been inverted by folding or where strata have been thrust over younger rocks. Where strata are overfolded the stratigraphical succession, as noted, is inverted. When fossils are present in the beds concerned their correct way up can be discerned. However, if fossil evidence is lacking the correct way up of the succession can be determined from evidence provided by 'way-up' structures. An extensive review of such structures has been provided by Shrock (1948). These include such features as graded bedding. As the name suggests graded bedding exhibits a gradual upward passage from coarser to finer material (Figure 3.9) and is characteristic of greywackes. Eroded cross bedding provides another indication of the way up, as do ripple marks (Figure 3.21). Other features which may provide some clue as to the top and bottom of a bed are swash marks, flow and groove casts, imprints such as made by rain drops or footprints, embedded boulders, mud cracks and fossils in their position of growth.

3.7 Unconformities

The record of geological history is incomplete, and the further back in time, the more incomplete the record becomes. It is often possible to derive evidence from relatively young rocks relating to their conditions of deposition and the palaeogeography. On the other hand older rocks are likely to have undergone a complex history and so are more difficult to interpret. Indeed because the evidence is slight, the interpretation of the first three-quarters of the Earth's history is very difficult.

Although a sequence of strata may exhibit no visible breaks in deposition it nonetheless is unlikely to represent a record of continuous deposition. Each bedding plane was once the surface of the sediment, that is, a bedding plane probably indicates a temporary pause in sedimentation. In addition it is highly likely that much sediment has never been lithified, having been removed by erosion, transported away and deposited, possibly many times over. Many sediments which have formed rocks no doubt have suffered the same fate. This can be demonstrated by comparing the length of time represented by any geological formation with known rates of sedimentation, which indicates that much of geological time is not represented by sediments. The oldest known sediments, which are about 3000 million years in age, are considerably younger than the age of the Earth (4500–5000 million years), and hence are not first generation material.

There are many breaks in the stratigraphical record. These are of differing size and importance, the major breaks representing significant periods of geological time. Locally conspicuous breaks have frequently been used to divide strata into series and stages. However, such surfaces are unlikely to be contemporaneous over large areas and consequently may not offer accurate time boundaries. Obviously stratigraphical divisions should be contemporaneous and these should be established, wherever possible, by the use of fossils.

An unconformity represents a break in the stratigraphical record and occurs when changes in the palaeogeographical conditions led to a cessation of deposition for a significant period of time. Such a break may correspond to a relatively short interval of geological time or a very long one. An unconformity normally means that uplift and erosion have taken place, resulting in some previously formed strata being removed. The beds above and below the surface of unconformity are described as unconformable.

The structural relationship between unconformable units allows four types of unconformity to be distinguished (Figure 3.22). In Figure 3.22a stratified rocks rest upon igneous or metamorphic rocks. This type of feature has frequently been referred to as a nonconformity (it has also been called a heterolithic unconformity, see Tomkeieff (1962)). An angular unconformity is shown in Figure 3.22b, where an angular discordance separates the two units of stratified rocks. In an angular unconformity the lowest bed in the upper sequence of strata usually rests on beds of differing ages. This is referred to as overstep. In a disconformity, as illustrated in Figure 3.22c, the beds lie parallel both above and below the unconformable surface but the contact between the two units concerned is an uneven surface of erosion. When deposition is interrupted for a significant period but there is no apparent erosion of sediments or tilting or folding, then subsequently formed beds are deposited parallel to those already existing. In such a case the interruption of sedimentation may be demonstrable only by the incompleteness of the fossil sequence. Such a feature has been called a non-sequence. Non-sequences cannot be proved in unfossiliferous rocks. This type of unconformity has also been termed a paraconformity by Dunbar and Rodgers (1957) (Figure 3.22d).

As long ago as 1917 Barrell concluded that unconformities were quite inadequate to reconcile the immensity of geological time with the evidence for rapid accumulation of sedimentary deposits and that innumerable short breaks due to non-deposition, or to the sublevation of loose sediments on the sea floor must account for most of the time which has elapsed. These small breaks he termed *diastems*. Diastems ordinarily are not susceptible to individual measurement, even qualitatively.

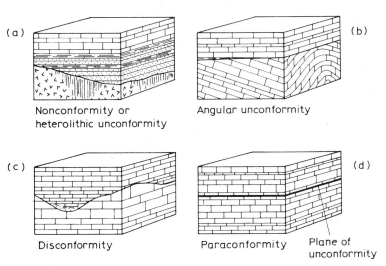

Figure 3.22 Types of unconformity

When traced laterally a single unconformity may appear at different locations in different forms. One of the reasons for this is that the uplift which causes strata to be elevated, and permits their erosion, is frequently not uniform over wide areas.

One of the most satisfactory criteria for the recognition of unconformities is evidence of an erosion surface between two formations. Such evidence may take the form of pronounced irregularities in the surface of the unconformity. Evidence may also take the form of weathered strata beneath the unconformity, weathering having occurred prior to the deposition of the strata above. Fossil soils provide a good example. The abrupt truncation of bedding planes (Figure 3.23), folds, faults, dykes, joints, etc. in the beds below the unconformity is characteristic of an unconformity, although large-scale thrusts will give rise to a similar structural arrangement. Post-unconformity sediments often commence with a conglomeratic deposit. The pebbles in the conglomerate may be derived from the older rocks below the

Figure 3.23 Carboniferous Limestone unconformably overlying rocks of Silurian age, the Horton Flags, at Helwith Bridge, Yorkshire

unconformity. The conglomerate also may contain *remanié* fossils, that is, fossils derived from previously existing rocks.

The relative importance of an unconformity can be demonstrated if the beds above and below contain fossils which enable them to be assigned to their correct positions in the stratigraphical column.

Like other surfaces, unconformities may become folded and overturned. In areas where this has occurred they may constitute a valuable piece of younging evidence, of value in interpreting the geometrical aspects of the structure, as well as throwing light on other aspects of the stratigraphical and structural history.

3.8 Stratigraphic units

The problem of the age of the Earth has fascinated thinking man from early times but because the subject was fraught with difficulties the answer alluded him until the discovery of radioactivity at the end of the nineteenth century. Radioactive decay proceeds at a constant rate and is not influenced by external factors. The original radioactive isotope is termed the parent and upon radioactive decay it gives rise to a stable isotope which is called the daughter. The interval of time required for half the parent material to decay to its daughter element is known as the half-life. After one half-life has passed, half the remaining parent material decays during the next half-life, and so on. Half-lives range in time with different atoms from minute fractions of a second to thousands of millions of years. If a radioactive substance is to be of use for age determination, then it must have a half-life within one or two orders of magnitude of the age of the elements from which the Earth is made. The more common systems used in geochronology are given in Table 3.4.

Table 3.4. Decay systems used in geochronology

Parent	Daughter	Half-life (years)
^{238}U	^{206}Pb	4.51×10^9
^{235}U	^{207}Pb	0.71×10^9
^{232}Th	^{208}Pb	$13.9 \ \times 10^9$
^{87}Rb	^{87}Sr	$50.0 \ \times 10^9$
^{40}K	^{40}Ar	$12.4 \ \times 10^9$

In order to determine the age of a mineral the ratio of the amount of radioactive parent material to the quantity of daughter element produced must be measured.

Some of the oldest known rocks have been recorded from South Africa, radioactive dating giving an approximate age of 4000 million years. It has therefore been assumed that the age of the Earth is probably between 4500 and 5000 million years.

Stratigraphy distinguishes rock units and time units. A rock unit, such as a stratum or a formation, possesses a variety of physical characteristics which enable it to be recognised as such, and so measured, described, mapped and analysed. Weller (1960) regarded groups, formations and members as rock units, a formation being a division of a group and a member is a division of a formation. He admitted, however, that there was no absolute distinction between these three types of rock unit. A rock unit is sometimes termed a lithostratigraphical unit.

A particular rock unit required a certain interval of time for it to form. Hence stratigraphy not only deals with strata but also deals with age, and the relationship between strata and age. Accordingly time units and time-rock units have been recognised. Time units are simply intervals of time, the largest of which are eons, although this term tends to be used infrequently. There are two eons, representing PreCambrian time and post-PreCambrian time. Eons are divided into eras, and eras into periods (Table 3.5). Periods are in turn divided into epochs and epochs into ages. Time units and time-rock units are directly comparable, that is, for each time unit there is a corresponding time-rock unit. For example, the time-rock unit corresponding to a period is a system. Indeed the time allotted to a time unit is determined from the rocks of the corresponding time-rock unit. Most time units in local exposures, except possibly some of the smallest, are incomplete.

A time-rock unit has been defined as a succession of strata bounded by theoretically uniform time planes, regardless of the local lithology of the unit. Fossil evidence usually provides the basis for the establishment of time planes. Ideal time-rock units would be bounded by completely independent time planes, however, practical time-rock units depend on whatever evidence is available.

The geological systems are time-rock units which are based on stratigraphical successions present in certain historically important areas. In other words in their type localities the major time-rock units are also rock units. The boundaries of major time-rock units generally are important structural or faunal breaks or are placed at highly conspicuous changes in lithology. Major unconformities frequently are chosen as boundaries. Away from their type areas major time-rock units may not be so distinctive or easily separated. In fact although systems are regarded as of global application there are large regions where the recognition of some of the systems has not proved satisfactory.

The formulation of the geological systems evolved gradually as the science of geology developed, being finally established towards the end of the nineteenth century. The division of the stratigraphical column into systems was largely a British achievement. Systems are divided into series. The type localities in which series are formulated usually have been chosen because the stratigraphical sections there are fully developed, fossiliferous, and well known as a result of early and continuing investigation. Normally series are applicable throughout major stratigraphical provinces but they are of less than world-wide extent. Series are divided into stages.

Biostratigraphical units are distinguished by their fossil content and without reference to their lithological character. Although time is not directly involved in the recognition of biostratigraphical units, the latter commonly have been considered to approximate time-rock units, if not to correspond to them exactly.

The term *zone* has been used in geology with a variety of meanings. It has been adopted in stratigraphy as a biostratigraphical unit, a zone being defined on a fossil basis. The simplest type of zone includes all those beds between the lowest and highest stratigraphic occurrence of any fossil, which

Table 3.5. The geological time-scale (after Holmes (1978))

Eras	Periods and systems	Derivation of names	Duration of period (Millions of years)	Total from beginning (Millions of years)
CAINOZOIC	QUATERNARY			
	Recent or Holocene*	Holos = complete, whole	2 or 3	2 or 3
	Glacial or Pleistocene*	Pleiston = most		
	TERTIARY			
	Pliocene*	Pleion = more	9 or 10	12
	Miocene*	Meion = less (i.e. less than in Pliocene)	13	25
	Oligocene*	Oligos = few	15	40
	Eocene*	Eos = dawn	20	60
	Paleocene*	Palaios = old	10	70
		'cene' from Kainos = recent		
	The above comparative terms refer to the proportions of modern marine shells occurring as fossils			
MESO-ZOIC	CRETACEOUS	Creta = chalk	65	135
	JURASSIC	Jura Mountains	45	180
	TRIASSIC	Threefold division in Germany	45	225
	(New Red Sandstone = desert sandstones of the Triassic Period and part of the Permian)			
PALAEOZOIC	PERMIAN	Permia, ancient kingdom between the Urals and the Volga	45	270
	CARBONIFEROUS	Coal (carbon)-bearing	80	350
	DEVONIAN	Devon (marine sediments)	50	400
	(Old Red Sandstone = land sediments of the Devonian Period)			
	SILURIAN	Silures, Celtic tribe of Welsh Borders	40	440
	ORDOVICIAN	Ordovices, Celtic tribe of North Wales	60	500
	CAMBRIAN	Cambria, Roman name for Wales	100	600
	PRECAMBRIAN ERA			
	ORIGIN OF EARTH			5000

* Frequently regarded as epochs or stages

may be a species, genus or some larger taxonomic group. Such range zones generally are not successive units because almost everyone is overlapped by others. Assemblage zones consist of strata characterised by the occurrence of two or more particular zonal fossils, commonly species or genera. They do not necessarily include beds equivalent to the entire range of any fossil and therefore they are generally more restricted stratigraphically than many range zones.

Zones are combined to form substages or stages. Although stages may be considered biostratigraphical units the object in their formation is to create time-rock units. Stages provide a means of long-range correlation extending far beyond the ranges of their constituent zones. They are developed by observing the lateral extent and overlapping relations of individual zones. Thus as one zone becomes unrecognisable, another adjacent one takes its place and correlation is carried on.

3.9 Correlation

The process by which the time relationships between strata in different areas is established is referred to as *correlation*. Correlation is therefore the demonstration of equivalency of stratigraphical units (see Krumbein and Sloss (1963). Unfortunately, however, equivalency may be expressed in terms of lithological and palaeontological, as well as chronological terms. Most stratigraphers use the term correlation when referring to lithostratigraphical equivalency as well as to time stratigraphical equivalency. In each situation it is necessary to identify the nature of the equivalency and the kind of correlation involved.

Correlation of one stratigraphical sequence with another is time consuming and involves investigation of as much evidence as possible provided by the rocks concerned. Any correlation must also be in accordance with the geological time scale. Palaeontological and lithological evidence are the two principal criteria used in correlation. However, there is no single criterion of correlation that is more reliable in all cases. Nonetheless, for approximate correlation of the larger stratigraphical units, fossils prove satisfactory. The difficulty appears when fine detail is needed.

The principle of physical continuity may be of some use in local correlation. In other words it can be assumed that a given bed, or bedding plane, is roughly contemporaneous throughout an outcrop of bedded rocks. Tracing of bedding planes laterally, however, is severely limited since individual beds or bedding planes die out, are interrupted by faults, are missing in places due to removal by erosion, are concealed by overburden, or merge with others laterally. Consequently outcrops are rarely good enough to permit an individual bed to be traced laterally over an appreciable distance. A more practicable procedure is to trace a member of a formation. However, this can also prove misleading if beds are diachronous (Figure 3.24).

Where outcrops are discontinuous physical correlation depends on lithological similarity, that is, on matching rock types across the breaks in the hope of identifying the beds involved. The lithological characters used to make comparison in such situations include gross lithology, subtle distinctions within one rock type such as a distinctive, heavy mineral suite or

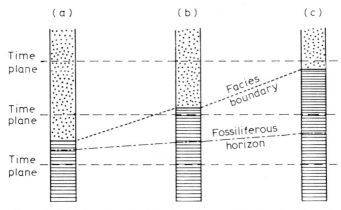

Figure 3.24 Diachronism of a facies boundary and the migration time of a fossil assemblage. The fossiliferous horizon may be regarded as a time plane if the localities A, B and C are not far distant. As a rule, time-planes cannot be identified

notable microscopic features, distinctive key beds, for example, the occurrence of manganese in the upper part of the Lower Cambrian of North Wales, or even the manner in which a rock weathers or the topographical expression it gives rise to. The greater the number of different, and especially unusual, characters that can be linked, the better are the chances of reliable correlation. Even so such factors must be applied with caution and wherever possible such correlation should be verified by the use of fossils.

If correlation can be made from one bed in a particular outcrop to one in another, it can be assumed that the beds immediately above and below are also correlative, provided, of course, that there was no significant break in deposition at either exposure. Better still, if two beds can be correlated between two local exposures then the intervening beds are presumably correlative, even if the character of the intervening rocks is different in the two outcrops. This again depends on there being no important break in deposition at either of the locations.

3.9.1 Drillhole logging techniques

Sedimentary rocks have widely differing physical properties, some of which can be used to distinguish different rock types. Drillhole logging techniques can be used to identify some of these physical properties and thereby act as an aid to correlation (Table 3.6; see also Anon (1981)). These techniques have also been used in site investigations and are therefore dealt with in more detail below.

The electrical resistivity of a rock depends primarily on the amount of fluid it contains, as well as on its own inherent resistivity. As the amount of fluid present is influenced by the porosity of a rock, the resistivity provides a measure of its porosity. For example, if two rocks have the same fluid content and one has a porosity of 10% and the other 30%, then the former is ten times as resistive as the latter. The electrical resistivity method makes use of various electrode configurations down-the-hole. As the instrument is raised from the bottom to the top of the hole it provides a continuous record

Table 3.6. Engineering uses of borehole geophysical logs (from Cratchley (1976))
(NOTE: Only Logs 2C, 4A, 4B and 4C can be used in cased holes)

Serial No. (a)	Type of log (b)	Parameters measured (c)	Application (d)
(1)	*Borehole geometry:*		
	A Caliper log	Continuous record of borehole diameter	Correction of other logs correlation with lithology
	B Verticality test	Point determination of inclination and azimuth of borehole	Correction of inter-borehole measurements and geological log measurements of ground movement
	C Television inspection	Visual examination and photography of borehole wall	Location of voids. Mapping of discontinuities
(2)	*Sonic logs:*		
	A Sonic or continuous velocity log (CVL)	Continuous record of (P wave) velocity in the borehole wall	Determination of rock quality, degree of fracturing, porosity
	B 3-D Sonic log	Continuous record of both P and S wave velocity in the borehole wall	As for 2A and determination of dynamic moduli in homogeneous material (with density log)
	C Interborehole sonic log	Record of apparent velocity between adjacent boreholes at discrete depths	Location of cavities and determination of rock quality between boreholes
(3)	*Electrical logs:*	*Continuous records of:*	
	A Single point resistivity log	Apparent resistivity between single electrode in borehole and ground surface	Lithostratigraphical correlation
	B Normal resistivity log	Apparent resistivity of borehole wall measured by electrode array	As 3A and rock quality/ porosity/groundwater salinity estimation
	C Induction (resistivity) log		
	D Laterolog (resistivity) log	Focused beyond borehole invaded zone	
	E Microlaterolog (resistivity)	Resistivity of invaded zone	Estimation of formation permeability
	F Spontaneous potential (SP) log	Natural potential differences in borehole wall	In conjunction with resistivity, lithological identification and correlation
(4)	*Radioactive logs:*		
	A Gamma log	Rapid sampling with continuous record of natural gamma-ray emission	Estimation of shale/clay content of rock
	B Gamma–gamma or Formation density log (FDL)	Gamma-ray intensity after bombardment by gamma-ray source	Determination of electron density and hence rock density
	C Neutron log	Quantity of hydrogen in rock around borehole, i.e. water and hydrocarbons	Determination of porosity and shale content

of the variations in resistivity of the wall rock. In the normal or standard resistivity configuration there are two potential and one current electrodes in the sonde. The depth of penetration of the electric current from the drillhole is influenced by the electrode spacing. In a short normal resistivity survey spacing is about 400 mm, whereas in a long normal survey, spacing is generally between 1.5 and 1.75 m. Unfortunately in such a survey, because of the influence of thicker adjacent beds, thin resistive beds yield resistivity values which are much too low, whilst thin conductive beds produce values which are too high. The microlog technique may be used in such situations. In this technique the electrodes are very closely spaced (25–50 mm) and are in contact with the wall of the drillhole. This allows the detection of small lithological changes so that much finer detail is obtained than with the normal electric log (Figure 3.25). A microlog is particularly useful in recording the position of permeable beds.

If, for some reason, the current tends to flow between the electrodes on

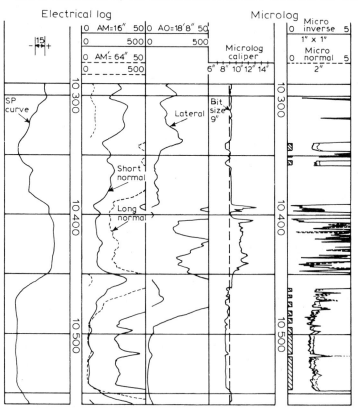

Figure 3.25 Microlog curves. Microresistivity curves are shown on the right. Permeable portions of the section penetrated are indicated (crosshatched bars) by extensions of the 50 mm micro-normal curve beyond the microinverse. Note that the diameter of the bore, as recorded by the microlog caliper, is smaller than bit size where a mud filter cake is formed at the position of permeable beds. A standard electrical log of the same stratigraphic interval is shown on the left for comparison (courtesy Schlumberger Inland Services, Inc.)

the sonde instead of into the rocks, then the laterolog or guard electrode is used. The laterolog 7 has seven electrodes in an array which focuses the current into the strata of the drillhole wall. The microlaterolog, a focused microdevice, is used in such a situation instead of the microlog.

Induction logging may be used when an electrical log cannot be obtained. In this technique the sonde sends electrical energy into the strata horizontally and therefore only measures the resistivity immediately opposite the sonde, unlike in normal electric logging where the current flows between electrodes. As a consequence the resistivity is measured directly in an induction log whereas in a normal electric log, since the current flows across the stratal boundaries, it is measured indirectly from the electric log curves. A gamma-ray log is usually run with an induction log in order to reveal the boundaries of stratal units.

A spontaneous potential (SP) log is obtained by lowering a sonde down a drillhole which generates a small electric voltage at the boundaries of permeable rock units and especially between such strata and less permeable beds. For example, permeable sandstones show large SPs, whereas shales are typically represented by low values. If sandstone and shale are interbedded then the SP curve has numerous troughs separated by sharp or rounded peaks, the widths of which vary in proportion to the thicknesses of the sandstones (Figure 3.26). SP logs are frequently recorded at the same time as resistivity logs. Interpretation of both sets of curves yields precise data on the depth, thickness and position in the sequence of the beds penetrated by the drillhole. The curves also enable a semi-quantitative assessment of lithological and hydrogeological characteristics to be made.

The sonic logging device consists of a transmitter–receiver system, transmitter(s) and receiver(s) being located at given positions on the sonde. The transmitters emit short high frequency pulses several times a second, and differences in travel times between receivers are recorded in order to obtain the velocities of the refracted waves. The velocity of sonic waves propagated in sedimentary rocks is largely a function of the character of the matrix. Normally beds with high porosities have low velocities, and dense rocks are typified by high velocities. Hence the porosity of strata can be assessed (see Wyllie *et al.* (1956)). In the 3-D sonic log one transmitter and one receiver are used at a time and this allows both compressional and shear waves to be recorded, from which, if density values are available, the dynamic elastic moduli of the beds concerned can be determined. As velocity values vary independently of resistivity or radioactivity, the sonic log permits differentiation amongst strata which may be less evident on the other types of log.

In interborehole acoustic scanning an enclosed electrical sparker, designed for use in a liquid-filled drillhole, produces a highly repetitive pulse shape. This signal is received by a hydrophone array in an adjacent drillhole, similarly filled with liquid. The array, therefore, only records compressional waves. Generally the source and receiver are at the same level in the two drillholes and are moved up and down together (see McCann *et al.* (1975)). Drillholes must be spaced closely enough to achieve the required resolution of detail and be within the range of the equipment. This, Grainger and McCann (1977) noted, is up to 400 m in Oxford Clay, 160 m in the Chalk and 80 m in sands and gravels. By contrast because soft organic clay is highly attenuating, transmission is only possible over a few metres. These

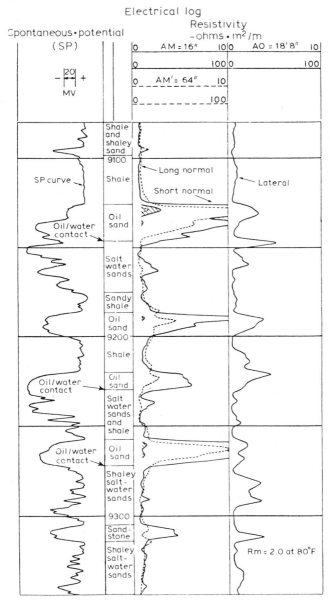

Figure 3.26 A typical electrical log combining the SP normal and lateral curves (courtesy Schlumberger Inland Services Inc.)

distances are for saturated material and the effective transmission range is considerably reduced in dry superficial layers.

Radioactive logs include gamma-ray or natural gamma, gamma-gamma or formation density, and neutron logs. They have the advantage of being obtainable through the casing in a drillhole. On the other hand, the various electric and sonic logs, with the exception of interborehole acoustic scanning, can only be used in uncased holes. The natural gamma log provides a record of the natural radioactivity or gamma radiation from elements such as potassium 40, and uranium and thorium isotopes, in the rocks. This radioactivity varies widely among sedimentary rocks, being generally high for clays and shales and lower for sandstones and limestones. Evaporites give very low readings. The gamma-gamma log uses a source of gamma-rays which are sent into the wall of the drillhole. There they collide with electrons in the rocks and thereby lose energy. The returning gamma-ray intensity is recorded, a high value indicating low electron density and hence low formation density. The neutron curve is a recording of the effects caused by bombardment of the strata with neutrons. As the neutrons are absorbed by atoms of hydrogen, which then emit gamma-rays, the log provides an indication of the quantity of hydrogen in the strata around the sonde. The amount of hydrogen is related to the water (or hydrocarbon) content and therefore provides another method of estimating porosity. Since carbon is a good moderator of neutrons, carbonaceous rocks are liable to yield spurious indications as far as porosity is concerned.

The caliper log measures the diameter of a drillhole. Different sedimentary rocks show a greater or lesser ability to stand without collapsing from the walls of the drillhole. For instance, limestones may present a relatively smooth face slightly larger than the drilling bit whereas soft shale may cave to produce a much larger diameter. A caliper log is obtained along with other logs to help interpret the characteristics of the rocks in the drillhole.

Dating, and therefore correlation, can be achieved by radioactive methods (see above). However, even under the best conditions, the ages so determined involve an error of several per cent. Indeed the degree of precision that these techniques afford is often less than that obtainable by fossils.

3.9.2 Use of fossils

Fossils provide evidence of the former existence of living organisms. If an organism is to be fossilised then it should be buried by sediment very soon after its death in order to reduce the likelihood of its decay. Since there are few places on land where deposition is taking place rapidly, terrestrial organisms generally occur as fossils with much greater rarity than do marine organisms. Fossils may occur as the complete remains of the organism, such as mammoths entombed in ice or insects in resin, but these are extremely rare; the shell or skeleton of the animal may be preserved unchanged; plants, and animals with chitinous exoskeletons may occur as carbonised remains; the mould of the shell or skeleton, or trunk of a tree may be preserved in the sediment; or lastly imprints, for example, of soft-bodied creatures such as jelly fish, and impressions may be left behind. Obviously the hard parts of an organism, if possessed, are the most easily preserved.

A rock may be barren of fossils for three reasons: first, it may have been deposited under conditions which did not support life, second, living forms may have existed but have failed to be preserved, and, third, living forms may have been fossilised but subsequent events may have led to their destruction. For example, their disappearance may have been attributable to removal by solution by groundwater or to metamorphism.

At the end of the eighteenth century William Smith formulated the *Law of Faunal Succession*, which states that strata of different ages are characterised by different fossils or suites of fossils. Smith demonstrated that each formation could be identified by its distinctive suite of fossils without the need of lateral tracing. In this way he developed the use of guide fossils as a method of recognising rocks of equivalent age. The recognition of strata by their fossil content depends on the fact that species and genera become extinct, with new ones replacing them. This has provided a constantly changing variety of living forms which has meant that there has been a progressive change from one age to another. Hence the Earth has been populated by a succession of fauna and flora, each distinct from that which preceded it. Many genera and species, however, were restricted to particular environments which they followed as the facies moved location. As a result their stratigraphical range in a given sequence is likely to have little relation to their total range in time. The lateral or geographical variation in a fossil assemblage complicates correlation. Generally geographical factors, especially climatic factors, influence the composition of floral and faunal assemblages. Species with a world-wide distribution are rare.

As far as correlation is concerned good fossils should have a wide geographical distribution and a limited stratigraphical range. In general groups of organisms which possessed complicated structures provide better guides for correlation than those that were simple. The usefulness of fossils is enhanced if the group concerned evolved rapidly, for where morphological changes take place rapidly, individual species are of short duration. These fossils provide a more accurate means of subdividing the geological column and therefore provide more precise correlation. Groups which were able to swim or float prove especially useful since they ranged widely and were little restricted in distribution by the conditions on the sea floor. However, sedentary fossils have also been used extensively for correlation purposes. This is possible because such animals pass through a free-swimming larval stage which also allows them a wide geographical distribution.

Ideally, for fossils to be of maximum value for correlation they should be independent of facies. This can never be the case since all living forms are adapted in some way to the environment in which they live. Some very highly successful species are particularly well adapted to their environment and can only tolerate small changes, significant environmental changes leading to their extinction. Fossils like these which have been closely tied to a particular environment are termed facies fossils. Facies fossils are of little use in correlation.

Some species have had a very long time range. For instance, *Lingula* has persisted from Cambrian times to the present day almost unchanged. Such organisms are useless as far as correlation is concerned. The stratigraphical range of a fossil can only be determined by experience. Even though the range of a fossil has been well established in one province, its range could be somewhat different in another.

The principal way in which fossils are used in correlation is based on the recognition of characteristic species in strata of a particular age. This method can be applied in two ways. First, index fossils can be established, which in turn allows a particular bed to be identified, and second, fossils may be used to distinguish zones. A zone may be defined as that strata which was laid down during a particular interval of time when a given fauna or flora existed. In some cases zones have been based on the complete fauna present whilst in other instances they have been based on the members of a particular phylum or class. Nonetheless, a zone is a division of time given in terms of rocks deposited. Although a faunal or floral zone is defined by reference to an assemblage of fossils, it is usually named after some characteristic species and this fossil is known as the zone fossil. Normally a faunal or floral zone is identifiable because certain species existed together for some time. It is assumed that these species have time ranges which are overlapping and that their time ranges are similar in different areas.

Stratal subdivision has frequently been made on the basis of the number of species contained or the percentage of species within a given stratigraphical unit, compared with the numbers in other units. For example, Lyell's (1870) subdivision of the Cenozoic era into Eocene, Miocene and Pliocene was based primarily upon the percentage of species still living, found in the respective faunal assemblages. Such calculations, however, may be misleading if some of the species involved have a longer range than that represented in the stratal sequence in question. Hence all species with a long range should be excluded from the calculation.

References

ALLEN, J. R. L. (1968). *Current Ripples*, North Holland Publishing Co., Amsterdam.

ALLEN, J. R. L. (1970). *Sedimentary Processes and Structures*, Allen & Unwin, London.

ANON. (1970). 'Working party report on the logging of cores for engineering purposes', *Q. J. Engg. Geol.*, **3**, 1–24.

ANON. (1981). 'Suggested methods for geophysical logging of boreholes. Commission on Standardization of Laboratory and Field Tests—International Society for Rock Mechanics', *Int. J. Rock Mech. Min. Sci. & Geomech. Abstr.*, **18**, 67–84.

BARRELL, J. (1917). 'Rhythms and the measurement of geologic time', *Bull. Geol. Soc. Am.*, **28**, 745–904.

BARTH, T. W. F., CORRENS, C. W. & ESKOLA, P. (1939). *Die Enstehung der Gesteine*, Springer-Verlag, Berlin.

BATHURST, R. G. C. (1975). *Carbonate Sediments and their Diagenesis*, Elsevier, Amsterdam.

BELL, F. G. (1982). 'The sedimentary petrography of the Fell Sandstones of the Alnwick-Rothburg District, Northumberland', *Proc. Geol. Ass.*, **92**

CAROZZI, A. V. (1960). *Sedimentary Petrology*, Wiley, New York.

CHOQUETTE, P. W. & PRAY, L. C. (1970). 'Geological nomenclature and classification of porosity in sedimentary carbonates', *Bull. Am. Ass. Petrol. Geologists*, **54**, 207–50.

CRATCHLEY, C. R. (1976). 'Geophysical methods (in *Manual of Applied Geology for Engineers*), Institution of Civil Engineers, London, 110–46.

DUNBAR, C. O. & RODGERS, J. (1957). *Principles of Stratigraphy*, Wiley, New York.

DUNHAM, R. J. (1962). 'Classification of carbonate rocks according to their texture', (in *Classification of Carbonate Rocks*), Ham, W. E. (ed.), Am. Ass. Petrol. Geologists, 108–121.

FOLK, R. L. (1951). 'Stages of textural maturity in sedimentary rock', *J. Sediment. Pet.*, **21**, 127–30.

FOLK, R. L. (1959). 'Practical petrographic classification of limestone', *Bull. Am. Ass. Petrol. Geologists*, **43**, 1–38.

FOLK, R. L. (1962). 'Spectral subdivision of limestone types' (in *Classification of Carbonate Rocks*), Ham, W. E. (ed.), Am. Ass. Petrol. Geologists, 62–85.

FOLK, R. L. (1973). 'Carbonate petrology in the Post-Sorbian age' (in *Evolving Concepts in Sedimentology*), Ginsburg, R. N. (ed.), Johns Hopkins University Press, Baltimore.

FOOKES, P. G. & HIGGINBOTTOM, I. E. (1975). 'The classification and description of near-shore carbonate sediments for engineering purposes', *Geotechnique*, **25**, 406–11.

FRIEDMAN, G. M. (1962). 'On sorting coefficients and the log normality of the grain size distribution of sandstones', *J. Geol.*, **70**, 737–53.

GOLDICH, S. S. (1938). 'A study of rock weathering', *J. Geol.*, **46**, 17–58.

GRABAU, A. W. (1904). 'On the classification of sedimentary rocks', *Am. Geol.*, **33**, 228–47.
GRAINGER, P. & McCANN, D. M. (1977). 'Interborehole acoustic measurements in site investigation', *Q. J. Engg. Geol.*, **10**, 241–56.
GRIFFITHS, J. C. (1967). *Scientific Method in the Analysis of Sediments*, McGraw-Hill, New York.
HARDIE, C. A. (1967). 'The gypsum-anhydrite equilibrium at one atmosphere pressure', *Am. Mineralogist*, **52**, 171–200.
HOLLINGWORTH, S. E. (1948). 'Evaporites', *Proc. Yorks. Geol. Soc.*, **27**, 192–8.
HOLMES, A. (1978). *Principles of Physical Geology*, 3rd edition, Van Nostrand Reinhold (UK).
INGRAM, R. L. (1953). 'Fissility of mudrocks', *Bull. Geol. Soc. Am.*, **64**, 869–78.
JONES, O. T. (1944). 'The compaction of muddy sediments', *Q. J. Geol. Soc.*, **100**, 137–60.
KAHN, J. S. (1956). 'The analysis and distribution of the properties of packing', *J. Geol.*, **64**, 385–95.
KRUMBEIN, W. C. (1938). 'Size frequency distribution of sediments and the normal phi curve', *J. Sediment. Pet.*, **8**, 84–90.
KRUMBEIN, W. C. & PETTIJOHN, F. J. (1938). *Manual of Sedimentary Petrography*, Appleton Century Crofts, New York.
KRUMBEIN, W. C. & SLOSS, L. L. (1963). *Stratigraphy and Sedimentation*, W. H. Freeman & Co., San Francisco.
KRYNINE, P. D. (1941). 'Differentiation of sediments during the life history of a landmass', *Bull. Geol. Soc. Am.*, **52**, 1915 (abstract).
KUENEN, Ph. H. (1953). 'Significant features of graded bedding', *Bull. Am. Ass. Petrol. Geologists*, **37**, 1044–66.
LYELL, C. (1870). *Principles of Geology*, Murray, London.
McCANN, D. M., GRAINGER, P. & McCANN, C. (1975). 'Interborehole acoustic measurements and their use in engineering geology', *Geophys. Prosp.*, **23**, 50–69.
MILNER, H. B. (1962). *Sedimentary Petrography*, 2 vols, Allen & Unwin, London.
PETTIJOHN, F. J. (1975). *Sedimentary Rocks*, Harper & Row, New York.
PETTIJOHN, F. J., POTTER, P. E. & SIEVER, R. (1972). *Sands and Sandstones*, Springer-Verlag, Berlin.
POWERS, M. C. (1953). 'A new measurement scale for sedimentary particles', *J. Sediment. Pet.*, **23**, 117–19.
RAISTRICK, A. & MARSHALL, C. E. (1938). *The Origin of Coal and the Coal Seams*, English Universities Press, London.
RILEY, N. A. (1941). 'Projection sphericity', *J. Sediment. Pet.*, **11**, 94–7.
ROBERTSON, E. C. (1965). 'Experimental consolidation of carbonate mud' (in *Dolomitization and Limestone Diagenesis*), *Soc. Econ. Pal. Min.*, Spec. Publ. 13, **170**.
ROBSON, D. A. (1958). 'A new technique for measuring the roundness of sand grains', *J. Sediment. Pet.*, **28**, 108–110.
SHROCK, R. R. (1948). *Sequence in Layered Rocks*, McGraw-Hill, New York.
SPEARS, D. A. (1980). 'Towards a classification of shale', *J. Geol. Soc.*, **137**, 125–30.
STEWART, F. H. (1963). 'Marine evaporites', *US Geol. Surv.*, Prof. Paper 440.
TAYLOR, J. M. (1950). 'Pore space reduction in sandstones', *Bull. Am. Ass. Petrol. Geologists*, **34**, 701–16.
TERZAGHI, R. D. (1940). 'Compaction of lime mud as a cause of secondary structure', *J. Sediment. Pet.*, **10**, 78–90.
TOMKEIEFF, S. I. (1962). 'Unconformity—an historical study', *Proc. Geol. Ass.*, **73**, 383–416.
TRASK, P. D. (1930). 'Mechanical analysis of sediments by centrifuge', *Econ. Geol.*, **25**, 581–99.
TRASK, P. D. (1931). 'Compaction of sediments', *Bull. Am. Ass. Petrol. Geologists*, **15**, 271–6.
WADELL, H. (1932). 'Volume, shape and roundness of rock particles', *J. Geol.*, **40**, 443–52.
WADELL, H. (1933). 'Sphericity and roundness of rock particles', *J. Geol.*, **41**, 310–32.
WELLER, J. M. (1960). *Stratigraphic Principles and Practice*, Harper & Row, New York.
WENTWORTH, C. K. (1922a). 'A scale of grade and class terms for clastic sediments', *J. Geol.*, **30**, 377–92.
WENTWORTH, C. K. (1922b). 'The shape of beach pebbles', *US Geol. Surv.*, Prof. Paper 131-C, 75–83.
WYLLIE, M. R. J., GREGORY, A. R. & GARDNER, L. W. (1956). 'Elastic wave velocities in heterogeneous and porous media', *Geophysics*, **21**, 41–70.

Chapter 4

Geological structures: folds and faults

The two most important features which are produced when strata are deformed by earth movements are folds and faults, that is, the rocks are buckled or fractured respectively. A fold is produced when a more or less planar surface is deformed to give a waved surface. On the other hand a fault represents a surface of discontinuity along which the strata on either side have been displaced relative to each other. Such deformation principally takes place due to movements along shearing planes. When these are small numerous flexuring and folding results, whilst if they are few and large, they cause faulting.

Whether a fold or a fault is developed depends on the nature of the stresses involved, as well as on the physical properties of the rocks, notably whether they are ductile or brittle. The ductility or brittleness of a rock is governed primarily by its composition and is influenced by its texture. But it is also influenced by the presence of pore liquids and the existing temperature–pressure conditions. Folding tends to be the result of ductile rock being deformed. This is likely to occur where rocks are subjected to slowly increasing stress under high confining pressures and elevated temperatures, as found deep within the Earth's crust. On the other hand, at lesser depths and with rapidly applied stress, brittle rocks in particular are sheared, thus giving rise to faulting.

FOLDS

4.1 Anatomy of folds

There are two important directions associated with folding, namely, *dip* and *strike*. True dip gives the maximum angle at which a bed of rock is inclined and should always be distinguished from apparent dip (Figure 4.1). The latter is a dip of lesser magnitude whose direction can run anywhere between that of true dip and strike. Strike is the trend of a fold and is orientated at right angles to the true dip, it has no inclination (Figure 4.1).

Folds are wave-like in shape and vary enormously in size (see Busk (1927)). Simple folds are divided into two types, anticlines and synclines

121

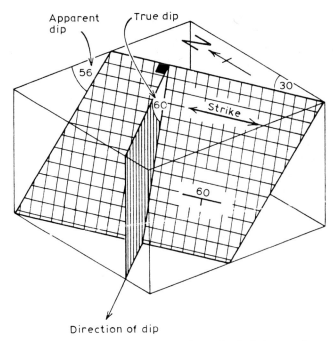

Figure 4.1 Dip and strike: orientation of cross-hatched plane can be
expressed as follows: strike 330° dip 60°W or dip 60° towards 240°

(Figure 4.2). In the former the beds are convex upwards, whereas in the latter
they are concave upwards. The crestal line of an anticline is the line which
joins the highest parts of the fold whilst the trough line runs through the
lowest parts of a syncline (Figure 4.3). The crest and trough planes are
planes which link the crests and troughs of each bed involved in the fold.
The amplitude of a fold is defined as the vertical difference between the crest
and the trough, whilst the length of a fold is the horizontal distance from
crest to crest or trough to trough. The inner part of a fold is referred to as
the core and it is surrounded by the envelope. The hinge of a fold is the
line along which the greatest curvature exists and it can be either straight
or curved. However, the axial line is another term which has been used to
describe the hinge line and, to add further confusion, some authors have used
the term axis in the same way as axial line. Hills (1964) used the term apex
for the highest point in the cross sectional profile of a fold where the rate
of change of dip is greatest. The limb of a fold occurs between the hinges,
all folds having two limbs, which if of unequal length, are defined as long
and short limbs.
   The axial plane of a fold is commonly regarded as the plane which bisects
the fold and passes through the hinge line. But this cannot be true in all
folds, for example, in disharmonic folding there is a different axial plane for
each bed. Consequently the more convenient term, axial surface, has been
introduced as not all surfaces which contain the fold axes are planar. The
axial surface has been defined as the locus of the hinges of all beds forming the fold. The axis of a fold has been used in different senses. Willis and

Figure 4.2 Anticline in Carboniferous Limestone, Ecton Hill, Derbyshire

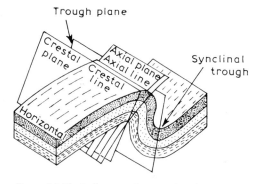

Figure 4.3 Block diagram of a non-plunging overturned anticline and syncline, showing various fold elements

Willis (1934) defined axis in the same way as crestal or trough line, that is, as the highest or lowest line of a fold respectively. According to Billings (1962) and Nevin (1949) the axis of a fold represents the intersection of the axial surface with a bed. McIntyre (1950) defined the axis as a line which generated a fold if moved parallel to itself. However, Hills (1964) has described the axis defined by McIntyre as the longitudinal axis and he has referred to the trace of the axial surface of a fold in transverse section as the transverse axis.

The inter-limb angle, which is the angle measured between the two projected planes from the limbs of the fold, can be used to assess the degree of closure of a fold. Fleuty (1964) recognised five degrees of closure based on the inter-limb angle. Gentle folds are those with an inter-limb angle of greater than 120°; open folds, 120–70°; close folds, 70–30°; tight folds are those with an inter-limb angle of less than 30° and finally in isoclinal folds the limbs are parallel and so the inter-limb angle is zero. He went on to express the attitude of a fold in terms of the dip of the axial plane.

Dip of axial plane		Terms
0°	Horizontal	Recumbent fold
1–10°	Sub-horizontal	
10–30°	Gentle	
30–60°	Moderate	Inclined fold
60–80°	Steep	
80–89°	Sub-vertical	Upright fold
90°	Vertical	

Folds are of limited extent and when one fades out the attitude of its axial line changes, that is, it dips away from the horizontal. This is referred to as the plunge or pitch of the fold (Figure 4.4). The angle of plunge is usually small, although steeply dipping plunges have been recorded in metamorphic terrains. The amount of plunge changes along the strike of a fold and a reversal of plunge direction can occur. The axial line is then waved, concave upwards areas being termed depressions whilst convex upwards areas are known as culminations. Fleuty (1964) also classified plunge in a similar manner to that in which he classified the attitude of folding.

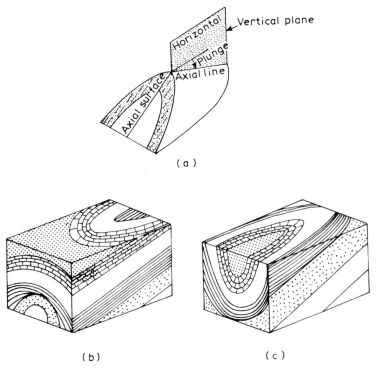

Figure 4.4 (a) Block diagram of an anticlinal fold of a single bed, illustrating plunge. (b) Plunging anticline. (c) Plunging syncline

Plunge		Terms
0°	Horizontal fold	
1–10°	Sub-horizontal	
10–30°	Gentle	
30–60°	Moderate	Plunging fold
60–80°	Steep	
80–89°	Sub-vertical	
90°	Vertical	Vertical fold

4.2 Types of folding

Anticlines and synclines are symmetrical if both limbs are equally arranged about the axial plane so that the dips on opposing flanks are the same, otherwise they are asymmetrical. In symmetrical folds the axis is vertical whilst in asymmetrical folds it is inclined. Folds which possess one limb which is appreciably longer than the other are referred to as inequant.

If a fold is eroded then the older strata are exposed in the centre of an anticline whilst the younger strata occupy the core of an eroded syncline. If a folded surface has a constant dip in all directions away from a single crest point or apex, or towards a single trough point or nadir, then the resultant feature is known as a dome or basin respectively. These folds do not possess an axial line. However, the two terms have not always been used in the strict sense as defined here, indeed, more often they have been used to describe structures akin to an upturned bowl or bowl of a spoon respectively, in which the beds dip away from or towards a plunging axis. The term pericline is sometimes used to refer to the former type of feature.

If beds which are horizontal or nearly so, suddenly dip at a high angle then the feature they form is termed a monocline (Figure 4.5). When traced along their strike, monoclines may eventually flatten out or pass into a normal fault, indeed they are often formed as a result of faulting at depth. A structural terrace may be regarded as the opposite of a monocline, it being a localised area of horizontal beds in dipping strata.

Isoclinal folds are those in which both the limbs and the axial plane are parallel (Figure 4.6a). A fan fold is one in which both limbs are overturned, in the anticlinal fan fold the two limbs dip towards each other whilst in the synclinal fan fold they dip away from each other (Figure 4.6b). Box folds have a box-shaped outline in cross section.

As folding movements become intensified, overfolds are formed in which both limbs are inclined, together with the axis, in the same direction but with different angles (Figure 4.6c). In a recumbent fold the beds have been completely overturned so that one limb is inverted, and the limbs, together with the axial plane, dip at an angle of less than 10° (Figure 4.6d).

Nappes are recumbent folds of large dimensions found in orogenic that is, mountainous regions. Such folds have been translated from their original positions by distances which range up to many kilometres. Those nappes whose places of origin cannot be determined are described as allochthonous, whilst those types whose origin *is* known are referred to as autochthonous. Many large nappes are displaced by thrusting.

An attempt to explain the mechanism of nappe movement has been ad-

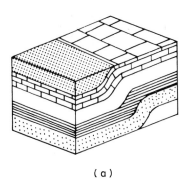

(a)

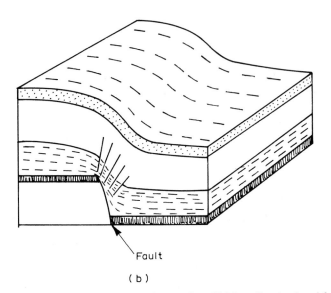

Fault

(b)

Figure 4.5 (a) Block diagram of monocline. (b) Monocline developed from a fault at depth

vanced by Hubbert and Rubey (1959). They contended that the weight of overburden in any region subjected to lateral stress is supported by the residual
stress of the rock and by the pore pressure of the fluids within it. If the pore
pressure is high enough the normal and shear stresses acting on the potential
shear planes in the rocks are reduced to almost zero, causing a very powerful
buoyancy effect to come into play. This means that the coefficient of friction
is reduced almost to zero, which in turn reduces the shear stress and/or angle
of slope required to move the overburden. Gravity gliding of large nappes
can thereby take place over lengthy distances down gentle slopes. Hubbert and
Rubey suggested that the build-up of pore pressure was greatest in
geosynclinal tracts, especially in rocks such as shale where loss by leakage is reduced.

An anticlinorium is an anticline of large dimensions upon which minor

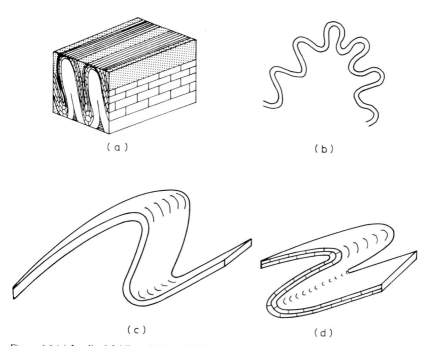

Figure 4.6 (a) Isoclinal folding. (b) Fan folding. (c) Overfold. (d) Recumbent fold

plications have been superimposed. Similarly a synclinorium is a large, composite downwarp possessing minor folds. A normal anticlinorium is one in which the axial surfaces of the minor folds converge downwards, whilst an abnormal anticlinorium is one in which the axial surfaces diverge downwards. A synclinorium can similarly be described as normal or abnormal if its axial surfaces converge or diverge upwards respectively.

4.3 Relationships of strata in folds

Parallel or concentric folds are those where the strata have been bent into parallel curves in which the thickness of the individual beds remains the same. Such folds can be produced by arching a pack of cards. In so doing differential slip occurs between individual cards; the same no doubt happens in parallel folding. From Figure 4.7a it can be observed that, because the thickness of the beds remains the same on folding, the shape of the folds changes with depth and in fact they fade out. Parallel folding occurs in competent (relatively strong) beds which may be interbedded with incompetent (relatively weak, plastic) strata. If the competent beds fracture and separate but still maintain the same fold geometry, then the feature is termed a disjunctive fold.

Similar folds are those which retain their shape with depth. This is accomplished by flowage of material from the limbs into the crestal and trough regions (Figure 4.7b). Although the thickness measured across in-

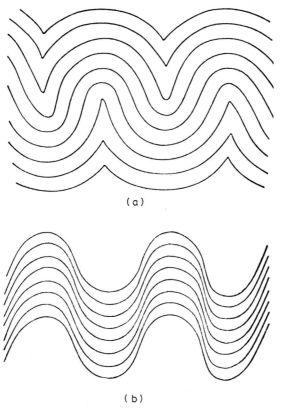

Figure 4.7 (a) Parallel folding. (b) Similar folding

dividual beds varies from the limbs to the crests and troughs, when it is measured parallel to the axial plane it is found to be constant. Similar folds are developed in incompetent strata. However, they are rare in nature for most so-called similar folds change their shape to some degree along the axial plane (see Ramsey (1962)). Most folds exhibit both the characteristics of parallel and similar folding, for example, thickening may occur in the hinge region without being accompanied by significant attenuation of the limbs.

Most folding is disharmonic in that the shape of the individual folds within the structure is not uniform, the fold geometry varying from bed to bed. Disharmonic folding occurs in interbedded competent and incompetent strata. Its essential feature is that incompetent horizons display more numerous and smaller folds than the more competent beds enclosing them. It is developed because competent and incompetent beds yield differently to stress and because secondary stresses may be generated by competent beds squeezing the incompetent, as in the cores of a fold and by the mass translation of incompetent beds which are moved against the competent.

Zig-zag or chevron folds have straight or nearly straight limbs with sharply curved or even pointed hinges (Figure 4.8). Wilson (1951) suggested that the term chevron should be restricted to folds in which the limbs are equal, whilst those folds with unequal limbs could be referred to as zig-zag

Figure 4.8 Chevron or zig-zag folding

folds. Such folds possess features which are characteristic of both parallel and similar folds in that the strata in their limbs remain parallel, beds may be thinned but they are never thickened, and the pattern of the folding persists with depth. These folds may be caused by rotation of their limbs about median planes. Some bedding slip occurs and gives rise to a small amount of distortion in the hinge regions. The planes about which the beds are sharply bent are called kink planes and their attitude governs the geometry of the fold. If one limb of a zig-zag fold dips more steeply than the other, the differential movement involved may have developed a fault which runs through the hinge of the fold. Chevron or zig-zag folds are characteristically found in thin-bedded rocks, especially where there is a rapid alternation of more rigid beds such as sandstones, with interbedded shales.

4.4 Minor structures associated with folding

Cleavage imparts to rocks the ability to split into thin slabs or micro-lithons along parallel or slightly sub-parallel planes of secondary origin. The distance between cleavage planes varies according to the lithology of the host rock, that is the coarser the texture, the further the cleavage planes are apart. For example, in psammitic rocks the cleavage planes may be separated by more than 10 mm, whereas in slates the spacings are only observed beneath the microscope.

Cleavage planes are formed by the combined effects of deformation and metamorphism (see Dieterich (1969)). For instance, they may be caused by mechanical means such as shear planes, kink planes or planes along which minerals have been flattened. On the other hand, cleavage planes may have been induced by low-grade metamorphism, associated with folding, which was responsible for recrystallisation in the plane of least stress of minerals of flaky habit. Accordingly, Leith (1905) recognised two principal types of cleavage, namely, fracture cleavage and flow cleavage.

The inclination of cleavage planes is usually independent of that of the bedding planes of rocks in which they occur. Where cleavage is developed in a series of beds of different lithologies, its attitude changes as it passes from one bed to another and it is described as being refracted. However, refraction does not occur in the hinges of folds but its dip increases as the dip of the limbs increases away from the hinge line. The greatest refraction is witnessed in steeply dipping beds.

Cleavage planes do not intersect although they may meet and branch. They are always roughly parallel to each other. In the most frequent instance the cleavage forms parallel to the axial planes of folds, having developed perpendicular to the direction of maximum principal stress. This type of cleavage

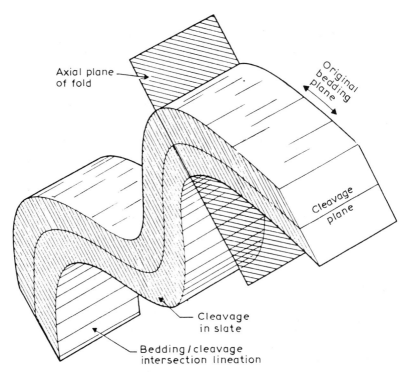

Axial plane
of fold

Original
bedding
plane

Cleavage
plane

Cleavage
in slate

Bedding / cleavage
intersection lineation

Figure 4.9 Relationships between axial-plane cleavage and original bedding.
Cleavage dips more steeply than bedding when the latter is the right way
up. Conversely, the bedding dips more steeply than the cleavage if the
bedding is overturned

has commonly been referred to as axial plane cleavage (Figure 4.9). In
open folds axial cleavage is transverse to the bedding planes, but when the
folds are isoclinal the cleavage runs parallel to the bedding except in the hinge
regions.

Flow cleavage occurs as a result of plastic deformation in which internal
readjustments involving gliding, granulation and the parallel reorientation
of minerals of flaky habit—such as micas, chlorite, graphite and hematite—
together with the elongation of quartz and calcite, take place. The cleavage
planes are commonly only a fraction of a millimetre apart and when the
cleavage is well developed the original bedding planes may have partially
or totally disappeared. Flow cleavage may develop in deeply buried rocks
which are subjected to simple compressive stress, in which case the cleavage
planes are orientated normal to the direction in which the stress was acting.
As a result the cleavage planes run parallel to the axial planes of the folds.
If flow cleavage develops as a result of a shearing couple then the cleavage
planes are inclined to the direction in which the couple was active.

Many authors equate flow cleavage with true slaty cleavage, which is
characterised by the parallel orientation of minerals of platy habit. There
are two schools of thought concerning the origin of slaty cleavage (see
Braddock (1970)). One adheres to the belief that the parallel orientation of

platy minerals is caused by the neocrystallisation of such minerals in planes perpendicular to the direction of maximum principal stress; the opposing school suggests that the parallel orientation is brought about by rotation of flaky minerals within rocks as a result of strain and that they lie in the plane of maximum shear.

Micro-shearing along individual cleavage planes or in narrow zones, together with elongation of parts of the rock in the direction of cleavage, are often associated with slaty cleavage (see Wilson (1946)). The former phenomenon is illustrated by the displacement of bedding in laminated slates and by the rotation of, and pressure fringes about, porphyroblasts. The effects of flattening, elongation and unilateral shear such as the extension of fossils, pebbles and concretions along cleavage planes, which are observed in slates,

Figure 4.10 Fracture cleavage developed in highly folded Horton Flags, Silurian, near Stainforth, Yorkshire. The inclination of the fracture cleavage is indicated by the near vertical hammer. The other hammer represents the dip of the bedding

are related to the bulk distortion and plastic flow of the rock under the influence of regionally applied forces and the local constraints caused by the competent and incompetent beds in folds.

According to Billings (1962) fracture cleavage can be regarded as closely spaced jointing, the distance between the planes being measured in millimetres or even in centimetres (Figure 4.10). Unlike flow cleavage there is no parallel alignment of minerals, fracture cleavage having been caused by shearing forces. It therefore follows the laws of shearing and develops at an angle of approximately 30° to the axis of maximum principal stress. However, fracture cleavage often runs almost normal to the bedding planes and in such instances it has been assumed that it is related to a shear couple. The external stress creates two potential shear fractures but since one of them trends almost parallel to the bedding it is unnecessary for fractures to develop in that direction. The other direction of potential shearing is that in which fracture cleavage ultimately develops and this is facilitated as soon as the conjugate shear angle exceeds 90°. Fracture cleavage is frequently found in folded incompetent strata which lie between competent beds. For example, where sandstone and shale are highly folded, fracture cleavage occurs in the shale in order to fill the spaces left between the folds of the sandstone. However, fracture cleavage need not be confined to the incompetent beds. Where it is developed in competent rocks it forms a larger angle with the bedding planes than it does in the incompetent strata. Some authors do not think that shearing offers a satisfactory explanation for the presence of fracture cleavage in the crestal regions of folds and have suggested that it may be developed there as a consequence of tension.

Because both folds and cleavage develop contemporaneously as a consequence of the same movements a definite relationship exists between them. For instance, where slaty cleavage is developed across the bedding it usually runs parallel to the axial planes of the folds. This arrangement allows a number of structural problems to be solved.

(1) Where the cleavage is vertical the axial plane is also vertical and the fold is symmetrical. The beds are the correct way up.
(2) The acute angle between the bedding and the cleavage indicates the direction in which the adjacent bed has slipped which allows the order of superposition to be determined.
(3) The attitude of the cleavage is always steeper than that of the bedding unless the latter approaches the vertical or is overturned.
(4) If the cleavage is horizontal then the axial planes of the folds are horizontal and in such a situation the correct sequence of the beds cannot be ascertained from the bedding–cleavage relationship.

If the cleavage–bedding relationship of slaty cleavage is considered in three dimensions then the plunge of the folds can be determined if:

(a) if the folds plunge then the strike of the cleavage is transverse to the bedding,
(b) the acute angle between the normal to the upper surface of the beds and the strike of the cleavage indicates the direction in which the folds plunge,

(c) the plunge of the fold is equal to the inclination of the bedding traces on the cleavage planes.

It must be noted that in some instances the attitude of the cleavage planes in the limbs of folds is not parallel to the axial planes but forms a fan-like pattern with either upwards or downwards convergence.

Lineation is due to some directional property within a rock and it has been

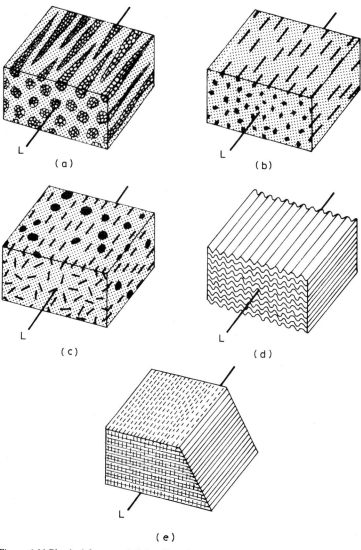

Figure 4.11 Physical features defining lineations (*L*) according to the concept of Turner and Weiss. (a) Lineation due to preferred orientation of linear bodies. (b) Lineation due to preferred orientation of acicular minerals. (c) Lineation due to preferred orientation of acicular and tabular minerals. (d) Linear fabric defined by penetrative folding. (e) Lineation defined by intersection of two planes such as bedding and cleavage) used with permission of McGraw-Hill Book Company)

defined as any striping which occurs on any rock surface (Figure 4.11). Although there are numerous ways in which lineations may be developed they are commonly associated with folding. Lineations related to folds are usually referred to three coordinates, a, b, and c, which are all orientated at right angles to each other. The a coordinate runs between the two limbs of the fold, the b along the hinge line and the c is normal to the two preceding coordinates.

According to De Sitter (1964) there are four important tectonic lineations which are caused by: an intersection of two planes, a tectonic rippling of a plane or minor folds or microfolds, the orientation of minerals and the elongation of rock fragments (Figure 4.11). The first two may merge and rippling or crenulation is often due to intersection. The most frequent type of intersection is that of the bedding planes with those of cleavage. Planes of cleavage or schistosity may be wrinkled into corrugations with an amplitude and a wavelength measured in millimetres. The lineation results from the parallel alignment of the crests of minute drag folds which were formed by bedding slip. The axial planes of the crenulations usually run parallel with those of the major folds, however, no such relationship may exist in rocks which have suffered tight compression and consequent differential movement. In axial plane cleavage the intersection with the original bedding is parallel to the axial plane of the associated fold and the attitude of the axial plane may be determined by the direction and plunge of this lineation.

Lineation may be imparted to a rock by the approximate parallel alignment in it of lath-shaped minerals such as hornblende or by some of the original minerals being granulated and the fragments thereof being strung in a linear arrangement. In folded rocks the long axes of deformed pebbles, fossils, oolites and even grains of quartz frequently possess an approximate parallelism.

According to Cloos (1946) boudinage is relatively common in folded strata in which competent and incompetent beds alternate, the boudinaged bed

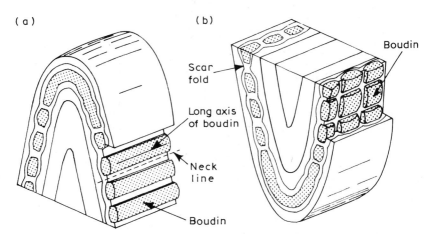

Figure 4.12 Diagrammatic representation of normal boudinage (a) and chocolate block boudinage (b), showing the relationship of both to associated folding

Figure 4.13 Mullion structure in bedding surface between sandstone and pelite, North Eifel, Germany (courtesy John Wiley & Sons Inc.)

always being composed of the competent rock. Boudinage is commonly ascribed to tension in one or two directions along the bedding planes or to compression normal to the bedding. In such cases the boudin lies parallel to the axial plane of the fold (Figure 4.12). However, if the boudinage is caused by stretching parallel to the axial line of a fold then the boudins are perpendicular to the axial line. When the competency of the boudins and the enclosing rock is not too dissimilar, boudins tend to break into small longitudinal sections which have rounded edges. This is because they do not possess the same capacity for stretching as the incompetent rocks which surround them.

Mullions are linear structures which are developed in competent beds due to compression occurring parallel to the bedding (see Wilson (1953)). They consist of a series of parallel columns (Figure 4.13). In typical mullions only one side of the competent bed is contorted.

Quartz rods are of frequent occurrence in severely crumpled metamorphic rocks (Figure 4.14). In their most extreme development they are long cylindrical rods of quartz, in their simplest form they are small concordant quartz veins. The veins grow by the segregation of quartzose material from the country rock into zones of minimum tension and also by recrystallisation along cleavage planes.

Figure 4.14 Quartz rods in Dalradian schists, Isle of Arran

Drag folds occur in association with major folds. (Figure 4.15). They may be slightly folded or highly compressed or even exhibit signs of overthrusting. Drag folds are usually attributed to bedding slip (see Ramberg (1963)).

When brittle rocks are distorted, tension gashes may develop as a result of stretching over the crest of a fold or of local extension caused by drag exerted when beds slip over each other (see Ramsey (1967)). Those tension gashes which are the result of bending of competent rocks usually appear as radial fractures concentrated at the crests of anticlines which are

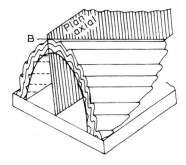

Figure 4.15 Drag folds on the flanks of an anticline. Note that the upper bed on each limb rides upwards towards the anticlinal crest relative to the lower beds. This creates a consistent directional asymmetry (courtesy of the Société Geologique de Belgique)

sharply folded. They represent failure following plastic deformation. Tension gashes formed by differential slip appear on the limbs of folds and are aligned approximately perpendicular to the local direction of extension. Tension gashes are distinguished from fracture cleavage and other types of fractures by the fact that their sides tend to gape. As a result they often contain lenticular bodies of vein quartz or calcite.

Tectonic shear zones were recorded in the compact clays of the Siwalik series at the Mangla dam site in Pakistan by Fookes (1965). The shear zones lie parallel to the bedding and appear to be due to displacements caused by concentric folding. Skempton (1966) found that such shear zones generally occurred in clay beds with high clay mineral contents. He noted that the shear zones ranged up to approximately 0.5 m in thickness and extended over hundreds of metres. Each shear zone exhibited a conspicuous principal slip which forms a gently undulating smooth surface. There are two other main displacement shears. The interior of a shear zone is dominated by displacement shears and slip surfaces lying *en echelon* inclined at 10–30° to the *ab* plane (*a* is the direction of movement, *b* lies in the plane of shear, and *c* is at right angles to this plane). These give rise to a complex pattern of shear lenses, the surfaces of which are slickensided. Relative movement between the lenses has been complicated, with many local variations. The principal slip surface showed a faint lineation directed along the *a* axis. Thrust shears, and possibly fracture cleavage, have also been noted in these shear zones.

Skempton (1966) showed that many of the slip surfaces lying *en echelon* formed at an early stage of folding, and that the principal displacement shears formed later, since they intersect the shear lenses. He maintained that because of the flat, polished nature of such slip surfaces the movement was concentrated along them. The clay is reduced to its residual strength along these slip surfaces.

FAULTS

Faults are fractures in crustal strata along which the adjacent rock has been displaced (Figure 4.16). The amount of displacement may vary from only a few tens of millimetres to several hundred kilometres. One of the most notable examples of lateral displacement is along the San Andreas fault in California, displacement totalling over 640 km. However, the apparent dis-

Figure 4.16 High-angled fault in rocks of Limestone Group, Carboniferous, Northumberland

placement of a fault need bear no relationship to the true displacement.

In many faults the fracture is a clean break but in others the displacement is not restricted to a simple fracture but is developed throughout a fault zone, differential movement having occurred along innumerable closely spaced fractures (see Badgley (1965)). If a fault surface is irregular it is usually grooved in the direction of slip.

The dip and strike of a fault plane can be described in the same way as are those of a bedding plane. The angle of hade is the angle enclosed between the fault plane and the vertical. The hanging wall of a fault refers to the upper rock surface along which displacement has occurred whilst the foot wall is the term given to that below. The vertical shift along a fault plane is called the throw whilst the term heave refers to the horizontal displacement. Where the displacement along a fault has been vertical then the terms downthrow and upthrow refer to the relative movement of strata on opposite sides of the fault plane.

4.5 Classification of faults

A classification of faults can be made on a geometrical or a genetic basis (see Gill (1971)). As far as a geometrical classification is concerned it can be based on the direction in which movement has taken place along the fault plane, on the relative movement of the hanging and foot walls, on the attitude of the fault in relation to the strata involved and on the fault pattern. If the direction of slippage along the fault plane is used to distinguish between faults, then three types may be recognised, namely, dip-slip faults, strike-slip faults, and oblique-slip faults. In the dip-slip fault the slippage occurred along the dip of the fault, in the strike-slip fault it took place along the strike and in the oblique-slip fault movement occurred diagonally across the fault plane

(Figure 4.17). When the relative movement of the hanging and foot walls is used as a basis of classification, then normal, reverse and wrench faults can be recognised. The normal fault is characterised by the occurrence of the hanging wall on the downthrown side whilst in the reverse fault the foot wall occupies the downthrown side. Reverse faulting involves a vertical duplication of strata, unlike normal faults where the displacement gives rise to a region of barren ground (Figure 4.17). In the wrench fault neither the foot nor the hanging wall have moved up or down in relation to one another (Figure 4.17). Considering the attitude of the fault to the strata involved, strike faults, dip (or cross) faults, and oblique faults can be recognised. A strike fault is one which trends parallel to the beds it displaces, a dip or cross fault is one which follows the inclination of the strata and an oblique fault runs at angle with the strike of the rocks it intersects. These three different classifications may be combined to give a fuller description of a fault (Figure 4.17). A classification based on the pattern produced by a number of faults does not take into account the effects on the rocks involved. Parallel faults, radial faults, peripheral faults and *en echelon* faults are among the patterns which have been recognised.

In areas which have not undergone intense tectonic deformation reverse and normal faults generally dip at angles in excess of 45°, whilst their low-angled equivalents, thrusts and lags, are inclined at less than that figure. Splay faults occur at the extremities of strike-slip faults and strike-slip faults are commonly accompanied by numerous smaller parallel faults. Sinistral and dextral strike-slip faults can be distinguished in the following manner: if, when looking across a fault plane, the displacement on the far side has been to the left then it is sinistral whereas if movement has been to the right the fault is described as dextral (Figure 4.18). Since both strike-slip faults and thrusts are formed as a result of lateral compressive movements they are often found associated with each other in the field.

Normal faults range in linear extension up to, occasionally, a few hundred kilometres in length. Generally the longer faults do not form single fractures throughout their entirety but consist of a series of fault zones. The net slip on such faults may total over a thousand metres. Normal faults are commonly quite straight in outline but sometimes they may be sinuous or irregular with abrupt changes in strike. When a series of normal faults run parallel to one another with their downthrows all on the same side, the area involved is described as being step faulted (Figure 4.19). Horsts and rift structures (graben) are also illustrated in Figure 4.19.

The strata involved in normal faulting are commonly stretched which led to the widely held belief that these faults were formed as a result of tension. However, tension may be indirectly a consequence of compressional, rotational or torsional forces. Differential vertical movements may cause stretching of the surface rocks but such movements may be due to the relaxation which is associated with the decline of compressive forces or couple. In fact Anderson (1951) suggested that all types of faults could be developed as a result of compression (Figure 4.18).

Billings (1962) suggested that the relative movement along a fault plane provided the most satisfactory basis for a genetic classification of faults. A gravity fault was one in which the movement of the hanging wall was downward in relation to the foot wall, it having formed as a result of tension.

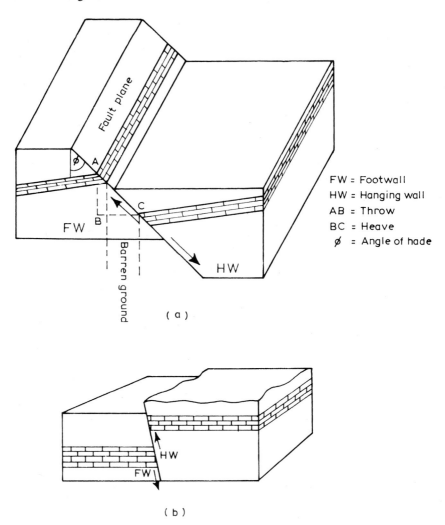

FW = Footwall
HW = Hanging wall
AB = Throw
BC = Heave
ø = Angle of hade

(a)

(b)

Thrust and reverse faults were believed by Billings to develop as a con-sequence of compression. Rift faults were presumed to owe their origin to shearing movements, the relative movement occurring along the strike of the fault and the fault itself following the regional trend of the strata involved. Tear faults were similarly formed by shearing movements but they cut trans-versely across the regional strike of the rocks in which they occur. Shear thrusts develop independently of bedding, their attitude being controlled by the physical properties of the strata and by the mechanics of deformation.

Where a resistant stratum is sheared at a flexure, one limb is pushed over the other along a break thrust (Figure 4.20a). Break thrusts are dependent upon the pre-existence of folding. The formation of tension fractures in the rocks involved in an asymmetrical anticline weakens the steeper limb and con-tinued earth movements may lead to it shearing to form a thrust. Where a

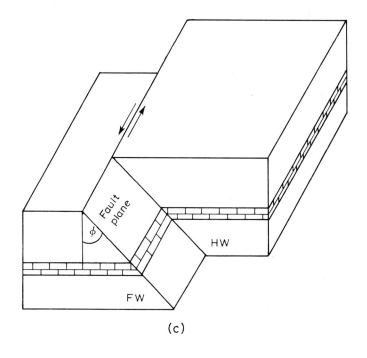

(c)

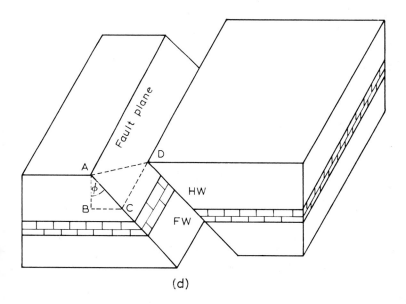

(d)

Figure 4.17 (a) Normal, dip-slip, strike fault. (b) Reverse fault. (c) Tear or strike-slip fault. Hanging wall and footwall do not move up or down in relation to one another. (d) An oblique-slip fault AB = throw; BC = heave; CD = lateral shift; AD = oblique slip

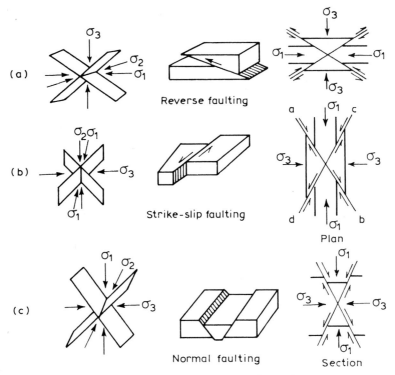

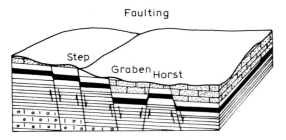

Figure 4.18 Initial stress distribution causing faulting (after Anderson (1951)) σ_1 maximum, σ_2 mean, σ_3 minimum (compressive) stress. In (a) ab = dextral (clockwise) and cd = sinistral (anticlockwise) displacement

Faulting

Figure 4.19 Block diagram illustrating step faulting and Horst and Graben structures

thrust emerges at the surface over which it then advances, the fault is re-ferred to as a surface thrust. Similar thrusts may arise when the crest of a rising anticline, composed of competent and incompetent strata, is removed by erosion (Figure 4.20b). This action permits the movement of the competent beds over an incompetent layer. If the thrust sheet is subjected to erosion as it advances over the incompetent layer then the sediments so formed may be deposited in front of it and eventually overridden. A stretch fault is the name given to a fault which, because of continued folding, has severed the overturned limb of a recumbent fold (Figure 4.20c).

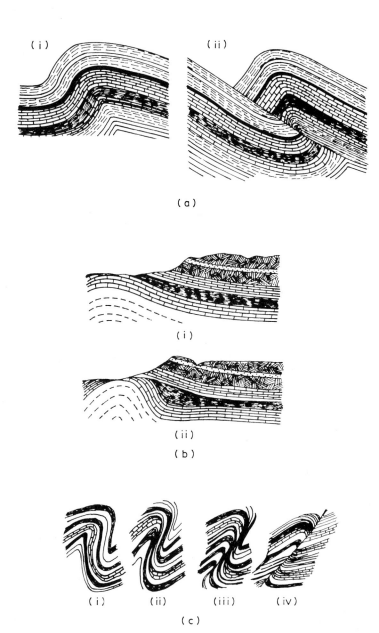

Figure 4.20 (a) Development of a break thrust. According to this concept, the strata are too competent to become overturned and stretched. The folding weakens the forelimb of the anticline by tension fracturing. Continued stress application causes a thrust to develop and it utilises the already existing fractures. (b) Sections illustrating the origin of an erosion thrust. The position of the overthrust shown in (ii) was determined by the land surface condition represented in (i), where a weak rock (dark shading) outcrops at the base of a scarp. The force acted from right to left. Erosion is supposed to have continued during the faulting. (c) Development of a stretch thrust from an overturned fold

An overthrust is a thrust fault which has an initial dip of ten degrees or less and its net slip measures several kilometres. Overthrusts may be folded or even overturned. As a consequence when they are subsequently eroded remnants of the overthrust rocks may be left as outliers surrounded by rocks which lay beneath the thrust. These remnant areas are termed klippe and the area which separates them from the parent overthrust is referred to as a fenster or window. The region which occurs in front of an overthrust is

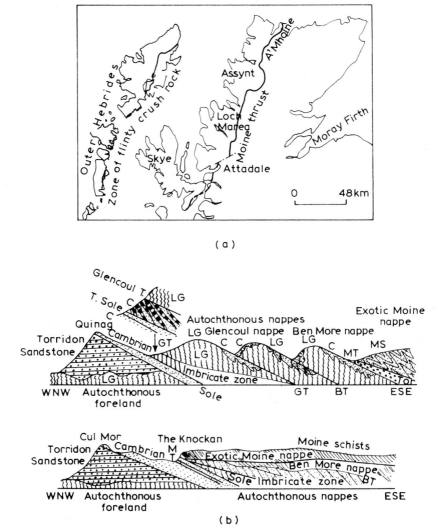

Figure 4.21 (a) The Moine thrust and related major structures. (b) Sketch sections of the Moine thrust zone, Scotland, showing the major low-angle thrust-planes, and imbricate structure in the overridden Cambrian beds. MS = Moine Schists, C = Cambrian, TOR = Torridonian, LG = Lewisham gneiss, GT = Glencoul thrust, BT = Ben More thrust, MA = Moine thrust. Inset, details of the imbricate structure (courtesy of Blackie & Son Ltd)

called the foreland. Small thrusts with a net slip amounting to thousands of metres may occur in the foreland but there are no overthrusts. The Moine thrust of the North West Highlands of Scotland provides a classic example of an overthrust (Figure 4.21). It is approximately 300 km in length and the translation exceeds 16 km, Moine Schists being carried over Cambrian and Torridonian strata. In some areas a clear-cut sole or single fault is observable but parallel faults also exist and in other places there is a wide fault zone. A set of small parallel thrusts, dipping more steeply, slice the overridden Cambrian rocks to form an imbricate structure.

With the initiation of a fault, fracturing causes an instantaneous and drastic alteration in the pre-existing stress system, accompanied by the introduction of a new internal boundary surface. Therefore any subsequent faulting must occur in accordance with the changed conditions. Pronounced changes take place in both direction and intensity of the stresses in the immediate vicinity of the fault plane but they rapidly diminish with distance. This is particularly true in the direction normal to the fault where alterations soon become insignificant. Near the centre of the fault the shear stress is greatly reduced which prevents further faulting occurring close by. However, there is a substantial increase in stress at both ends of the fault, indicating a tendency for lateral extension, once the process has begun. Differential movement along a fault plane tends to increase the minimum principal stress whilst causing a decrease in the maximum principal stress. Consequently a normal fault brings about an increase in the horizontal stress normal to the strike of the fault whereas a thrust causes the horizontal stress to diminish.

The local relief of stress provided by faults tends to diminish due to the tendency of the regional stress system to re-establish itself later. This may cause renewed faulting in the region already fractured. Hence the process becomes more and more complicated. Fault systems which are produced during one geological epoch usually possess a parallel alignment and it has therefore been inferred that stress systems over large areas of the crust often have a general uniformity in both direction and intensity.

A recent review of the mechanics of fault formation has been given by Murrell (1977).

4.6 Criteria for the recognition of faults

The abrupt ending of one group of strata against another may be caused by the presence of a fault, but abrupt changes also occur at unconformities and intrusive contacts. Nevertheless it is usually a matter of no great difficulty to distinguish between these three relationships. Repetition of strata may be caused by faulting, that is, when the beds are repeated in the same order and dip in the same direction, whereas when they are repeated by folding they recur in the reverse order and may possess a different inclination (Figure 4.22a). Omission of strata suggests that faulting has taken place although such a feature could again occur as a result of unconformity (Figure 4.22b).

Many features are characteristically associated with faulting and consequently when found, indicate the presence of a fault. Cleavage and shear and tension joints are frequently associated with major faults (see

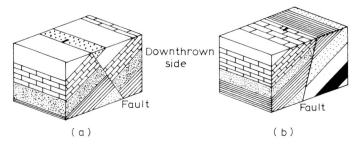

Figure 4.22 (a) Repetition of bed at surface (fault parallel with strike and hading against dip). (b) Omission of bed at surface (fault parallel with strike and hading with dip)

Louderback (1950)). Structures of this nature were simulated experimentally by Riedel (1929) who covered two adjacent boards with clay and then moved one board, so generating shearing stresses in the clay above. Tension fractures formed at angles of between 45 and 47° to the direction of shearing and as the movement progressed the fractures rotated and opened to form an angle of about 60°. The plasticity of the clay was reduced by spraying a film of water onto its surface. If, however, this was not done then shearing planes developed in the deformed zone, one dominant set usually occurring which formed an angle of 12 to 17° with the direction of relative displacement. The acute angle between the tension gashes or the shearing planes and the direction of relative displacement points towards the direction in which movement has occurred. The clay did not shear along a single plane parallel to the direction of relative displacement, tension fractures and shearing planes being developed *en echelon* in a zone between the blocks (Figure 4.23). Shear

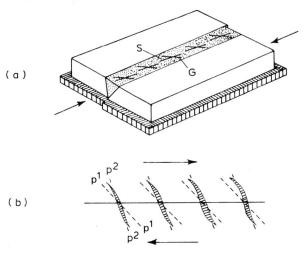

Figure 4.23 (a) Development of shearing planes (S) and tension gashes (G) in clay which has been displaced by moving two boards as shown. The zone of deformation within which the fractures occur is shown stippled. (b) Tension fractures which originally formed parallel to p^1 rotated on displacement to p^2, as shown by arrows. Fractures on rotation become tension gashes.

and tension joints formed along a fault are frequently referred to as feather joints because of their barb-like appearance. Feather joints may be sub-divided into pinnate shear joints and pinnate tension joints. Where pinnate shear planes are closely spaced and involve some displacement, then fracture cleavage is developed.

Slickensides are polished striated surfaces which occur on a fault plane and are caused by the frictional effects generated by its movement (see Tija (1964)). Only slight movements are required to form slickensides and their presence has been noted along shear joints. Indeed Hancock (1968) observed no displacements along two conspicuously slickensided sets of fractures in the Cotswold Hills. Nevertheless, where movement has occurred the striations illustrate the general direction of movement. Very low scarps, sometimes less than a millimetre high, occur perpendicular to the striations and represent small accumulations of material formed as a consequence of the drag effect created by the movement of the opposing block. The shallow face of the scarp points in the direction in which the block moved. Sometimes two or more sets of slickensides, which usually intersect at an acute angle, may be observed indicating successive movements in slightly different directions or a sudden deviation in the movement during one displacement.

Intraformational shears, that is, zones of shearing parallel to bedding are associated with faulting. Salehy *et al.* (1977) noted their presence in clays, mudstones and shales of Carboniferous age and that they generally occurred at the contact of such beds with overlying sandstones. They pointed out that these shear zones tended to die out when traced away from the faults concerned and suggested that such zones are probably formed as a result of flexuring of strata adjacent to faults. A shear zone may consist of a single polished or slickensided shear plane, a more complex shear zone may be up to 300 mm in thickness. Intraformational shear zones are not restricted to argillaceous rocks, for instance, they occur in chalk. Their presence means that the strength of the rock along the shear zone has been reduced to its residual value.

As a fault is approached the strata involved frequently exhibit flexures which suggest that the beds have been dragged into the fault plane by the frictional resistance generated along it. Indeed along some large dip-slip faults the beds may be inclined vertically. A related effect is seen in faulted gneisses and schists where a pre-existing foliation is strongly turned into the fault zone and a secondary foliation results.

If the movement along a fault has been severe then the rocks involved may have been crushed, sheared or pulverised (see Sibson (1977)). Where shales or clays have been faulted the fault zone may be occupied by clay gouge. Fault breccias, which consist of a jumbled mass of angular fragments containing a high proportion of voids, come into being when more competent rocks are faulted. They are formed mainly by the continued opening of tension gashes where the normal stresses across the fault are low. Accordingly fault breccias are confined to the uppermost region of the crust and are associated with normal faults. Crush breccias develop when rocks are sheared by a regular pattern of fractures, the individual fragments being bounded by intersecting shear surfaces. If the movements are prolonged then the corners of the fragments may be rounded with the result that crush conglomerates are formed. The material between the lenses forms a fine-grained matrix. Crush conglomerates and breccias indicate that strong normal stresses were operational across the fault. These rocks are squeezed as well as stretched and so they are elongated along a plane of flattening which bisects intersecting shear surfaces, the plane usually lying in the acute bisectrix. Although crush conglomerates may occur in association with any type of fault, they most commonly accompany reverse faults.

Movements of even greater intensity are responsible for the occurrence of mylonite along a fault zone, this may be regarded as a micro-breccia and shearing planes may be observed within it. If, because of the severity of the crushing, no individual fragments can be distinguished within the material then it is described as an ultramylonite. Mylonites are principally found in association with low-angle thrusts and are thought to be developed at depths where the confining pressures are sufficient to maintain rock coherence. The ultimate stage in the intensity of movement is reached with the formation of pseudo-tachylite. This looks like glass and many authorities have suggested that the movements responsible for its formation must have been so severe as to melt the rock material involved. Pseudo-tachylite is formed upon cooling. Large blocks called horses or slices may be found stranded within a fault zone.

Although a fault may not be observable its effects may be reflected in the topography (Figure 4.24). For example, if blocks are tilted by faulting then a series of scarps are formed. If the rocks on either side of a fault are of

Figure 4.24 Giggleswich Scar, Yorkshire, formed of Carboniferous Limestone, which more or less follows the trend of the Dent fault. The area to the right of the road represents the downthrown side

different hardness then a scarp may form along the fault as a result of differential erosion. If the scarp is the product of movement along the fault plane and differential erosion, then it is termed a composite fault scarp. Triangular facets occur along a fault scarp associated with an upland region. They represent the remnants left behind after swift flowing rivers have cut deep valleys into the scarp. Such deeply carved rivers deposit alluvial cones over the fault scarp. Scarplets are indicative of active faults and are found near the foot of mountains where they run parallel to the base of the range. They rarely exceed 30 m in height and are usually confined to unconsolidated deposits. On the other hand natural escarpments may be offset by cross faults. Stream profiles may be similarly interrupted by faults, or, in a region of recent uplift, their courses may in fact be relatively straight due to them following faults. Springs often occur along faults. A lake may form if a fault intersects the course of a river and the downthrown block is tilted upstream. Faults may be responsible for the formation of waterfalls in the path of a stream. Sag pools may be formed if the downthrown side settles different amounts along the strike of a fault. However, it must be emphasised that the physiographical features noted above may be developed without the aid of faulting and consequently they do not provide a fool-proof indication of such stratal displacement.

Faults provide a path of escape and they are therefore frequently associated with mineralisation, silicification and igneous phenomena. For example, dykes are often injected along faults.

FAULTS AND EARTH MOVEMENT

In almost all known instances of historic fault breaks the fracturing has occurred along a pre-existing fault. Whilst it seems probable that a given fault would break again at the same location as the last break this cannot be concluded with certainty. However, the likelihood of a new fault interfering with an engineering structure is so remote that it can be reasonably neglected except in unusual situations such as near the tip of the wedge of an active thrust fault.

4.7 Earthquakes

Although earthquakes have been reported from all parts of the world they are primarily associated with areas of recent mountain building and with the global rift system, in other words, with the edges of the plates which form the Earth's crust (Figure 4.25). The Earth's crust is being slowly displaced at the margins of the plates, presumably by convection currents in the upper mantle. Differential displacements give rise to elastic strains which eventually exceed the strength of the rocks involved and a fault then develops. The strain is partly or wholly dissipated. Hence an earthquake occurs. From time to time earthquakes of minor significance occur in the stable areas of the world, witness the shocks which are generated by movements along the Great Glen and Highland Boundary Fault zones in Scotland.

The focus or hypocentre of an earthquake is the name given to the location

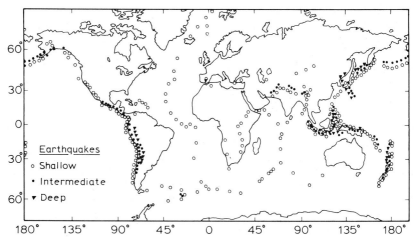

Figure 4.25 Distribution of earthquake epicentres

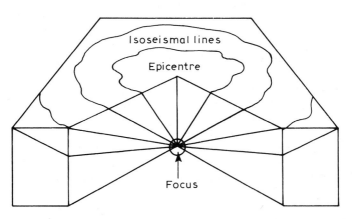

Figure 4.26 Block diagram showing isoseismal lines and their relation to the epicentre and to the wave paths radiating from the focus of an earthquake. The Roman numerals indicate isoseismal grading

of origin. The epicentre of an earthquake is located on the surface of the Earth immediately above the focus and shock waves radiate from the focus in all directions (Figure 4.26). Earthquake foci are confined within a limited zone of the upper Earth, the lower boundary occurring at 700 km depth from the surface. No earthquakes are known to have originated below this level. Moreover, earthquakes rarely originate at the Earth's surface. In fact most earthquakes originate within the upper 25 km of the Earth. Because of its significance, the depth of foci has been used as the basis of a threefold classification of earthquakes; those occurring within the upper 70 km are referred to as shallow, those located between 70 and 300 km as intermediate, and those between 300 and 700 km as deep. Seventy per cent of all earthquakes are of shallow type. Within active tectonic regions shallow earthquakes appear to align themselves to one side whilst the deep earthquakes occur towards the opposing side (Figure 4.25).

An earthquake propagates three types of shock wave. The first pulses that are recorded are termed primary or P waves. Sometimes they are referred to as push and pull waves since oscillation occurs to and fro in the path of the wave. P waves are also called compression waves or longitudinal waves. The next pulses recorded are the S waves, sometimes referred to as secondary or shake waves. These waves oscillate at right angles to the path of propagation. S waves usually have a larger amplitude than the P waves but the latter travel almost three times as fast as the former. The third type of vibration is known as the L wave. These waves travel from the focus of the earthquake to the epicentre above, and from there they radiate over the Earth's surface. Two types of L or surface wave can occur in a solid medium, namely, Rayleigh waves and Love waves. In the former, surface displacement occurs partly in the direction of propagation and partly in the vertical plane. They can only be generated in a uniform solid. Love waves occur in non-uniform solids and oscillate in a horizontal plane, normal to the path of propagation. L waves are recorded after S waves.

As P waves travel faster than S waves, the further they travel from the focus of an earthquake the greater is the time interval between them. Thus the distance of a recording station from the epicentre of an earthquake can be calculated from this time interval. With three such determinations the epicentre can be located, it occurring where three circles, each centred on a recording station and with radii equal to their respective distances from the epicentre, intersect.

P waves are not as destructive as S or L waves. This is because they have a smaller amplitude and the force their customary vertical motion creates rarely exceeds the force of gravity. On the other hand, S waves may develop violent tangential vibrations strong enough to cause great destruction. The intensity of an earthquake depends upon the amplitude and frequency of wave motion. S waves commonly have a higher frequency than L waves, nevertheless the latter may be more powerful because of their larger amplitude.

The velocities of P and S waves vary with the density and elastic constants of the rocks through which they travel. As a result, when they pass from one material of different composition or state to another, they are subject to reflection and refraction. The Earth's surface also reflects waves.

Large earthquakes cause shock waves which travel throughout the world. Near the epicentre the shocks are only felt for a matter of seconds or minutes but with increasing distance from the epicentre the disturbance lasts for a progressively longer time. The vibrations of a shallow earthquake decrease rapidly in intensity away from the epicentre. On the other hand, although the shocks of deep-seated earthquakes are usually weak when they reach the surface, they extend over a much wider area.

As noted above, it is assumed that most earthquakes are caused by faulting. Faulting is probably due to a gradual build-up of shearing stress which results in rocks fracturing when their yield point is exceeded. Initially movement may occur over a small area of the fault plane, to be followed later by a slippage over a much larger surface. Such initial movements account for the foreshocks which precede an earthquake. These are followed by the main movement, but complete stability is not restored immediately. The shift of rock masses involved in faulting relieves the main stress but develops new stresses in adjacent areas. Because stress is not relieved evenly everywhere, minor adjustments may arise along the fault plane and thereby generate aftershocks. The decrease in strength of the aftershocks is irregular and occasionally they may continue for a year or more.

There is a rough relationship between the length of a fault break and the amount of displacement involved, and both are related to the magnitude (M) of the resultant earthquake. Displacements range from a few millimetres up to 11 m, which was the vertical displacement on the Chedrang fault during the Assam earthquake (1897; $M = 8.7$). The largest horizontal displacement was on the Bogdo fault after the Mongolian earthquake (1957; $M = 8.3$). As expected the longer fault breaks have the greater displacements and generate the larger earthquakes. On the other hand, it has been shown by Ambraseys (1969) and Bonilla (1970) that the smaller the fault displacement, the greater the number of observed fault breaks. They also found that for the great majority of fault breaks the maximum displacement was less than 6 m and that the average displacement along the length of the fault was less than 50% of the maximum. The length of the fault break during a particular earthquake is generally only a fraction of the true length of the fault. Individual fault breaks during simple earthquakes have ranged in length from less than a kilometre to several hundred kilometres. What is more, fault breaks do not only occur in association with large and infrequent earthquakes but also occur in association with small shocks and continuous slow slippage known as fault creep. Fault creep may amount to several millimetres per year and progressively deforms buildings located across such faults (see Steinbrugge and Zacher (1960)).

There is little information available on the frequency of breaking along active faults. All that can be said is that some master faults have suffered repeated movements—in some cases it has recurred in less than 100 years. By contrast much longer intervals, totalling many thousands of years, have occurred between successive breaks. Therefore, because movement has not been recorded in association with a particular fault in an active area, it cannot be concluded that the fault is inactive.

Earthquakes resulting from displacement and energy release on one fault can sometimes trigger small displacements on other unrelated faults many kilometres distant. Breaks on subsidiary faults have occurred at distances as great as 25 km from the main fault, but with increasing distance from the main fault the amount of displacement decreases. For example, displacements on branch and subsidiary faults located more than 3 km from the main fault break are generally less than 20% of the main fault displacement.

Some areas are affected by earthquake swarms, that is, by a series of nearly equally large shocks accompanied by small ones.

Bolt (1970) suggested that between earthquakes the fault rupture in basement rocks would be welded by the large lithostatic forces normal to the fault plane. Consequently, regional deformation in the basement rocks would occur as elastic strain.

Major volcanic eruptions may cause extremely violent local earthquakes, such as that due to Vesuvius which affected the island of Ischia in the Bay of Naples in 1883. This earthquake was not particularly intense on the adjoining mainland.

The seismograph is an instrument which records the occurrence of an earthquake, whereas a seismometer measures the vibrations. When recording an earthquake, it is essential that some part of the instrument remains stationary whilst the rest is shaken by the earth tremors. The usual way of doing this is to suspend a mass, with a minimum of attachment to the ground, and to rely upon the inertia of the mass to hold it motionless while the ground vibrates. One of the commonest seismometers consists of a weighted pendulum which swings in the horizontal plane and is suspended from and pivoted against a firmly anchored support. The latter vibrates during an earthquake whilst the pendulum remains stationary. Seismometers usually measure earth tremors in one direction. Consequently two instruments are placed at right angles to each other in the horizontal plane whilst a third is arranged vertically, so that movements in all three directions are measured. Modern practice is to have an array of as many as 100 seismometers laid out in an L- or T-shaped arrangement to record shock waves. An accurate record of time must be kept during an earthquake so that its intensity can be assessed.

4.8 Intensity and magnitude of earthquakes

Whilst the severest earthquakes wreak destruction over areas of 2500 km² or more, most only affect tens of square kilometres. Earthquake intensity scales depend on human perceptibility and the destructivity of earthquakes. Whereas the degree of damage may be estimated correctly and objectively, the perceptibility of an earthquake depends on the location of the observer and his sensibility. Several earthquake intensity scales have been proposed. The one given in Table 4.1 is the Mercalli scale (see Wood and Neumann (1931)) which was slightly modified by Richter (1956).

The magnitude of an earthquake is an instrumentally measured quantity. Richter (1935) devised a logarithmic scale for comparing the magnitudes of Californian earthquakes. His method has since been widely extended and developed. He related the magnitude of a tectonic earthquake to the total amount of elastic energy released when overstrained rocks suddenly rebound and so generate shock waves. The relationship is given by the expression

$$\log E = a + bM$$

where M is the magnitude and E the energy released expressed in ergs. The values of the constants a and b have been modified several times as data have accumulated. One of the most recent versions of the equation is (see Hazzard (1978))

$$\log E = 9.4 + 2.14\,M - 0.05\,M^2$$

The largest earthquakes have had a magnitude of 8.9 and these release about 700 000 times as much energy as earthquakes at the threshold of damage. Earthquakes of magnitude 5.0 or greater generate sufficiently severe ground motions to be potentially damaging to structures (Figure 4.27). It has been estimated that in a typical year the Earth experiences two earthquakes over magnitude 7.8, 17 with magnitudes between 7 and 7.8, and about a 100 between 6 and 7.

Many seismologists believe that the duration of an earthquake is the most

Table 4.1. Modified Mercalli scale, 1956 version, with Cancani's equivalent acceleration. (These are not peak accelerations as instrumentally recorded.)

Degrees	Description	Acceleration (mm/s²)
I	Not felt. Only detected by seismographs	Less than 2.5
II	Feeble. Felt by persons at rest, on upper floors, or favourably placed	2.5–5.0
III	Slightly felt indoors. Hanging objects swing. Vibration like passing of light trucks. Duration estimated. May not be recognised as earthquake	
IV	Moderate. Hanging objects swing. Vibration like passing of heavy trucks, or sensation of a jolt like a heavy ball striking the walls. Standing motor cars rock. Windows, dishes, doors rattle. Glasses clink. Crockery clashes. In the upper range of IV wooden walls and frames creak	10–25
V	Strong. Felt outdoors, direction estimated. Sleepers wakened. Liquids disturbed, some spilled. Small unstable objects displaced or upset. Doors swing, close, open. Shutters and pictures move. Pendulum clocks stop, start change rate.	25–50
VI	Strong. Felt by all. Many frightened and run outdoors. Persons walk unsteadily. Windows, dishes, glassware broken. Ornaments, books, etc. fall off shelves. Pictures fall off walls. Furniture moved or overturned. Weak plaster and masonry cracked. Small bells ring (church, school). Trees, bushes shaken visibly or heard to rustle	50–100
VII	Very strong. Difficult to stand. Noticed by drivers of motor cars. Hanging objects quiver. Furniture broken. Damage to masonry D, including cracks. Weak chimneys broken at roof line. Fall of plaster, loose bricks, stones, tiles, cornices, also unbraced parapets and architectural ornaments. Some cracks in masonry C. Waves on ponds, water turbid with mud. Small slides and caving in along sand or gravel banks. Large bells ring. Concrete irrigation ditches damaged	100–250
VIII	Destructive. Steering of motor cars affected. Damage to masonry C, partial collapse. Some damage to masonry B, none to masonry A. Fall of stucco and some masonry walls. Twisting, fall of chimneys, factory stacks, monuments, towers, elevated tanks. Frame houses moved on foundations if not bolted down, loose panel walls thrown out. Decayed piling broken off. Branches broken from trees. Changes in flow or temperature of springs and wells. Cracks in wet ground and on steep slopes	250–500
IX	Ruinous. General panic. Masonry D destroyed, masonry C heavily damaged, sometimes with complete collapse, masonry B seriously damaged. General damage to foundations. Frame structures, if not bolted, shifted off foundations. Frames cracked, serious damage to reservoirs. Underground pipes broken. Conspicuous cracks in ground. In alluviated areas sand and mud ejected, earthquake fountains, sand craters	500–1000
X	Disastrous. Most masonry and frame structures destroyed with their foundations. Some well-built wooden structures and bridges destroyed. Serious damage to dams, dykes, embankments. Large landslides. Water thrown on banks of canals, rivers, lakes, etc. Sand and mud shifted horizontally on beaches and flat land. Rails bent slightly	1000–2500

Degrees	Description	Acceleration (mm/s²)
XI	Very disastrous. Rails bent greatly. Underground pipelines completely out of service	2500–5000
XII	Catastrophic. Damage nearly total. Large rock masses displaced. Lines of sight and level distorted. Objects thrown into the air	Over 5000

important factor as far as damage or failure of structures, soils and slopes are concerned. What is important in hazard assessment is the prediction of the duration of seismic shaking above a critical ground acceleration threshold. The magnitude of an earthquake affects the duration much more than it affects the maximum acceleration, since the larger the magnitude, the greater the length of ruptured fault. Hence the more extended the area from which the seismic waves are emitted. With increasing distance from the fault the duration of shaking is longer and the intensity of shaking is less, the higher frequencies being attenuated more than the lower ones. The attenuation, or reduction of acceleration amplitude, which occurs as waves travel from the causative fault, is the result of decreasing seismic energy attributable to dispersion that takes place as the outward propagating waves occupy an increasing space. Dispersion effects tend to increase the duration of strong shaking as the distance from the causative fault increases.

4.9 Ground conditions and seismicity

The physical properties of the soils and rocks through which seismic waves travel, as well as the geological structure, also influence surface ground motion. For example, if a wave traverses vertically through granite overlain by a thick uniform deposit of alluvium, then theoretically the amplitude of the wave at the surface should be double that at the alluvium–granite contact. According to Ambraseys (1974) maximum acceleration within an earthquake source area may exceed 200% g for competent bedrock. On the other hand, normally consolidated clays with low plasticity are incapable of transmitting accelerations greater than 10–15% g to the surface. Clays with high plasticity allow accelerations of 25–35% g to pass through. Saturated sandy clays and medium dense sands may transmit 50–60% g, and in clean gravel and dry dense sand accelerations may reach much higher values.

The response of structures on different foundation materials has proved surprisingly varied. In general, structures not specifically designed for earthquake loadings have fared far worse on soft saturated alluvium than on hard rock. This is because motions and accelerations are much greater on deep alluvium than on rock. By contrast a rigid building may suffer less on alluvium than on rock. The explanation is attributable to the alluvium having a cushioning effect and the motion may be changed to a gentle rocking. This is easier on such a building than the direct effect of earthquake motions experienced on harder ground. Nonetheless alluvial ground beneath any kind of poorly constructed feature facilitates its destruction.

(a)

(b)

Figure 4.27 Earthquake damage at the Saada Hotel, Turkey (courtesy of
the American Iron and Steel Institute)

Intensity attenuation on rock is very rapid whereas it is extremely slow on soft formations and speeds up only in the fringe area of the shock. Hence the character of intensity attenuation in any shock will depend largely on the surface geology of the shaken area.

Ground vibrations caused by earthquakes often lead to compaction of cohesionless soil and associated settlement of the ground surface. Loosely packed saturated sands and silts tend to lose all strength and behave like fluids during strong earthquakes (see Seed (1970); Martin (1978)). When such materials are subjected to shock, a densification occurs. During the relatively short time of an earthquake, drainage cannot be achieved and this densification therefore leads to the development of excessive pore water pressures. These cause the soil mass to act as a heavy fluid with practically no shear strength, that is, a quick condition occurs (Figure 4.28). Water moves upward from the voids to the ground surface where it emerges to form sand boils. An approximately linear relationship has been shown to occur between the density index of sands and the stress required to cause initial liquefaction in a given number of stress cycles. If liquefaction occurs in a sloping soil mass, the entire mass will begin to move as a flow slide. Such slides develop in loose, saturated cohesionless materials during earthquakes. Loose saturated silts and sands often occur as thin layers underlying firmer materials. In such instances liquefaction of the silt or sand during an earthquake may cause the overlying material to slide over the liquefied layer. Structures on the main slide are frequently moved without suffering damage. However, a graben-like feature often forms at the head of the slide and buildings located in this area are subjected to large differential settlements and often are destroyed. Buildings near the toe of the slide are commonly heaved upwards or are even pushed over by the lateral thrust. Major slides during earthquakes can result from failure in clay deposits.

Figure 4.28 Tilting of apartment buildings on liquefied sands, Niiagata, 1964 (courtesy of Professor H. Bolton Seed)

Clay soils do not undergo liquefaction when subjected to earthquake activity. However, it has been shown that under repeated cycles of loading large deformations can develop, although the peak strength remains about the same. Nonetheless these deformations can reach the point where, for all practical purposes, the soil has failed.

It can be concluded from the foregoing paragraphs that macroseismic observations can be used to establish sesimic intensity increments for the basic categories of ground. Such an idea was first advanced by Reid (1908) who, after the San Francisco (1906) earthquake, introduced the concept of foundation coefficients for several major soil and rock types. The coefficient for the type of foundation that produced the least vibrational force as revealed by observed earthquake damage was designated unity. Estimates of the probable accelerations associated with these coefficients also were noted.

Although it is now recognised, from the complex nature of strong motion accelerograph records, that acceleration itself has little meaning, unless the frequency is given also, there is reason to believe that these coefficients provide a true picture of the relative earthquake intensities experienced on the types of foundation cited. The basic categories of ground subsequently recognised by Medvedev (1965) are granites, limestones and sandstones, moderately firm ground, coarse fragmental ground, sandy ground, clayey ground and fill. By comparing the specific intensity changes for the basic categories of ground, the total intensity increment for each category with reference to a single standard (granite) can be found. All changes then become positive quantities or increments. The values of the intensity increments are given in Table 4.2. In order to determine the rated seismic intensity it is necessary to relate the results of observation to the physical characteristics of the ground. One of the chief physical characteristics of the ground in the evaluation of its influence on seismic intensity is the seismic rigidity, which is the product of the rate of propagation of longitudinal seismic waves (V_c) and density (ρ).

Table 4.2. Seismic intensity increments for the basic categories of ground, from macroseismic data (after Medvedev (1965))

Ground	Medvedev	Reid	V_c (km/s)	ρ (Mg/m³)	$V_c \times \rho$
(1) Granites	0	0	5.6	2.9	16.2
(2) Limestones, sandstones, shales	0.2–1.3	0–1.2	4.5–2.5	2.8–2	12.6–5
(3) Gypsum, marl	0.6–1.4	1–1.5	3–1.7	2.4–1.7	7.2–2.9
(4) Coarse-fragmental ground (rubble, pebble, gravel)	1–1.6	1.2–2.1	2.1–0.9	2–1.6	4.2–1.4
(5) Sandy ground	1.2–1.8	1.2–2.1	1.6–0.6	1.9–1.6	3.1–1
(6) Clayey ground	1.2–2.1	1.5–2	1.5–0.6	2–1.6	3–1
(7) Fill	2.3–3	2.1–3.4	0.6–0.2	1.5–1.3	0.9–0.26
(8) Moist ground (gravel, sand, clay)	1.7–2.8	2.3			
(9) Moist fill and soil ground (bog)	3.3–3.9	3.5			

4.10 Earthquake prediction and seismic zoning

It has been found that the properties of rock change as it undergoes
dilation under stress immediately prior to failing and generating an earthquake (see Pakiser *et al.* (1969); Scholtz *et al.* (1973)). Most important is the
variation in seismic wave velocity. The P waves slow down due to the
minute fracturing in rocks during dilation but then increase as water
occupies microfractures prior to an earthquake. Hence a continuous record
of such changes in shock wave velocity may be of value in earthquake
prediction as a quake may occur a day or so after the wave velocities return
to normal. Furthermore the ratio of the P and S wave velocities decreases
by up to 20% before a major quake. The period of time covered by these
changes indicates the size of the earthquake that can be expected. For
example, an earthquake of magnitude 5.4 may be preceded by a period of
four months of lowered velocities, of magnitude 4 by only two months. Thus
if the period of changes lasts for 14 years it suggests a potentially violent earthquake of magnitude 7.

The dilatancy of rocks leads to an increase in their volume and results in
minor tilting of the ground surface near an active fault. This ground movement can be measured by very accurate surveying. Around Palmdale, near
Los Angeles, an area of over 12800 km² gently rose over the period
1959–74, the centre having lifted by 0.45 m. The domed area straddles the
San Andreas fault, which locally has not moved since 1932.

Other features brought about by rock dilation include a decrease in electrical
resistivity and a change in magnetic susceptibility of the rock concerned as
well as the release of small quantities of radon. This increases the radon
content in the groundwater. Monitoring these subtle changes provides further
evidence for earthquake prediction.

At the present time earthquake prediction is in its infancy and is not
precise enough to be a significant help. Nevertheless, protection against earthquake damage is necessary and so the concept of seismic zoning has been
developed.

Seismic evidence is obtained instrumentally and from the historical record.
Maps can be drawn, by using these data, which indicate the epicentral areas
of earthquakes and these are then zoned according to activity. The longer
the history on which these maps are based, the better they are likely to be.
A seismic zoning map, therefore, shows the zones of different seismic danger
in a particular area. Hence it provides a broad picture of the earthquake risk
that can be involved in seismic regions.

According to Medvedev (1968) seismic zoning of a region is based upon
a study of earthquakes occurring in the region concerned, an investigation
of the laws governing the occurrence of earthquakes of different intensity,
an analysis of the geological conditions under which earthquakes occur and
the investigation of special features accompanying the occurrence of earthquakes. The production of maps of seismic zoning in the Soviet Union therefore takes account of the depth of earthquake foci, the relationship between
energy at the focus and epicentral intensity, the correlation between magnitude and intensity, and information concerning attenuation of intensity
with distance.

Specific conditions in each seismic region, peculiarities of the seismic

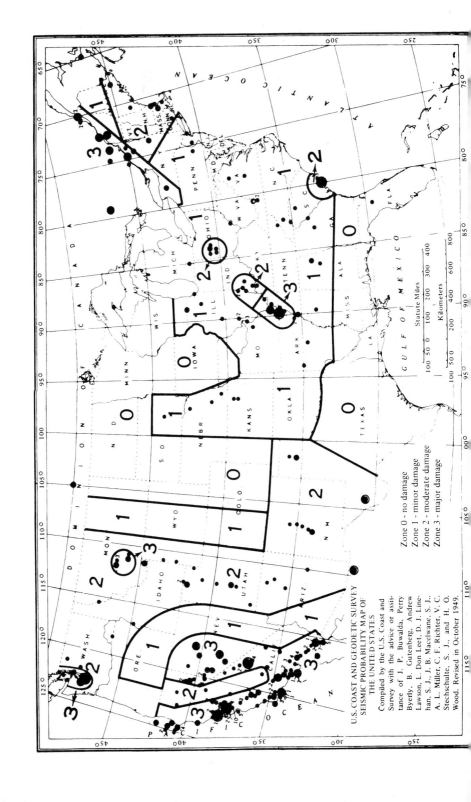

U.S. COAST AND GEODETIC SURVEY
SEISMIC PROBABILITY MAP OF
THE UNITED STATES

Compiled by the U.S. Coast and
Survey with the advice or assis-
tance of J. P. Buwalda, Perry
Byerly, B. Gutenberg, Andrew
Lawson, L. Don Leet, D. J. Line-
han, S. J., J. B. Macelwane, S. J.,
A. L. Miller, C. F. Richter, V. C.
Stechschulte, S. J., and H. O.
Wood. Revised in October 1949.

Zone 0 - no damage
Zone 1 - minor damage
Zone 2 - moderate damage
Zone 3 - major damage

regime, different forms of seismo-geological linking and the extent to which the area has been studied, mean that there can be no rigorously standard way of using the information for zoning in all regions. What is more in the compilation of these maps all engineering data concerning the surface manifestations of earthquakes must refer to identical ground conditions as they can have a strong influence on intensity. Geological investigation may be able to throw light upon the state of strain within the crust and can delimit areas where faulting is widespread. However, geological data can only be used to give a qualitative assessment of seismic risk within an area. To obtain a quantitative expression the geological data must be examined in conjunction with engineering and seismic data.

Historical and geological studies have been used to distinguish regions at different levels of risk within the USA. A map zoning the USA was prepared by the US Coast and Geodetic Survey with the cooperation of seismologists in all parts of the country (Figure 4.29a). This map was incorporated in the Uniform Building Code, of which it became legally a part. In 1959 Richter published a zoning map of the USA, based on seismo-statistical and geological data. Zoning was according to the modified Mercalli scale. The map distinguishes regions of intensities 6, 7, 8 and 9, and it is assumed that earthquakes with intensities exceeding 9 are possible in the western zone (Figure 4.29b). However, Richter's map may be regarded as an overestimation. This is especially true of the eastern and central parts of the country.

Seismic zoning has led to a reduction of earthquake risk. Buildings in quake-prone areas must be made as safe as possible even though in practice the earthquake-proof building does not exist. Earthquake-protected buildings are constructed of reinforced concrete, with a steel frame and deep founda-

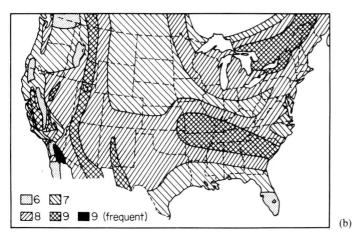

□6 ◪7
▨8 ▩9 ■9 (frequent)   (b)

Figure 4.29 (a) Seismic probability map of the USA. This map is commonly used to establish seismic design criteria, the following maximum ground accelerations are associated with the zones. Zone 3, 33% g; Zone 2, 16% g; Zone 1, 8% g; Zone 0, 4% g. In Zone 3 close to a major active fault the maximum ground acceleration is estimated to be approx. 50% g. (b) Seismic zoning of the USA. 6–9 intensity ratings

tions. Even with good construction practice it is difficult to avoid the threat of resonance in high buildings, when slow earthquake waves coincide with the natural period of motion of the building.

4.11 Tsunamis

One of the more terrifying side-effects of earthquakes which occurs along coastal regions is inundation by large masses of water called tsunamis. Most tsunamis originate as a result of fault movement on the sea floor, though they can also be developed by submarine landslides or volcanic activity. Seismic tsunamis are usually formed where submarine faults have a significant vertical movement, and faults of this type most commonly occur along the coasts of South America, the Aleutian Islands and Japan. Horizontal fault movements, such as occur along the Californian coast, do not result in tsunamis.

In the open ocean tsunamis normally have a very long wavelength and their amplitude is hardly noticeable. Successive waves may be from five minutes to an hour apart. They travel at speeds of around 650 km/h. However, their speed is proportional to the depth of the water which means that in coastal areas the waves slow down and increase in height, rushing onshore as highly destructive breakers. Waves have been recorded up to nearly 20 m in height above normal sea level. Large waves are most likely when tsunamis move into narrowing islets. Such waves cause terrible devastation and can wreck entire villages. For example, if the wave is breaking as it crosses the shore, it can destroy houses merely by the weight of water. The subsequent backwash may carry many buildings out to sea and may remove several metres depth of sand from dune coasts.

Structures which offer protection against potential damage caused by tsunamis include breakwaters, revetments, etc. Design against tsunami damage involves raising buildings off the ground and orienting shear walls normal to the direction of the wavefront. Reinforced concrete structures having adequate foundations do not usually suffer noticeable damage.

Historical data provide the basis for prediction of tsunamis in Hawaii (see Adams (1970)). The effects of tsunamis in Hawaii have been recorded for more than 150 years and allow an estimation of the events following a tsunamigenic earthquake to be made. Usually the first wave is like a very rapid change in tide, for example, the sea level may change 7 or 8 m in 10 min. A bore occurs where there is a concentratiion of wave energy by funnelling, as in bays, or by convergence, as on points. A steep front rolls over relatively quiet water. Behind the front the crest of such a wave is broad and flat, the wave velocity is about 30 km/h. Along rocky coasts large blocks of material may be dislodged and moved shoreward.

The Pacific Tsunami Warning System (PTWS) is a communications network covering all the countries bordering the Pacific Ocean and is designed to give advance warning of dangerous tsunamis. Clearly the PTWS cannot provide a warning of an impending tsunami to those areas which are very close to the earthquake epicentre which is responsible for the generation of the tsunami. On the other hand, waves generated off the coast of Japan will take 10 h to reach Hawaii. In the latter case the system is important.

Furthermore the atmosphere is disturbed by the same forces that cause the tsunamis but the sound waves travel faster. Consequently the detection of sound waves could provide an additional warning mechanism for the future. In the same way, the ionosphere is disturbed by tsunami earthquakes and variations in reflected radio waves could offer yet more warning.

Evacuation of an area depends on estimating just how destructive any tsunami will be when it arrives on a particular coast. As has been shown the PTWS in some instances can provide a few hours for evacuation to take place if it appears that it is necessary. The resettlement of coastal populations away from possible danger zones, however, is not a feasible proposition.

References

ADAMS, W. M. (1970). 'Tsunami effects and risk at Kahuku Point, Uahu, Hawaii', *Engg Geol. Case Histories*, No. 8, Geol. Soc. Am., 63–70.

AMBRASEYS, N. N. (1969). 'Maximum intensity of ground movements caused by faulting', *Proc. 4th World Conf. Earthquake Engg Chile*, **1**, 154–62.

AMBRASEYS, N. N. (1974). Notes on engineering seismology, (in *Engineering Seismology and Earthquake Engineering*), Solnes, J. (ed.), NATO Advanced Studies Institute Series, Applied Sciences, No. 3, 33–54.

ANDERSON, E. M. (1951). *Dynamics of Faulting and Dyke Formation*, Oliver & Boyd, Edinburgh.

BADGLEY, P. C. (1965). *Structural and Tectonic Principles*, Harper & Row, New York.

BILLINGS, M. P. (1962). *Structural Geology*, Prentice-Hall, Englewood Cliffs, New Jersey.

BOLT, A. B. (1970). 'Causes of earthquakes', (in *Earthquake Engineering*), Weigel, R. L. (ed.), Prentice-Hall, Englewood Cliffs, New Jersey, 21–46.

BONILLA, M. G. (1970). 'Surface faulting and related effects', (in *Earthquake Engineering*), Weigel, R. L. (ed.), Prentice-Hall, Englewood Cliffs, New Jersey, 47–74.

BRADDOCK, W. A. (1970). 'The origin of slaty cleavage: evidence from the PreCumbrian rocks in Colorado', *Bull. Geol. Soc. Am.*, **81**, 589–600.

BUSK, H. G. (1927). *Earth Flexures*, Cambridge University Press, London.

CLOOS, E. (1946). 'Lineation', *Geol. Soc. Am. Mem.*, **18**.

DE SITTER, L. U. (1964). *Structural Geology*, McGraw-Hill, New York.

DIETERICH, J. H. (1969). 'Origin of cleavage in folded rock', *Am. J. Sci.*, **257**, 155–65.

FLEUTY, M. J. (1964). 'The description of folds', *Proc. Geol. Ass.*, **75**, 461–89.

FOOKES, P. G. (1965). 'Orientation of fissures in stiff fissured, overconsolidated clay of the Silawik system', *Geotechnique*, **15**, 195–206.

GILL, J. E. (1971). 'Continued confusion in the classification of faults', *Bull. Geol. Soc. Am.*, **82**, 1389–92.

HANCOCK, P. L. (1968). 'The relation between folds and late formed joints in south Pembrokeshire', *Geol. Mag.*, **101**, 174–84.

HAZZARD, A. O. (1978). 'Earthquake and engineering: structural design', *The Consulting Engineer*, 14–23, May.

HILLS, E. S. (1964). *Elements of Structural Geology*, Methuen, London.

HUBBERT, M. K. & RUBEY, W. W. (1959). 'The role of fluid pressure in mechanics of overthrust faulting', *Bull. Geol. Soc. Am.*, **70**, 115–206.

LEITH, C. K. (1905). 'Rock cleavage', *US Geol. Surv. Bull.*, **239**

LOUDERBACK, G. D. (1950). Faults and engineering geology, (*in Application of Geology to Engineering Practice*), Berkey Volume, Geol. Soc. Am., 125–50.

McINTYRE, D. B. (1950). 'Notes on lineation, boudinage and recumbent folds in the Struan Flags (Moine)', *Geol. Mag.*, **87**, 427–32.

MARTIN, G. R. (1978). 'Soil stability in strong earthquakes', *The Consulting Engineer*, 30–35, May.

MEDVEDEV, S. V. (1965). *Engineering Seismology*, Israel Program for Scientific Translations, Keter Publishing House, Jerusalem.

MEDVEDEV, S. V. (1968). 'Measurement of ground motion and structural vibrations caused by earthquake', *Proc. Int. Sem. Earthquake Engg Skopje*, UNESCO, 35–38.

MURRELL, S. A. F. (1977). 'Natural faulting and the mechanics of brittle shear failure', *J. Geol. Soc.* **133**, 175–90.

NEVIN, C. M. (1949). *Principles of Structural Geology*, Wiley, New York.

PAKISER, L. C., EATON, J. P., HEALY, J. H. & RALEIGH, L. B. (1969). 'Earthquake prediction and control', *Science*, **166**, 1467–74.

RAMBERG, H. (1963). 'Evolution of drag folds', *Geol. Mag.*, **100**, 97–106.

RAMSEY, J. G. (1962). 'The geometry and mechanics of formation of "similar" type folds', *J. Geol.*, **60**, 309–27.

RAMSAY, J. G. (1967). *Folding and Fracturing of Rocks*, McGraw-Hill, London.

REID, H. F. (1908). *The Californian Earthquake of April 18, 1906*, Report of the State Earthquake Investigation Commission Vol. 2, *The Mechanics of the Earthquake*, Carnegie Institute, Washington DC.

RICHTER, C. F. (1935). 'An instrumental earthquake scale', *Bull. Seis. Soc. Am.*, **25**, 1–32.

RICHTER, C. F. (1956). *Elementary Seismology*, W. H. Freeman & Co., San Francisco.

RIEDEL, W. (1929). 'Zur Mechanik Geologischer Brucherscheinaugen', *Cb Min. Abt B*, 354–68.

SALEHY, M. R., MONEY, M. S. & DEARMAN, W. R. (1977). 'The occurrence and engineering properties of intraformational shears in carboniferous rocks', *Proc. Conf. Rock Engg Newcastle University*, **1**, 311–28.

SCHOLTZ, C. H., SYKES, L. R. & AGGARWAL, T. (1973). 'Earthquake prediction: a physical basis', *Science*, **181**, 803–10.

SEED, H. B. (1970). 'Soil problems and soil behaviour', (in *Earthquake Engineering*), Weigel, R. L. (ed.), Prentice-Hall, Englewood Cliffs, New Jersey, 227–52.

SIBSON, R. H. (1977). 'Fault rocks and fault mechanisms', *J. Geol. Soc.*, **133**, 191–214.

SKEMPTON, A. W. (1966). 'Some observations on tectonic shear zones', *Proc. 1st Int. Conf. Rock Mech., Lisbon*, **1**, 329–55.

STEINBRUGGE, K. V. & ZACHER, E. G. (1960). 'Creep on the San Andreas fault: fault creep and property damage', *Bull. Seis. Soc. Am.*, **50**, 389–98.

TIJA, H. D. (1964). 'Slickensides and fault movements', *Bull. Geol. Soc. Am.*, **75**, 683–6.

WILLIS, R. & WILLIS, B. (1934). *Geologic Structures*, McGraw-Hill, New York.

WILSON, G. (1946). 'The relationship of slaty cleavage and kindred structure to tectonics', *Proc. Geol. Ass.*, **57**, 263–302.

WILSON, G. (1951). 'The tectonic significance of small-scale structures and their importance to the geologist in the field', *Ann. Soc. Geol. Belgique*, **84**, 423–548.

WILSON, G. (1953). 'Mullion and rodding structures in the Moine series of Scotland', *Proc. Geol. Ass.*, **64**, 118–51.

WOOD, N. O. & NEUMANN, F. (1931). 'The modified Mercalli intensity scale of 1931', *Bull. Seis. Soc. Am.*, **21**, 277–83.

Chapter 5

Discontinuities

As far as design and practice in engineering are concerned, rock masses may be grouped into two categories, simply those which are unweathered and those which are weathered. In the case of unweathered rock masses interest centres mainly on the incidence and character of the discontinuities since these adversely affect engineering performance. This is not to say that discontinuities are of no consequence in weathered rocks, indeed weathering tends to be concentrated along discontinuities.

A discontinuity represents a plane of weakness within a rock mass across which the rock material is structurally discontinuous. Although discontinuities are not necessarily planes of separation, most in fact are and they possess little or no tensile strength. Discontinuities vary in size from small fissures on the one hand to huge faults on the other. The most common discontinuities are joints and bedding planes (Figure 5.1). Other important discontinuities are planes of cleavage and schistosity, fissures and faults.

5.1 Nomenclature of joints

Joints are fractures along which little or no displacement has occurred and are present within all types of rock. At the surface, joints may open as a consequence of denudation, especially weathering.

A group of joints which run parallel to each other are termed a *joint set* whilst two or more joint sets which intersect at a more or less constant angle are referred to as a *joint system*. A conjugate joint system describes two sets of joints which intersect symmetrically about some other structural plane or line. In a complementary joint system two sets of shear joints of the same age, and initiated by the same stress field, intersect at an angle of about 60°. All complementary joint systems are conjugate but the converse is not true.

If joints are planar and parallel or sub-parallel they are described as systematic, conversely, when they are irregular they are termed non-systematic. If one set of joints is dominant then the joints are known as primary joints, the other set or sets of joints being termed secondary. Joints commonly occur in narrow zones where each joint is replaced *en echelon* by another. Cross joints, which are irregular in shape, may cut the rock

(a)

(b)

(c)

Figure 5.1 (a) Joints and bedding planes in sandstone of Coal Measures age, near Ripley, Derbyshire. Note the plumose structure on the joint surface in the left-centre of the exposure. (b) Two diagonally opposed joint systems developed in argillaceous shaly limestone in the Limestone Group of the Lower Carboniferous of Northumberland near Cullernose Point. (c) Discontinuity surface in Horton Flags, Silurian, along which rocks have slipped, Arcow Quarry, Helwith Bridge, Yorkshire

within these zones. Hodgson (1961) contended that cross joints do not intersect systematic joints or well developed bedding planes.

The attitude of joints may be referred to that of the enclosing bedding planes when no predominant trend is discernible in the region. Consequently strike joints are parallel to the strike of the bedding planes and dip joints run in the direction of the dip of the bedding. It has been suggested that joints can be classified in the same way as faults, that is, according to the direction in which any movement has taken place (if it can be discerned) along the joint plane. Joints can be grouped as normal dip-slip joints, reverse dip-slip joints and strike-slip joints. In some cases, because of the heavy disruption of strata, joints can only be compared with these groupings when the strata has been restored to horizontal. In such situations these joints are therefore referred to as normal dip-slip equivalent joints, reverse dip-slip equivalent joints and strike-slip equivalent joints. According to Hancock (1968) normal dip-slip and normal dip-slip equivalent joints are rare in all types of structural settings. Reverse dip-slip and reverse dip-slip equivalent joints are locally developed in regions of intense folding and reversed folding. Strike-slip and strike-slip equivalent joints are common in all structural settings.

On a basis of size, joints can be divided into master joints which penetrate several rock horizons and persist for hundreds of metres; major joints which are smaller joints but which are still well defined structures and minor joints which do not transcend bedding planes. Lastly, minute fractures occasionally occur in finely bedded sediments and such micro-joints may only be a few millimetres in size. Master joints are not usually found in the folds of thin interbedded competent and incompetent rocks. In such instances major and minor joints are generally restricted to the competent rocks.

Joints may be associated with folds and faults, having developed towards the end of an active tectonic phase or when such a phase has subsided. However, joints do not appear to form parallel to other planes of shear failure such as normal and thrust faults. The orientation of joint sets in relation to folds depends upon their size, the type and size of the fold, and the thickness and competence of the rocks involved. At times the orientation of the joint sets can be directly related to the folding and may be defined in terms of the *a*, *b* and *c* axes of the 'tectonic cross' (Figure 5.2). Those joints which cut the fold at right angles to the axis are called *ac* or cross joints. The *bc* or longitudinal joints are perpendicular to the latter joints and diagonal or oblique joints make an angle with both the *ac* and the *bc* joints. Diagonal joints are classified as shear joints whereas *ac* and *bc* joints are regarded as tension joints.

Joint faces are often irregular so that the adjacent walls are completely interlocking. Figure 5.3 shows the primary surface structures on a joint plane. When the joint is unweathered the dominant feature is the main joint face which possesses a rather rough surface on which there are plumose markings. Generally the axes of the plumes are parallel to the bedding planes. At times the main joint face is separated from the surrounding fringe by a pronounced shoulder, which is often formed by an abrupt termination of the fine

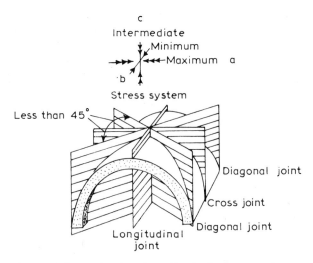

Figure 5.2 Geometric orientation of longitudinal, cross and diagonal joints relative to fold axis and to principal stress axes (from Willis, B. & Willis, R. (1934). *Geologic Structures*, McGraw-Hill, New York)

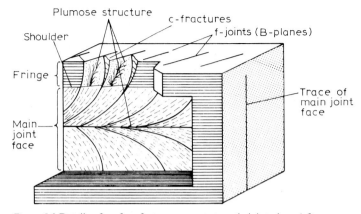

Figure 5.3 Details of surface features on a systematic joint plane (after Hodgson (1961))

ridges of the plumose structure against small joints arranged *en echelon* in the fringe. These fringe joints (*f* joints) commonly make an angle of 5 to 25° with the main joint face. Plumose structures may develop on the *f* joints and run at high angles to the bedding planes. The main joint is usually a shear fracture whilst the *f* joints may be either minor tension fractures or complementary shear fractures. Occasionally small cross fractures may run between the *f* joints. Concentric rib marks and stylolites may also occur on joint surfaces.

5.2 Origins of joints

Joints are formed through failure in tension, in shear, or through some combination of both. Rupture surfaces formed by extension tend to be clean and rough with little detritus. They tend to follow minor lithological variations. If such surfaces are sheared, the resulting load deformation curve displays a peak, rising well above the residual strength which is only reached at large values of displacement. Simple surfaces of shearing are generally smooth and contain considerable detritus. They are unaffected by local lithological changes. Shearing along this type of discontinuity does not yield as great a contrast between peak and residual strength as does a tension joint. The material along shear joints is commonly much more susceptible to alteration than that along tension joints.

Price (1966) contended that the majority of joints are post-compressional structures, formed as a result of the dissipation of residual stress after folding has occurred. Some spatially restricted small joints associated with folds, such as radial tension joints, are probably initiated during folding. The shearing stresses at the time of joint formation probably measure a few hundred megapascals. Such stresses usually can be dissipated by a movement amounting to a millimetre or so along a shear plane. But such movement can only dissipate the residual stresses in the immediate neighbourhood of a joint plane so that a very large number of joints need to form in order to dissipate the stresses throughout a large area. Hobbs (1968) proposed that joint frequency was proportional to the inverse of the bed thickness, the inverse of the square root of Young's modulus of the bed and the square root of the shear modulus of the surrounding beds. Formerly Price (1966) had suggested that joint frequency was related to the lithology of the rock type, the dimensions of the rock unit and the degree of tectonic deformation. On the last point he quoted the work of Harris *et al.* (1960) who found that the highest joint frequencies were associated with areas where the structures exhibited maximum curvature.

Price (1959) suggested that rocks could retain residual strain energy and that the residual stresses associated with the residual strain energy were modified during uplift in such a way as to give rise to tension joints or, in some cases, to both tension and shear joints. During the principal tectonic phase the resistance to the deformation of the rocks grows until it is in equilibrium with the maximum available tectonic stresses. When these conditions are attained there are no further changes in strain and deformation. It was assumed that after tectonic deformation there was usually a phase when the rocks were uplifted and surface material was removed by erosion. As a consequence gravitational loading and lateral stresses were reduced. Moreover when rocks are uplifted they are subjected to tensile stress and Price estimated that the latter would be about 70 MPa if they were elevated some 6000 m. He further argued that since the gravitational load increases by approximately 7 kPa for every 0.3 m increase in depth of overburden then in an uplift of 6000 m the gravitational load would decrease by about 140 MPa. As a result the tensile stresses which develop when rocks are elevated is equal to approximately half the change in the gravitational load. As uplift continues the tensile strain increases until one of the tensile stresses reaches the tensile strength of the rock concerned and a joint then forms across the bed at its weakest point. In discontinuous strata the initiation of a joint in one bed does not necessarily lead to rupture in the surrounding beds but it does give rise to an increase in stress in the latter beds near the boundaries of the joint. Because of the differences in the elastic properties between the jointed bed and the neighbouring rocks the elastic displacements differ. Shear stresses are therefore produced on planes parallel to the tensile strain. Across the face of the joint the tensile stress is zero.

With the initiation of shear joints a certain amount of residual stress is liberated and as they develop the maximum principal stress gradually decreases, whilst the least principal stress increases. Eventually the vertical load due to gravity may assume the role of principal maximum stress. If the rocks undergo continuous uplift then they become subjected to tension and tension joints are formed. Shear joints are usually markedly planar fractures which are not affected by local changes in lithology. On the other hand tension joints sometimes present irregular surfaces which may follow the outline of and are deflected by minor changes in lithology such as pebbles in conglomerates.

In horizontal beds which have suffered little tectonic compression two sets of tension joints may be developed whereas in those rocks which have been subjected to considerable tectonic compression, but have remained unfolded, two sets of shear joints may be formed. If uplift follows compression then two sets of tension joints may be developed subsequent to the shear joints.

These structures are related to one cycle of subsidence, compression and up-lift. When there are an abundant number of joint sets present in horizontally bedded rocks they have probably undergone several cycles of subsidence, compression and uplift.

Joints are also formed in other ways. For example, joints develop within igneous rocks when they initially cool down, and in wet sediments when they dry out. The most familiar of these are the columnar joints in lava flows, sills and some dykes. The cross joints, longitudinal joints, diagonal joints and flat-lying joints associated with large granitic intrusions have been described in Chapter 1. Sheet or mural joints have a similar orientation to flat-lying joints. When they are closely spaced and well developed they impart a pseudo-stratification to the host rock. It has been noted that the frequency of sheet jointing is related to the depth of overburden, in other words, the thinner the rock cover the more pronounced the sheeting. This suggests a connection between removal of overburden by denudation and the development of sheeting. Indeed such joints have often developed suddenly during quarrying operations. It may well be that some granitic intrusions contain considerable residual strain energy and that with the gradual removal of load the associated residual stresses are dissipated by the formation of sheet joints.

Chapman (1958) maintained that other sets of joints developed after the formation of sheet joints and that they were contained between the sheeting surfaces. He showed that the orientation and degree of development of these joints was related to topography. For instance, in Arcadia National Park, Maine, Chapman distinguished six sets of joints in roadside cuttings, yet only a short distance away the number of joint sets was reduced. Observations suggested that the best developed joints formed a pattern where two sets were mutually perpendicular, to which another two sets were diagonally arranged. It seemed that the two perpendicular sets were related to the direction of slope and contour of the topography and Chapman suggested that they may be formed as a result of gravity sliding of rock sheets. The diagonal joints, it was tentatively suggested, may represent the opening of incipient joint sets which were already formed.

5.3 Description of jointed rock masses

5.3.1 Incidence of discontinuities

The shear strength of a rock mass and its deformability are very much influenced by the discontinuity pattern, its geometry and how well it is developed. Observation of discontinuity spacing, whether in a field exposure or in a core stick, aids appraisal of rock mass structure. In sedimentary rocks bedding planes are usually the dominant discontinuity and the rock mass can be described as shown in Table 5.1. The same boundaries can be used to describe the spacing of joints (see Anon (1977)).

The mechanical behaviour of a rock mass is strongly influenced by the number of sets of discontinuities which intersect, since this governs the amount of deformation that the rock mass will undergo. The number of sets also affects the degree of overbreak which occurs on excavation, they there-fore may be an important factor in rock slope stability (Figure 5.4).

Table 5.1. Description of bedding plane and joint spacing (cf. Table 12.4)

Description of bedding plane spacing	Description of joint spacing (cf. Barton (1978))	Limits of spacing	Mass factor (j)
Very thickly bedded	Extremely wide	Over 2 m	0.8–1.0
Thickly bedded	Very wide	0.6–2 m	0.5–0.8
Medium bedded	Wide	0.2–0.6 m	0.2–0.5
Thinly bedded	Moderately wide	60 mm–0.2 m	0.1–0.2
Very thinly bedded	Moderately narrow	20–60 mm	Less than 0.1
Laminated	Narrow	6–20 mm	
Thinly laminated	Very narrow	Under 6 mm	

Systematic sets should be distinguished from non-systematic sets when recording the discontinuities in the field. Barton (1978) suggested that the number of sets of discontinuities at any particular location could be described in the following manner.

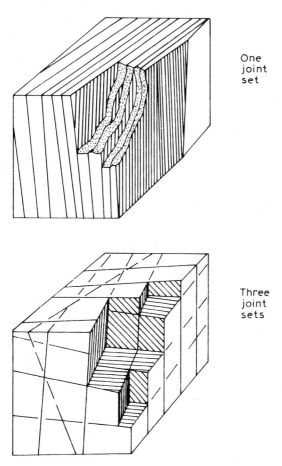

One joint set

Three joint sets

Figure 5.4 Effect of the number of joint sets on the mechanical behaviour and appearance of a rock mass (after Barton (1978))

(1) Massive, occasional random joints
(2) One discontinuity set
(3) One discontinuity set plus random
(4) Two discontinuity sets
(5) Two discontinuity sets plus random
(6) Three discontinuity sets
(7) Three discontinuity sets plus random
(8) Four or more discontinuity sets
(9) Crushed rock, earth-like.

5.3.2 Geometry of discontinuities

As joints represent surfaces of weakness, the larger and more closely spaced they are, the more influential they become in reducing the effective strength of the rock mass. The persistence of a joint plane refers to its continuity. This is one of the most difficult properties to quantify since joints frequently continue beyond the rock exposure and consequently in such instances it is impossible to estimate their continuity. Nevertheless Barton (1978) suggested that the modal trace lengths measured for each discontinuity set can be described as follows

Very low persistence	Less than 1 m
Low persistence	1–3 m
Medium persistence	3–10 m
High persistence	10–20 m
Very high persistence	Greater than 20 m

Simple sketches and block diagrams help to indicate the relative persistence of the various sets of discontinuities.

Block size provides an indication of how a rock mass is likely to behave, since block size and interblock shear strength determine the mechanical performance of a rock mass under given conditions of stress. The following descriptive terms have been recommended for the description of rock masses in order to convey an impression of the shape and size of blocks of rock material (see Barton (1978)).

(1) Massive—few joints or very wide spacing
(2) Blocky—approximately equidimensional (Figure 5.5)
(3) Tabular—one dimension considerably shorter than the other two
(4) Columnar—one dimension considerably larger than the other two
(5) Irregular—wide variations of block size and shape
(6) Crushed—heavily jointed to 'sugar cube'.

The orientation of the short or long dimensions should be specified in the columnar and tabular blocks respectively. In addition it may be useful to note the ratio of the orthogonal dimensions, for example, 1 vertical: 2 north: 6 east. The block size may be described by using the terms given below (see Anon (1977)).

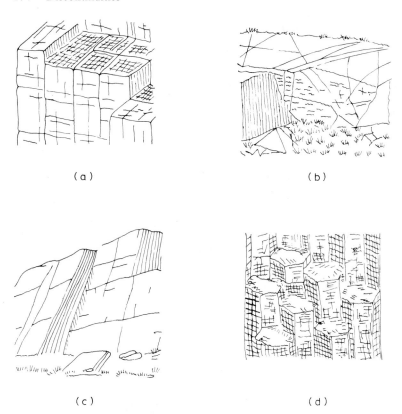

(a)

(b)

(c)

(d)

Figure 5.5 Sketches of rock masses illustrating (a) blocky, (b) irregular, (c) tabular, (d) columnar block shapes

Term	Block size	Equivalent discontinuity spacings in blocky rock	Volumetric joint count (J_v)* (joints/m^3)
Very large	Over 8 m^3	Extremely wide	Less than 1
Large	0.2–8 m^3	Very wide	1–3
Medium	0.008–0.2 m^3	Wide	3–10
Small	0.0002–0.008 m^3	Moderately wide	10–30
Very small	Less than 0.0002 m^3	Less than moderately wide	Over 30

* After Barton (1978)

Discontinuities, especially joints, may be open or closed. How open they are (Table 5.2) is of importance in relation to the overall strength and permeability of a rock mass and this often depends largely on the amount of weathering which the rocks have suffered. Furthermore, where the effects of weathering have penetrated deeply into a joint, a wide weak zone may be present. Some joints may be partially or completely filled. The type and amount of filling not only influence the effectiveness with which the opposing joint surfaces are bound together, thereby affecting the strength of the rock

Table 5.2. Description of the aperture of discontinuity surfaces

Anon (1977)			Barton (1978)	
Description	*Width of aperture*		*Description*	*Width of aperture*
Tight	Zero		Very tight	Less than 0.1 mm
Extremely narrow	Less than 2 mm	Closed	Tight	0.1–0.25 mm
Very narrow	2–6 mm		Partly open	0.25–0.5 mm
Narrow	6–20 mm		Open	0.5–2.5 mm
Moderately narrow	20–60 mm	Gapped	Moderately wide	2.5–10 mm
Moderately wide	60–200 mm		Wide	Over 10 mm
Wide	Over 200 mm		Very wide	10–100 mm
		Open	Extremely wide	100–1000 mm
			Cavernous	Over 1 m

mass, but also influence permeability. If the infilling is sufficiently thick, for example, over 100 mm, the walls of the joint will not be in contact and hence the strength of the joint plane will be that of the infill material. Materials such as clay or sand may have been introduced into a joint opening. Mineralisation is frequently associated with joints. This may effectively cement a joint, however, in other cases the mineralising agent may have altered and weakened the rocks along the joint conduit.

Infill occupying discontinuities may possess a wide range of physical properties, especially with regard to its shear strength, deformability and permeability. Its short-term and long-term behaviour may differ appreciably. The range of behaviour is influenced by the mineralogy of the infill, its particle size distribution, its water content and permeability, its over-consolidation ratio, any previous shear displacement, width of aperture, and roughness and state of wall rock. The infill may be assessed by using the same method(s) as used to assess the wall rock (see below).

5.3.3 Surfaces of discontinuities

The nature of the opposing joint surfaces also influences rock mass behaviour as the smoother they are, the more easily can movement take place along them. However, joint surfaces are usually rough and may be slickensided. Hence the nature of a joint surface may be considered in relation to its waviness, roughness and the condition of the walls. Waviness and roughness differ in terms of scale and their effect on the shear strength of the joint. Waviness refers to first order asperities which appear as undulations of the joint surface and are not likely to shear off during movement. Therefore the effects of waviness do not change with displacements along the joint surface. Waviness modifies the apparent angle of dip but not the frictional properties of the discontinuity. On the other hand, roughness refers to second order asperities which are sufficiently small to be sheared off during movement. Increased roughness of the discontinuity walls results in an increased effective friction angle along the joint surface. These effects diminish or disappear when infill matter is present. The procedure for measuring joint roughness in the field has been given by Barton (1978). (See also pages 183–184.)

The following visual classification of roughness can be used when quantitative measurements are not made (see Anon (1977)).

Category	Degree of roughness
1	Polished
2	Slickensided
3	Smooth
4	Rough
5	Defined ridges
6	Small steps
7	Very rough

This classification only has meaning when the direction of the irregularities on the surface is in the least favourable direction to resist sliding. As a consequence it is necessary to record the trend of the lineation on the joint surface in relation to the direction of shearing. Uniformity of assessment may be obtained by identifying and photographing each category at the site in question.

An alternative set of descriptive terms has been suggested by Barton (1978) and these should be based upon two scales of observation, namely, small scale (several centimetres) and intermediate scale (several metres). The intermediate scale of roughness is divided into stepped, undulating and planar, and the small scale of roughness, superimposed upon the former, includes rough (or irregular), smooth and slickensided categories. The direction of the slickensides should be noted as shear strength may vary with direction. Barton recognised the following classes (Figure 5.6).

(1) Rough (or irregular), stepped
(2) Smooth, stepped
(3) Slickensided, stepped
(4) Rough (irregular), undulating
(5) Smooth, undulating   Increasing
(6) Slickensided, undulating   shear
(7) Rough (irregular) planar   strength
(8) Smooth, planar
(9) Slickensided, planar

The compressive strength of the rock comprising the walls of a discontinuity is a very important component of shear strength and deformability, especially if the walls are in direct rock to rock contact, as in the case of unfilled joints. Weathering (and alteration) frequently is concentrated along the walls of discontinuities, thereby reducing their strength. The weathered material can be assessed in terms of its grade (see Table 5.3) and manual index tests (see Tables 9.2 and 12.2). Alternatively, a Schmidt hammer (see Chapter 11) can be used to obtain an idea of the compressive strength of the material concerned (see Barton (1978)). Samples of wall rock can be tested in the laboratory, not just for strength, but if they are highly weathered, also for swelling and durability.

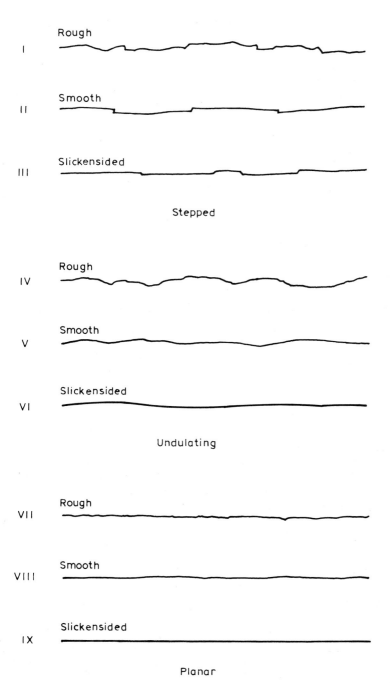

Figure 5.6 Typical roughness profiles and suggested nomenclature. The length of each profile is in the range 1–10 m. The vertical and horizontal scales are equal (after Barton (1978))

5.3.4 Flow of water and discontinuities

Seepage of water through rock masses usually takes place via the discontinuities, although in some sedimentary rocks seepage through the pores may also play an important role. The prediction of groundwater levels, probable seepage paths and approximate water pressures frequently provides an indication of stability or construction problems. Barton (1978) suggested that seepage from open or filled discontinuities could be assessed according to the following descriptive scheme.

Seepage rating	*A. Open discontinuities* Description	*B. Filled discontinuities* Description
(1)	The discontinuity is very tight and dry, water flow along it does not appear possible.	The filling material is heavily consolidated and dry, significant flow appears unlikely due to very low permeability.
(2)	The discontinuity is dry with no evidence of water flow.	The filling materials are damp but no free water is present.
(3)	The discontinuity is dry but shows evidence of water flow, i.e. rust staining, etc.	The filling materials are wet, occasional drops of water.
(4)	The discontinuity is damp but no free water is present.	The filling materials show signs of outwash, continuous flow of water (estimate l/min).
(5)	The discontinuity shows seepage, occasional drops of water but no continuous flow.	The filling materials are washed out locally, considerable water flow along outwash channels (estimate l/min and describe pressure, i.e. low, medium, high).
(6)	The discontinuity shows a continuous flow of water (estimate l/min and describe pressure, i.e. low, medium, high).	The filling materials are washed out completely, very high water pressures are experienced, especially on first exposure (estimate l/min and describe pressure).

5.4 Strength of jointed rock masses and its assessment: a review

Joints in a rock mass reduce its effective shear strength at least in a direction parallel with the discontinuities. Hence the strength of jointed rocks is highly anisotropic. Joints offer no resistance to tension whereas they offer high resistance to compression. Nevertheless they may deform under compression if there are crushable asperities, compressible filling or apertures along the joint or if the wall rock is altered.

Where discontinuities dip into a rock face, they only impose a direct mechanical instability on the face when they are of the same scale. They do, however, allow the ingress of water into the rock mass and as a result facilitate an increase in pore water pressure. This reduces the effective strength of the rock mass. Conversely, when discontinuities daylight into a rock face they adversely affect stability the more closely they run parallel to the face. In this case the slope of the face is to a greater or lesser extent controlled by the discontinuities.

5.4.1 Some earlier views

John (1965) considered that when a jointed rock mass failed by sliding along one joint or a set of joints, the limiting stress ratios could be determined based on the parameters of the joints and the confining pressure. In other words, where a load is applied in a direction parallel or sub-parallel to the joint direction the shear strength depends on the shearing resistance along the joint surfaces. At low normal pressures shearing stresses along a joint with relatively smooth asperities produce a tendency for one block to ride up onto and over the asperities of the other, whereas at high normal pressures shearing takes place through the asperities. When a jointed rock mass undergoes shearing this may be accompanied by dilation especially at low pressures and small shear displacements probably occur as shear stress builds up.

The shearing strength along irregular joint planes was investigated by Patton (1966), who performed a series of shear tests on the horizontal surfaces of specimens cast from plaster, which contained irregular asperities. These specimens were tested at a number of loadings to obtain the maximum shear strengths. It was found that if displacements continued after the initial failure, then a residual shearing resistance could be obtained. Both sets of results were then plotted on a Mohr diagram (Figure 5.7). The maximum shearing strengths are shown by the line OAB whilst the residual shearing strength is represented by the line OC. The line OA was obtained at low loadings and can be represented by the following expression

$$S = \sigma \tan (\phi_u + i)$$

where S is the angle between the asperities and the shearing surface, σ is the normal loading, ϕ_u is the angle of frictional sliding resistance of the planar surface, and i is the irregularity. Line AB represents failure at high normal loads, failure having taken place through the base of the asperities. The ver-

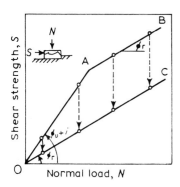

Figure 5.7 Failure envelopes for multiple inclined surfaces (after Patton (1966))

tical distance between the two lines represents the amount of shearing resistance lost with displacement. It shows that the internal cohesive strength of the asperities plays an important part at all loads other than zero, reaching a maximum value when they are sheared (that is, at A), even though there is no cohesion intercept on the Mohr diagram. For OA the cohesion mobilised is directly proportional to the normal loading whereas for AB it is independent of normal loading. At stresses less than about 20.5 MPa the maximum shear strength envelope for joints is given by the line OA from which the apparent angle of shearing resistance (ϕ) can be obtained ($\phi = \phi_u + i$). For practical purposes ϕ_u is almost equal to ϕ_r (the angle of residual shearing resistance) and thus i can be simply derived by subtracting ϕ_r from ϕ. According to Patton the above expression offers a method of interpreting the results of laboratory and field tests on jointed rocks but its practical application requires an evaluation of i in the field. Field measurements by Patton demonstrated that a value of i from 10–15° is reasonable for the component strength due to the irregularities in discontinuities. Irregularities with the highest i values are the first to be sheared off because they have the narrowest bases, hence progressive failure is likely to occur along a joint surface. Tests on rock joints have shown a similar behaviour to the idealised tests carried out by Patton.

Bock (1979) further developed the work of Patton (1966) and proposed a failure criterion for rough joints and compound shear surfaces which depended on the angle of friction (ϕ), the cohesion of the intact rock (c) and the inclination (γ) of the gently inclined parts of the asperities which are dipping against the shear direction.

A number of model experiments were conducted by Lajtai (1969) in order to investigate the relative contributions to shear strength of fundamental shear strength, internal friction and joint friction. He pointed out that the shear strength (τ) of rocks was regarded as a function of normal stress (σ) which could be expressed in the Coulomb equation

$$\tau = c + \sigma \tan \phi$$

where c is the cohesion and ϕ is the angle of internal friction. If the relationship was applied to a completely smooth continuous joint then the resistance against shear could be represented by

$$\tau_j = \sigma \tan \phi_j$$

where τ_j and ϕ_j are the joint friction and angle of joint friction respectively. However, if the joint surface was not continuous then the ultimate fracture plane would have to cut through the asperities and the total shear strength would be much higher. It could perhaps be given by

$$\tau_j = c + (1 - \kappa)\sigma \tan \phi + \kappa \sigma \tan \phi_j$$
where
$$\kappa = \frac{\text{Area of open joint}}{\text{Total area}}, \text{ i.e. the degree of separation.}$$

This equation assumes that the fundamental shear strength and internal friction of the asperities, together with the joint friction along the discontinuities, are mobilised at the same time. If neither of the two frictional resistances are mobilised before failure then this gives the lowest strength possible and is significant in foundation design. In the latter case failure occurs when the minimum principal stress exceeds the tensile resistance of the asperities. In fact Lajtai found that in most cases only a small fraction of the total joint friction could be mobilised prior to tension failure in the asperities.

Brown (1970) also used models to try to assess the importance of joints in a rock mass. He subjected prismatic samples, $100 \times 100 \times 200$ mm, to triaxial compression testing. The samples were composed of closely toleranced blocks of high strength gypsum plaster arranged in different ways in different samples so as to give different joint patterns. It was noticed that there was a development of stick-slip oscillations in all tests except those carried out at zero confining pressure. Post-peak sliding took place on some failure planes with the development of some residual strength in samples tested at confining pressures of 1.4, 3.5 and 7 MPa. The deformation modulus increased with increasing confining pressure. Brown identified the following seven modes of failure

(1) axial cleavage of the plaster at low confining pressures
(2) shear failure of the plaster along an approximately planar surface independent of the joints
(3) collapse at low confining pressures as a consequence of block movement which involved joint opening and dilation of the sample
(4) formation of one composite shear plane, partly along the joints and partly through the plaster
(5) the development of complex non-planar shear failure surfaces, again partly along the joints and partly through the plaster
(6) the development of multiple conjugate shear planes through the plaster at high confining pressures
(7) the formation of multiple conjugate shear planes partly through the plaster and partly through the joints.

He concluded that from the engineering point of view it should be noted that modes of failure other than those commonly recognised are likely to occur in block-jointed rock masses. For instance, failure to consider the possibility of axial cleavage fracture around tunnel openings in jointed brittle rock could have serious practical consequences. Another important point engineering-wise was the consistently low strength of block-jointed samples recorded at low confining pressures. Brown suggested that the difference in strength between jointed samples in which failure took place solely through the plaster and those which were unjointed was due to differences in the stress distribution in the two samples.

The application of the finite element method to the problem of jointed rocks was considered by Goodman et al. (1968). They maintained that for a realistic analysis the jointed rock mass should be treated as an aggregate of massive rock blocks separated by joints, rather than as a continuum. In their analysis they distinguished three distinct joint parameters, namely, the unit stiffness across the joint (k_n), the unit stiffness along the joint (k_s) and the shear strength along the joint (S). The value of k_n depends upon the contact area ratio between the joint surfaces, the perpendicular aperture distribution and amplitude, and the relevant properties of the filling materials when present. The factors which control the value of k_s include the roughness of the joint surfaces which is determined by the distribution, size and inclination of the asperities and the tangential aperture distribution, and amplitude. Lastly, the value of S is governed by the friction along the joint, the cohesion due to interlocking, and the strength of the filling material when present. The moisture content influences all three parameters; however, the authors regarded the joint water pressures as a separate variable whose effect was analogous to that of pore pressures in soil. Joints with a high stiffness have negligible joint displacements compared with the elastic displacements of the rock blocks, whilst those which have a low stiffness value undergo displacements of very much greater magnitude than the elastic displacements of the blocks. A moderate joint stiffness corresponds to displacements of joints to the same extent as the elastic displacements in the blocks. The joint strength was considered as high, moderate or low in accordance with whether the joints played a negligible, participating or dominant role in the strength of the rock mass.

The stiffness properties of a discontinuity are determined by measuring the loads and displacements on a rock specimen in both the shear and normal directions. Rosso (1976) compared the results of joint stiffness measurements obtained by direct shear and triaxial compression with those obtained *in situ*. The results of various tests were found to compare quite well.

5.4.2 Recent views on joint strength

Barton (1976) proposed the following empirical expression for deriving the shear strength (τ) along joint surfaces

$$\tau = \sigma_n \tan (JRC \log_{10}(JCS/\sigma_n) + \phi_b)$$

where σ_n is the effective normal stress, JRC is the joint roughness coefficient, JCS is the joint wall compressive strength and ϕ_b is the basic friction angle. According to Barton, the values of the joint roughness coefficient range from 0–20, from the smoothest to the roughest surface (Figure 5.8). The joint wall compressive strength is equal to the unconfined compressive strength of the rock if the joint is unweathered. This may be reduced by up to 75% when the walls of the joints are weathered. Both these factors are related as smooth-walled joints are less affected by the value of JCS, since failure of asperities plays a less important role. The smoother the walls of the joints, the more significant is the part played by its mineralogy (ϕ_b). The experience gained from rock mechanics indicates that under low effective normal stress levels, such as occur in engineering, the shear strength of joints can vary within relatively wide limits. The maximum effective normal stress acting across joints considered critical for stability lies, according to Barton, in the range 0.1–2.0 MPa.

Tse and Cruden (1979) pointed out that fairly small errors in estimating the joint roughness coefficient could produce serious errors in estimating the peak shear strength (τ) from the equation

$$\tau = \sigma_n \tan (JRC \log_{10}(JCS/\sigma_n) + \phi_b)$$

especially if the ratio JCS/σ_n was large. They therefore recommended a numerical method of checking of the value of JRC, based on a detailed

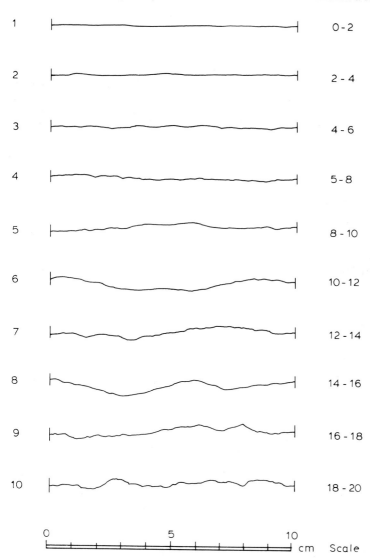

1 ├────────────────────┤   0 - 2

2 ├────────────────────┤   2 - 4

3 ├────────────────────┤   4 - 6

4 ├────────────────────┤   5 - 8

5 ├────────────────────┤   8 - 10

6 ├────────────────────┤   10 - 12

7 ├────────────────────┤   12 - 14

8 ├────────────────────┤   14 - 16

9 ├────────────────────┤   16 - 18

10 ├────────────────────┤   18 - 20

0   5   10
├─┼─┼─┼─┼─┼─┼─┼─┼─┼─┤ cm   Scale

Figure 5.8 Roughness profiles and corresponding range of JRC values
associated with each one (after Barton (1976))

profiling and analysis. Weissbach (1978) developed a profilograph for
measuring the roughness of joints.

Previously Barton and Choubey (1977) had suggested that tilt and push
tests provided a more reliable means of estimating the joint roughness coefficient than comparison with typical profiles. Barton and Bandis (1980) also
supported the use of such tests, particularly in heavily jointed rock masses,
when three joint sets are present. In a tilt test, two immediately adjacent
blocks are extracted from an exposure and the upper is laid upon the lower in
the exact same position as it was in a rock mass. Both are then tilted and the

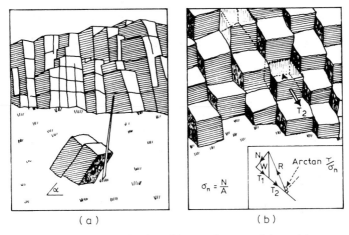

Figure 5.9 Two extremely simple and inexpensive ways of determining an accurate scale-free value of JRC. (a) Tilt test. (b) Pull test (after Barton and Bandis (1980))

angle (α) at which sliding occurs is recorded (Figure 5.9a). The JRC is estimated from

$$JRC = \frac{\alpha - \phi_r}{\log_{10}(JCS/\sigma_{no})}$$

where $\sigma_{no} = \gamma H \cos^2 \alpha$ (i.e. normal stress induced by self-weight of block), γ = unit weight, H = thickness of upper block, and ϕ_r = residual friction angle.

In a pull test an external shearing force (T_2) is applied via a bolt grouted into the block in question (Figure 5.9b). The value of JRC is given by

$$JRC = \frac{\arctan\left[\dfrac{T_1 + T_2}{N}\right] - \phi_r}{\log_{10}(JCS.A/N)}$$

where A is the joint area and N is the normal and tangential components of the self-weight of the upper block. In both cases the joint wall compression strength (JCS) and the residual friction angle (ϕ_r) can be estimated by using a Schmidt hammer (see Barton and Choubey (1977))

$$\log_{10} JCS = 0.00088 \gamma_d R + 1.01 \tag{1}$$

where γ_d = dry unit weight, R = Schmidt hammer rebound number.

$$\phi_r = (\phi_b - 20°) + 20 (r/R) \tag{2}$$

where ϕ_b = basic friction angle, r = Schmidt hammer rebound number on wet joint surface, R = Schmidt hammer rebound number on dry unweathered sawn surface.

In practice it is found that JRC is only a constant for a fixed joint length. Generally longer profiles (of the same joint) have lower JRC values.

Indeed Barton and Bandis (1980) suggested that mobilisation of peak strength along a joint surface seems to be a measure of the distance the joint has to be displaced in order that asperities are brought into contact. This distance increases with increasing joint length. Consequently when testing, longer samples tend to give lower values of peak shear strength. Barton and Choubey (1977) suggested that blocks defined by intersecting discontinuities probably provided the best size of samples for shear testing or joint surface analysis.

Barton and Bandis (1980) emphasised the influence of scale on the JRC. They also noted that the shear force-displacement curves change with increasing scale, that is, that behaviour along a joint on shearing changes from brittle to plastic and the shear stiffness is reduced. The changing shape of the curves was attributed to the progressive damage that occurs to larger and larger asperities as the scale is increased.

5.4.3 Testing jointed rocks

Krahn and Morgenstern (1979) carried out a series of direct shear tests on natural and artificially produced discontinuities in limestone in order to demonstrate how the ultimate shearing resistance was influenced by surface structure and roughness. They found that except at low normal stresses there was essentially no difference between peak and ultimate or residual resistance for relatively smooth and flat artificial surfaces. All the natural discontinuities, however, showed a significant drop from peak to ultimate strength. The ultimate friction angle (ϕ_u) varied from 14–32° for natural discontinuities while the cohesion intercept varied between 55 and 82 kPa. The major reason for the wide variation in the shearing resistance was attributable to the shearing resistance of the discontinuities. In other words the roughest surfaces offered the greatest resistance. They concluded that the ultimate frictional resistance of jointed hard unweathered rock depends on the initial surface roughness along the joint and the type of surface alteration which occurs during shearing. Each ultimate resistance reflects the characteristic roughness of the particular type of discontinuity. In order to determine the ultimate frictional resistance of a rock mass they suggested that the differing geological histories of the various discontinuities must be recognised, and sampling and testing should proceed accordingly.

Although closely jointed rock is extremely difficult to test, if it is tested under triaxial conditions then the larger the diameter of the sample the better the results. This is because measurements on small samples give values relating to the intact rock and do not take account of the jointing. Jaeger (1970) described large-scale triaxial tests carried out on closely jointed rock. He suggested that one approach in dealing with such rock was to regard it as randomly jointed on a small scale and to apply soil mechanics theory to it. Earlier Bray (1967) had advanced a similar idea as a method of investigating fractured rock. In such an instance the normal (σ_n) and shear (τ) stresses at failure are related by

$$\tau = c + \mu\sigma_n$$

where the coefficient of friction (μ) and the cohesion (c) are those of the individual joints. Jaeger found that under confining conditions, adjacent blocks interlocked and the strength of the rock mass was thereby increased.

In the sample he tested, movement took place along a large number of planes and some barrelling, and occasional tension gashes, developed. Generally movement on one particular plane tended to become dominant as strain increased. This final surface of shear was usually heavily slickensided.

There is no single method of testing highly fractured rock which will give the values of strength and deformability directly. Raphael and Goodman (1979), however, have shown how the interpretation of results from a selection of test procedures can permit reasonable values to be confined between high and low estimates. For example, the modulus of elasticity of a rock mass with one set of discontinuities spaced at a regular distance, S, can be expressed by

$$E = \frac{1}{(1/E_r) + (1/K_n S)}$$

where E_r is the modulus of elasticity of the intact rock (MPa), and K_n is the normal stiffness of each fracture (MPa/m).

5.5 Discontinuities and rock quality indices

Several attempts have been made to relate the numerical intensity of fractures to the quality of unweathered rock masses and to quantify their effect on deformability. For example, the concept of rock quality designation (RQD) was introduced by Deere (1964). It is based on the percentage core recovery when drilling rock with NX (57.2 mm) or larger diameter diamond core drills. Assuming that a consistent standard of drilling can be maintained, the percentage of solid core obtained depends on the strength and degree of discontinuities in the rock mass concerned. The RQD is the sum of the core sticks in excess of 100 mm expressed as a percentage of the total length of core drilled. However, the RQD does not take account of the joint opening and condition, a further disadvantage being that with fracture spacings greater than 100 mm the quality is excellent irrespective of the actual spacing (Table 5.3). This particular difficulty can be overcome by using the fracture spacing index as suggested by Franklin et al. (1971). This simply refers to the frequency with which fractures occur within a rock mass (Table 5.3).

Table 5.3. Classification of rock quality in relation to the incidence of discontinuities

Quality classification	RQD (%)	Fracture frequency per metre	C factor	Mass factor (j)	Velocity ratio (V_{cf}/V_{cl})
Very poor	0–25	Over 15	0.00–0.15		0.0–0.2
Poor	25–50	15–8	0.15–0.30	Less than 0.2	0.2–0.4
Fair	50–75	8–5	0.30–0.45	0.2–0.5	0.4–0.6
Good	75–90	5–1	0.45–0.65	0.5–0.8	0.6–0.8
Excellent	90–100	Less than 1	0.65–1.00	0.8–1.0	0.8–1.0

The concept of fissuration factor was introduced by Hansagi (1974). Like the RQD it refers to the fragmentation of rock cores. However, it is based not only on the total length of the intact fragments and the average length of total core recovery, but also on the number of cylindrical pieces obtained (the lower limit of which is linked with the core diameter and the lengths of core required for strength determination). Furthermore, it is not the total length of all the core fragments obtained which is included in the calculation but a fixed length of the drillhole. Hence the core loss occurring during the drilling process is also taken into account. The fissuration factor (C) is derived from

$$C = 1/2S(pH + Kn)$$

where S = one investigated unit length of drillhole (this is dependent on the diameter of the core and the rock strength), p = the number of cylindrical samples which can be obtained from cores corresponding to length S, H = the height of the cylindrical sample used for compression testing, K = the total length of the core fragments with cylindrical lengths greater than the core diameter, n = the number of these core fragments.

The purpose of determining the C factor is to record the variations in fissuration and strength along the drillhole at each investigation unit at a maximum distance of 1 m. Hansagi maintained that the C factor was more sensitive to changes in the quality of rock than the RQD because the lower limit for the determination of the latter was 100 mm.

An estimate of the numerical value of the deformation modulus of a jointed rock mass can be obtained from various *in situ* tests (see Bell (1978)). The values derived from such tests are always smaller than those determined in the laboratory from intact core specimens and the more heavily the rock mass is jointed, the larger the discrepancy between the two values. Thus, if the ratio between these two values of deformation modulus is obtained from a number of locations on a site, the engineer can evaluate the rock mass quality. In this context the concept of the rock mass factor (j) has been introduced by Hobbs (1975). He defined the rock mass factor as the ratio of the deformability of a rock mass within any readily identifiable lithological and structural component to that of the deformability of the intact rock comprising the component. Consequently it reflects the effect of discontinuities on the expected performance of the intact rock (Table 5.3). The value of j depends upon the method of assessing the deformability of the rock mass, and the value beneath an actual foundation will not necessarily be the same as that determined even from a large-scale field test. According to Hobbs, the greatest difficulties which occur in a jointed rock mass in relation to foundation design are experienced when the fracture spacing falls within a range of about 100–500 mm, in as much as small variations in fracture spacing and condition result in exceptionally large changes in j-value.

The effect of discontinuities in a rock mass can be estimated by comparing the *in situ* compressional wave velocity with the laboratory sonic velocity of an intact core sample obtained from the rock mass. The difference in these two velocities is caused by the structural discontinuities which exist in the field. The velocity ratio, V_{cf}/V_{cl}, where V_{cf} and V_{cl} are the compressional wave velocities of the rock mass *in situ* and of the intact specimen respectively, was first proposed by Onodera (1963). For a high-quality massive rock with only a few tight joints, the velocity ratio approaches unity. As the degree of jointing and fracturing becomes more severe, the velocity ratio is reduced (Table 5.3). The sonic velocity is determined for the core sample in the laboratory under an axial stress equal to the computed overburden stress at the depth from which the rock material was taken, and at a moisture content equivalent to that assumed for the *in situ* rock. The field seismic velocity preferably is determined by uphole or crosshole seismic measurements in drillholes or test adits, since by using these measurements it is possible to explore individual homogeneous zones more precisely than by surface refraction surveys.

5.6 Recording discontinuity data

5.6.1 Direct discontinuity surveys

Before a discontinuity survey commences the area in question must be mapped geologically to determine rock types and delineate major structures. It is only after becoming familiar with the geology that the most efficient and accurate way of conducting a discontinuity survey can be devised. A comprehensive review of the procedure to be followed in a discontinuity survey has been provided by Barton (1978).

One of the most widely used methods of collecting discontinuity data is simply by direct measurement on the ground. A direct survey can be carried out subjectively in that only those structures which appear to be important are measured and recorded. In a subjective survey the effort can be concentrated on the apparently significant joint sets. Nevertheless, there is a risk of overlooking sets which might be important. Conversely, in an objective survey all structures intersecting a fixed line or area of the rock face are measured and recorded.

Several methods have been used for carrying out direct discontinuity surveys. Halstead *et al.* (1968) used the fracture set mapping technique by which all discontinuities occurring in 6 by 2 m zones, spaced at 30 m intervals along the face, were recorded. Knill (1971) also suggested using an area sampling method on the rock face concerned. On the other hand, Piteau (1971) and Robertson (1971) maintained that using a series of line scans provides a satisfactory method of joint surveying. The technique involves extending a metric tape across an exposure, levelling the tape and then securing it to the face. Two other scan lines are set out as near as possible at right angles to the first, one more or less vertical, the other horizontal. The distance along a tape at which each discontinuity intersects is noted, as is the direction of the pole to each discontinuity (this provides an indication of the dip direction). The dip of the pole from the vertical is recorded as this is equivalent to the dip of the plane from the horizontal. The strike and dip directions of discontinuities in the field can be measured with a compass and the amount of dip with a clinometer. Measurement of the length of a discontinuity provides information on its continuity. It has been suggested that measurements should be taken over distances of about 30 m, and to ensure that the survey is representative the measurements should be continuous over that distance. The line scanning technique yields more detail on the incidence of discontinuities and their attitude than other methods (see Priest and Hudson (1981)). A minimum of at least 200 readings per locality is recommended to ensure statistical reliability.

Terzaghi (1965) pointed out that the number of observations of joints of any one set is a function of the angle of intersection between that set and the face under examination. Hence joints intersecting the face at low angles are poorly represented in any one joint survey on that face. She demonstrated, however, that it was possible to correct data for this effect but suggested that the simpler remedy was to choose several exposures with different orientations.

Priest and Hudson (1976) described a line scanning technique to record discontinuities in a tunnel in order to determine the stability of the rock mass involved, in this case the Lower Chalk. Similar work was done by Young and Fowell (1978) in the Kielder experimental tunnel. In each case a grid of vertical and horizontal lines was laid out over each tunnel face exposed and the intersection of the discontinuities recorded. This enabled a contoured plan of the face to be drawn which showed the spatial density of the discontinuities (Figure 5.10).

Hudson and Priest (1979) have pointed out that where discontinuities occur in sets, the discontinuity frequency along a scanline is a function of scanline orientation. They showed that the spacing distributions of discontinuities is a negative exponential distribution with the mean spacing of discontinuities being the reciprocal of the average number of discontinuities per metre (λ). This value can simply be calculated by dividing the number of scanline intersections by the total scanline length. According to these two

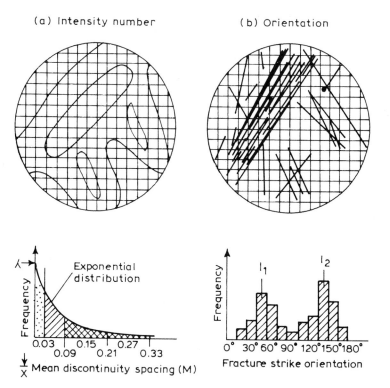

Figure 5.10 Recording of discontinuities for one cycle of tunnel advance (after Young and Fowell (1978))

Table 5.4. Recording of joint data (adapted from Richards *et al.* (1978))

Property		Method of measurement
Orientation		Direct measurement on surface of discontinuity, using specially designed compass/clinometer. Measurement of dip with clinometer, dip direction estimated by comparison with a known direction, preferably instrumentally, if magnetic compass cannot be used. Estimated by visual alignment with compass/clinometer. Measurement from oriented core in oriented hole. Determination of orientation from stereophotographs taken by phototheodolite. Ambiguous or involved coding should be avoided. Dip/dip direction is preferred to strike notations.
Position	Absolute	Measured position of feature with respect to surveyed or well-established reference on known global or site axes
	Relative	Measured position with respect to other, similar features such as members of the same joint set, to give apparent spacings and intensity of discontinuities. Obtained by direct measurement or abstracted from geotechnical plans and sections. Measured position with respect to excavated faces. Some discontinuities may have been formed at a different time from others in which case it may be possible to work out the age relationships by noting how one set of discontinuities intersects with another
Continuity		Visual estimation or measurement of two-dimensional continuity in near-planar surface outcrops, with horizontal or vertical extrapolation to nearby exposures showing features with similar characteristics. Three-dimensional continuity assessed using evidence from vertical and horizontal exposures, boreholes and adits, if available. Qualitative and quantitative descriptions related to extent of proposed excavations. In a weak rock there may be fewer continuous joints and more smaller shear discontinuities and slickensides, the frequency and orientation of which influence the development of any shear failure
Openness		Whether a discontinuity is open or not and if it is open by how much? If a discontinuity is open, is the wall rock altered, if so to what degree?
Factors affecting shear strength	Infill	Details obtained from freshly excavated outcrops, or from drill core only if good quality recovery and if logged before deterioration. Estimation of amount of infill along discontinuity and determination of its character. The following types of discontinuities were recognised by Brekke and Howard (1972) in relation to stability in excavations. (1) Joints, seams and sometimes even minor faults may be sealed through precipitation from solution of quartz or calcite. 'Welded' discontinuities, however, may be fractured by later movements.

Property	Method of measurement
	(2) Clean discontinuities are those where the wall rock is fresh.
	(3) Calcite fillings may, particularly when they are porous or flaky, dissolve during the lifetime of the structure.
	(4) Coatings or fillings of chlorite and graphite give rise to very slippery, low-strength joints, seams or faults especially when wet.
	(5) Inactive clay material in seams and faults naturally represents a very weak material that may squeeze or be washed out.
	(6) Swelling clay may cause serious problems through free swell and consequent loss of strength, or through considerable swelling pressure when confined. Brekke and Selmer-Olson (1965) maintained that the swelling capacity of montmorillonite was one of the most important causes of instability in excavated rock.
	(7) Material that has been altered to a more cohesionless material (sand-like) may run or flow following excavation.
	(8) Chemical and mineralogical analysis may be necessary. Determination of physical properties of gouge by laboratory testing. Correct soils description of gouge desirable
Waviness	Description of the deviation of the surface from a true plane. Surface irregularities that are unlikely to shear off and that provide a keying effect. Quantitative description using approximation to wavelength and amplitude of the irregularity. Direct observation of large outcrops only
Roughness	Description of surface asperities that may shear off, but which affect the frictional properties of the surface. Obtained from fresh and occasionally weathered rock exposures and from drill core. Situations on opposing faces of discontinuities may indicate a shear genesis and may be interpreted in terms of relative directional movement
Frictional Properties	Obtained by small-scale shear tests on specially selected rock samples containing joints, or from rock core. Difficult to avoid disturbing the surface of the discontinuity

If a face is very large and contains too many discontinuities to measure individually, it may be necessary to resort to statistical methods of analysis. In such instances the number of discontinuities which are exposed on a face are sampled and consequently the field work and subsequent analysis must ensure that the data which are sampled are representative of the conditions in the field. If statistical or idealised methods are used it is essential that isolated, 'non-systematic' or random variations are also examined to find out whether or not they affect stability. A 'randomly oriented' discontinuity may provide the release surface to allow failure.

DISCONTINUITY SURVEY DATA SHEET

GENERAL INFORMATION

Seq. no. [ ] Site [ ] Date (Day Month Year) [ ] Operator [ ] Discontinuity data sheet No. [ ] of [ ]

NATURE AND ORIENTATION OF DISCONTINUITY

Chainage or No.	Type	Dip	Dip direction	Persistence	Aperture	Nature of infilling	Consistency of infilling	Surface roughness	Trend of lineation	Waviness wavelength	Waviness amplitude	Water/flow	Remarks

Type
0. Fault zone
1. Fault
2. Joint
3. Cleavage
4. Schistosity
5. Shear
6. Fissure
7. Tension crack
8. Foliation
9. Bedding

Dip, Dip direction and trend of lineation
(Expressed in degrees)

Persistence
(Expressed in metres)

Aperture
1. Wide (>200 mm)
2. Mod. wide (60-200 mm)
3. Mod. narrow (20-60 mm)
4. Narrow (6-20 mm)
5. Very narrow (2-6 mm)
6. Ext. narrow (<2 mm)
7. Tight

Nature of infilling
1. Clean
2. Surface staining
3. Non-cohesive
4. Inactive clay or clay matrix
5. Swelling clay or clay matrix
6. Cemented
7. Chlorite, talc or gypsum
8. Others—specify

Compressive strength of infilling
1. Very soft (<40 kPa)
2. Soft (40-80 kPa)
3. Firm (80-150 kPa)
4. Stiff (150-300 kPa)
5. Very stiff (300-600 kPa)
6. Hard/v. weak (600-1250 kPa)
7. Weak (1.25 MPa)
8. Mod. weak (5-12.5 MPa)
9. Mod. strong (12.5-50 MPa)
10. Strong (50-100 MPa)
11. Very strong (100-200 MPa)
12. Ext. strong (>200 MPa)

Roughness
1. Polished
2. Slickensided
3. Smooth
4. Rough
5. Defined ridges
6. Small steps
7. Very rough

Waviness
Express wavelength and amplitude in metres

Water
1. Dry
2. Seepage

Flow
3. <10 ml/s
4. 10-100 ml/s
5. 0.1-1 l/s
6. 1-10 l/s
7. 10-100 l/s
8. >100 l/s

Figure 5.11 Discontinuity survey data sheet (after Anon (1977))

workers the relationship between the RQD and the average number of discontinuities per metre is given by the following expression

$$RQD = 100e^{-0.1\lambda} (0.1\lambda + 1)$$

In addition, Hudson and Priest showed how the distributions of block areas, for most locations, can be predicted adequately from discontinuity frequency measurements made along scanlines, and how to derive cumulative frequency curves for block volumes from scanline data. A summary of the other details which should be recorded about discontinuities is given in Table 5.4 and Figure 5.11.

In their joint survey of sediments of Cretaceous age in south-east England, Fookes and Denness (1969) used the cavity technique, that is, they excavated blocks of material from the faces of exposures, and measurements were taken in the area vacated by the blocks. This meant that only fresh discontinuities were recorded. Skempton *et al.* (1969) carried out a similar procedure in the London Clay at Wraysbury and Edgware. In addition to recording the dip and strike data for the joints and fissures they also measured the height, and, where possible, the length of each joint. A large orientated block was excavated for examination in the laboratory.

5.6.2 Drillholes and discontinuity surveys

This information gathered by any of the above methods can be supplemented with data from orientated cores from drillholes. The value of the data depends in part on the quality of the rock concerned, in that poor quality rock is likely to be lost during drilling. However, it is impossible to assess the persistence, degree of separation or the nature of the joint surfaces. What is more, infill material, especially if it is soft, is not recovered by the drilling operations.

Core orientation can be achieved by using the Craelius core orientator or by integral sampling (see Rocha (1971) and Figures 5.12a and 5.12b, respectively).

Drillhole inspection techniques include the use of drillhole periscopes, drillhole cameras or closed-circuit television. The drillhole periscope affords direct inspection and can be orientated from outside the hole. However, its effective use is limited to about 30 m. The drillhole camera can also be orientated prior to photographing a section of the wall of a drillhole. The television camera provides a direct view of the drillhole and a recording can be made on videotape. These three systems are limited in that they require relatively clear conditions and so may be of little use below the water table, particularly if the water in the drillhole is murky.

The televiewer produces an acoustic picture of the drillhole wall. One of its advantages is that drillholes need not be flushed prior to its use.

Snow (1968) demonstrated that the discharges from water injection or packer tests in jointed crystalline rocks provide estimates of the spatial frequency of joints, and the mean and variance of the size distributions of the apertures, which are log normal. He also maintained that decreasing permeability in rock masses with depth is more a result of decreasing fracture openings than of decreasing fracture spacing.

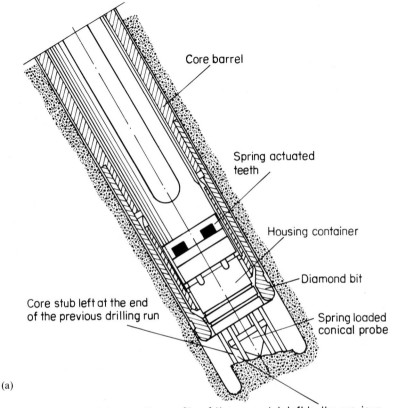

Core barrel

Spring actuated teeth

Housing container

Diamond bit

Spring loaded conical probe

Core stub left at the end of the previous drilling run

(a)

Probes which take up the profile of the core stub left by the previous drilling run and which are locked in position when the spring loaded cone is released

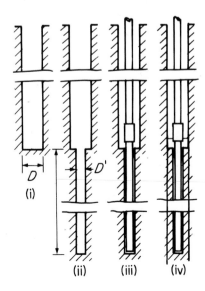

D

D'

(i)

(ii)   (iii)   (iv)

(b)

5.6.3 Photographs and discontinuity surveys

Many data relating to discontinuities can be obtained from photographs of exposures. Photographs may be taken looking horizontally at the rock mass from the ground or they may be taken from the air looking vertically, or occasionally obliquely, down at the outcrop. These photographs may or may not have survey control. Uncontrolled photographs are taken using hand-held cameras, stereo-pairs being obtained by taking two photographs of the same face from positions about 5% of the distance of the face apart, along a line parallel to the face. Delineation of major discontinuity patterns and preliminary subdivision of the face into structural zones can be made from these photographs. Unfortunately, data cannot be transferred with accuracy from them onto maps and plans. Conversely discontinuity data can be accurately located on maps and plans by using controlled photographs. Controlled photographs are obtained by aerial photography with complementary ground control or by ground-based phototheodolite surveys. Aerial and ground-based photography are usually done with panchromatic film but the use of colour and infrared techniques is becoming more popular. Aerial photographs, with a suitable scale, have proved useful in the investigation of discontinuities. Photographs taken with a phototheodolite can also be used with a stereo-comparator which produces a stereoscopic model. Measurements of the locations or points in the model can be made with an accuracy of approximately 1 in 5000 of the mean object distance. As a consequence, a point on a face photographed from 50 m can be located to an accuracy of 10 mm. In this way the frequency, orientation and continuity of discontinuities can be assessed. Such techniques prove particularly useful when faces which are inaccessible or unsafe have to be investigated.

Moore (1974) outlined a stereo-photogrammetric method by which he recorded the position of major discontinuities in the Oxford Clay as they were exposed during excavation in brick pits. Photographs of the faces were taken with a phototheodolite from three reinforced concrete pillars specially constructed for the purpose. The photographs enabled a series of contours,

---

Figure 5.12 (a) Details and method of operation of the Craelius core orientor. The teeth clamp the instrument in position in the inside of the core barrel until released by pressure on the conical probe. The housing contains a soft aluminium ring against which a ball bearing is indented by pressure from the conical probe thus marking the bottom of the hole position. The probe is released by pressure against the core stub and, when released, locks the probe in position and releases the clamping teeth to allow the instrument to ride up inside the barrel ahead of the core entering the barrel. (Craelius Diabor AB). (b) Stages of the integral sampling method. A drillhole (diameter D) is drilled to a depth where the integral sample is to be obtained, then another hole (diameter D^1) coaxial with the former and with the same length as the required sample is drilled, into which a reinforcing bar is placed. The bar is then bonded to the rock mass. Drilling of the drillhole then is resumed to obtain the integral samples. The method has been used with success in all types of rock masses, from massive to highly weathered varieties, and provides complete information on the spacing and orientation as well as the opening and infilling of discontinuities

at 2 m vertical intervals on the face, to be plotted on a plan with a scale of 1:250. Each contour was marked where it was intersected by a joint. Contours on successive faces were referenced to the same eastings line and northings line. A model was constructed from the data in order to illustrate the three-dimensional nature of the joint system.

5.6.4 Recording discontinuity data

The simplest method of recording discontinuity data is by using a histogram on which the frequency is plotted along one axis and the strike direction along the other. Directional information, however, is more effectively represented on a rose diagram. This provides a graphical illustration of the angular relationships between joint sets. The strikes of the joints and their frequencies are represented by the directions on each rose diagram, the lengths of the vectors being plotted either on a half or full-circle. Directions are usually plotted for data contained in 5° arcs, while magnitudes are plotted to scale (Figure 5.13).

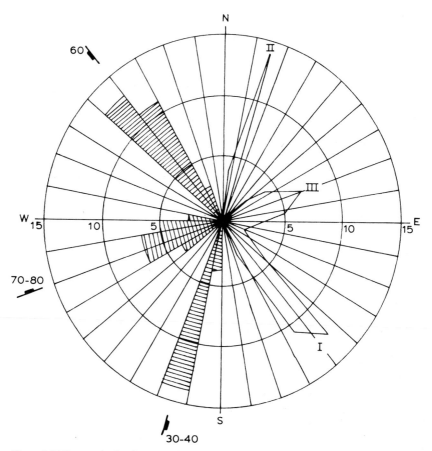

Figure 5.13 Two methods of representing orientation data on a joint rosette (courtesy of Pergamon Press Inc.)

Data from a discontinuity survey are now, however, usually plotted on a stereographic projection. The use of spherical projections, commonly the Schmidt or Wulf net, means that traces of the planes on the surface of the 'reference sphere' can be used to define the dips and dip directions of discontinuity planes. In other words the inclination and orientation of a particular plane is represented by a great circle or a pole, normal to the plane, which are traced on an overlay placed over the stereonet. The method whereby great circles or poles are plotted on a stereogram has been explained by Hoek and Bray (1978). When recording field observations of the amount and direction of dip of discontinuities it is convenient to plot the poles rather than the great circles. The poles can then be contoured in order to provide an expression of orientation concentration. This affords a qualitative appraisal of the influence of the discontinuities on the engineering behaviour of the rock mass concerned (Figure 5.14). The orientation of discontinuities can also be illustrated by using a block diagram (Figure 5.15).

5.7 Fissures in clay

5.7.1 Character of fissures

Ward *et al.* (1959) noted the presence of fissures in the London Clay, which they defined as small fractures confined to a bed or horizons within a bed. Unfortunately bedding is not always readily discernible in some clays.

In addition to fissures, Skempton *et al.* (1969) recorded joints in the London Clay. These were planar structures, the surfaces of which had a matt texture, although a thin layer of clay gouge occurred along some joints. Joints were predominantly normal to the horizontal bedding, with a pronounced trend in two orthogonal directions. They ranged in height up to 2.6 m and in length up to 6.0 m. Skempton *et al.* also recognised smooth surfaces of moderate size (over 0.03 m² in area) which dipped at angles varying between 5 and 25°. These surfaces they termed sheeting. Sheeting usually had a slightly undulating shape but sometimes it was planar. These authors noted that the fissures in the uppermost 13 m of the London Clay may be curved or planar, that they are rarely more than 150 mm in size and that they exhibit no preferred orientation (cf. Fookes and Parrish (1968)), although they tend to be inclined at low angles to the bedding planes. Fissures usually have a matt surface texture. The mean size of the fissures decreases and the number per unit volume correspondingly increases as the upper surface of the clay is approached, suggesting that stress release and weathering play an important role in fissure formation. Similarly Ward *et al.* (1965) found that at Ashford the more weathered brown London Clay was more fissured than the blue below. At depth the high pressures tend to keep the fissures closed, at least in the sense that they are not visibly open fractures. No appreciable relative movements have taken place along fissures. However, 5% of all fissures in the London Clay are slickensided with polished and striated surfaces. The slickensides are due to shearing consequent upon minor internal distortions of the clay mass. The polished nature of the surface indicates a considerable degree of particle orientation.

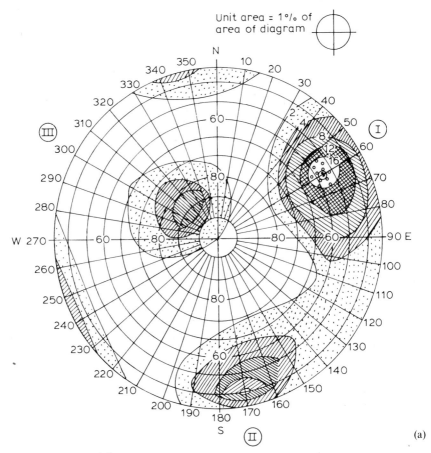

(a)

5.7.2 Origins of fissures

Although weathering and stress release may give rise to fissures, it has been suggested that most are more likely to be original features which are opened on exposure. For instance, Fookes and Parrish (1968) noted that the orientation of many fissures in the London Clay is very closely similar to the orientation of joints in adjacent rock. This suggests that fissures in stiff clay and joints in semi-brittle rocks, when in the same stress field, develop with more or less similar orientations. The authors therefore concluded that the preferred orientation of many of the fissures in the London Clay is related to the stress field as indicated by the regional structure.

Fookes and Denness (1969) suggested that the intensity of fissuring in some sediments of Cretaceous age was influenced by near surface desiccation cracks and that bedding planes appeared to be the major factor governing the fissure patterns developed. They claimed that vertical stress release seemed to have little influence except when almost parallel to the bedding direction. But when this does happen they supposed that dilation occurs within the upper metre or so of the clay mass due to elastic expansion of clay material and separation of individual lithological units. Fookes and Denness maintained that fissuring sympathetic to the bedding planes might have developed as a

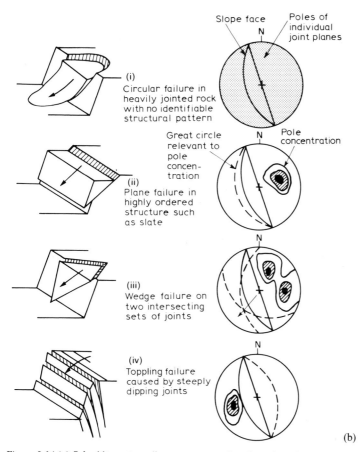

Figure 5.14 (a) Schmidt contour diagram representing the orientation of three sets of joints plotted on a polar equal-area net. The main sets I and II are approximately normal to each other and the minor set III is nearly horizontal (after Barton (1978)). (b) Representation of structural data concerning four possible slope failure modes plotted on equal-area nets as poles and great circles (after Hoek & Bray (1977))

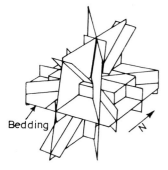

Figure 5.15 Block diagram of discontinuities

result of differential settlement between neighbouring particles shortly after deposition, leading to their separation with the formation of discontinuities approximately normal to the bedding. Minor changes in lithology do not appear to influence the attitude of fissuring but do influence the size of fissures and the intensity of fissuring. Generally with increasing time after exposure the proportion of non-planar to planar fissures increases since the former can develop from the latter by extension from their extremities. They concluded that stress release tended to generate more non-planar than planar

It would appear from the above that fissures in clay tend to form in a number of ways. Fookes and Denness (1969) therefore proposed the following summary of the possible modes of fissure formation

(1) formed or modified by tectonic stresses,
(2) inherited from underlying rocks,
(3) formed during deposition or soon after syneresis and/or changes in the chemistry of the pore fluids,
(4) formed or modified by *in situ* physico-chemical changes due to such agencies as groundwater, weathering or ion exchange,
(5) formed or modified by non-diastrophic processes such as hill creep, rebound on unloading, or stress release during erosion.

Table 5.5. Classification of fissures (from Fookes and Denness (1969))
(a) Fissure classification (general)

Type	Description	Occurrence	Size	Orienta-tion	Intensity	Restrictions
Discon-tinuity	Break or interruption of the mechanical properties of a solid	In any material	Any	Any	Any	None
Fissure	Discontinuity divid-ing an otherwise continuous material without separation of units	In any material	Any	Any	Any	None
True fissure	Fissure dividing an otherwise con-tinuous material of essentially uniform strength	In any material	Any	Any	Any	None
Apparent fissure	Fissure dividing a soft material when harder units are present in the same mass or separating soft material from hard	In any material	Any	Any	Any	Weathered zones

Note: An apparently continuous combination of true and apparent fissures is collectively termed an apparent fissure.

(b) Size classification of fissures

Type	Size (area)
Very large fissure	$\geqslant 100$ m^2
Large fissure	$1-100$ m^2
Normal fissure	$0.01-1$ m^2
Small fissure	$1-100$ cm^2
Very small fissure	$\leqslant 1$ cm^2

(c) Surface geometry classification of fissures

Type	Description	Occurrence	Size	Orienta-tion	Frequency	Restrictions
	L/R range (see Figure 5.16)					
Planar	$\leqslant \pi/8$	Ubiquitous	Any	Any	Very common	None
Semi-curved	$\pi/8 - \pi/4$	Ubiquitous	Any	Any	Common	None
Curved	$\geqslant \pi/4$	Ubiquitous	Any	Any	Common	None
Hinged	Combination of planar and semi-curved/curved	Ubiquitous	Any	Any	Rare	None
Semi-undulose	Combination of two or more alternately convex and concave semi-curved	Ubiquitous	Any	Any	Fairly common	None
Undulose	Combination of two or more alternately convex and concave curved	Ubiquitous	Any	Any	Fairly common	None
Conchoidal	As a conchoid	Ubiquitous	< 1 m^2	Any	Rare	None

Note: An alternately convex and concave formation of curved and semi-curved surfaces is grouped according to whether most of the undulations are curved or semi-curved. If they are curved and semi-curved equally, a common occurrence, the surface is classified as undulose.

(d) Surface markings (roughness) classification of fissures

Type	Description		Occurrence	Size	Orienta-tion	Frequency	Restrictions
	Sandpaper grade	D/L range (see Figure 5.17)					
Slicken-sided	<00 (shiny when dry, some directional features)	—	Mainly in shear surfaces	Any	Parallel to shear zones	Rare	Fine-grained rocks
Very smooth	<00	—	Ubiquitous	Any	Any	Common	Fine-grained rocks
Smooth	$00-01$	—	Ubiquitous	Any	Any	Common	None
Slightly rough	$01-02$	—	Ubiquitous	Any	Any	Common	None
Rough	$02-03$	—	Ubiquitous	Any	Any	Common	None
Very rough	$03-04$	—	Ubiquitous	Any	Any	Common	None
Pock marked	>04	$\leqslant \frac{1}{10}$	Ubiquitous	Any	Any	Common	None
Pitted	—	$\frac{1}{10} - \frac{1}{2}$	Ubiquitous	Any	Any	Common	None

Note: Any fissure with $D/L \geqslant \frac{1}{2}$ is not considered to be continuous and is treated as a fissure or fissures excluding the feature which causes $D/L \geqslant \frac{1}{2}$.

Table 5.5. cont'd
(e) Fabric classification of fissures

Classification	Description
Major girdle	Fissure set with one axis within every plane common to all, no fissure having the same dip as another unless the common axis is vertical, and none having the same strike unless the common axis is horizontal (see Figure 5.18) e.g. around the circumference of a stereogram. Its solid envelope of revolution is such that it occupies the whole of finite space.
Minor girdle	Fissure set with no common axis but whose members are all tangential to one cone (see Figure 5.18), e.g. the south-east quadrant of a stereogram. Its solid envelope of revolution is such that it occupies the whole of finite space except the interior of the cone.
Uni-planar set	Fissure set which is condensed to a single plane (see Figure 5.18), e.g. at the centre of a stereogram.

(f) Intensity classification of fissured material

Intensity type	Area of fissures per unit volume (m^2/m^3)	Average size of intact blocks
Very low	$\leqslant 3$	$\geqslant 1\ m^3$
Low	3–10	$0.027–1\ m^3$
Moderate	10–30	$0.001–0.027\ m^3$
High	30–100	$27–1000\ cm^3$
Very high	100–300	$1–27\ cm^3$
Excessive	$\geqslant 300$	$\leqslant 1\ cm^3$

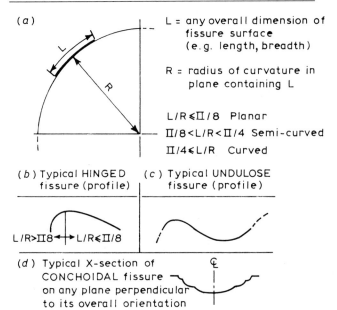

(a)

L = any overall dimension of fissure surface (e.g. length, breadth)

R = radius of curvature in plane containing L

$L/R \leqslant \Pi/8$ Planar
$\Pi/8 < L/R < \Pi/4$ Semi-curved
$\Pi/4 \leqslant L/R$ Curved

(b) Typical HINGED fissure (profile)

$L/R > \Pi8 \longleftrightarrow L/R < \Pi/8$

(c) Typical UNDULOSE fissure (profile)

(d) Typical X-section of CONCHOIDAL fissure on any plane perpendicular to its overall orientation

Figure 5.16 Surface geometry classification of fissures (after Fookes and Denness (1969))

(a) Plan of fissure surface

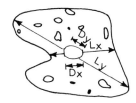

(b) Any section on inset (a)

D/L < 10   POCK-MARKED (unless
subject to a smoother classification)
1/10 ≤ D/L ≤ 1/2   PITTED
1/2 < D/L   2 separate fissures

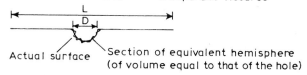

Actual surface   Section of equivalent hemisphere
(of volume equal to that of the hole)

Figure 5.17 Surface marking (roughness) classification of fissures (after
Fookes and Denness (1969))

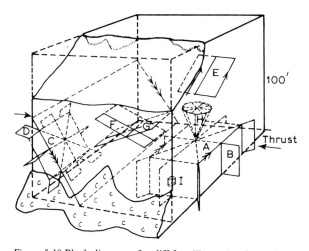

Figure 5.18 Block diagram of a cliff face illustrating fissure fabric
classification. Bedding is horizontal. A = Uniplanar set parallel to bedding
(>>); B = Major girdle right angles to bedding; C = Uniplanar set parallel
to fault zone (>>>); D = Major girdle right angles to fault zone;
E = Uniplanar set parallel to exposure slope; F, G = Conjugate pair of
uniplanar sets in diamond pattern with long axis parallel to thrust
direction; H = Minor girdle related to bedding: I = Sample site for small-
scale fissure observations. NB, E may be accompanied by a major girdle
perpendicular to it near old exposure slopes

5.7.3 Classification of fissures

One of the most extensive classifications of fissures was proposed by Fookes and Denness (1969), although they admitted that the system would need refining as further data on fissures became available. Their classification included the type of fissure, its size, surface geometry, surface roughness and fabric on the one hand, and the intensity of fissures on the other (Table 5.5). Not only does a classification based on size allow differentiation between large and small fissures but the size of fissures probably reflects the amount of residual energy which has been released (see Price (1966)). This presumably also applies to the intensity of fissuring. Unfortunately it is difficult to calculate fissure areas and therefore it may be necessary to rely on estimates.

The mechanism of fissure formation may have some influence on the roughness of their surfaces (see Price (1966)) and, more important, affects the resistance to shear failure. Measurements of surface geometry are tedious and such classifications are not particularly useful unless the geometrical parameters can be related to the shear strength characteristics of the discontinuities (undulatory characteristics also would have to be considered in the context of normal pressures to which they are subjected since any shearing motion would either be over the protrusions or through them). Application of such a classification would involve measuring some length dimension of the discontinuity and expressing it in terms of the characteristic radius of curvature in the plane of the dimension concerned.

Classification of fissure fabric affords a means whereby the general orientation of fissures within a unit volume of material with respect to each other and a certain datum, such as bedding, may be established.

The classification of the intensity of fissures is determined by counting the number of fissures in a given volume and dividing by that volume. Before any relationship can be postulated between intensity and genesis of fissures the total area of the fissures per unit volume must be obtained. Fookes and Denness (1969) maintained that the mechanical properties of the sediment, its degree of induration and weathering, had an influence on fissure intensity. Consistent grouping of fissure orientations suggests a constant influence during their formation.

5.7.4 Strength of fissured clay

Fissures play an extremely important role in the failure mechanism of fissured clay. Indeed many clays are seriously weakened by the presence of a network of fissures. Terzaghi (1936) provided the first quantitative data relating to the influence of fissures and joints on the strength of clays, pointing out that such features are characteristic of overconsolidated clays. He maintained that fissures in normally consolidated clays have no significant practical consequences and that they appeared to be due to a process comparable with syneresis. On the other hand fissures can have a decisive influence on the engineering performance of an overconsolidated clay, in that the overall strength of such fissured clay can be as low as one-tenth that of the intact clay. For instance, it appears that the average shearing resistance of stiff fissured clay at the moment of sliding usually ranges between 15 and

30 kPa whereas the initial shearing resistance of such clays ranges between 100 and 300 kPa. In addition to allowing clay to soften, fissures and joints allow concentrations of shear stress which locally exceed the peak strength of clay thereby giving rise to progressive failure. Under stress the fissures in clay seem to propagate and coalesce in a complex manner.

Skempton (1948) attributed the reduction in strength of the London Clay exposed in cuttings to softening along fissures which opened as a result of small movements consequent upon the removal of lateral support on excavation. The ingress of water into fissures means that the pore water pressure in the clay concerned increases, which in turn means that its strength is reduced. Skempton and La Rochelle (1965) investigated slope failures in the London Clay at Bradwell, Essex. These began to occur on cut slopes a few days after excavation, even though the slopes had been designed on strengths lower than the undrained parameters for the intact clay. The investigation revealed that a large proportion of the reduction in the overall strength below these design parameters was attributed to the fissures. For example, it was noted that blocks of intact clay which were found on the failed slopes showed no signs of deformation or failure.

Skempton and La Rochelle considered that where unfavourable orientation of fissures exists the major part of a failure plane may follow fissures. As a result the clay may be near its residual strength. The overall strength of the clay could be further reduced by the separation of the walls of closed fissures. Whereas intact clay has a low tensile strength, there is no resistance to the opening of a fissure and once open there is no shear resistance along the fissure itself. If regular fissure patterns with an unfavourable orientation occur at a site they recommended that an attempt should be made to estimate the influence of these fissures on the overall strength of the clay mass. It is therefore necessary to determine the average area of a potential failure plane which would pass through open fissures, closed fissures and intact clay. It is also necessary to establish the stress distribution in the clay mass in order to obtain the correct value of the overall residual strength for design purposes.

Skempton et al. (1969) summarised the shear strength parameters of the London Clay in terms of effective stress as follows

(1) peak strength of intact clay: $c' = 31$ kPa, $\phi' = 20°$,
(2) 'peak' strength on fissure and joint surfaces: $c' = 6.9$ kPa, $\phi' = 18.5°$,
(3) residual strength of intact clay: $c'_r = 1.4$ kPa, $\phi'_r = 16°$.

Thus the strength along joints or fissures in clay is only slightly higher than the residual strength of the intact clay. Similar results previously had been obtained by Marsland and Butler (1967) who carried out a series of large-scale shear box tests on stiff fissured Barton Clay. They found that the strength developed along a closed fissure has hardly more than the residual value in the drained or undrained conditions. They also concluded that if a plane of failure develops along fissures which facilitate more effective drainage, then the drained strength parameters should be used in a stability analysis.

It can be concluded that the upper limit of the strength of fissured clay is represented by its intact strength whilst the lower limit corresponds to the strength along the fissures. The operational strength, which is somewhere between the two, is, however, often significantly higher than the fissure strength.

Ward *et al.* (1959) recommended that tests to determine the mechanical behaviour of fissured clay should be made on as large a scale as possible. The effect of size of the test specimen on the undrained strength of a stiff fissured clay subsequently was demonstrated by Bishop and Little (1967). For example, they showed that the strength of the London Clay, when determined by an *in situ* shear box, 600 × 600 mm, was only 55% of that obtained in the laboratory by testing unconsolidated–undrained samples, 37.5 mm in diameter, in triaxial conditions. The greater the curvature and complexity of the fissures, the greater the strengths obtained, particularly those measured on small specimens.

Lo (1970) showed that as the size of the test specimen of fissured clay increased, so would the number of fissures it contained. In addition the probability of the test specimen possessing larger fissures would also increase, as would the probability of these having a critical orientation. Lastly, there would be a greater likelihood of coalescing adjacent fissures corresponding with the potential plane of failure.

Fissures open fairly rapidly once fissured clay is exposed. This has an important effect upon the properties measured both in the field and in the laboratory. For instance, Ward *et al.* (1965) showed that the strength of the London Clay measured 4–8 h and 2.5 days after excavation was 85% and 75% of that obtained 0.5 h after excavation. This was attributed to the gradual extension of fissures and microcracks in the clay with time.

Marsland (1971) also emphasised the importance of fissures on the properties of stiff clay and, more importantly, the dangers of using data obtained from laboratory tests on small samples in the design of both temporary and permanent works. He pointed out that for an individual clay the strength measured in the field or laboratory depended on the intensity of the fissuring, their size and spacing; the changes in stress consequent upon boring, excavation, sampling and testing of the clay; the length of time the clay remains under each stress condition and the size of the specimen tested in relation to the fissures. Different clays have different basic properties and previous stress histories which vary with location and depth. Marsland felt that larger tests in which long periods were allowed for equilibrium to be reached would reduce the variation to more tolerable limits. Wherever possible these studies should be made in conjunction with measurements to determine the stability or instability in the field. He recommended that each type of fissured clay should be regarded as a new material and that it should be studied by all means possible.

5.7.5 *Fissure patterns*

The fissure patterns in till at Hurlford, Ayrshire, have been investigated by McGown *et al.* (1974). They used the cavity technique to determine the fissure patterns and orientations of stones in the till since it proved extremely difficult to extract large undisturbed block samples because of the presence of boulders. Small and normal sized fissures were recorded, the former having slightly lower angles of dip than the larger which were semi-vertical to vertical. The intensity of the fissures was found to increase as the surface of the till was approached. This was attributated to greater stresses caused by ice movement and to the effects of weathering. Although the area per unit volume may provide an indication of the intensity, it gives no indication of the space between the fissures. Accordingly McGown *et al.* measured the distance of intersection of each fissure (or projected fissure) on one of three axes originating from one corner of the cavity. When a fissure intersected more than one axis the intersection nearest the origin was recorded.

McGown *et al.* noted a very definite preferred orientation of fissures in the till. These influenced its shear strength and therefore its stability. These authors showed that the opening of fissures sympathetically orientated to cut slopes, and softening of till along fissures as a result of weathering, were responsible for small slip failures. In other words, these two factors gave rise to a rapid reduction of undrained shear strength along the fissures. They found that the operational strength around a potential slip surface was directionally dependent and may be considered less than that measured by large diameter, vertically orientated samples on which conventional designs are based. Factors of safety based on conventional designs may therefore greatly overestimate the overall stability of a slope excavated in this till.

McGown *et al.* (1977) found that the undrained shear strength of fissures in till may be as little as one-sixth that of the intact soil. They emphasised that the distinction between the nature of the various fissure coatings (sand, silt or clay-size material) is of critical importance in determining the shear strength behaviour of the fissured soil mass. Deformation and permeability are also controlled by the nature of the fissure surface and coatings. Fissures and laminations have very similar influences on the behaviour of a soil mass. However, fissures tend to be much more variable, in nature, spacing, orientation and areal extent. They therefore usually give rise to a greater degree of variability in the soil mass.

After an investigation of stiff fissured clays in South Africa, Williams and Jennings (1977) suggested that the fissuring was attributable to tensile failure brought about by shrinkage on drying. Tension failure occurred near the surface and gave rise to more or less vertical cracking. With increasing depth in such shrinking soil tensile failure gave way to shear failure. Shear failure occurs where there is an appreciable difference between the major and minor principal stresses. In such a situation the major principal stress is compressive or positive in a vertical direction and the minor principal stress in the horizontal direction reaches a negative value sufficient to cause failure. The near-surface fissures may have rough surfaces, whereas at lower horizons in particular, fissures with shiny surfaces occur. These, the authors referred to as slickensides. Slickensides were recorded down to depths of 15 m from the surface. Measurements of slickensides showed that their average length is around 200 mm, with a maximum of up to 750 mm. They generally occur within 150 and 500 mm of each other. Williams and Jennings assumed that these slickensides were formed when reversal of movement occurred, due to expansion and shrinkage in the soil, brought about by seasonal changes of climate. They showed that slickensides occur in those fissured clays which have a plasticity index greater than 30% and a clay fraction in excess of 30%. This indicates that the slickensides are associated with expansive clay deposits, the activities of which are above 0.7. These authors also found that the shear strength along a slickensided fissure was close to the residual value of the intact clay. They suggested that the latter should therefore be regarded as the operational strength of such clay deposits.

References

ANON. (1977). 'The description of rock masses for engineering purposes', Working Party Report, *Q. J. Engg Geol.*, **10**, 355–88.

BARTON, N. (1976). 'The shear strength of rock and rock joints', *Int. J. Rock Mech. Min. Sci. & Geomech. Abstr.*, **13**, 255–79.

BARTON, N. (1978). 'Suggested methods for the quantitative description of discontinuities in rock masses', ISRM Commission on Standardization of Laboratory and Field Tests, *Int. J. Rock Mech. Min. Sci. & Geomech. Abstr.*, **15**, 319–68.

BARTON, N. & BANDIS, S. (1980). 'Some effects of scale on the shear strength of joints', *Int. J. Rock Mech. Min. Sci. & Geomech. Abstr.*, **17**, 69–76.

BARTON, N. & CHOUBEY, V. (1977). 'The shear strength of rock joints in theory and practice', *Rock Mechanics*, **10**, 1–54.

BELL, F. G. (1978). '*In-situ* testing and geophysical surveying', (in *Foundation Engineering in Difficult Ground*), Bell, F. G. (ed.), Butterworths, London, 233–80.

BISHOP, A. W. & LITTLE, A. L. (1967). 'The influence of size and orientation of the sample on the apparent strength of London Clay at Maldon, Essex', *Proc. Geot. Conf., Oslo*, **1**, 89–96.

BOCK, H. (1979). 'A simple failure criterion for rough joints and compound shear surfaces', *Engg Geol.*, **14**, 241–54.

BRAY, J. W. (1967). 'A study of jointed and fractured rock. Part I—Fracture patterns and their failure characteristics. Part II—Theory of limiting equilibrium', *Rock Mech. Engg Geol.*, **5**, 117–36, 197–216.

BREKKE, T. L. & HOWARD, T. R. (1972). 'Stability problems caused by seams and faults', *Proc. 1st N. Am. Tunneling Conf.* AIME, New York, 25–41.

BREKKE, T. L. & SELMER-OLSON, R. (1965). 'Stability problems in underground constructions caused by montmorillonite-carrying joints and faults', *Engg Geol.*, **1**, 3–19.

BROWN, E. T. (1970). 'Strength models of rocks with intermittent joints', *Proc. ASCE, Div. Soil Mech. Foundation Engg*, **96**, 1935–49.

CHAPMAN, C. A. (1958). 'Control of jointing by topography', *J. Geol.*, **66**, 552–68.

DEERE, D. (1964). 'Technical description of cores for engineering purposes', *Rock Mech. Engg Geol.*, **1**, 18–22.

FOOKES, P. G. & DENNESS, B. (1969). 'Observational studies on fissure patterns in Cretaceous sediments of south-east England', *Geotechnique*, **19**, 453–77.

FOOKES, P. G. & PARRISH, D. G. (1968). 'Observations on small-scale structural discontinuities in the London Clay and their relationship to regional geology', *Q.J. Engg Geol.*, **1**, 217–40.

FRANKLIN, J. L., BROCH, E. & WALTON, G. (1971). 'Logging the mechanical character of rock', *Trans Inst. Min. Metall.*, **81**, Mining Section, A1–9.

GOODMAN, R. E., TAYLOR, R. L. & BREKKE, T. L. (1968). 'A model for the mechanics of jointed rocks', *Proc. ASCE Div. Soil Mech. Foundation Engg*, **94**, 637–59.

HALSTEAD, P. N., CALL, P. D. & RIPPERE, K. H. (1968). 'Geological structural analysis for open pit slope design, Kimberley pit, Ely, Nevada', Reprint: *Annual AIME Meeting*, New York.

HANCOCK, P. L. (1968). 'Joints and faults: the morphological aspects of their origins', *Proc. Geol. Ass.*, **79**, 141–51.

HANSAGI, I. A. (1974). 'Method of determining the degree of fissuration of rock', *Int. J. Rock Mech. Min. Sci. & Geomech. Abstr.*, **11**, 379–88.

HARRIS, J. F., TAYLOR, G. L. & WALPER, J. L. (1960). 'Relation of deformational features in sedimentary rocks and regional and local structure', *Bull Am. Ass. Petrol. Geologists*, **44**, 1853–73.

HOBBS, D. W. (1968). 'The formation of tension joints in sedimentary rocks', *Geol. Mag.*, **104**, 550–56.

HOBBS, N. B. (1975). 'Factors affecting the prediction of settlement of structures on rocks with particular reference to the Chalk and Trias', (in *Settlement of Structures*), Brit. Geotech. Soc., Pentech Press, London, 579–610.

HODGSON, R. A. (1961). 'Classification of structures on joint surfaces', *Am. J. Sci.*, **259**, 493–507.

HOEK, E. & BRAY, J. W. (1978). *Rock Slope Engineering*, Inst. Min. Metall., London.

HUDSON, J. A. & PRIEST, S. D. (1979). 'Discontinuities and rock mass geometry', *Int. J. Rock Mech. Min. Sci. & Geotech. Abstr.*, **16**, 339–62.

JAEGER, J. C. (1970). 'Behaviour of closely jointed rock', *Proc. 11th Symp. Rock Mech., Berkeley*, Pergamon Press, New York, 56–68.

JOHN, K. W. (1965). 'Civil engineering approach to evaluate strength and deformability of regularly jointed rock', *Rock Mech.*, **1**, 69–80.

KNILL, J. L. (1971). *Collecting and Processing of Geological Data for Purposes of Rock Engineering. The Analysis and Design of Rocks Slopes*, University of Alberta, Edmonton.

KRAHN, J. & MORGENSTERN, N. R. (1979). 'The ultimate frictional resistance of rock discontinuities', *Int. J. Rock Mech. Min. Sci. & Geomech. Abstr.*, **16**, 127–33.

LAJTAI, E. Z. (1969). 'Shear strength of weakness planes in rocks', *Int. J. Rock Mech., Min. Sci.*, **5**, 499–515.

LO, K. Y. (1970). 'The operational strength of fissured clays', *Geotechnique*, **20**, 57–74.

McGOWN, A., RADWAN, A. M. & GABR, A. W. A. (1977). 'Laboratory testing of fissured and laminated soils', *Proc. 9th Int. Conf. Soil Mech. Foundation Engg., Tokyo*, **1**, 205–10.

McGOWN, A., SALDIVAR-SALI, A. & RADWAN, A. M. (1974). 'Fissure patterns and slope failures in till at Hurlford, Ayrshire', *Q. J. Engg Geol.*, **7**, 1–26.

MARSLAND, A. (1971). 'The shear strength of stiff fissured clays', *Build. Res. Stn.*, Current Paper, 21/71, Watford.

MARSLAND, A. & BUTLER, M. E. (1967). 'Strength measurements on stiff fissured Barton Clay from Fawley, Hampshire', *Proc. Geot. Conf., Oslo*, **1**, 139–46.

MOORE, J. F. A. (1974). 'Mapping of major joints in the Lower Oxford Clay using terrestrial photogrammetry', *Q. J. Engg Geol.*, **7**, 57–67.

ONODERA, T. F. (1963). 'Dynamic investigation of foundation rocks', *Proc. 5th Symp. Rock Mech., Minnesota*, Pergamon Press, New York, 517–33.

PATTON, F. D. (1966). 'Multiple modes of shear failure in rock', *Proc. 1st Int. Cong. Rock Mech., Lisbon*, **1**, 509–14.

PITEAU, D. R. (1971). 'Geological factors significant to the stability of slopes cut in rock', *Symp. Planning Open Pit Mines*, Johannesburg, Balkema, Amsterdam, 43–53.

PRICE, N. L. (1959). 'Mechanics of jointing in rocks', *Geol. Mag.*, **96**, 149–60.

PRICE, N. L. (1966). *Fault and Joint Development in Brittle and Semi-Brittle Rock*, Pergamon Press, London.

PRIEST, S. D. & HUDSON, J. A. (1976). 'Discontinuity spacings in rock', *Int. J. Rock Mech. Min. Sci. & Geomech. Abstr.*, **13**, 135–48.

PRIEST, S. D. & HUDSON, J. A. (1981). 'Estimation of discontinuity spacing and trace length using scanline surveys', *Int. J. Rock Mech. Min. Sci. & Geomech. Abstr.*, **18**, 183–97.

RAPHAEL, J. M. & GOODMAN, R. E. (1979). 'Strength and deformability of highly fractured rock', *Proc. ASCE J. Geot. Engg Div.*, **105**, GT11, Paper 14988, 1285–300.

RICHARDS, L. R. LEG, G. M. M. & WHITTLE, R. A. (1978). 'Appraisal of stability conditions in rock slopes', (in *Foundation Engineering in Difficult Ground*), Bell, F. G. (ed.), Butterworths, London, 449–512.

ROBERTSON, A. M. (1971). 'The interpretation of geological factors for use in slope theory', *Symp. Planning Open Pit Mines*, Johannesburg, Balkema, Amsterdam, 55–71.

ROCHA, M. (1971). 'Method of integral sampling', *Rock Mech.*, **3**, 1–12.

ROSSO, R. S. (1976). 'A comparison of joint stiffness measurements in direct shear triaxial compression and *in situ*', *Int. J. Rock Mech. Min. Sci. & Geomech. Abstr.*, **13**, 167–73.

SKEMPTON, A. W. (1948). 'The rate of softening in stiff fissured clay with special reference to the London Clay', *Proc. 2nd Int. Conf. Soil Mech. Foundation Engg, Rotterdam*, **2**, 50–53.

SKEMPTON, A. W. & LA ROCHELLE, P. (1965). 'The Bradwell slip: a short-term failure in London Clay', *Geotechnique*, **15**, 221–41.

SKEMPTON, A. W., SCHUSTER, R. L. & PETLEY, D. J. (1969). 'Joints and fissures in the London Clay at Wraysbury and Edgware', *Geotechnique*, **19**, 205–17.

SNOW, D. T. (1968). Rock fracture spacing opening and porosities', *Proc. ASCE Div. Soil Mech. Foundation Engg*, **94**, (SM1), 73–91.

TERZAGHI, K. (1936). 'Stability of slopes of natural clay', *Proc. 1st Int. Conf. Soil Mech. Foundation Engg, Cambridge, Mass.*, **1**, 161–5.

TERZAGHI, R. D. (1965). 'Sources of error in joint surveys', *Geotechnique*, **15**, 287–304.

TSE, R. & CRUDEN, D. M. (1979). 'Estimating joint roughness coefficient', *Int. J. Rock Mech. Min. Sci. & Geomech. Abstr.*, **16**, 303–7.

WARD, W. H., MARSLAND, A. & SAMUELS, S. G. (1965). 'Properties of the London Clay at the Ashford Common shaft: *In situ* and undrained strength tests', *Geotechnique*, **15**, 321–44.

WARD, W. H., SAMUELS, S. G. & BUTLER, M. E. (1959). 'Further studies of the properties of the London Clay', *Geotechnique*, **9**, 33–8.

WEISSBACH, G. (1978). 'A new method for the determination of the roughness of joints in the laboratory', *Int. J. Rock Mech. Min. Sci. & Geomech. Abstr.*, **15**, 131–4.

WILLIAMS, A. B. & JENNINGS, J. E. (1977). 'The *in situ* shear behaviour of fissured soils', *Proc. 9th Int. Conf. Soil Mech. Foundation Engg, Tokyo*, **2**, 243–6.

YOUNG, R. P. & FOWELL, R. J. (1978). 'Assessing rock discontinuities', *Tunnels and Tunnelling*, **10**, No. 5, June, 45–8.

Chapter 6

Groundwater

6.1 The origin and occurrence of groundwater

The principal source of groundwater is meteoric water, that is, precipitation (rain, sleet, snow and hail). However, two other sources are very occasionally of some consequence. These are juvenile water and connate water. The former is derived from magmatic sources whilst the latter represents the water in which sediments were deposited. This was trapped in the pore spaces of sedimentary rocks as they were formed and has never been expelled.

The amount of water that infiltrates into the ground depends upon how precipitation is dispersed, namely, on what proportions are assigned to immediate run-off and to evapotranspiration, the remainder constituting the proportion allotted to infiltration/percolation (Figure 6.1). Water which enters the ground can be estimated if the amount of run-off and evapotranspiration are known, by subtracting these from the total precipitation.

Infiltration refers to the seepage of surface water into the ground, percolation being its subsequent movement, under the influence of gravity, to the zone of saturation. In reality one cannot be separated from the other. The infiltration capacity is influenced by the rate at which rainfall occurs (which also affects the quantity of water available), the vegetation cover, the porosity of the soils and rocks, their initial moisture content and the position of the zone of saturation. Gentle rainfall is more effective than heavy rainfall as far as infiltration is concerned, since most heavy rain runs off the surface. If a soil is relatively dry when rain begins there is a strong capillary action in the subsurface layers which acts in the same direction as gravity and hence generates high infiltration values. As water infiltrates and the surface layer becomes semi-saturated, these capillary forces decline and so the infiltration capacity is reduced.

The retention of water in a soil depends upon the capillary force and the molecular attraction of the particles. As the pores in a soil become thoroughly wetted the capillary force declines so that gravity becomes more effective. In this way downward percolation can continue after infiltration has ceased but as the soil dries, so capillarity increases in importance. No further percolation occurs after the capillary and gravity forces are balanced. Thus water percolates into the zone of saturation when the retention capacity is

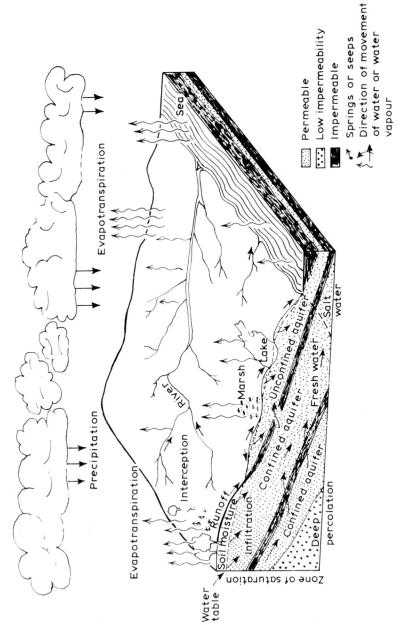

Precipitation

Evapotranspiration

Evapotranspiration

Interception

River

Marsh

Lake

Sea

Water table

Zone of saturation

Soil moisture

Runoff

Infiltration

Unconfined aquifer

Confined aquifer

Confined aquifer

Fresh water

Salt water

Deep percolation

Permeable
Low impermeability
Impermeable
Springs or seeps
Direction of movement
of water or water
vapour

Figure 6.1 The hydrologic cycle

satisfied. This means that the rains which occur after the deficiency of soil moisture has been catered for are the ones which count as far as supplementing groundwater is concerned.

6.1.1 The water table

The pores within the zone of saturation are filled with water, generally referred to as *phreatic water*. The upper surface of this zone is therefore known as the phreatic surface but is more commonly termed the *water table*. Above the zone of saturation is the zone of aeration in which both air and water occupy the pores. The water in the zone of aeration is commonly referred to as *vadose water*. Meinzer (1942) divided this zone into three belts, those of soil water, the intermediate belt and the capillary fringe (Figure 6.2). The uppermost or soil water belt discharges water into the atmosphere in perceptible quantities by evapotranspiration. In the capillary fringe, which occurs immediately above the water table, water is held in the pores by capillary action. An intermediate belt occurs when the water table is far enough below the surface for the soil water belt not to extend down to the capillary fringe. The degree of saturation decreases from the water table upwards, saturation occurring only in the immediate neighbourhood of the water table. However, where the water table is at shallow depth and the maximum capillary rise is large, moisture is continually attracted from the water table due to evaporation from the ground surface. Hence the soil is saturated or nearly so. The height to which moisture can rise due to capillary action depends upon the type of soil but generally it is not more than 2.5 m (see below).

The geological factors which influence percolation not only vary from one rock outcrop to another but may do so within the same one. This, together with the fact that rain does not fall evenly over a given area, means that the contribution to the zone of saturation is variable. This in turn in-

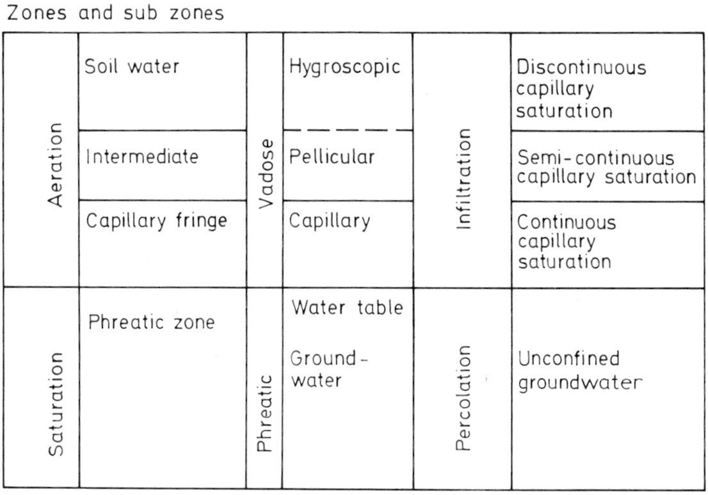

Zones and sub zones

Aeration	Soil water	Vadose	Hygroscopic	Infiltration	Discontinuous capillary saturation
	Intermediate		Pellicular		Semi-continuous capillary saturation
	Capillary fringe		Capillary		Continuous capillary saturation
Saturation	Phreatic zone	Phreatic	Water table Ground-water	Percolation	Unconfined groundwater

Figure 6.2 Zones and sub-zones of groundwater

fluences the position of the water table, as do the points of discharge. A rise in the water table as a response to percolation is partly controlled by the rate at which water can drain from the area of recharge. Accordingly it tends to be greatest in areas of low transmissivity (see below). Mounds and ridges form in the water table under the areas of greatest recharge. Superimpose upon this the influence of water draining from lakes, streams and wells, and it can be seen that a water table is continually adjusting towards equilibrium. Because of the low flow rates in most rocks this equilibrium is rarely, if ever, attained before another disturbance occurs. By using measurements of groundwater levels obtained from wells and by observing the levels at which springs discharge, it is possible to construct groundwater

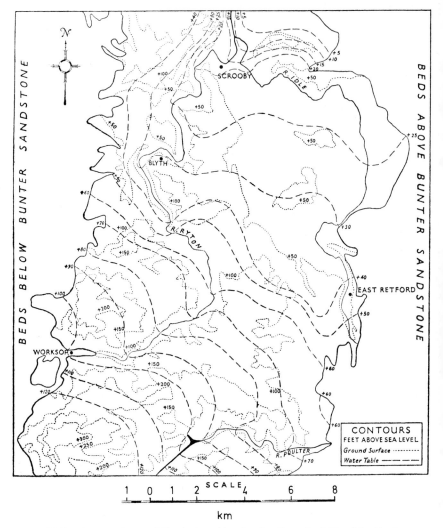

Figure 6.3 Sketch map of part of Nottinghamshire showing the water table of the Bunter Sandstone

contour maps showing the form and elevation of the water table (Figure 6.3).

As pointed out, the water table fluctuates in position, particularly in those climates where there are marked seasonal changes in rainfall. Thus permanent and intermittent water tables can be distinguished, the former marking the level beneath which the water table does not sink whilst the latter is an expression of the fluctuation. Usually water tables fluctuate within the lower and upper limits rather than between them, especially in humid regions, since the periods between successive recharges are small. The position at which the water table intersects the surface is termed the spring line. Intermittent and permanent springs similarly can be distinguished.

A perched water table is one which forms above a discontinuous impermeable layer such as a lens of clay in a formation of sand, the clay impounding a water mound.

6.1.2 Aquifers, aquicludes and aquitards

An *aquifer* is the term given to a rock or soil mass which not only contains water but from which water can be readily abstracted in significant quantities. The ability of an aquifer to transmit water is governed by its permeability. Indeed the permeability of an aquifer usually is in excess of 10^{-5} m/s (see Table 6.2). As far as yield is concerned an aquifer must also have an adequate storage capacity. The storage coefficient of an aquifer is defined as the volume of water taken into or released from storage in each column of the aquifer having a base of one square metre and a height equal to the full thickness of the aquifer, when the head is lowered one metre.

By contrast, a formation with a permeability of less than 10^{-9} m/s is one which, in engineering terms, is regarded as impermeable and is referred to as an *aquiclude*. For example, clays and shales are aquicludes. Even when such rocks are saturated they tend to impede the flow of water through stratal sequences.

According to De Wiest (1967) an aquitard is a formation which transmits water at a very slow rate but which, over a large area of contact, may permit the passage of large amounts of water between adjacent aquifers which it separates. Sandy clays provide an example.

An aquifer is described as unconfined when the water table is open to the atmosphere, that is, the aquifer is not overlain by material of lower permeability (Figure 6.1). Conversely a confined aquifer is one which is overlain by impermeable rocks (Figure 6.1). Confined aquifers may have relatively small recharge areas as compared with unconfined aquifers and therefore may yield less water. Very often the water in a confined aquifer is under piezometric pressure, that is, there is an excess of pressure sufficient to raise the water above the base of the overlying bed when the aquifer is penetrated by a well. Piezometric pressures are developed when the buried upper surface of a confined aquifer is lower than the water table in the aquifer at its recharge area. Where the piezometric surface is above ground level, then water overflows from a well. Such wells are described as artesian. A synclinal structure is the commonest cause of artesian conditions (Figure 6.4a). The London Basin is a much quoted example, although excessive abstraction has reduced its present piezometric level below the ground surface. Other geological structures which give rise to artesian conditions are illustrated in Figure

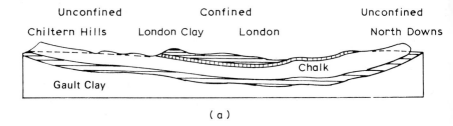

Unconfined   Confined   Unconfined

Chiltern Hills   London Clay   London   North Downs

Chalk

Gault Clay

(a)

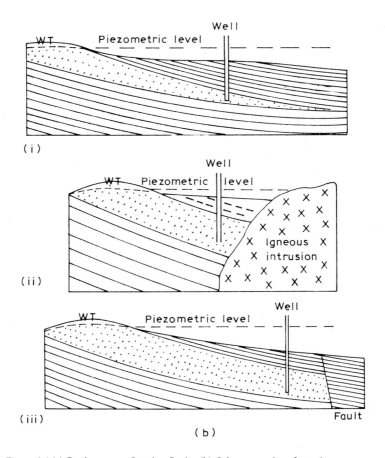

Well

WT   Piezometric level

(i)

Well

WT   Piezometric level

X Igneous X
X intrusion X

(ii)

Well

WT   Piezometric level

(iii)   Fault

(b)

Figure 6.4 (a) Section across London Basin. (b) Other examples of artesian conditions (permeable layer, stippled, sandwiched between impermeable beds)

6.4b. The term *subartesian* is used to describe those conditions in which the water is not under sufficient piezometric pressure to rise to the ground surface.

Where a stream channel is in contact with an unconfined aquifer, water may flow from the stream into the ground or *vice versa*, depending upon the relative water levels. On this basis two types of stream can be distinguished. An effluent stream receives water *from* an aquifer, whereas an influent stream supplies water *to* an aquifer. Most streams are effluent in humid areas whilst influent streams are characteristic of arid areas. A stream may be influent in one location and effluent in another. Moreover, changes can occur with time as stream stages, relative to nearby groundwater levels, change. The rate of recharge from influent streams is largely governed by the permeability of their beds. If these are covered by silt then the stream may lose little water except perhaps during times of flood. During flood groundwater levels are temporarily raised near a stream channel by inflow from the stream. This is referred to as bank storage. Conversely groundwater discharging into a stream forms the base-flow of the stream.

6.2 Capillary movement in soil

Capillary movement in a soil refers to the movement of moisture through the minute pores between the soil particles which act as capillaries. It takes place as a consequence of surface tension, therefore moisture can rise from the water table. This movement, however, can occur in any direction, not just vertically upwards. It occurs whenever evaporation takes place from the surface of the soil, thus exerting a 'surface tension pull' on the moisture, the forces of surface tension increasing as evaporation proceeds. Accordingly capillary moisture is in hydraulic continuity with the water table and is raised against the force of gravity, the degree of saturation decreasing from the water table upwards. Equilibrium is attained when the forces of gravity and surface tension are balanced.

The boundary separating capillary moisture from the gravitational water in the zone of saturation is, as would be expected, ill-defined and cannot be determined accurately. That zone immediately above the water table which is saturated with capillary moisture is referred to as the closed capillary fringe, whilst above this, air and capillary moisture exist together in the pores of the open capillary fringe. The depth of the capillary fringe is largely dependent upon the particle size distribution and density of the soil mass, which in turn influence pore size. In other words the smaller the pore size, the greater is the depth. For example, capillary moisture can rise to great heights in clay soils (Table 6.1) but the movement is very slow. In soils

Table 6.1. **Capillary rises and pressures in soils (after Jumikis (1968))**

Soil	Capillary rise (mm)	Capillary pressure (kPa)
Fine gravel	Up to 100	Up to 1.0
Coarse sand	100–150	1.0–1.5
Medium sand	150–300	1.5–3.0
Fine sand	300–1000	3.0–10.0
Silt	1000–10 000	10.0–100.0
Clay	Over 10 000	Over 100.0

which are poorly graded the height of the capillary fringe generally varies whereas in uniformly textured soils it attains roughly the same height. Where the water table is at shallow depth and the maximum capillary rise is large, moisture is continually attracted from the water table, due to evaporation from the surface, so that the uppermost soil is near saturation. For instance, under normal conditions peat deposits may be assumed to be within the zone of capillary saturation. This means that the height to which the water can rise in peat by capillary action is greater than the depth below ground to which the water table can be reduced by drainage. The coarse fibrous type of peat, containing appreciable sphagnum, may be an exception.

Drainage of capillary moisture cannot be effected by the installation of a drainage system within the capillary fringe as only that moisture in excess of that retained by surface tension can be removed, but it can be lowered by lowering the water table. The capillary ascent, however, can be interrupted by the installation of impermeable membranes or layers of coarse aggregate. These two methods can be used in the construction of embankments, or more simply the height of the fill can be raised.

Below the water table the water contained in the pores is under normal hydrostatic load, the pressure increasing with depth. Because these pressures exceed atmospheric pressure they are designated positive pressures. On the other hand the pressures existing in the capillary zone are less than atmospheric and so are termed negative pressures. Thus the water table is usually regarded as a datum of zero pressure between the positive pore pressure below and the negative above.

At each point where moisture menisci are in contact with soil particles the forces of surface tension are responsible for the development of capillary or suction pressure (Table 6.1). The air and water interfaces move into the smaller pores. In so doing the radii of curvature of the interfaces decrease and the soil suction increases. Hence the drier the soil, the higher is the soil suction.

Soil suction is a negative pressure and indicates the height to which a column of water could rise due to such suction. Since this height or pressure may be very large, a logarithmic scale has been adopted to express the relationship between soil suction and moisture content, the latter is referred to as the pF value

pF value	Equivalent suction	
	(mm water)	(kPa)
	mm water	kPa
0	10	0.1
1	100	1.0
2	1 000	10.0
3	10 000	100.0
4	100 000	1 000.0
5	1 000 000	10 000.0

Soil suction tends to force soil particles together and these compressive stresses contribute towards the strength and stability of the soil. There is a particular suction pressure for a particular moisture content in a given soil, the magnitude of which is governed by whether it is becoming wetter or drier. In fact as a clay soil dries out the soil suction may increase to the order of several thousands of kilopascals. However, the strength of a soil attributable to soil suction is only temporary and is destroyed upon saturation. At that point soil suction is zero.

6.3 Porosity and permeability

Porosity and permeability are the two most important factors governing the accumulation, migration and distribution of groundwater. However, both may change within a rock or soil mass in the course of its geological evolution. Furthermore it is not uncommon to find variations in both porosity and permeability per metre of depth beneath the ground surface.

6.3.1 Porosity

The porosity of a rock can be defined as the percentage pore space within a given volume. Total or absolute porosity is a measure of the total void volume and is the excess of bulk volume over grain volume per unit of bulk volume. It is usually determined as the excess of grain density (the same as specific gravity, now referred to as relative density) over dry density, per unit of grain density, and can be obtained from the following expression

$$\text{Absolute porosity} = \left(1 - \frac{\text{Dry density}}{\text{Grain density}}\right) \times 100$$

The effective, apparent or net porosity is a measure of the effective void volume of a porous medium and is determined as the excess of bulk volume over grain volume and occluded pore volume. It may be regarded as the pore space from which water can be removed.

The factors affecting the porosity of a rock include particle size distribution, sorting, grain shape, fabric, degree of compaction and cementation, solution effects, and lastly mineralogical composition, particularly the presence of clay particles. In experiments with packing arrangements, Frazer (1935) found that for a given mode of packing of equal sized spheres porosity was independent of size. Rhombohedral packing (Figure 6.5a) was the tightest form and produced a porosity of 25.9% whilst the loosest type of packing gave rise to a porosity of 87.5%. However, in natural assemblages as grain sizes decrease so friction, adhesion and bridging become more important because of the higher ratio of surface area to volume. Therefore as the grain size decreases the porosity increases. For example, in coarse sands it ranges from 39–41%, medium 41–48% and fine 44–49%. Whether the grain size is uniform or non-uniform is of fundamental importance with respect to porosity (Figure 6.5b). The highest porosity is commonly attained when all the grains are the same size. The addition of grains of different size to such an assemblage lowers its porosity and this is, within certain limits, directly proportional to the amount added. As would be expected the skewness of the size distribution influences porosity. For example, sands with a negative skewness, that is, an excess of coarse particles in relation to fines, tend to have higher porosities.

Irregularities in grain shape result in a larger possible range of porosity,

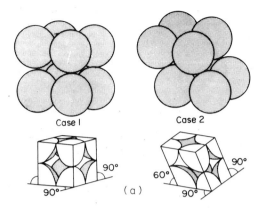

Figure 6.5 (a) Packing of spherical grains. Unit cells of cubic and rhombohedral packing (after Fraser (1935))

as irregular forms may theoretically be packed either more tightly or more loosely than spheres. Similarly angular grains may either cause an increase or a decrease in porosity, although the only type of angularity which has been found experimentally to produce a decrease is that in which the grains are mildly and uniformly disc-shaped.

After a sediment has been buried and indurated several additional factors help determine its porosity. The chief amongst these are closer spacing of grains, deformation and granulation of grains, recrystallisation, secondary growth of minerals, cementation and, in some cases, solutioning. For instance, when chemical cements are present in sandstones in large amounts their influence on porosity is dominant and masks the control of other factors. Thus, two types of porosity may be distinguished, original and secondary. Original porosity is an inherent characteristic of the rock in that it was determined at the time the rock was formed. The process by which a given sediment has accumulated affects its porosity in two ways. First, the nature and variety of the materials deposited affects the entire deposit by controlling the range and uniformity of the sizes present, as well as their degree of rounding. Second, is by the manner in which the material is packed. Hence the original porosity results from the physical impossibility of packing grains in such a way as to exclude interstitial voids of a conjugate nature (an account of packing is given in Chapter 3). On the other hand, secondary porosity results from later changes undergone by the rock which may either increase or decrease its original porosity.

The porosity can be determined experimentally by using either the standard saturation method (see Franklin (1970)) or an air porosimeter (see Ramana and Venkatanarayana (1971). Both tests give an effective value of porosity, although that obtained by the air porosimeter may be somewhat higher because air can penetrate pores more easily than can water.

6.3.2 *Permeability*

Permeability may be defined as the ability of a rock to allow the passage of fluids into or through it without impairing its structure. In ordinary hydraulic usage a substance is termed permeable when it permits the passage of a measurable quantity of fluid in a finite period of time and impermeable when the rate at which it transmits that fluid is slow enough to be negligible

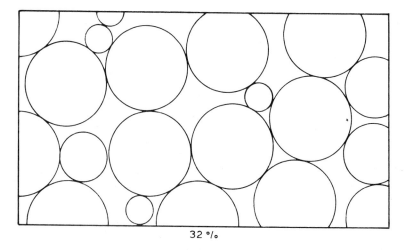

32 %

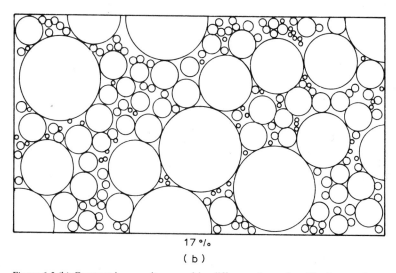

17 %

(b)

Figure 6.5 (b) Contrast in porosity caused by difference in sorting. Numbers under diagrams indicate percent porosity in each diagram

under existing temperature-pressure conditions (Table 6.2). The permeability of a particular material is defined by its coefficient of permeability or hydraulic conductivity (the term *hydraulic conductivity* is now used in place of coefficient of permeability.) The transmissivity or flow in litres per day through a section of aquifer one metre wide under a hydraulic gradient of unity is sometimes used as a convenient quantity in the calculation of groundwater flow instead of the hydraulic conductivity.

Permeability is assessed in the laboratory by using either a constant head or falling head permeameter. A constant head permeameter (Figure 6.6a) is

Table 6.2. Relative values of soil permeabilities

Degree of permeability	Range of hydraulic conductivity (k) (m/s)	Soil type
High	10^{-3}	Medium and coarse gravel
Medium	$10^{-3} - 10^{-5}$	Fine gravel; coarse, medium and fine sand; dune sand; clean sand–gravel mixtures
Low	$10^{-5} - 10^{-7}$	Very fine sand; silty sand; loose silt, loess, well-fissured clays
Very low	$10^{-7} - 10^{-9}$	Dense silt, dense loess, clayey silt, poorly fissured clays
Impermeable	10^{-9}	Unfissured clays

used to measure the permeability of granular materials such as gravels and sands (see Akroyd (1964); BS1377 (1975)). A sample is placed in a cylinder of known cross sectional area (A) and water is allowed to move through it under a constant head. The amount of water discharged (Q) in a given period of time (t) together with the difference in head (h) over a given length of sample (l), measured by means of manometer tubes, is obtained. The results are substituted in the Darcy expression (see Section 6.4.1) and the hydraulic conductivity (k) thereby derived.

$$Q/t = (Ak)h/l = Aki$$

where i is the hydraulic gradient.

Determination of the permeability of fine sands and silts, as well as many rock types, is made by using a falling head permeameter (Figure 6.6b). The sample is placed in the apparatus which is then filled with water to a certain height (h_1) in the standpipe. Then the stopcock is opened and the water infiltrates through the sample, the height of the water in the standpipe falling to h_2. The times at the beginning (t_1) and end (t_2) of the test are recorded. These, together with the cross sectional area (A) and length of sample (l) are then substituted in the following expression, which is derived from Darcy's law, to obtain the hydraulic conductivity (k)

$$k = \frac{2.3026al}{A(t_2 - t_1)} \times (\log_{10}h_i - \log_{10}h_2)$$

where a is the cross sectional area of the standpipe. The permeability of clay cannot be measured by using a permeameter, it must be determined indirectly, for example, from the consolidation test.

Bernaix (1969) examined the variations of permeability in rocks under stress by using a radial percolation test. He used a cylindrical specimen 60 mm in diameter and 150 mm in length in which an axial hole, 12 mm diameter and 125 mm in length, was drilled. The specimen is placed in the radial percolation cell, which can contain water under pressure, and the central cavity is in contact with atmospheric pressure; or water can be injected under pressure into the cavity (Figure 6.7). The flow is radial over almost the whole height of the sample and is convergent when the water pressure is applied to the outer faces of the specimen, and divergent when the water is under

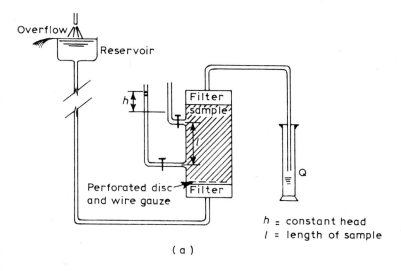

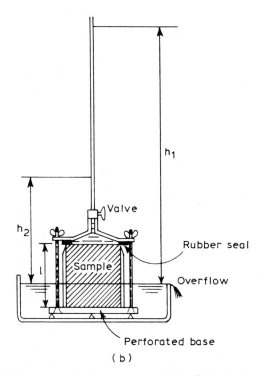

Figure 6.6 (a) The constant head permeameter. (b) The falling head permeameter

pressure inside the specimen. For radial flow from a cylinder of unfractured rock with interconnected pores the hydraulic conductivity (k) is given by

$$k = \frac{q\gamma_w}{2\pi l \Delta p} \ln \frac{R_2}{R_1}$$

in which R_1 and R_2 are the radii of the inner and outer surfaces respectively q is the flow discharge, γ_w is the unit weight of water, l is the length of cylinder over which flow is occurring and Δp is the difference between external and internal water pressures and is positive for convergent flow.

It has been shown by Bernaix (1969) that porous rocks remain more or less unaffected by pressure changes. On the other hand, he found that fissured rocks exhibit far greater permeability in divergent flow than in convergent flow. Moreover the fissured rocks which were examined exhibited a continuous increase in permeability as the pressure attributable to divergent flow was increased. Indeed it would appear that some amount of hydraulic fracturing occurs in divergent flow. Goodman and Sundaram (1980) noted the same type of behaviour when they tested tuff, schist, sandstone and limestone.

The flow through a unit cross section of material is modified by temperature, hydraulic gradient and the hydraulic conductivity. The latter is affected by the uniformity and range of grain size, shape of the grains, stratification, the amount of consolidation and cementation undergone, and the presence and nature of discontinuities. Temperature changes affect the flow rate of a fluid by changing its viscosity. The rate of flow is commonly assumed to be directly proportional to the hydraulic gradient but this is not always so in practice.

Permeability and porosity are not necessarily as closely related as would be expected, for instance, very fine textured sandstones frequently have a higher porosity than coarser ones, though the latter are more permeable.

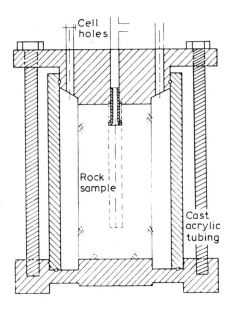

Figure 6.7 Radial permeameter cell

From experiments on uniformly sorted sands of known sizes it has been found that, other factors being equal, permeability varies roughly as the square of the diameter of the grains (see Krumbein and Monk (1943)). This follows from the fact that when the diameter of the grains is doubled, the throat-plane area between them undergoes a fourfold increase, the rate of flow being dependent on the size of this channel. Any departure from the spherical shape affects the permeability by varying the size and shape of the interstices and by causing looser or tighter packing. At equal grain diameters and porosity, the hydraulic conductivity decreases with increasing uniformity of pore spaces. Hence as the form of the grains departs from that of a true sphere the permeability increases.

As can be inferred from above, the permeability of a clastic material is also affected by the interconnections between the pore spaces. If these are highly tortuous then the permeability is accordingly reduced. Consequently tortuosity figures importantly in permeability, influencing the extent and rate of free water saturation. It can be defined as the ratio of the total path covered by a current flowing in the pore channels between two given points to the straight line distance between them. Incidentally, the difference between the absolute and effective porosity provides an indirect measure of assessing tortuosity. The tortuosity (T_o) may be calculated from the following expressions

$$T_o = l_1/l \text{ or } T_o = \sqrt{(n_p \times n_f)}$$

where l_1 is the length of the current path in the sample (this can be measured by passing an electric current between two electrodes), l is the length of the sample, n_p is the porosity parameter, and n_f is the porosity expressed as a proper fraction. The relative electrical resistivity is defined as the ratio of the resistance offered by the rock (R_p) when saturated with a mineralising solution, to the electrical resistance of the solution (R_s). It depends upon the effective porosity and the specific surface attributable to the tortuosity, hence

$$n_p = R_p/R_s$$

Stratification in a formation varies within limits both vertically and horizontally. It is frequently difficult to predict what effect stratification has on the permeability of the beds. Nevertheless in the great majority of cases where a directional difference in permeability exists, the greater permeability is parallel to the bedding. For example, the Permo-Triassic sandstones of the Mersey and Weaver Basins are notably anisotropic as far as permeability is concerned, the flow parallel to the bedding being higher than across it. Ratios of 5:1 are not uncommon and occasionally values of 100:1 have been recorded where fine marl partings occur.

The term pervious is frequently used to describe a rock mass which is traversed by discontinuities which can hold water and which allow it to percolate through, even though the rock itself may have extremely low porosity. In fact the flow of water through such rock masses is very much dependent upon discontinuities and also upon any impermeable boundaries. Basaltic lava flows which are intersected by cooling joints provide a typical example. As expected the frictional resistance to flow through such joint systems is frequently much lower than that offered by a porous medium, hence appreciable quantities of water may be transmitted. Another example is provided by the massive limestones of Lower Carboniferous age of the Pennine area, the permeability of an intact sample being much lower than that obtained by field tests (10^{-16}–10^{-11} m/s, and 10^{-5}–10^{-1} m/s respectively). The significantly higher permeability found in the field is attributable to the joint systems and bedding planes which have been opened by solutioning. The mass permeability of sandstones is also very much influenced by the discontinuities. For instance, the average laboratory permeability for the Fell Sandstone Group from Shirlawhope Well near Longframlington, Northumberland, is 17.4×10^{-7} m/s (see Bell (1978)). This compares with an estimated value of 2.4×10^{-3} m/s obtained from field tests. From the foregoing examples it can be concluded that as far as the assessment of flow through pervious rock masses is concerned, field tests (see below) provide far more reliable results than can be obtained from testing intact samples in the laboratory. However, the walls of discontinuities are invariably irregular and this has a retarding effect upon flow movement. Moreover discontinuities tend to close with depth. Indeed joints, when exposed at the surface, have usually been opened up by weathering processes. Dissipation of residual stress on removal of overburden also aids joint development.

Dykes often act as barriers to groundwater flow so that the water table on one side may be higher than on the other. Fault planes occupied by clay gouge may have a similar effect. Conversely they may act as conduits where the fault plane is not sealed. The movement of water across a permeable boundary which separates aquifers of different permeabilities leads to deflection of flow, the bigger the difference the larger the deflection. When groundwater meets an impermeable boundary it flows along it and, as noted previously, in some situations, such as the occurrence of a dyke, may be impounded. The nature of a rock mass also influences whether flow is steady or unsteady. Generally it is unsteady since it is usually due to discharge from storage.

As the water table falls so the rate at which water drains away declines and *vice versa*. Thus the flow rate is influenced by recharge and time.

The direction of groundwater flow can be assessed by monitoring the movement of radioactive tracers such as tritium or carbon-14 (see Mather *et al.* (1973)), or by using dyes. Tracers are introduced into the groundwater via boreholes. Koerner *et al.* (1979) reviewed various other methods used for detecting groundwater seepage including temperature sensing, infrared sensing, microwave sensing, acoustic emission control and seismic and electric methods, as well as the use of tracers. Their summary of the advantages and disadvantages of these methods is given in Table 6.3.

6.4 Flow through soils and rocks

Water possesses three forms of energy, namely, potential energy attributable to its height, pressure energy owing to its pressure, and kinetic energy due to its velocity. The latter can usually be discounted in any assessment of flow through soils. Energy in water is usually expressed in terms of head. The head possessed by water in soils or rocks is manifested by the height to which water will rise in a standpipe above a given datum. This height is usually referred to as the piezometric level and provides a measure of the

Table 6.3. Advantages and disadvantages of methods of monitoring groundwater flow (from Koerner et al. (1979))

Method (1)	Primary use (2)	Advantages (3)	Disadvantages (4)
Tracers (non-radioactive)	Seepage	Inexpensive Defines flow path Fluorescent dyes are detectable at low concentrations Most widely used	Source of seepage is required Often difficult to place Absorption is a problem Dilution is common
Tracers (radioactive)	Seepage	Easily detectable Widely used	Difficult to place Expensive Health hazard Environmental hazard
Temperature	Seepage	Uses natural phenomenon	Lengthy readout time Requires 0.01°C sensitivity Not widely used Still in research
Infrared	Groundwater	Uses natural phenomenon Covers large area	Limited detail in complicated topography Expensive Not widely used Still in research
Microwave (pulsed)	Groundwater and seepage	Traces surface of water Good penetration depth Continuous data for contouring	Expensive Needs sharp interface Upstream detection not possible
Microwave (continuous)	Groundwater and seepage	Traces surface of water Continuous data for contouring	Expensive Needs sharp interface Upstream detection not possible Still in research Limited penetration depth
Geophysical (seismic)	Groundwater	Refraction method can identify type of liquid Relatively common use Technique well established	Expensive Refraction needs dense lower layer
Geophysical (electric)	Groundwater	Cost is less than seismic Relatively common use Identify type of liquid SP gives direction and relative magnitude of flow	Salts and metal troublesome Depth limited SP method not widely used
Acoustic emission	Seepage	Traces seepage flow and relative magnitude Inexpensive	Background noise troublesome Not widely used Still in research

total energy of the water. If at two different points within a continuous area of water there are different amounts of energy, then there will be a flow towards the point of lesser energy and the difference in head is expended in maintaining that flow. Other things being equal, the velocity of flow between two points is directly proportional to the difference in head between them. The hydraulic gradient (i) refers to the loss of head or energy of water flowing through the ground. This loss of energy by the water is due to the friction resistance of the ground material, and this is greater in fine- than coarse-grained soils. Thus, in a given engineering project there is no guarantee that the rate of flow will be uniform, indeed this is exceptional. However, if it is assumed that the resistance to flow is constant, then for a given difference in head the flow velocity is directly proportional to the flow path.

6.4.1 Darcy's law

Before any mathematical treatment of groundwater flow can be attempted certain simplifying assumptions have to be made, namely, that the material is isotropic and homogeneous, that there is no capillary action, and that a steady state of flow exists. Since rocks and soils are anisotropic and heterogeneous, since they may be subject to capillary action, and as flow through them is characteristically unsteady, any mathematical assessment of flow must be treated with caution.

The basic law concerned with flow is that enunciated by Darcy (1856) which states that the rate of flow (v) per unit area is proportional to the gradient of the potential head (i) measured in the direction of flow

$$v = ki$$

and for a particular rock or soil or part of it, of area (A)

$$Q = vA = Aki$$

where Q is the quantity in a given time. The ratio of the cross sectional area of the pore spaces in a soil to that of the whole soil is given by $e/(1 + e)$. Hence a truer velocity of flow, that is, the seepage velocity (v_s), is

$$v_s = [(1 + e)/e]ki$$

where e is the void ratio and k is the hydraulic conductivity. Darcy's law is valid as long as a laminar flow exists. Departures from Darcy's law therefore occur when the flow is turbulent. They also occur when the velocity of flow is high. Such conditions exist in very permeable media, normally when the Reynolds number* can attain values above four. Accordingly it is usually accepted that this law can be applied to those soils which have finer textures than gravels. Furthermore Darcy's law probably does not accurately represent the flow of water through a porous medium of extremely low permeability, because of the influence of surface and ionic phenomena and the presence of gases.

Apart from an increase in the mean velocity, the other factors which cause deviations from the linear laws of flow include, first, the non-uniformity of pore spaces, since differing porosity gives rise to differences in the seepage rates through pore channels. A second factor is an absence of a running-in section where the velocity profile can establish a steady state parabolic distribution. Lastly, such deviations may be developed by perturbations due to jet separation from wall irregularities.

---

*Reynolds number (N_R) is commonly used to distinguish between laminar and turbulent flow and is expressed as follows

$$N_R = \rho \frac{vR}{\mu}$$

where ρ is density, v is mean velocity, R is hydraulic radius and μ is viscosity. Flow is laminar for small values of Reynolds number.

Darcy omitted to recognise that permeability also depends upon the density (ρ) and viscosity of the fluid (μ) involved, and the average size (D_n) and shape of the pores in a porous medium. In fact permeability is directly proportional to the unit weight of the fluid concerned and is inversely proportional to its viscosity. The latter is very much influenced by temperature. The following expression attempts to take these factors into account.

$$k = CD_n^2 \, \rho_x / \mu$$

where C is a dimensionless constant or shape factor which takes note of the effects of stratification, packing, size distribution and porosity. It is assumed in this expression that both the porous medium and the water are mechanically and physically stable, but this may never be true. For example, ion exchange on clay and colloid surfaces may bring about changes in mineral volume which in turn affect the shape and size of the pores. Moderate to high groundwater velocities will tend to move colloids and clay particles. Solution and deposition may result from the pore fluids. Small changes in temperature and/or pressure may cause gas to come out of solution which may block pore spaces.

It has been argued that a more rational concept of permeability would be to express it in terms that are independent of the fluid properties. Thus the intrinsic permeability (k_i) characteristic of the medium alone has been defined as

$$k_i = CD_n^2$$

However, it has proved impossible to relate C to the properties of the medium. Even in uniform spheres it is difficult to account for the variations in packing arrangement. In this context a widely accepted relationship for laminar flow through a permeable medium is that given by Fair and Hatch (1935)

$$k = \cfrac{1}{m\left[\dfrac{(1-n)^2}{n^3}\left(\dfrac{\theta}{100}\Sigma\dfrac{p}{D_m}\right)^2\right]}$$

where n = porosity, m = packing factor found by experiment to have a value of 5, θ = particle shape factor varying from 6.0 for spherical to 7.7 for angular grains, p = percentage of particles by weight held between each pair of adjacent sieves, D_m = geometric mean opening $(D_1 D_2)^{1/2}$ of the pair.

6.4.2 Fissure flow

Generally it is the interconnected systems of discontinuities which determine the permeability of a particular rock mass. Indeed the permeability of a jointed rock mass is usually several orders higher than that of intact rock. According to Serafim (1968) the following expression can be used to derive the filtration through a rock mass intersected by a system of parallel-sided joints with a given opening (e) separated by a given distance (d).

$$k = \frac{e^3 \gamma_w}{12d\mu}$$

where γ_w is the unit weight of water and μ its viscosity. The velocity of flow (v) through a single joint of constant gape is expressed by

$$v = \left(\frac{e^2 \gamma_w}{12\mu}\right) i$$

where i is the hydraulic gradient. Subsequently Wittke (1973) suggested that where the spacing between discontinuities is small in comparison with the dimensions of the rock mass, it is often admissible to replace the fissured rock, with regard to its permeability, by a continuous anisotropic medium, the permeability of which can be described by means of Darcy's law. He also provided a resumé of procedures by which three-dimensional problems of flow through rocks under complex boundary conditions could be solved.

Lovelock *et al.* (1975) suggested that the contribution of the fissures (T_f) to the transmissivity of an idealised aquifer can be approximated from the following expression

$$T_f = \frac{g}{12\mu_k} \sum_{x=1}^{n} b_x^3$$

where b_x is the effective aperture of the xth of n horizontal, parallel-sided, smooth-walled openings, g is the acceleration due to gravity, μ_k is the kinematic viscosity of the fluid, and flow is laminar. The third power relationship means that a small variation in effective aperture gives rise to a large variation in fissure contribution. A fuller account of the theory of groundwater flow can be obtained from De Wiest (1967), Verruijt (1970) or Skeat (1969).

6.4.3 Flow nets

Flow nets provide a graphical representation of the flow of water through the ground and indicate the loss of head involved (Figure 6.8). They also provide data relating to the changes in head velocity and effective pressure

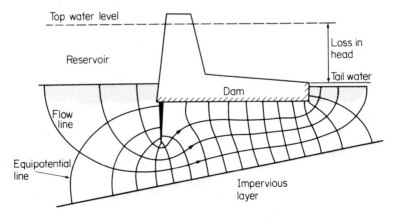

Figure 6.8 Flow net beneath concrete gravity dam, with cut-off at the heel, showing seventeen equipotential drops and four flow channels

which occur in a foundation subjected to flowing groundwater conditions. For example, where the flow lines of a flow net move closer together this indicates that the flow increases, although their principal function is to indicate the direction of flow. The equipotential lines indicate equal losses in head or energy as the water flows through the ground, so that the closer they are, the more rapid is the loss in head. Hence a flow net can provide quantitative data related to the flow problem in question, for example, seepage pressures can be determined at individual points within the net.

It is possible to estimate the amount of water flowing through a soil from a flow net. If the total loss of head and the permeability of the soil are known, then the quantity of water involved can be calculated by using Darcy's law. However, it is not really as simple as that for the area through which the water flows usually varies, as does the hydraulic gradient, since the flow paths vary in length. By using the total number of flow paths (f), the total number of equipotential drops (d) and the total loss of head (i_t) together with the permeability (k) in the following expression

$$Q = ki_t(f/d)$$

the quantity of water flow can be estimated.

6.5 Pore pressures, total pressures and effective pressures

Subsurface water is normally under pressure which increases with increasing depth below the water table to very high values. Such water pressures have a significant influence on the engineering behaviour of most rock and soil masses and their variations are responsible for changes in the stresses in these masses, which affect their deformation characteristics and failure.

The efficiency of a soil in supporting a structure is influenced by the effective or intergranular pressure, that is, the pressure between the particles of the soil which develops resistance to applied load. Because the moisture in the pores offers no resistance to shear, it is ineffective or neutral and therefore pore pressure has also been referred to as neutral pressure. Since the pore or neutral pressure plus the effective pressure equals the total pressure, reduction in pore pressure increases the effective pressure. Reduction of the pore pressure by drainage consequently affords better conditions for carrying a proposed structure.

The effective pressure at a particular depth is simply obtained by multiplying the unit weight of the soil by the depth in question and subtracting the pore pressure for that depth. In a layered sequence the individual layers may have different unit weights. The unit weight of each should then be multiplied by its thickness and the pore pressure involved subtracted. The effective pressure for the total thickness involved is then obtained by summing the effective pressures of the individual layers (Figure 6.9). Water held in the capillary fringe by soil suction does not affect the values of pore pressure below the water table. However, the weight of water held in the capillary fringe does increase the weight of overburden and so the effective pressure.

Volume changes brought about by loading compressive soils depend upon the level of effective stress and are not affected by the area of contact. The latter may also be neglected in saturated or near-saturated soils.

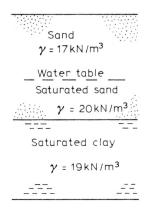

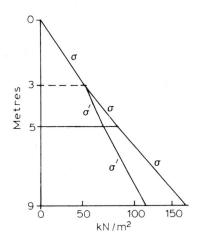

γ = unit weight

Depth (m)	Total pressure (σ)		Pore pressure (u)		Effective pressure (σ') $\sigma' = \sigma - u$
3	3×17	$= 51$	0		51
5	$(3 \times 17)+(2 \times 20)$	$= 91$	2×9.8	$= 19.6$	71.4
9	$(3 \times 17)+(2 \times 20)+(4 \times 19)$	$= 167$	6×9.8	$= 58.8$	108.2

Figure 6.9 Pressure diagram and example of calculation of total and effective overburden pressures

There is some evidence which suggests that the law of effective stress as used in soil mechanics, in which the pore pressure is subtracted from all direct stress components, holds true for some rocks. Those with low porosity may, at times, prove the exception. However, Serafim (1968) suggested it appeared that pore pressures have no influence on brittle rocks. This is probably because the strength of such rocks is mainly attributable to the strength of the bonds between the component crystals or grains.

The changes in stresses and the corresponding displacements due, for example, to construction work influence the permeability of a rock mass. For instance, with increasing effective shear stress the permeability increases along discontinuities orientated parallel to the direction of shear stress, whilst it is lowered along those running normal to the shear stress. Consequently the imposition of shear stresses, and the corresponding strains, lead to an anisotropic permeability within joints.

6.6 Assessment of *in situ* permeability and pore water pressures

Accurate recording of groundwater conditions is important, particularly if excavation level extends beneath the water table. Not only should the water levels be observed in boreholes, say twice daily, but at least one standpipe should be installed for long-term observation. Piezometers may be installed in boreholes, and *in situ* permeability tests carried out. In some instances the groundwater may contain substances in great enough quantity to affect concrete adversely, the sulphate content and pH value being of particular interest. A chemical analysis of the groundwater is then required to assess the need for special precautions.

An initial assessment of the magnitude and variability of the *in situ* hydraulic conductivity can be obtained from tests carried out in boreholes as the hole is advanced. By artificially raising the level of water in the borehole (falling head test) above that in the surrounding ground, the flow rate from the borehole can be measured. However, in very permeable soils it may not be possible to raise the level of water in the borehole. Conversely the water level in the borehole can be artificially depressed (rising head test) so allowing the rate of water flow into the borehole to be assessed. Wherever possible a rising and a falling head test should be carried out at each required level and the results averaged. Once a steady state has been attained the permeability (k) can be determined by measuring the rate of flow of water (q) under a constant applied change in head (H) by using the following expression of Hvorslev (1951)

$$k = q/FH$$

where F is a shape factor.

In a rising or falling head test in which the piezometric head varies with time, the permeability is determined from the expression

$$k = A/F(t_2 - t_1) \times \ln(H_1/H_2)$$

where H_1 and H_2 are the piezometric heads at times t_1 and t_2 respectively and A is the inner cross sectional area of the casing in the borehole. The test procedure involves observing the water level in the casing at given times, then a graph of water level against time is constructed (Figure 6.10).

In cohesive soils of low permeability it is preferable to carry out the test over an unlined section of borehole. However, in soft cohesive soils, because the walls of the borehole may collapse, a gravel filter needs to be installed over the test length. Casing can be used when testing in sands and gravels and the borehole must be kept topped up with water to prevent piping.

The permeability of an individual bed of rock can be determined by a water injection or packer test carried out in a drillhole. This is done by sealing off a length of uncased hole with packers and injecting water under pressure into the test section (Figure 6.11). Usually because it is more convenient, these permeability tests are carried out after the entire length of a hole has been drilled. Two packers are used to seal off selected test lengths and the tests are performed from the base of the hole upwards. The hole must be flushed to remove sediment prior to a test being performed. Water is generally pumped into the test section at steady pressures for periods of 15 min, readings of water absorption being taken every 5 min. The test usually consists of five cycles at successive pressures of 6, 12, 18, 12 and 6 kPa for every metre depth of packer below the surface (see Dixon and Clarke (1975)). The permeability from packer tests can be derived from

$$k = q/C_s HR$$

where q is the steady flow rate under an effective applied head (H), C_s is

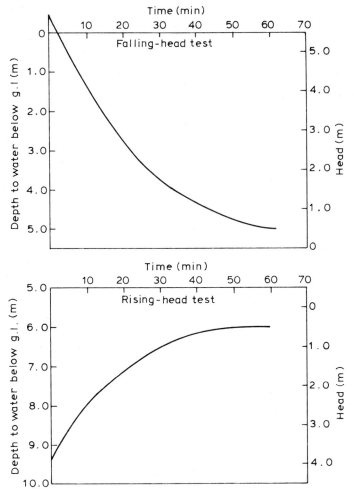

Figure 6.10 Rising and falling head permeability tests

a constant depending upon the length and diameter of the test section and *R* is the radius of the drillhole.

Piezometers are installed in the ground in order to monitor and obtain accurate measurements of pore water pressures (Figure 6.12). Observations should be made regularly so that changes due to external factors such as excessive precipitation, tides, the seasons etc., are noted, it being most important to record the maximum pressures which have occurred. Standpipe piezometers allow the determination of the position of the water table and the permeability (see Sherrell (1976)). For example, the water level can be measured with an electric dipmeter or with piezometer tips which have leads going to a constant head permeability unit enabling the rate of flow through the tip to be measured. Hydraulic piezometers can be installed at various depths in a borehole where it is required to determine water pressures. They are connected to a manometer board which records the changes in pore water

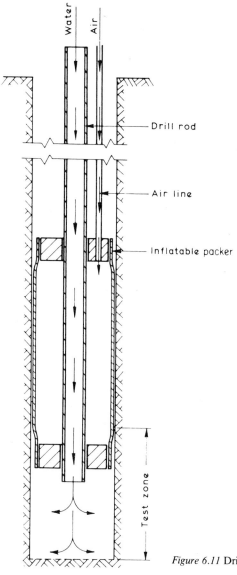

Figure 6.11 Drillhole 'packer' test arrangement

pressure. Usually simpler types of piezometer are used in the more permeable soils. When a piezometer is installed in a borehole it should be surrounded with a filter of clean sand. The sand should be sealed both above and below the piezometer to enable the water pressures at that particular level to be measured with a minimum of influence from the surrounding strata, since the latter may contain water at different pressures. The response to piezometers in rock masses can be very much influenced by the incidence and geometry of the discontinuities so that the values of water pressure obtained may be misleading if due regard is not given to these structures.

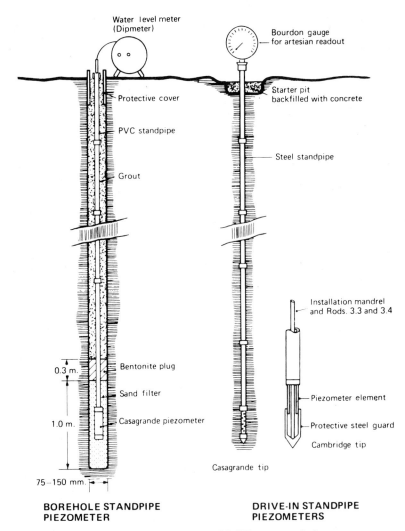

Figure 6.12 Standard piezometers (courtesy of Soil Instruments Ltd)

6.7 Critical hydraulic gradient, quick conditions and hydraulic uplift phenomena

As water flows through the soil and loses head, its energy is transferred to the particles past which it is moving, which in turn creates a drag effect on the particles. If the drag effect is in the same direction as the force of gravity, then the effective pressure is increased and the soil is stable. Indeed the soil tends to become more dense. Conversely if water flows towards the surface then the drag effect is counter to gravity thereby reducing the effective pressure between particles. If the velocity of upward flow is sufficient it can buoy up the particles so that the effective pressure is reduced to zero. This represents a critical condition where the weight of the submerged soil is balanced by the upward acting seepage force. The critical hydraulic gradient (i_c) can be calculated from the following expression

$$i_c = \frac{G_s - 1}{1 + e}$$

where G_s is the relative density (specific gravity) of the particles and e is the void ratio. A critical condition sometimes occurs in silts and sands. If the upward velocity of flow increases beyond the critical hydraulic gradient a quick condition develops.

Quicksands, if subjected to deformation or disturbance, can undergo a spontaneous loss of strength. This loss of strength causes them to flow like viscous liquids. Terzaghi (1925) explained the quicksand phenomenon in the following terms. First, the sand or silt concerned must be saturated and loosely packed. Secondly, on disturbance the constituent grains become more closely packed which leads to an increase in pore water pressure, reducing the forces acting between the grains. This brings about a reduction in strength. If the pore water can escape very rapidly the loss in strength is momentary. Hence the third condition requires that pore water cannot escape readily. This is fulfilled if the sand has a low permeability and/or the seepage path is long. Casagrande (1936) demonstrated that a critical porosity existed above which a quick condition could be developed. He maintained that many coarse-grained sands, even when loosely packed, have porosities approximately equal to the critical condition whilst medium- and fine-grained sands, especially if uniformly graded, exist well above the critical porosity when loosely packed. Accordingly fine sands tend to be potentially more unstable than coarse grained varieties. It must also be remembered that the finer sands have lower permeabilities.

Quick conditions brought about by seepage forces are frequently encountered in excavations made in fine sands which are below the water table, as for example, in cofferdam work. As the velocity of the upward seepage force increases further from the critical gradient the soil begins to boil more and more violently. At such a point structures fail by sinking into the quicksand. Liquefaction of potential quicksands may be caused by sudden shocks such as the action of heavy machinery (notably pile driving), blasting and earthquakes. Such shocks increase the stress carried by the water, the neutral stress, and give rise to a decrease in the effective stress and shear strength of the soil. There is also a possibility of a quick condition developing in a layered soil sequence where the individual beds have different permeabilities. Hydraulic conditions are particularly unfavourable where water initially flows through a very permeable horizon with little loss of head, which means that flow takes place under a great hydraulic gradient.

There are several methods which may be employed to avoid the development of quick conditions. One of the most effective techniques is to prolong the length of the seepage path thereby increasing the frictional losses and so reducing the seepage force. This can be accomplished by placing a clay blanket at the base of an excavation where seepage lines converge. If sheet piling is used in excavation of critical soils then the depth to which it is sunk determines whether or not quick conditions will develop. Consequently it should be sunk deep enough to avoid a potential critical condition occurring at the base level of the excavation. The hydrostatic head also can be reduced by means of relief wells and seepage can be intercepted by a wellpoint system placed about the excavation. Furthermore a quick condition may be prevented by increasing the downward acting force. This may be brought about by laying a load on the surface of the soil where seepage is discharging. Gravel filter beds may be used for this purpose. Suspect soils also can be densified, treated with stabilising grouts, or frozen.

When water percolates through heterogeneous soil masses it moves preferentially through the most permeable zones and it issues from the ground as springs. Piping refers to the erosive action of some such springs, where sediments are removed by seepage forces, so forming subsurface cavities and tunnels. In order that erosion tunnels may form, the soil must have some cohesion, the greater the cohesion, the wider the tunnel. In fact fine sands and silts are most susceptible to piping failures. Obviously the danger of piping occurs when the hydraulic gradient is high, that is, when there is a rapid loss of head over a short distance. This may be indicated on a flow net by a close network of squares where the flow is upward. As the pipe develops by backward erosion it nears the source of water supply so that eventually the water breaks into and rushes through the pipe. Ultimately the hole, so produced, collapses from lack of support. Piping has been most frequently noted downstream of dams, the reservoir providing the water source (see Penman (1977)). Leaking drains can also give rise to piping.

Subsurface structures should be designed to be stable with regard to the highest groundwater level that is likely to occur. Structures below groundwater level are acted upon by uplift pressures. If the structure is weak this pressure can break it and, for example, cause a blow-out of a basement floor or collapse of a basement wall. If the structure is strong but light it may be lifted, that is, subjected to heave. Uplift can be taken care of by adequate drainage or by resisting the upward seepage force. Continuous drainage blankets are effective but should be designed with filters to function without clogging. The entire weight of structure can be mobilised to resist uplift if a raft foundation is used. Anchors, grouted into bedrock, can provide resistance to uplift.

Moore and Longworth (1979) recorded a failure in a brick pit, 29 m in depth, excavated in Oxford Clay. The failure was brought about by a build-up of hydrostatic pressure in an underlying aquifer (either the Cornbrash or Blisworth Limestone located at depths of 6 and 11 m respectively below the surface of the pit). This initially gave rise to a heave of some 150 mm, and then ruptured the surface clay, thereby allowing the rapid escape of approximately 7000 m³ of water. The floor of the pit then settled up to 100 mm.

Hydraulic uplift phenomena have previously been reported by Rowe (1968) and occur under a wide range of geological conditions and at differing scales. Horswill and Horton (1976) referred to a case of hydraulic disruption at the base of a shaft, 11 m in diameter, which was sunk in the Upper Lias Clay at Empingham Dam site. The uplift took place when the shaft had reached a point of 15 m above the Marlstone Rock Bed, the offending limestone aquifer. Within a few hours the clay in the shaft was deformed and fractured, facilitating the inflow of water, thereby flooding the shaft.

If water flowing under pressure through the ground is confined between two impermeable horizons then it is termed artesian water. Artesian conditions are commonly developed in synclinal structures where an aquifer is sandwiched between two impermeable layers but outcrops at a higher elevation than the position at which the pressure is measured (see Figure 6.4). They can cause serious trouble in excavations and both the position of the water table and the piezometric pressures should be determined before work commences. Otherwise excavations which extend close to strata under artesian pressure may be severely damaged due to blow-outs taking place in the floors. Slopes may also fail. Indeed such sites may have to be abandoned. Artesian pressures have been controlled by sinking bleeder wells (see Wade and Taylor (1979)).

6.8 Control of groundwater

All excavations below the water table encounter groundwater, which, besides being a nuisance, can have undesirable effects. Groundwater lowering processes which depend on pumping are essentially temporary remedies, but if the terrain is suitable, permanent local control of the water table can be achieved by drainage ditches, counterfort drains, or drainage adits, the water being discharged into a stream at a lower elevation.

The method adopted for dewatering excavations depends upon the permeability of the soil and its variation within the stratal sequence, the depth of base level below the water table, piezometric conditions in underlying horizons, the method of providing support to the sides of the excavations, and on safeguarding neighbouring structures. Surface water or shallow subsoil water can be dealt with by diversionary ditches or drains. However, the toes of ditch slopes in sandy soils may be suspect in that they develop quick conditions. In such instances a graded filter should be placed in the bottom of the ditch.

6.8.1 Wellpoints and bored wells

The simplest method of groundwater lowering is to pump it from a sump within the excavation. This can generally be done in excavations in rock and in gravels where the rate of inflow does not cause instability of the sides or base of the excavation. However, in silts and sands the rate of inflow is often effective enough to cause erosion and slumping of the sides, and 'boiling' at the base. In these conditions a wellpoint system (Figure 6.13) or bored wells, installed about the perimeter of the excavated face, is required. Such groundwater lowering techniques depend on excessive pumping which lowers the water table and thereby develops a cone of exhaustion. The radius of the cone of exhaustion at the withdrawal points depends upon the rates of pumping and recharge. The amount of discharge (Q) which is necessary to lower the water table through a given depth can be estimated by using the Dupuit equation, which for the gravity well condition is as follows

$$Q = \frac{\pi k (H^2 - h_0^2)}{\ln(R/r_0)}$$

where H is the elevation of the original water table above an impermeable

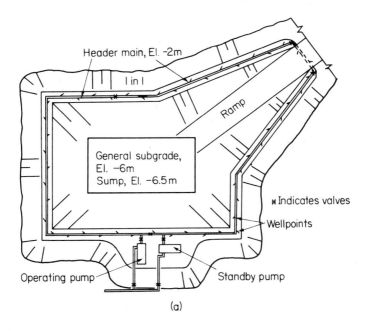

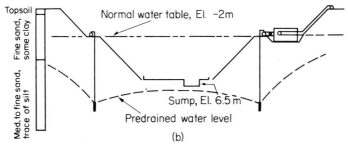

Figure 6.13 (a) Typical layout of a wellpoint system. (b) Section through layout

horizon, h_0 is the elevation of the operating level of the pumping well above this horizon, R is the radius of the area of influence, r_0 is the radius of the well and k is the hydraulic conductivity (Figure 6.14a).

The equation for a confined aquifer under artesian pressure is

$$Q = \frac{2\pi kb(H - h_0)}{\ln (R/r_0)}$$

where b is the thickness of the confined layer (Figure 6.14b). Because drawdown is a linear function whereas soil volume is a cubic function, increased pumping from one point of withdrawal soon becomes inefficient.

The installation of wellpoints is rapid; individual wellpoints can be placed in a matter of minutes in some soils, and the flexibility of the system allows for rearranging their spacing according to the rate of inflow (see Cashman (1975)). The radii of influence of the individual wellpoints overlap and they

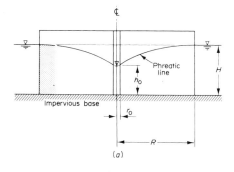

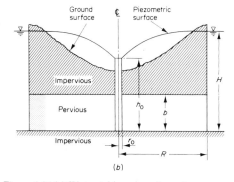

Figure 6.14 (a) Water table or gravity well condition. (b) Confined aquifer or artesian well condition.

are laid out so as to lower the water table by approximately a metre below the base level of the excavation.

A bored filter well consists of a perforated tube surrounded by an annulus of filter media and the operational depth may, in theory, be unlimited. The successful operation of deep bored filter wells is largely dependent on the grading of the filter. Bored wells are preferable to wellpointing for deep excavations where the area of the excavation is small in relation to the depth. They are also preferable in ground containing cobbles and boulders where wellpoint installation is difficult. Deep wells are particularly suited to variable soils and multi-layer aquifers, as well as to the control of groundwater under artesian or subartesian conditions. For example, they are the appropriate method of dewatering an aquifer at depth beneath an impermeable stratum, in order to prevent a blow-out in an excavation terminating within the impermeable material. A bored shallow well system can be used on a highly permeable site where pumping is required for several months, rather than wellpointing where risers at close centres could hinder construction operations.

Whenever the phreatic or piezometric surface is lowered, the effective load on the soil is increased, causing compression and consequent settlement. However, any resulting settlement due to the abstraction of water from clean sands is likely to be insignificant unless the sand was initially very loosely packed. On the other hand pumping from an aquifer containing layers of soft clay, peat or other compressible soils or from a confined aquifer overlain by compressible soils may cause significant settlements. The amount of settlement undergone depends on the thickness of the compressible layers and their compressibility, as well as on the amount of groundwater lowering. The permeability of the soil and the length of the pumping period influence the rate of settlement. As a result it may be necessary to limit the radius of influence by the use of groundwater recharge methods. An example of preventing settlement of adjacent buildings due to groundwater lowering has been given by Zeevaart (1957).

6.8.2 Electro-osmosis

Electro-osmosis has been used as a dewatering method to stabilise soils of low permeability such as silts and clays which cannot be easily dewatered by the above-mentioned techniques. Basically electro-osmosis consists of placing electrodes into the ground to a depth of about 2 m below excavation base level, and passing a direct electric current between them. The electric current induces a flow of water from the anodes to the cathodes, the latter acting as wellpoints from which the water can be removed by pumping. According to Casagrande (1947) the coefficient of electro-osmotic permeability is similar for sands, silts and clays and therefore for practical purposes most soils may be assumed to have a value of about 0.5×10^{-3} mm/s for a gradient of 0.1 volt/mm. However, the mass permeability of stiff fissured clays is frequently similar to their electro-osmotic permeability which means that electro-osmotic drainage methods are inappropriate since cheaper methods of stabilisation as, for example, sand drains, are more effective. Potential gradients in excess of 0.05 volt/mm between electrodes should not be entertained for long-term applications since they result in high energy losses in the form of ground heating.

According to Farmer (1975) electro-osmosis suffers a fundamental drawback when used solely for dewatering in that it is a decelerating process, in that as the soil dries out around the anode, so its electrical resistance increases. As a result electro-osmosis becomes progressively less efficient as the soil water content is reduced. This in fact means that the water content in soils is rarely reduced sufficiently to achieve complete stabilisation and even where it is, subsequent re-hydration may often rapidly reverse the process. It also means that electro-osmosis only operates efficiently in saturated or almost saturated soils. Electro-osmosis also consumes a high amount of energy, it is therefore by no means a common method of ground treatment.

In the cases where electro-osmosis has been successful, this has generally been due to the introduction of a chemical into the soil, either through anode solution or by direct electrolyte replacement. In other words the introduction of metal ions from the anode leads to ionic replacement in clay minerals or to partial cementation of pore spaces, thereby enhancing soil stability. The engineering properties of a clay are related to its mineralogy and more particularly to the type of cations associated with the clay minerals present. Montmorillonite provides the best example. For example, Na and Li montmorillonites can absorb large quantities of water and so have very low shearing strengths. On the other hand Ca, Mg, Fe or Al montmorillonites absorb appreciably less water and so have more stable structures. Because montmorillonite is prone to base exchange Na and Li may be replaced by Ca, Mg, Fe or Al, thereby improving the engineering performance of the soil. The technique of ionic replacement to improve soil properties is known as electrochemical stabilisation. It can increase the shear strength of clay soils by up to 60%.

6.8.3 Exclusion techniques

Where surrounding property has to be safeguarded it will usually be more appropriate to provide a barrier about the excavation in order to prevent inflow of water, whilst maintaining the surrounding water table at its normal level, rather than to adopt a dewatering technique. Methods of forming such a barrier, in a rough order of relative cost (cheapest first) include steel sheet piling, concrete diaphragm walls, contiguous bored pile walls, bentonite cut-off walls, cement or clay-cement grout curtains and frozen soil barriers (see Bell (in press)). The economy of providing a barrier to exclude ground-water depends on the existence of an impermeable stratum beneath the excavation to form an effective cut-off for the barrier. If this stratum does not exist or if it lies at too great a depth to be practicable to use as a cut-off, then upward seepage occurs which may give rise to instability at excavation level. In these circumstances the barrier will not be effective unless it can be extended horizontally beneath the excavation. The only methods of forming a horizontal barrier are by grouting or freezing.

6.9 Subsidence due to the abstraction of groundwater

The removal of water from sediments reduces the pore pressures and as a consequence the effective pressures are increased. Penman (1978) observed that for every 1 m the water table is lowered an increased vertical effective stress equivalent of 1 Mg/m^2 is imposed on the substrata. This in turn leads to consolidation, the degree of which depends on the compressibility of the material involved. The net result is surface subsidence. Peat is the most compressible of materials and is highly porous, indeed its water content may range up to 2000% (see Chapter 10). Accordingly drainage of peat leads invariably to subsidence, the Fenlands providing a classic example. There peat has been drained for over 400 years. In some parts of the Fens the thickness of peat has almost been halved as a result. For example, in the years between 1848 and 1932 a total subsidence of 2.7 m was recorded by the Holme Post, the original thickness of peat being 6.7 m (Figure 6.15).

Carillo (1948) revealed that subsidence in parts of Mexico City occurred at a rate of 1 mm/day. This was due to the abstraction of water from several sand aquifers located in very soft clay of volcanic origin. The aquifers extend under the city from an approximate depth of 50 m below ground surface to well below 500 m. Water has been abstracted for over 100 years. Carillo recorded that in 1944 the rate of abstraction was $7 \text{ m}^3/\text{s}$ and that the drop in the static head of the wells ranged from 0.4–2.05 m/year. This gave rise

Figure 6.15 The Holme Post, a cast-iron pillar erected in 1851 on the south west edge of Whittlesey Mere. It replaced the wooden posts which were erected in 1848 to indicate peat shrinkage caused by drainage. The post was driven 7 m through peat into clay until its top was flush with the ground. Within 10 years ground level had fallen 1.5 m through shrinkage. A second post was erected in 1957 with its top at the same level as that of the original post (right-hand side). Between 1850 and 1970 the ground surface subsided some 4 m

to accelerating subsidence in the central area of the city with serious consequences for both drainage and buildings. For instance, the sewer system, which formerly worked by gravity, now requires pumps. Buildings on end-bearing piles rose above the ground. Negative skin friction caused heavy overloading on the piles which led to some sudden differential settlements. By 1959 most of the old city had suffered on average at least 4 m of subsidence, and in the north-east part, as much as 7.5 m had been recorded (Figure 6.16). However, a prohibition order was imposed in 1953 whereby no more wells were sunk in the Valley of Mexico and subsequently there has been a slow decline in the rate of abstraction. The installation of piezometers in boreholes indicated that the piezometric level was some 30 m beneath the old water table, this corresponded to an increase in vertical effective stress of about 300 kPa.

Abstraction of water from the Chalk over the past 150 years has caused subsidence in excess of 0.3 m in some areas of London (see Wilson and Grace (1942)). In 1820 the artesian head in the Chalk was approximately +9.1 m AOD, but by 1936 this had declined in some places to −90 m AOD. The decline in artesian head has been accompanied by under-drainage in the London Clay. Between 1865 and 1931 subsidence averaged between 60 and 180 mm throughout much of London (Figure 6.17). Several other examples of subsidence, brought about by the abstraction of fluids, including oil, have been quoted by Poland and Davis (1963).

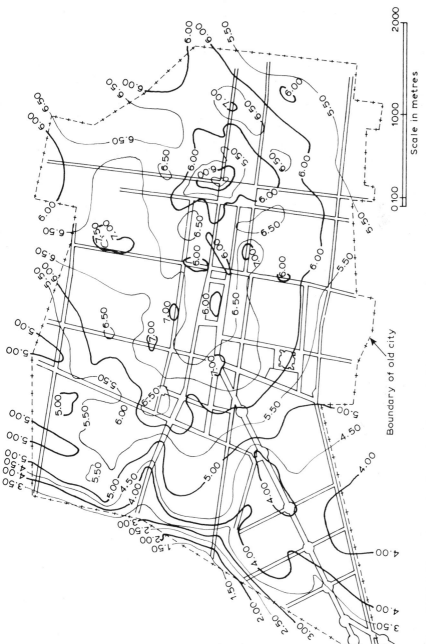

Figure 6.16 Subsidence, due to the abstraction of water, in Mexico City 1891–1959. Lines of equal subsidence are given in half-metre intervals (from Poland and Davis (1963))

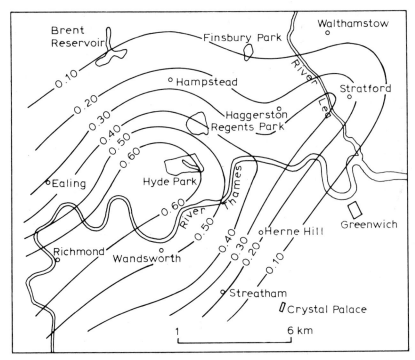

Figure 6.17 Lines of equal subsidence (1865–1931) due to abstraction of water from the Chalk beneath London, contours are in tenths of a foot (after Wilson and Grace (1942)) (Crown copyright, reproduced with the permission of the Controller of Her Majesty's Stationery Office)

6.10 Frost action in soil

Frost action in a soil is influenced by the initial temperature of the soil, as well as the air temperature, the intensity and duration of the freeze period, the depth of frost penetration, the depth of the water table, and the type of ground and exposure cover. If frost penetrates down to the capillary fringe in fine-grained soils, especially silts, then, under certain conditions, lenses of ice may be developed. The formation of such ice lenses may, in turn, cause frost heave and frost boil which may lead to the break-up of roads, the failure of slopes, etc. Shrinkage, which gives rise to polygonal cracking in the ground, presents another problem when soil is subjected to freezing. The formation of these cracks is attributable to thermal contraction and desiccation. Water which accumulates in the cracks is frozen and consequently helps increase their size. This water also may aid the development of lenses of ice.

6.10.1 Classification of frozen soil

According to Thomson (1980) ice may occur in frozen soil as small disseminated crystals whose total mass exceeds that of the mineral grains. It may also occur as large tabular masses which range up to several metres thick, or as ice wedges. The latter may be several metres wide and may extend to 10 m or so in depth. As a consequence frozen soils need to be described and classified for engineering purposes. A recent method of classifying frozen soils involves the identification of the soil type and the character of the ice (see Andersland and Anderson (1978)). First, the character of the actual soil is classified according to the Unified Soil Classification System (see Table 9.4a). Second, the soil characteristics consequent upon freezing are added to the description. Frozen soil characteristics are divided into two basic groups based on whether or not segregated ice can be seen with the naked eye (Table 6.4). Third, the ice present in the frozen soil is classified, this refers to inclusions of ice which exceed 25 mm in thickness (Table 6.4).

The amount of segregated ice in a frozen mass of soil depends largely upon the intensity and rate of freezing. When freezing takes place quickly no layers of ice are visible whereas slow freezing produces visible layers of ice of

Table 6.4. Description and classification of frozen soils (from Andersland and Anderson (1978))

I: Description of soil phase (independent of frozen state)	Classify soil phase by the unified soil classification system				
	Major group		*Subgroup*		
	Description	*Designation*	*Description*		*Designation*
	Segregated ice not visible by eye	N	Poorly bonded or friable		Nf
			Well bonded	No excess ice	Nb — n
II: Description of frozen soil				Excess ice	e
	Segregated ice visible by eye (ice 25 mm or less thick)	V	Individual ice crystals or inclusions		Vx
			Ice coatings on particles		Ve
			Random or irregularly oriented ice formations		Vr
			Stratified or distinctly oriented ice formations		Vs
III: Description of substantial ice strata	Ice greater than 25 mm thick	ICE	Ice with soil inclusions		ICE + soil type
			Ice without soil inclusions		ICE

various thicknesses. Ice segregation in soil also takes place under cyclic freezing and thawing conditions.

6.10.2 Mechanical properties of frozen soil

The presence of masses of ice in a soil means, that as far as engineering is concerned, the properties of both have to be taken into account. Ice has no long-term strength, that is, it flows under very small loads. If a constant load is applied to a specimen of ice, instantaneous elastic deformation occurs. This is followed by creep, which eventually develops a steady state. Instantaneous elastic recovery takes place on removal of the load, followed by recovery of the transient creep.

The mechanical properties of frozen soil are very much influenced by the grain size distribution, the mineral content, the density, the frozen and unfrozen water contents, and the presence of ice lenses and layering. The strength of frozen ground develops from cohesion, interparticle friction and particle interlocking, much the same as in unfrozen soils. However, cohesive forces include the adhesion between soil particles and ice in the voids, as well as the surface forces between particles. More particularly, the strength of frozen soils is sensitive to particle size distribution, particle orientation and packing, impurities (air bubbles, salts or organic matter) in the water-ice matrix, temperature, confining pressure and the rate of strain. Obviously the difference in the strength between frozen and unfrozen soils is derived from the ice component.

The density index influences the behaviour of frozen granular soils, especially their shearing resistance, in a manner similar to that when they are unfrozen (see Chapter 10). The cohesive effects of the ice matrix are superimposed on the latter behaviour and the initial deformation of frozen sand is dominated by the ice matrix. Sand in which all the water is more or less frozen exhibits a brittle type of failure at low strains, for example, at around 2% strain. However, the presence of unfrozen films of water around particles of soil, not only means that the ice content is reduced, but leads to a more plastic behaviour of the soil during deformation. For instance, frozen clay, as well as often containing a lower content of ice than sand, has layers of unfrozen water (of molecular proportions) around the clay particles. These molecular layers of water contribute towards a plastic type of failure.

Lenses of ice are frequently formed in fine-grained soils frozen under a directional temperature gradient. The lenses impart a laminated appearance to the soil. In such situations the strength of the bond between soil particles and ice matrix is greater than between particles and adjacent ice lenses. Under very rapid loading the ice behaves as a brittle material, with strengths in excess of those of fine-grained frozen soils. By contrast the ice matrix deforms continuously when subjected to long-term loading, with no limiting long-term strength. The laminated texture of the soil in rapid shear possesses the greatest strength when the shear zone runs along the contact between ice lens and frozen soil.

When loaded, stresses at the point of contact between soil particles and ice bring about pressure melting of the ice. Because of differences in the surface tension of the melt water, it tends to move into regions of lower stress, where it refreezes. The process of ice melting and the movement of unfrozen water are accompanied by a breakdown of the ice and the bonding with the grains of soil. This leads to plastic deformation of the ice in the voids and to a rearrangement of particle fabric. The net result is time-dependent deformation of the frozen soil, namely, creep (see Eckardt (1979); Takegawa *et al.* (1979)). Frozen soil undergoes appreciable deformation under sustained loading, the magnitude and rate of creep being governed by the composition of the soil, especially the amount of ice present, the temperature, the stress and the stress history.

The creep strength of frozen soils is defined as the stress level, after a given time, at which rupture, instability leading to rupture or extremely large deformations without rupture occur (see Andersland *et al.* (1978)). Frozen fine-grained soils can suffer extremely large deformations without rupturing at temperatures near to freezing point. Hence the strength of these soils must be defined in terms of the maximum deformation which a particular structure can tolerate. As far as laboratory testing is concerned, axial strains of 20%, under compressive loading, are frequently arbitrarily considered as amounting to failure. The creep strength is then defined as the level of stress producing this strain after a given interval of time.

When strain is plotted against time, three stages of creep are apparent under uniform load (Figure 6.18). At first, strain increases quickly, but then settles at a uniform minimal rate of increase in its second stage. A third, plastic stage is eventually reached during which complete loss of resistance occurs. This feature is well demonstrated in clay at temperatures near the freezing point of water. Sanger and Kaplar (1963) tested an organic silty clay, and showed that failure (in the tertiary stage of creep) occurred at 550 kPa after 17 h when the sample was maintained a 0°C, but no failure point had been reached after 60 h when tested at −2°C.

In fine grained sediments the intimate bond between the water and the clay particles results in a significant proportion of soil moisture remaining unfrozen at temperatures as cold as −25°C. The more clay material in the

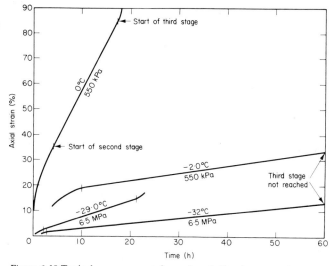

Figure 6.18 Typical creep curves of samples of silty clays at various applied stresses and temperatures (after Sanger and Kaplar (1963))

soil, the greater is the quantity of unfrozen moisture. Lovell (1957) measured the unconfined compressive strength of frozen clays and demonstrated that there was a dramatic increase in structural strength with decreasing temperature. In fact, it appears to increase exponentially with the relative proportion of moisture frozen. Using silty clay as an example, the amount of moisture frozen at $-18°C$ is only 1.25 times that frozen at $-5°C$, but the increase in compressive strength is more than four-fold.

By contrast, the water content of granular soils is almost wholly converted into ice at a very few degrees below freezing point. Hence frozen granular soils exhibit a reasonably high compressive strength only a few degrees below freezing, and there is justification for using this parameter as a design index of their performance in the field, provided that a suitable factor of safety is incorporated. The order of increase in compressive strength with decreasing temperature is shown in Figure 6.19.

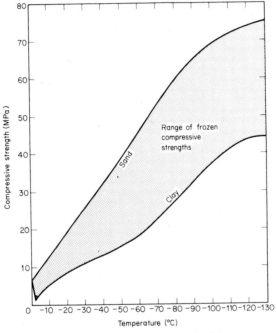

Figure 6.19 Increase in compressive strength with decreasing temperature

Uniaxial compression tests carried out by Parameswaran (1980) on cylindrical specimens of frozen Ottawa Sand containing about 20% by weight, of water, indicated that strength increases with increasing strain rates and decreasing temperatures. The initial tangent modulus (E_i) also increased with increasing strain rates and decreasing temperature. The considerably lower values of modulus and strength at $-2°C$, as compared with those at lower temperatures, were probably due to larger amounts of water remaining unfrozen at $-2°C$.

Because frozen ground is more or less impermeable this increases the problems due to thaw by impeding the removal of surface water. What is more, when the thaw occurs the amount of water liberated may greatly exceed that originally present in the melted out layer of the soil (see below). As the soil thaws downwards the upper layers become saturated, and since water cannot drain through the frozen soil beneath, they may suffer a complete loss of strength. Indeed under some circumstances excess water may act as a transporting agent thereby giving rise to soil flows.

Settlement is associated with thawing of frozen ground. As ice melts, settlement occurs, water being squeezed from the ground by overburden pressure or by any applied loads. Excess pore pressures develop when the rate of ice melt is greater than the discharge capacity of the soil. Since excess pore pressures can lead to the failure of slopes and foundations, both the rate and amount of thaw settlement should be determined. Pore pressures should also be monitored.

Further consolidation, due to drainage, may occur on thawing. If the soil was previously in a relatively dense state, then the amount of consolidation is small. This situation only occurs in coarse grained frozen soils containing very little segregated ice. On the other hand, some degree of segregation of ice is always present in fine-grained frozen soils. For example, lenses and veins of ice may be formed when silts have access to capillary water. Under such conditions the moisture content of the frozen silts significantly exceeds the moisture content present in their unfrozen state. As a result when such ice-rich soils thaw under drained conditions they undergo large settlements under their own weight. Methods of predicting the amount and rate of settlement due to thawing of frozen ground have been discussed by Nixon and Ladanyi (1978).

The thaw-settlement phenomena produced by melting of permafrost layers differ only in scale from those which occur due to melting ice in seasonally frozen ground. In both instances the amount of settlement is mainly governed by whether or not the frozen soil is ice-rich and/or contains layers of ice. Large depressions, known as thermokarst pits, may be developed at the surface when large wedges of ice in the soil melt. The hollows generally fill with water.

Drainage patterns may be affected when thaw occurs in permafrost. Melt water, because it cannot penetrate the frozen ground, tends to follow any porous paths thawed out in the permafrost. This action is at times responsible for serious soil erosion. Furthermore in permafrost regions ice may form at the base of shallow or slow-flowing streams, as well as at their surfaces. The development of ice at both the surface and base of a stream impedes or even stops its flow within its channel. Hence water is forced over the banks or into porous formations adjacent to the stream. This overflow water is frozen to form sheets, which gradually produce successive layers across the area surrounding the stream. By the end of the winter such layers of ice (aufeis), may extend up to 4 m in thickness and cover several square kilometres.

Chamberlain and Gow (1979) showed that significant structural changes can be brought about in the fabric of fine-grained soils when they are subjected to alternate freezing and thawing. Their experiments indicated that a reduction took place in the void ratio of the soil and that its vertical permeability was increased. The latter increase was most significant in soils with large plasticity indices, and usually the increase was smaller, the higher the level of applied stress. They attributed the increase in vertical permeability in clayey soils to the formation of polygonal shrinkage cracks, whereas in coarse grained soils they suggested that it was due to the reduction in the amount of fines in the pores of the coarse fraction.

6.10.3 Frost heave

Jumikis (1968) listed the following factors as necessary for the occurrence of frost heave, namely, capillary saturation at the beginning and during the freezing of the soil, a plentiful supply of subsoil water and a soil possessing fairly high capillarity together with moderate permeability. According to Kinosita (1979) the ground surface experiences an increasingly larger amount of heave, the higher the initial water table. Indeed it has been suggested that frost heave could be prevented by lowering the water table (see Andersland and Anderson (1978)).

Grain size is another important factor influencing frost heave. For example, gravels, sands and clays are not particularly susceptible to heave whilst silts definitely are. The reason for this is that silty soils are associated with high capillary rises but at the same time their voids are large enough to allow moisture to move quickly enough for them to become saturated rapidly. If ice lenses are present in clean gravels or sands then they simply represent small pockets of moisture which have been frozen. Indeed Taber (1930) gave an upper size limit of 0.007 mm, above which, he maintained, layers of ice do not develop. However, Casagrande (1932) suggested that the particle size critical to heave formation was 0.02 mm. If the quantity of such particles in a soil is less than 1%, no heave is to be expected, but considerable heaving may take place if this amount is over 3% in non-uniform soils and over 10% in very uniform soils.

Croney and Jacobs (1967) suggested that under the climatic conditions experienced in Britain well-drained cohesive soils with a plasticity index exceeding 15% could be looked upon as non-frost susceptible. Where the drainage is poor and the water table is within 0.6 m of formation level they suggested that the limiting value of plasticity index should be increased to 20%. In addition in experiments with sand they noted that as the amount of silt added was increased up to 55% or the clay fraction up to 33%, increase in permeability in the freezing front was the overriding factor and heave tended to increase. Beyond these values the decreasing permeability below the freezing zone became dominant and progressively reduced the heave. This indicates that the permeability below the frozen zone was principally responsible for controlling heave. These two authors also suggested that the permeability of soft chalk is sufficiently high to permit very serious frost heave but in the harder varieties the lower permeabilities minimise or prevent heaving.

Horiguchi (1979) demonstrated, from experimental evidence, that the rate of frost heave increased as the rate of heat removal from the freezing front was increased. However, removal of heat does not increase the rate of heave indefinitely—it reaches a maximum, after which it declines. The maximum rate of heave was shown to be influenced by particle size distribution in that it increased in soils with finer grain size. Horiguchi also found that when the particles were the same size in different test specimens, then the maximum rate of heave depended upon the types of exchangeable cations present in the soil. The rate of heave is also influenced by the thickness of overburden. For instance, Penner and Walton (1979) indicated that the maximum rate of ice accumulation at lower overburden pressures occurred at temperatures nearer to 0°C than at higher overburden pressures. However, it appears that the rate of heave for various overburden pressures tends to converge as the temperature below freezing is lowered. As the overburden pressure increases, the zone over which heaving takes place becomes greater and it extends over an increasingly larger range of temperature.

Maximum heaving, according to Jumikis (1956), does not necessarily occur at the time of maximum depth of penetration of the 0°C line, there being a lag between the minimum air temperature prevailing and the maximum penetration of the freeze front. In fact soil freezes at temperatures slightly lower than 0°C.

As heaves amounting to 30% of the thickness of the frozen layer have frequently been recorded, moisture, other than that initially present in the frozen layer, must be drawn from below, since water increases in volume by only 9% when frozen. In fact when a soil freezes there is an upward transfer of heat from the groundwater towards the area in which freezing is occurring. The thermal energy, in turn, initiates an upward migration of moisture within the soil. The moisture in the soil can be translocated upwards either in the vapour or liquid phase or by a combination of both. Moisture diffusion by the vapour phase occurs more readily in soils with larger void spaces than in fine grained soils. If a soil is saturated, migration in the vapour phase cannot take place.

In a very dense, closely packed soil where the moisture forms uninterrupted films throughout the soil mass, down to the water table, then, depending upon the texture of the soil, the film transport mechanism becomes more effective than the vapour mechanism. The upward movement of moisture due to the film mechanism, in a freezing soil mass, is slow. Nonetheless a considerable amount of moisture can move upwards as a result of this mechanism during the winter. What is more, in the film transport mechanism the water table is linked by the films of moisture to the ice lenses.

Before freezing, soil particles develop films of moisture about them due to capillary action. This moisture is drawn from the water table. As the ice lens grows, the suction pressure it develops exceeds that of the capillary attraction of moisture by the soil particles. Hence moisture moves from the soil to the ice lens. But the capillary force continues to draw moisture from the water table and so the process continues. Jones (1980) suggested that if heaving is unrestrained, the heave (H) can be estimated as follows

$$H = 1.09kit$$

where k is permeability, i is the suction gradient (this is difficult to derive) and t is time.

A recent investigation into moisture migration in frozen soils has been carried out by Mageau and Morgenstern (1980) who demonstrated that induced moisture migration does occur in frozen soils. They suggested that the rate of migration seems to be governed by the apparent permeability of the soil and the suction force within the frozen fringe. As ice develops in a soil, this reduces its permeability to a critical value at which an ice lens begins to form. It appears that the rate of water uptake by an ice lens is not influenced by the frozen soil above it. A dominant suction pressure develops in the frozen fringe between the warmest ice lens and the frozen–unfrozen interface. Moisture migrates via the interconnected zones of free water within the pores and through the unfrozen films in the frozen soil, under the influence of a temperature gradient. Mageau and Morgenstern concluded that the rate of heave is controlled principally by the frozen fringe of soil between the warmest ice lens and the frozen–unfrozen interface. The amount of unfrozen water in the soil governs the extent of migration. The amount of unfrozen soil is, in turn, controlled by soil type and temperature. For example, Mageau and Morgenstern found that moisture migration in clayey silt was reduced to an insignificant level at a soil temperature of about $-2.0°C$.

The frost heave test allows the prediction of frost heave (see Jacobs (1965)). A critical review of this test has been provided by Jones (1980), who claimed that it gives poor reproducibility of results. Furthermore, this type of test is unfortunately time consuming and so a rapid freeze test has been developed by Kaplar (1971). Approximate predictions of frost heave have also been based on grain size distribution. However, in a recent discussion of frost heaving, Reed et al. (1979) noted that such predictions failed to take account of the fact that soils can exist at different states of density and therefore porosity, yet they have the same grain size distribution. What is more, pore size distribution controls the migration of water in the soil and hence, to a large degree, the mechanism of frost heave. They accordingly derived expressions, based upon pore space, for predicting the amount of frost heave (Y) in mm/day.

$$Y = 581.1(X_{3.0}) - 5.46 - 29.46(X_{3.0})/(X_0 - X_{0.4})$$

where $X_{3.0}$ = cumulative porosity for pores >3.0 μm but <300 μm, X_0 = total cumulative porosity, $X_{0.4}$ = cumulative porosity for pores >0.4 μm but <300 μm.

A simpler expression based on pore diameters rather than cumulative porosity, but which was somewhat less accurate, was as follows

$$Y = 1.694(D_{40}/D_{80}) - 0.3805$$

where D_{40} and D_{80} are the pore diameters whereby 40% and 80% of the pores are larger respectively.

Where there is a likelihood of frost heave occurring it is necessary to esti- mate the depth of frost penetration (see Jumikis (1968)). Once this has been done, provision can be made for the installation of adequate insulation or drainage within the soil and to determine the amount by which the water table may need to be lowered so that it is not affected by frost penetration. The base of footings should be placed below the estimated depth of frost penetration as should water supply lines and other services. Frost-susceptible soils may be replaced by gravels. The addition of certain chemicals to soil can reduce its capacity for water absorption and so can influence frost sus- ceptibility. For example, Croney and Jacobs (1967) noted that the addition of calcium lignosulphate and sodium tripolyphosphate to silty soils were both effective in reducing frost heave. The freezing point of the soil may be lowered by mixing in solutions of calcium chloride or sodium chloride, in concentrations of 0.5–3.0% by weight of the soil mixture. The heave of non-cohesive soils containing appreciable quantities of fines can be reduced or prevented by the addition of cement or bituminous binders. Addition of cement both reduces the permeability of a soil mass and gives it sufficient tensile strength to prevent the formation of small ice lenses as the freezing isotherm passes through.

References

AKROYD, T. N. W. (1964). *Laboratory Testing in Soil Engineering*, Soil Mechanics Limited, Bracknell.

ANDERSLAND, O. B. & ANDERSON, D. M. (eds). (1978). *Geotechnical Engineering for Cold Regions*, McGraw-Hill, New York.

ANDERSLAND, O. B., SAYLES, F. H. & LADANYI, B. (1978). 'Mechanical properties of frozen ground', (in *Geotechnical Engineering for Cold Regions*), McGraw-Hill, New York. 216–75.

BELL, F. G. (1978). 'Some petrographic factors relating to porosity and permeability in the Fell Sandstone of Northumberland', *Q. J. Engg Geol.*, **11**, 113–26.

BELL, F. G. (In press). *Ground Treatment*, Butterworths, London.

BERNAIX, J. (1969). 'New laboratory methods of studying the mechanical properties of rocks', *Int. J. Rock Mech. Min. Sci.*, **6**, 43–90.

BRITISH STANDARDS INSTITUTION (1975). *Methods of Test for Soils for Civil Engineering Purposes*, BS 1377, British Standards Institution, London.

CARILLO, N. (1948). 'Influence of artesian wells on the sinking of Mexico City', *Proc. 2nd Int. Conf. Soil. Mech. Foundation Engg*, Rotterdam, **2**, 156–9.

CASAGRANDE, A. (1932). 'Discussion on frost heaving', *Proc. Highway Res. Board*, Bull., No. 12, 169, Washington DC.

CASAGRANDE, A. (1936). 'Characteristics of cohesionless soils affecting the stability of slopes and earth fills', *J. Boston Soc. Civil Engrs*, **23**, 3–32.

CASAGRANDE, L. (1947). 'The application of electo-osmosis to practical problems in foundations and earthworks', *Build. Res. Tech. Paper*, No. 30, DSIR, Watford.

CASHMAN, P. M. (1975). 'Control of groundwater by groundwater lowering', (In *Methods of Treatment of Unstable Ground*), Bell, F. G. (ed.), Butterworths, London, 12–25.

CHAMBERLAIN, E. J. & GOW, A. J. (1979). 'Effect of freezing and thawing on the permeability and structure of soils', *Engg Geol.*, *Spec. Issue on Ground Freezing*, **13**, 73–92.

CRONEY, D. & JACOBS, J. C. (1967). 'The frost susceptibility of soils and road materials', *Trans Road Res. Lab.*, Report LR90, Crowthorne.

DARCY, H. (1856). *Les Fontaines Publiques de la Ville de Dijon*, Dalmont, Paris.

DE WIEST, R. J. H. (1967). *Geohydrology*, Wiley, New York.

DIXON, J. C. & CLARKE, K. B. (1975). 'Field investigation techniques', (in *Site Investigations in Areas of Mining Subsidence*), Bell, F. G. (ed.), Butterworths, London, 40–74.

ECKARDT, H. (1979). 'Creep behaviour of frozen soils in uniaxial compression tests', *Engg Geol.*, *Spec. Issue on Ground Freezing*, **13**, 185–96.

FAIR, G. M. & HATCH, L. P. (1935). 'Fundamental factors governing the streamline flow of water through sand', *J. Am. Water Works Ass.*, **25**, 1151–65.

FARMER, I. W. (1975). 'Electro-osmosis and electrochemical stabilization', (in *Methods of Treatment of Unstable Ground*), Bell, F. G. (ed.), Butterworths, London, 385–426.

FRANKLIN, J. L. (1970). *Index Properties of Rocks. Part I—Suggested Methods for Determination of Water Content, Porosity, Density and Related Properties*, Imperial College, London.

FRAZER, H. J. (1935). 'Experimental study of porosity and permeability of clastic sediments', *J. Geol.*, **43**, 910–1010.

GOODMAN, R. E. & SUNDARAM, P. N. (1980). 'Permeability and piping in fractured rocks', *Proc. ASCE, J. Geol. Engg Div.*, **106**, GT5, Paper 15433, 485–98.

HORIGUCHI, K. (1979). 'Effect of rate of heat removal on rate of frost heaving', *Engg Geol.*, *Spec. Issue on Ground Freezing*, **13**, 63–72.

HORSWILL, P. & HORTON, A. (1976). 'Cambering and valley bulging in the Gwash Valley at Empingham, Rutland', *Phil. Trans R. Soc. Ser. A.*, **283**, 427–62.

HVORSLEV, M. J. (1951). 'Time lag and soil permeability in ground water observations', *Bull. No. 36*, Waterways Experimental Station, Vicksburg.

JACOBS, J. C. (1965). 'The Road Research Laboratory frost heave test', *Trans. Road Res. Lab.*, Lab. note LN/766/JCJ, Crowthorne.

JONES, R. H. (1980). 'Frost heave of roads', *Q. J. Engg Geol.*, **13**, 77–86.

JUMIKIS, A. R. (1956). 'The soil freezing experiment'. *Highway Res. Board, Bull. No. 135*, Factors Affecting Ground Freezing, Nat. Acad. Res. Coun., Pub. 425, Washington DC.

JUMIKIS, A. R. (1968). *Soil Mechanics*, Van Nostrand, Princeton.

KAPLAR, C. W. (1971). 'Experiments to simplify frost susceptibility testing of soils', *US Army Corps Engrs*, Cold Regions Res. and Engg Lab., Hanover NH, Tech. Report 223.

KINOSITA, S. (1979). 'Effects of initial soil-water conditions on frost heaving characteristics', *Engg Geol., Spec. Issue on Ground Freezing*, **13**, 53–62.

KOERNER, R. M., REIF, J. S. & BURLINGAME, M. J. (1979). 'Detection methods for location of subsurface water and seepage', *Proc. ASCE J. Geot. Engg Div.*, **105**, GT11, Paper 14989, 1301–16.

KRUMBEIN, W. C. & MONK, C. D. (1943). 'Permeability as a function of the size parameters of unconsolidated sand', *Trans. Am. Inst. Min. Engrs*, **151**, 153–63.

LOVELL, C. W. (1957). 'Temperature effects on phase composition and strength of a partially frozen soil', *Highway Res. Board*, Bull. No. 168, Washington DC.

LOVELOCK, P. E. R., PRICE, N. & TATE, T. K. (1975). 'Ground water conditions in the Penrith Sandstone at Cliburn, Westmorland', *J. Inst. Water Engrs*, **29**, 157–74.

MAGEAU, D. W. & MORGENSTERN, N. R. (1980). 'Observations on moisture migration in frozen soils', *Canadian Geotech. J.*, **17**, 54–60.

MATHER, J. D., GRAY, D. A., ALLEN, R. A. & SMITH, D. B. (1973). 'Ground water recharge in the Lower Greensand of the London Basin—results from tritium and carbon-14 determination', *Q. J. Engg Geol.*, **6**, 141–52.

MEINZER, O. (1942). 'Occurrence, origin and discharge of ground water', (in *Hydrology*), Meinzer, O. (ed.), Dover, New York, 385–443.

MOORE, J. F. A. & LONGWORTH, T. (1979). 'Hydraulic uplift at the base of a deep excavation in Oxford Clay', *Geotechnique*, **29**, 35–46.

NIXON, J. F. & LADANYI, B. (1978). 'Thaw consolidation', (in *Geotechnical Engineering for Cold Regions*), Andersland, D. B. & Anderson, D. M. (eds.) McGraw-Hill, New York, 164–215.

PARAMESWARAN, V. R. (1980). 'Deformation behaviour and strength of frozen sand', *Canadian Geotech. J.*, **17**, 74–88.

PENMAN, A. D. M. (1977). 'The failure of the Teton Dam', *Ground Engg*, **10**, No. 6, Dec. 18–27.

PENMAN, A. D. M. (1978). 'Ground water and foundations', (in *Foundation Engineering in Difficult Ground*), Bell, F. G. (ed.), Butterworths, London, 204–25.

PENNER, E. & WALTON, T. (1979). 'Effects of temperature and pressure on frost heaving', *Engg Geol., Spec. Issue on Ground Freezing*, **13**, 29–40.

POLAND, J. F. & DAVIS, G. H. (1963). 'Land subsidence due to the removal of fluids', *Reviews in Engg Geol.*, Geol. Soc. Am., 190–269.

RAMANA, Y. V. & VENKATANARAYANA, B. (1971). 'An air porosimeter for the porosity of rocks', *Int. J. Rock Mech. Min. Sci.*, **8**, 29–53.

REED, M. A., LOVELL, C. W., ALTSCHAEFFL, A. G. & WOOD, L. E. (1979). 'Frost heaving rate predicted from pore size distribution', *Canadian Geotech. J.*, **16**, 463–72.

ROWE, P. W. (1968). 'Failure of foundations and slopes on layered deposits in relation to site investigation practice', *Proc. Inst. Civil Engrs Supp. Vol.*, Paper No. 70575, 73–132.

SANGER, F. J. & KAPLAR, C. W. (1963). 'Plastic deformation of frozen soils', *Proc. Permafrost Int. Conf.*, Lafayette, Ind. NAS-NRC, Publication No. 1281, Washington DC, Nov., 305–15.

SERAFIM, J. L. (1968). 'Influence of interstitial water on rock masses', (In *Rock Mechanics in Engineering Practice*), Stagg, K. G. & Zienkiewicz, O. C. (eds), Wiley, London, 55–97.

SHERRELL, F. W., (1976). 'Engineering geology and ground water', *Ground Engineering*, **9**, No. 4, 21–7.

SKEAT, W. D. (ed.). (1969). *Manual of British Water Engineering Practice*, Vol. 2, Institution of Water Engineers, London.

TABER, S. (1930). 'Mechanics of frost heaving', *J. Geol.*, **38**, 303–17.

TAKEGAWA, K., NAKAZAWA, A., RYOKAI, K. & AKAGAWA, S. (1979). 'Creep characteristics of frozen soils', *Engg Geol., Spec. Issue on Ground Freezing*, **13**, 197–206.

TERZAGHI, K. (1925). *Erdbaumechanik auf Bodenphysikalischer Grundlage*, Deuticke, Vienna.

THOMSON, S. (1980). 'A brief review of foundation construction in the western Canadian Arctic', *Q.J. Engg Geol.*, **13**, 67–76.

VERRUIJT, A. (1970). *The Theory of Ground Water Flow*, Macmillan, London.

WADE, N. H. & TAYLOR, H. (1979). 'Control of artesian pressure by bleeder wells', *Canadian Geotech. J.*, **16**, 488–96.

WILSON, G. & GRACE, H. (1942). 'The settlement of London due to underdrainage of the London Clay', *J. Inst. Civil Engrs*, **19**, 100–27.

WITTKE, W. (1973). 'Percolation through fissured rock', *Bull. Int. Ass. Engg Geol.*, No. 7, 3–28.

ZEEVAART, L. (1957). 'Foundation design and behaviour of the Tower Latino Americano in Mexico City', *Geotechnique*, **7**, 115–33.

Chapter 7

Geomorphological processes I

All landmasses are continually being worn away or denuded by weathering
and erosion, the agents of erosion being the sea, rivers, wind and ice. The
detrital products resulting from denudation are transported by water, wind,
ice and the action of gravity and are ultimately deposited. In this manner
the surface features of the Earth are gradually, but constantly, changing.
As landscapes are continually developing it is possible to distinguish successive stages in their evolution. These stages have been termed youth, maturity
and senility. However, the form of landscape which arises during any one
of these stages is conditioned partly by the processes of denudation to which
the area is subjected, and partly by the structure of the rocks on which the
landforms are being developed. Thus, in the words of W. M. Davis (1909),
'landscape is a function of structure, process and stage'. Earth movements
and type of climate also play a significant role in landscape development.

WEATHERING

Weathering of rocks is brought about by physical disintegration, chemical
decomposition and biological activity. The agents of weathering, unlike those
of erosion, do not themselves provide for the transportation of debris from
a rock surface. Therefore unless this rock waste is otherwise removed it
eventually acts as a protective blanket, preventing further weathering taking
place. If weathering is to be continuous, fresh rock exposures must be constantly revealed, which means that the weathered debris must be removed
by the action of gravity, running water, wind or moving ice. During its
transportation weathered material acts as a tool for the agents of erosion,
enhancing their destructive power.

7.1 Rate of weathering

The type and rate of weathering varies from one climatic regime to another
(see Ollier (1969)). In humid regions chemical and chemico-biological pro-

cesses are generally much more significant than those of mechanical disintegration. The degree and rate of weathering in humid regions depends on the temperature, the amount of moisture and organic matter available, and the relief. If the temperature is high then weathering is extremely active and it has been calculated that an increase of 10°C more than doubles the rate at which chemical reactions occur. The higher the moisture content in the soil mantle, the more readily are silicates and aluminium silicates hydrolysed and substances removed in solution. When organic matter is dissolved by leaching waters carbon dioxide is liberated. Thus the greater the amount of organic matter contributed to the soil, the more active the weathering. In order to allow chemical weathering to proceed the surface layers of rock must remain in place or be removed at a rate which does not inhibit the change of the rock debris from an alkaline to an acid state and impede the removal of soluble material. Such a condition is only found on plains. As the relief increases so mechanical disintegration intensifies and ultimately a point is reached where the rate of slope wash is greater than that of chemical weathering.

The rate at which weathering proceeds depends not only upon the vigour of the weathering agent but also on the durability of the rock mass concerned. This, in turn, is governed by the mineralogical composition, texture and porosity of the rock on the one hand, and the incidence of discontinuities within the rock mass on the other.

The alteration products which develop when the primary rock-forming minerals are weathered are shown in Figure 7.1. The inherent stability of a mineral is influenced by the environment in which it was formed. For example, those minerals which crystallise from a magma at very high temperatures and pressures are relatively unstable when exposed to atmospheric

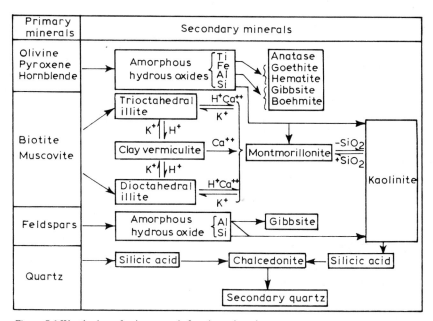

Figure 7.1 Weathering of primary rock-forming minerals

conditions. Goldich (1938) arranged the principal igneous rock-forming minerals in the following order of susceptibility to chemical attack

Most susceptible	Olivine
(High-temperature formation)	Calcic plagioclase
	Pyroxene
	Calci-sodic plagioclase
	Amphibole
	Soda plagioclase
	Biotite
(Low-temperature formation)	Muscovite
Least susceptible	Quartz

It can be seen from the above list that the least stable minerals occur in ultrabasic and basic igneous rocks such as dunites, peridotites, basalts and gabbros. Hence such rocks offer less resistance to weathering than acidic igneous rocks, which are commonly composed of potash feldspar, muscovite and quartz. Muscovite and quartz, in particular, can survive intense weathering and withstand more than one cycle of erosion. They are accordingly found in very mature muds and sands and their lithified equivalents. Most sedimentary rocks are the products of denudation, transportation and deposition and so they frequently contain high proportions of minerals which are stable under atmospheric conditions. Clay minerals are notable examples; they are the stable end products which form when most of the minerals listed above decompose.

Generally coarse-grained rocks weather more rapidly than do fine-grained types of similar mineralogical composition. The degree of interlocking

Figure 7.2 Spheroidal weathering developed in basalt lavas along the shore near Corrie, Isle of Arran

between minerals is a particularly important textural factor. Obviously the more strongly a rock is bonded together, the greater is its resistance to weathering. The closeness of the interlocking of grains governs the porosity of a rock. This in turn determines the amount of water a rock can hold. Not only are the more porous rocks more susceptible to chemical attack, but they are more prone to frost action than the less porous varieties.

Discontinuities represent planes of weakness along which weathering action is concentrated. Perhaps the most familiar features produced by weathering along discontinuities are the clints and grykes which characterise limestone pavements. The bedding plane spacing in sedimentary rocks affects the rate at which they weather, massive bedded types offering much more resistance than those which are laminated. Weathering which has taken place along the cooling joints in the Cornish granites has led to the development of 'tors'. Also as weathering reveals the flat-lying joints in these granites, it imparts to them a pseudo-stratification. The spheroidal weathering frequently seen in rotting basalts and dolerites, is also caused by weathering along joint planes (Figure 7.2). Intense weathering may be associated with fault zones and cases have been reported where it has extended to several hundred metres in depth. Cleavage, schistosity and foliation in metamorphic rocks also influence the rate of weathering.

7.2 Mechanical weathering

Mechanical or physical weathering is particularly effective in climatic regions which experience significant diurnal changes of temperature. This does not necessarily imply a large range of temperature, as frost and thaw action can proceed where the range is limited.

As far as frost susceptibility is concerned the porosity, pore size and degree of saturation all play an important role. When water turns to ice it increases in volume by up to 9% thus giving rise to an increase in pressure within the pores. This action is further enhanced by the displacement of pore water away from the developing ice front. Once ice has formed the ice pressures rapidly increase with decreasing temperature, so that at approximately $-22°C$ ice can exert a pressure of 200 MPa (see Winkler (1973)). Usually coarse grained rocks withstand freezing better than fine grained types. Indeed the critical pore size for freeze-thaw durability appears to be about 0.005 mm. In other words rocks with larger mean pore diameters allow outward drainage and escape of fluid from the frontal advance of the ice line and are therefore less frost susceptible. Fine grained rocks which have 5% sorbed water are often very susceptible to frost damage whilst those containing less than 1% are very durable. Alternate freeze-thaw action causes cracks, fissures, joints and some pore spaces to be widened. As the process advances, angular rock debris is gradually broken from the parent body.

The mechanical effects of weathering are well displayed in hot deserts, where wide diurnal ranges of temperature cause rocks to expand and contract. Because rocks are poor conductors of heat these effects are mainly localised in their outer layers where alternate expansion and contraction create stresses which eventually rupture the rock. In this way flakes of rock break away from the parent material, the process being termed exfoliation. The effects of exfoliation are concentrated at the corners and edges of rocks so that their outcrops gradually become rounded.

7.3 Chemical and biological weathering

Chemical weathering leads to mineral alteration and the solution of rocks. Alteration is principally effected by oxidation, hydration, hydrolysis and carbonation whilst solution is brought about by acidified or alkalised waters. Chemical weathering also aids rock disintegration by weakening the rock fabric and by emphasizing any structural weaknesses, however slight, that it possesses. When decomposition occurs within a rock the altered material frequently occupies a greater volume than that from which it was derived and in the process internal stresses are generated. If this swelling occurs in the outer layers of a rock then it causes them to peel off from the parent body.

In dry air rocks decay very slowly. The presence of moisture hastens the rate tremendously, first, because water is itself an effective agent of weathering and, second, because it holds in solution substances which react with the component minerals of the rock. The most important of these substances are free oxygen, carbon dioxide, organic acids and nitrogen acids.

Free oxygen is an important agent in the decay of all rocks which contain oxidisable substances, iron and sulphur being especially suspect. The rate of oxidation is quickened by the presence of water; indeed it may enter into the reaction itself, as for example, in the formation of hydrates. However, its role is chiefly that of a catalyst. Carbonic acid is produced when carbon dioxide is dissolved in water and it may possess a pH value of about 5.7. The principal source of carbon dioxide is not the atmosphere but the air contained in the pore spaces in the soil where its proportion may be a hundred or so times greater than it is in the atmosphere. An abnormal concentration of carbon dioxide is released when organic material decays. Furthermore humic acids are formed by the decay of humus in soil waters; they ordinarily have pH values between 4.5 and 5.0 but occasionally they may be under 4.0. The nitrogen acids, HNO_3 and HNO_2, are formed by organic decay or bacterial action in soils. They play only a minor part in weathering. In volcanic regions and in the oxidised zones of sulphide deposits, the sulphur acids, H_2SO_3 and H_2SO_4, become important and in some localities their pH value may be lowered below 1.0.

The simplest reactions which take place on weathering are the solution of soluble minerals and the addition of water to substances to form hydrates. Solution commonly involves ionisation, for example, this takes place when salt and gypsum deposits and carbonate rocks are weathered (see Chapter 13). Hydration and dehydration take place amongst some substances, a common example being gypsum and anhydrite

$$CaSO_4 + 2H_2O = CaSO_4 \cdot 2H_2O$$
(anhydrite)   (gypsum)

The above reaction produces an increase in volume of approximately 6% and accordingly causes the enclosing rocks to be wedged further apart. These reactions are slow but those involving ferric oxides and hydrates are even slower. Iron oxides and hydrates are conspicuous products of weathering, usually the oxides are a shade of red and the hydrates yellow to dark brown.

Sulphur compounds are readily oxidised by weathering. Because of the hydrolysis of the dissolved metal ion, solutions forming from the oxidation of sulphides are acidic. For instance, when pyrite is initially oxidised, ferrous sulphate and sulphuric acid are formed. Further oxidation leads to the formation of ferric sulphate. Very insoluble ferric oxide or hydrated oxide is formed if highly acidic conditions are produced

$$FeS_2 + H_2O + 7(O) = FeSO_4 + H_2SO_4 \tag{1}$$
(pyrite)

$$2FeSO_4 + (O) + H_2SO_4 = Fe_2(SO)_4 + H_2O \tag{2}$$
$$2FeS_2 + 15(O) + 4H_2O = F_2O_3 + 4H_2SO_4 \tag{3}$$

Perhaps the most familiar example of a rock prone to chemical attack is limestone. Limestones are chiefly composed of calcium carbonate and they are suspect to acid attack because CO_3 readily combines with H to form the stable bicarbonate HCO_3.

$$CaCO_3 + H_2CO_3 = Ca(HCO_3)_2$$

In water with a temperature of 25°C the solubility of calcium carbonate ranges from 0.01–0.05 g/l, depending upon the degree of saturation with carbon dioxide. Dolostone is somewhat less soluble than limestone. When a limestone is subject to dissolution any insoluble material present in it remains behind. An account of the weathering of the Carboniferous Limestone of northern England has been given by Sweeting (1964).

Weathering of the silicate minerals is primarily a process of hydrolysis. Much of the silica which is released by weathering forms silicic acid but where it is liberated in large quantities some of it may form colloidal or amorphous silica. As noted above, mafic silicates usually decay more rapidly than felsic silicates and in the process they release magnesium, iron and lesser amounts of calcium and alkalies. Olivine is particularly unstable, decomposing to form serpentine, which on further weathering forms talc and carbonates. Chlorite is the commonest alteration product of augite (the principal pyroxene) and of hornblende (the principal amphibole).

When subjected to chemical weathering feldspars decompose to form clay minerals, the latter are consequently the most abundant residual products. The process is effected by the hydrolysing action of weakly carbonated waters which leach the bases out of the feldspars and produce clays in colloidal form. The alkalies are removed in solution as carbonates from orthoclase (K_2CO_3) and albite (Na_2CO_3), and as bicarbonate from anorthite ($Ca(HCO_3)_2$). Some silica is hydrolysed to form silicic acid. Although the exact mechanism of the process is not fully understood the equation given below is an approximation towards the truth

$$2KAlSi_3O_6 + 6H_2O + CO_2 = Al_2Si_2O_5(OH)_4 + 4H_2SiO_4 + K_2CO_3$$
(orthoclase)   (kaolinite)

The colloidal clay eventually crystallises as an aggregate of minute clay minerals. Deposits of kaolin are formed when percolating acidified waters decompose the feldspars contained in granitic rocks.

Clays are hydrated aluminium silicates and when they are subjected to severe chemical weathering in humid tropical regimes they break down to form laterite or bauxite. The process involves the removal of siliceous material and this is again brought about by the action of carbonated waters. Intensive leaching of soluble mineral matter from surface rocks takes place during the wet season. During the subsequent dry season groundwater is drawn to the surface by capillary action and minerals are precipitated there as the water evaporates. The minerals generally consist of hydrated peroxides of iron, and sometimes of aluminium, and very occasionally of manganese. The precipitation of these insoluble hydroxides gives rise to an impermeable lateritic soil (see Chapter 10). When this point is reached the formation of laterite ceases as no further leaching can occur. As a consequence lateritic deposits are usually less than 7 m thick.

High-grade aluminium laterites are known as bauxites. Bauxite, like laterite, is found in sheet-like deposits near the surface. It is formed by the decomposition of clays or igneous rocks rich in aluminium silicates

$$Al_2Si_2O_5(OH)_4 + nH_2O = Al_2O_3.nH_2O + 2H_2SiO_3$$
$$\text{(kaolinite)} \qquad\qquad \text{(bauxite)}$$

Often the upper part of a deposit is hardened by iron oxides and bauxite may have a pisolitic structure.

Plants and animals play an important role in the breakdown and decay of rocks, indeed their part in soil formation is of major significance. Tree roots penetrate cracks in rocks and gradually wedge the sides apart whilst the adventitious root system of grasses breaks down small rock fragments to particles of soil size. Burrowing rodents also bring about mechanical disintegration of rocks. The action of bacteria and fungi is largely responsible for the decay of dead organic matter. Other bacteria are responsible, for example, for the reduction of iron or sulphur compounds. It has also been suggested that bacterial action plays an important part in the formation of residual deposits such as laterites and bauxites.

7.4 Breakdown of the more susceptible rocks: shales and mudstones

Many shales and mudstones are particularly susceptible to weathering. After a study of the disintegration of shales in water, Badger *et al.* (1956) concluded that this was brought about by two main processes, namely, air breakage and the dispersion of colloid material. It was noted that the former process only occurred in those shales which were mechanically weak whilst the latter appeared to be a general cause of disintegration. They also observed that the degree of disintegration of a shale when it was immersed in different liquids was governed by the manner in which those liquids affected air breakage and ionic dispersion forces. For example, in a liquid with a low dielectric constant little disintegration took place as a result of ionic forces because of the suppression of ionic dissociation from the shale colloids. It was found that the variation in disintegration of different shales in water was not usually connected with their total amount of clay colloid or the variation in the types of clay minerals present. It was rather controlled by the type of exchangeable cations attached to the clay particles and the accessibility of the latter to attack by water which, in turn, depended on the porosity of the shale. Air breakage could assist this process by presenting new surfaces of shale to water. It was also suggested that like coal, shale may have a rank and that low rank shales disintegrated more easily.

Nakano (1967) found that although some mudstones from Japan, when immersed in water swelled slowly and underwent a consequent decrease in bulk density and strength, they did not disintegrate even after immersion for a lengthy period of time. However, if they were dried and then wetted they disintegrated rapidly into small pieces. After conducting a series of slaking tests in vacuum, as well as in air, Nakano concluded that air breakage was not a significant mechanism in the breakdown of mudstones since he noticed no difference in the results of the slaking tests. He attributed the weakening of mudstone to chemical dissolution, by hydrogen bonding of originally adsorbed water molecules around clay particles with newly adsorbed ones. He assumed that part of the free energy, which was evolved by water molecules when adsorbed around clay particles, acted as the destructive force in slaking. It was observed that mudstones started to slake when drying in a relative humidity of 98%. In another instance, the relative humidity was 94%. In each case the drier the mudstone was, compared with its natural state, the greater was the intensity of disintegration in water. This means that such mudstones (the clay fraction in these mudstones consisted of montmorillonite) may deteriorate readily in the zone of fluctuating water table or water vapour pressure. This does not mean that beneath these zones mudstones will not be affected, they can be softened, especially if heavily fractured.

After an exhaustive investigation of the breakdown of Coal Measures rocks, Taylor and Spears (1970) concluded that the disintegration of sandstones and siltstones was governed by their fracture pattern and that after a few months of weathering the resulting debris was greater than cobble size. After that, the degradation to component grains took place at a very slow rate. However, shales and seat earths are rapidly broken down to a gravel sized aggregate. The polygonal fracture pattern characteristic of many mudstones and the laminations of shales, together with joints, contribute towards their degradation within a matter of months. Listric surfaces (small shear surfaces) in seat earths may mean that they disintegrate within a few wetting and drying cycles. In fact some rocks like the Brooch and Park seat earths, after desiccation, disintegrated very rapidly in water (the former was literally 'explosive' and the latter broke down in less than 30 min). Although the expandable clay content in these two rocks is high, and leads to intra-particle swelling, the authors maintained that this alone was not responsible for their rapid disintegration. In contrast to the work done by Nakano (1967), Taylor and Spears found that breakdown could be arrested by the removal of air from the samples under vacuum. Thus they concluded that air breakage was a principal disintegration mechanism in the weaker rocks. They suggested that during dry periods evaporation from the surfaces of rock fragments gives rise to high suction pressures which result in increased shearing resistance. With extreme desiccation most of the voids are filled with air, which, on immersion in water, become pressurised by the capillary pressures developed in the outer pore spaces. The mineral fabric may then fail along its weakest plane exposing an increased surface area to the same process. Taylor and Spears did not dismiss the physico-chemical ideas of Badger *et al.* (1956) or those of Nakano (1967), but suggested that such processes became progressively more important with time. They supported the concept of rank in shales.

7.5 Assessment of the degree of weathering and its classification

7.5.1 Tests of weatherability

A number of tests have been designed to assess certain aspects of weatherability, an example being the freeze-thaw test. Freeze-thaw action can quickly break down weaker rocks such as shales, marls and soft chalk.

The slake-durability test estimates the resistance to wetting and drying of a rock sample, particularly mudstones and rocks which exhibit a certain degree of alteration. In this test the sample, which consists of ten pieces of rock, each weighing about 40 g, is placed in a test drum, oven dried and then weighed. After this, the drum, with sample, is half immersed in a tank of water and attached to a rotor arm which rotates the drum for a period of 10 min at 20 rev/min (Figure 7.3a). The cylindrical periphery of the drum is formed of 2 mm sieve mesh so that broken-down material can be lost whilst the test is in progress. After slaking the drum and the material retained are dried and weighed. The slake-durability index is then obtained by dividing the weight of the sample retained by its original weight and expressing the answer as a percentage. Franklin and Chandra (1972) found a general qualitative correlation between slake-durability indices, rate of weathering and the stable slope angles of the quarries and pits from which they collected their samples. They provided a grading system for the slake-durability test (Figure 7.3b).

Swelling of rocks is associated with weathering; shales, mudstones and marls being particularly prone. Small amounts of swelling have been recorded in some sandstones. The clay mineral content plays a significant role in these rocks. For example, kaolinite is not expansive whilst montmorillonite is, Na montmorillonite being able to expand to many times its original volume. The swelling is principally due to the ingress of water, so that the rock must be porous or fractured. When rocks which are prone to swelling imbibe water they may exert considerable pressures against a confining structure, drainage is therefore important. However, where possible, it is better to prevent the ingress of water. On some sites in swelling rocks, because of the drastic changes which may take place in their character upon wetting, resort to covering the offending strata may be necessary. If a rock has an intact unconfined compressive strength exceeding 40 MPa, it is not subject to swelling.

Failure of consolidated and poorly cemented rocks occurs during saturation when the swelling pressure (or internal saturation swelling stress, σ_s),

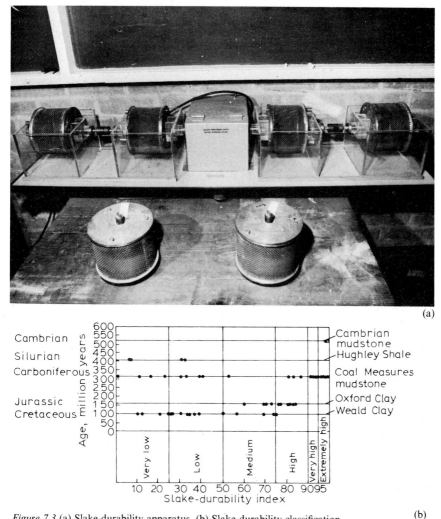

(a)

(b)

Figure 7.3 (a) Slake-durability apparatus. (b) Slake-durability classification and the variation in durability of rocks of differing age (after Franklin and Chandra (1972))

developed by capillary suction pressures exceeds their tensile strength. An estimate of σ_s can be obtained from the modulus of deformation (E):

$$E = \sigma_s/\varepsilon_D$$

where ε_D is the free-swelling coefficient. The latter is determined by a sensitive dial gauge recording the amount of swelling of an oven-dried core specimen per unit height, along the vertical axis during saturation in water for 12 h (see Duncan *et al.* (1968)), ε_D being obtained as follows

$$\varepsilon_D = \frac{\text{Change in length after swelling}}{\text{Initial length}}$$

7.5.2 Engineering classifications of weathering

Olivier (1976) developed a rock durability classification based on the internal saturation pressure and uniaxial compressive strength. However, as a means of routine field assessment, which could be rapidly derived, it suffered a major limitation, namely, the length of time required to obtain the modulus of deformation. Olivier (1979) therefore proposed the geodurability classification which was simply based on the free-swelling coefficient and uniaxial compressive strength (Figure 7.4). This classification was developed primarily to assess the durability of mudrocks and poorly cemented sandstones during tunnelling operations, since the tendency of such rocks to deteriorate after exposure governs the stand-up time of tunnels. Olivier suggested that the classification could be used in decisions relating to primary tunnel support, particularly where and when shotcrete should be applied.

Several attempts have been made to devise an engineering classification of weathered rock. The problem can be tackled in two ways. One method

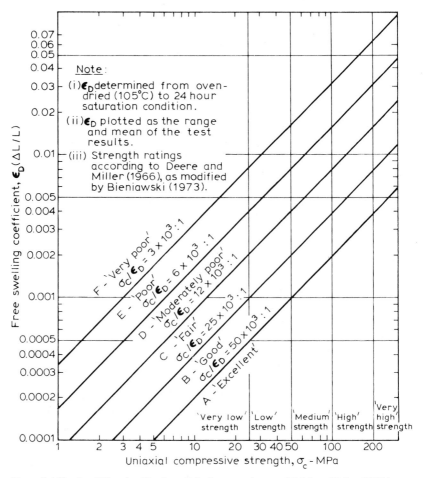

Figure 7.4 Geodurability classification of the intact rock material (after Olivier (1979))

is to attempt to assess the grade of weathering by reference to some simple index test. Such methods provide a quantitative, rather than a qualitative, answer. When coupled with a grading system, this means that the disadvantages inherent in these simple index tests are largely overcome.

Hamrol (1961) devised a quantitative classification of weatherability in which he first distinguished two weathering types. Type I weathering excluded cracking of any kind, whilst Type II weathering consisted entirely of cracking. This represents a division between chemical and physical weathering respectively, but such distinction can be extremely difficult to make. In Type I the void ratio increases as weathering progresses which means that the saturation moisture content increases and the dry density decreases. These two parameters therefore were used as the basis of an index test, the numerical value (i_1) of which was expressed as the weight of water absorbed by an oven-dried rock when it is saturated for a limited period, divided by its dry weight and expressed as a percentage. This is referred to as a quick absorption test. When considering Type II weathering, Hamrol distinguished between unfilled and filled cracks, the index being

$$i_{II} = (x + y + z) \times 100$$

where x, y and z were the dimensions of the crack along three orthogonal axes. Further indices could be obtained by relating the change in the degree of weathering (j) to a given time (Δt), hence

$$\hat{j}_1 = \Delta i_1 / \Delta t \qquad \text{and} \qquad j_{II} = \Delta i_{II} / \Delta t$$

Unfortunately Hamrol gave no scale to his indices so that their meaning in terms of engineering performance is lacking (he did mention that with an $i_1 = 10$, a weathered granite would crumble in the fingers, but little else).

Onodera et al. (1974) also used the number and width of microcracks (cracks less than 1 mm in width which occur in the rock fabric) as an index of the physical weathering of granite. They found a linear relationship between effective porosity (n_e) and density of microcracks (determined with the aid of a petrological microscope), defined as

$$n_e = 100 \times (\text{Total width of cracks/Length of measured line})$$

They also found that the mechanical strength of granite decreased rapidly as the density of microcracks increased from about 1.5 to 4%.

Lumb (1962) defined a quantitative index, Xd, related to the weight ratio of quartz and feldspar in decomposed granite from Hong Kong, as follows

$$Xd = (N_q - N_{qo})/(1 - N_{qo})$$

where N_q is the weight ratio of quartz and feldspar in the soil sample, and N_{qo} is the weight ratio of quartz and feldspar in the original rock. For fresh rock $Xd = 0$, whilst for completely decomposed rock $Xd = 1$.

Iliev (1967) developed a coefficient of weathering (K) for granitic rock. This coefficient was based upon the ultrasonic velocities of the rock material according to the expression

$$K = (V_u - V_w)/V_u$$

where V_u and V_w are the ultrasonic velocities of the fresh and weathered rock respectively. A quantitative index indicating the grade of weathering as determined from the ultrasonic velocity and the corresponding coefficient of weathering is as follows (cf. Table 7.3)

Grade of weathering	Ultrasonic velocity (m/s)	Coefficient of weathering
Fresh	Over 5000	0
Slightly weathered	4000–5000	0–0.2
Moderately weathered	3000–4000	0.2–0.4
Strongly weathered	2000–3000	0.4–0.6
Very strongly weathered	Under 2000	0.6–1.0

After an extensive testing programme Irfan and Dearman (1978a) concluded that the quick absorption, Schmidt hammer, and point load strength tests prove reliable field tests for the determination of a quantitative weathering index for granite (Table 7.1). This index can be related to the various grades of weathering recognised by visual determination and given in Table 7.3.

Table 7.1. Weathering indices for granite (after Irfan and Dearman (1978a))

Type of weathering	Quick absorption ($I\%$)	Bulk density (t/m^3)	Point load strength (MPa)	Unconfined compressive strength (MPa)
Fresh	Less than 0.2	2.61	Over 10	Over 250
* Partially stained	0.2–1.0	2.56–2.61	6–10	150–250
* Completely stained	1.0–2.0	2.51–2.56	4–6	100–150
Moderately weathered	2.0–10.0	2.05–2.51	0.1–4	2.5–100
Highly/completely weathered	Over 10	Less than 2.05	Less than 0.1	Less than 2.5

* Slightly weathered

As mineral composition and texture influence the physical properties of a rock, petrographic techniques can be used to evaluate successive stages in mineralogical and textural changes brought about by weathering. Accordingly Irfan and Dearman (1978b) developed a quantitative method of assessing the grade of weathering of granite in terms of its megascopic and microscopic petrography. The megascopic factors included an evaluation of the amount of discoloration, decomposition and disintegration shown by the rock. The microscopic analysis involved assessment of mineral composition and degree of alteration by modal analysis and a microfracture analysis. The latter involved counting the number of clean and stained microcracks and voids under the microscope in a 10 mm traverse across a thin section. The types of microfracture recognised included stained grain boundaries, open grain boundaries, stained microcracks in quartz and feldspar, infilled microcracks in quartz and feldspar, clean transgranular microcracks crossing the grains, filled or partially infilled microcracks, and pores in plagioclase. The data are used to derive the micropetrographic index, I_p, as follows

$$I_p = \frac{\% \text{ sound or primary minerals}}{\% \text{ unsound constituents}}$$

The unsound constituents are the secondary minerals together with micro-

Table 7.2. Stages of weathering of rock material in terms of microscopical properties (after Irfan and Dearman (1978b))

Stage 1:   No penetration of brown iron-staining. Microcracks are very short, fine, intragranular and structural. The centres of plagioclases are clouded and slightly sericitised. Altered minerals <6%; microcrack intensity <0.5%; micropetrographic index >12.

Stage 2:   Three substages are recognised depending on the amount of discoloration and type and amount of microfracturing.

    (i)   The rock is iron-stained only along the joint faces. No penetration of iron-staining.

    (ii)   Penetration of iron-staining (brown) inwards from the joint faces along the microcracks. Formation of simple, branched microcracks; tight and partially stained. Slight alteration of the centres of plagioclases. Occasional staining along quartz-quartz and quartz-feldspar grain boundaries. Grain boundaries are sharp.

    (iii)   More inward penetration of brown iron-staining along microcracks and partial staining of plagioclases. Microfracturing of feldspars and quartz by mainly intragranular, but some transgranular microcracks.

    Unstained core: Altered minerals 6–9%; microcrack intensity 0.5–1.0%; micropetrographic index 9–12.

    Stained rims: Altered minerals 9–12%; microcrack intensity 1.0–2.0%; micropetrographic index 6–9.

Stage 3:   Complete discoloration of rock by deep brown iron-staining. Partial alteration of plagioclases to sericite and gibbsite (?). Formation of single pores in plagioclases due to leaching. Potash feldspars are unaltered. Slight loss of pleochroism and bleaching of biotite. Grain boundaries are tight but stained brown by iron-oxide. The rock fabric is highly microfractured by complex branched, transgranular microcracks. Altered minerals 12–15%; microcrack intensity 2.0–5.0%; micropetrographic index 4–6.

Stage 4:   Nearly complete alteration of plagioclase to sericite and gibbsite and formation of nearly opaque areas in plagioclases. Very slight alteration of potash feldspar. Interconnected pores are formed in plagioclase feldspars due to removal and leaching of alteration products. Some solution of silica forming diffused quartz grain boundaries. Intense microfracturing of the rock fabric by a complex branched and dendritic pattern of microcracks. The whole of the rock is iron-stained. Altered minerals 15–20%; microcrack intensity 5.0–10.0%; micropetrographic index 2–4.

Stage 5:   Complete alteration of plagioclases. Potash feldspar is partially altered, but highly microfractured. Biotite is partially and muscovite slightly altered; expansion of biotite. Quartz is reduced in grain size and amount by microfracturing and solution. Almost all the grain boundaries are open. The fabric is intensely microfractured by a dendritic pattern of micro- and macrocracks. Parallel sided, partially filled or clean macrocracks are formed. Highly bleached, highly porous. Rock texture is intact. Altered minerals 20%; microcrack intensity 10%; micropetrographic index 2.

cracks and voids (cf. Mendes *et al.* (1966); Chapter 11). Irfan and Dearman were able to identify five stages and three substages of weathering in granite (Table 7.2).

7.5.3 Descriptive classifications

A second method of assessing the grade of weathering is based on a simple description of the geological character of the rock concerned as seen in the field, the description embodying different grades of weathering which are related to engineering performance (Table 7.3). This approach was first developed by Moye (1955), who proposed a grading system for the degree of weathering found in granite at the Snowy Mountains scheme in Australia.

Table 7.3. Engineering grade classification of weathered rock

Grade	Degree of decomposition	Field recognition (after Little (1969); Fookes et al. (1972); Dearman (1974))				Engineering properties	
		Soils (and soft rocks)	Rocks (mainly chemical decomposition)	Rocks (physical disintegration)*	Carbonate rocks (solution)	After Little (1969)	After Hobbs (1975)
VI	Soil	The original soil is completely changed to one of new structure and composition in harmony with existing ground surface conditions	The rock is discoloured and is completely changed to a soil in which the original fabric of the rock is completely destroyed. There is a large volume change	The rock is changed to a soil by granular disintegration and/or grain fracture. The structure of the rock is destroyed and the soil is a residuum of minerals unaltered from the original rock		Unsuitable for important foundations. Unstable on slopes when vegetation cover is destroyed and may erode easily unless hard cap is present. Requires selection before use as fill	In completely weathered rock and residual soil it may be possible to obtain fair quality samples depending upon the parent rock type and the consistency of the product. Generally the samples will tend to be less disturbed than when taken in the same rock in the highly weathered state. The bearing capacity and settlement characteristics of rock in these extreme states can be assessed using the usual methods for testing soils
V	Completely weathered	The soil is discoloured and altered with no trace of original structures	The rock is discoloured and is wholly decomposed and friable, but the original fabric is mainly preserved. The properties of the rock mass depend in part on the nature of the parent rock. In granitic rocks feldspars are completely kaolinised	The rock is changed to a soil by granular disintegration and/or grain fracture. The structure of the rock is preserved	Grades V and VI cannot occur. These grades can be applied to interbedded soluble and insoluble rocks. Void size should be recorded	Cannot be recovered as cores by ordinary rotary drilling methods. Can be excavated by hand or ripping without the use of explosives. Unsuitable for foundations of concrete dams or large structures. May be suitable for foundation of earth dams and for fill. Unstable in high cuttings at steep angles. New joint patterns may have formed. Requires erosion protection	

Table 7.3—cont.

IV	Highly weathered†	The soil is mainly altered with occasional small lithorelicts of original soil. Little or no traces of original structures	The rock is discoloured; discontinuities may be open and have discoloured surfaces (eg. stained by limonite) and the original fabric of the rock near the discontinuities is altered; alteration penetrates deeply inwards, but corestones are still present. The rock mass is partially friable. Less than 50% rock	More than 50% and less than 100% of the rock is disintegrated by open discontinuities or spheroidal scaling spaced at 60 mm or less and/or by granular disintegration. The structure of the rock is preserved	More than 50% of the rock has been removed by solution. A small residuum may be present in the voids	Similar to Grade V. Sometimes recovered as core by careful rotary drilling. Unlikely to be suitable for foundations of concrete dams. Erratic presence of boulders makes it an unreliable foundation for large structures	In highly weathered rock difficulties will generally be encountered in obtaining undisturbed samples for testing. If samples are obtained the strength and modulus will generally be underestimated, frequently by large margins, even with apparently undisturbed samples. In such rocks in situ tests with either the Menard pressure meter or the plate should be carried out to determine the bearing capacity and settlement characteristics. The greatest difficulties in assessing bearing capacity and settlement are likely to be encountered in highly weathered rocks, in which the rock fabric becomes increasingly disintegrated or increasingly more plastic

Table 7.3—cont.

		Field recognition (after Little (1969); Fookes et al. (1972); Dearman (1974)			Engineering properties	
Grade	Degree of decomposition†	Soils (and soft rocks)	Rocks (mainly chemical decomposition)	Rocks (physical disintegration)*	Carbonate rocks (solution)	After Little (1969)	After Hobbs (1975
III	Moderately weathered†	The soil is composed of large discoloured lithorelicts of original soil separated by altered material. Alteration penetrates inwards from the surfaces of discontinuities	The rock is discoloured; discontinuities may be open and have greater discoloration with the alteration penetrating inwards; the intact rock is noticeably weaker, as determined in the field than the fresh rock. The rock mass is not friable. 50–90% rock	Up to 50% of the rock is disintegrated by open discontinuities or by spheroidal scaling spaced at 60 mm or less and/or by granular disintegration. The structure of the rock is preserved	Up to 50% of the rock has been removed by solution. A small residuum may be present in the voids. The structure of the rock is preserved	Possessing some strength—large pieces (e.g. NX drill core) cannot be broken by hand. Excavated with difficulty without the use of explosives. Mostly crushes under bulldozer tracks. Suitable for foundations of small concrete structures and rock fill dams. May be suitable for semipervious fill. Stability in cuttings depends on structural features especially joint attitudes	In moderately weathered rock the intact modulus and strength can be very much lower than in the fresh rock and thus the j-value will be higher than in the fresh state, unless the joints and fractures have been opened by erosion or softened by the accumulation of weathering products. The intact modulus and strength can be measured in the laboratory and the bearing capacity assessed, in the same way as for fresh rock. Triaxial tests may be more appropriate than uniaxial tests, and it would be advisable to adopt conservative values for the factor of safety

Table 7.3—cont.

II	Slightly weathered	The material is composed of angular blocks of fresh soil, which may or may not be discoloured. Some altered material starting to penetrate inwards from discontinuities separating blocks	100% rock: discontinuities open and spaced at more than 60 mm	The rock may be slightly discoloured, particularly adjacent to discontinuities which may be open and have slightly discoloured surfaces; the intact rock is not noticeably weaker than the fresh rock. Some decomposed feldspar in granites. Over 90% rock	100% rock; discontinuity surfaces open. Very slight solution etching of discontinuity surfaces may be present	Requires explosives for excavation. Suitable for concrete dam foundations. Highly permeable through open joints. Often more permeable than the zones above or below. Questionable as concrete aggregate	In faintly and slightly weathered rock it is possible that the j-value, owing to the reduction in stiffness of the joints as a result of penetrative weathering alone, will show a fairly sharp decrease compared with that of the same rock in the fresh state. The intact modulus, by definition, is unaffected by penetrative weathering. The safe bearing capacity is not therefore affected by faint weathering, and may be only slightly affected by slight weathering
I	Fresh rock	The parent soil shows no discoloration, loss of strength or other effects due to weathering	100% rock; discontinuities closed	The parent rock shows no discoloration, loss of strength or other effects due to weathering	100% rock; discontinuities closed	Staining indicates water percolation along joints; individual pieces may be loosened by blasting or stress relief and support may be required in tunnels and shafts	

* Discontinuity spacing should be recorded.

† The ratio of the original soil or rock to altered material should be estimated where possible.

Similar classifications were advanced by Kiersch and Treacher (1955) and by Knill and Jones (1965). These classifications were mainly based on the degree of chemical decomposition exhibited by a rock mass and were primarily directed towards weathering in granitic rocks. However, local conditions and particular rock types have a marked influence on the development of weathering profiles and therefore influence the criteria used to select weathering zones. Subsequently Dearman (1974) suggested descriptions which could be used to establish the grade of mechanical weathering, and that of solution weathering of relatively pure carbonate rock (Table 7.3). Others, working on different rock types, have proposed modified classifications of weathering grade. For example, Lovegrove and Fookes (1972) made slight variations in their identification of grades of weathering of volcanic tuffs and associated sediments in Fiji. Classifications of weathered chalk and weathered marl have been developed by Ward *et al.* (1968) and Chandler (1969) respectively (see Chapter 13).

Nevertheless Dearman (1974) maintained that an ideal profile of weathering, which was irrespective of rock type, would be worthwhile in that each grade would provide an indication of the general engineering properties of the material concerned (Figure 7.5). Usually the grades will lie one above the other in a weathered profile developed from a single rock type, the highest grade being at the surface. But this is not necessarily the case in complex geological conditions (see Knill and Jones (1965)). Even so the concept of grade of weathering can still be applied. Such a classification can be used to produce maps showing the distribution of the grade of weathering at particular engineering sites.

MOVEMENT OF SLOPES

7.6 Soil creep and valley bulging

Movements of slopes can range in magnitude from soil creep on the one hand to instantaneous and colossal landslides on the other. Sharpe (1938) defined creep as the slow downslope movement of superficial rock or soil

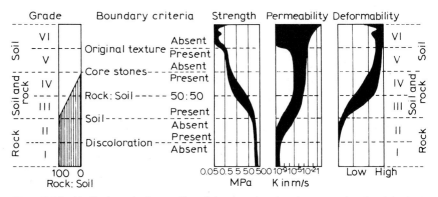

Figure 7.5 An idealised weathering profile and the general engineering properties of each horizon (after Dearman (1974))

debris, which is usually imperceptible except by observations of long duration. Creep is a more or less continuous process, which is a distinctly surface phenomenon and occurs on slopes with gradients somewhat in excess of the angle of repose of the material involved. Like landslip its principal cause is gravity although it may be influenced by seasonal changes in temperature and by swelling and shrinkage in surface rocks. Other factors which contribute towards creep include interstitial rain washing, ice crystals heaving stones and particles during frost, and the wedging action of rootlets. The liberation of stored strain energy in the weathered zone, particularly of over-consolidated clays with strong diagenetic bonds, is another contributory cause of creep (see Bjerrum (1967)). Although creep movement is exceedingly slow there are occasions on record when it has carried structures with it.

Evidence of soil creep may be found on almost every soil-covered slope. For example, it occurs in the form of small terracettes, downslope tilting of poles, the curving downslope of trees and soil accumulation on the uphill sides of walls. Indeed walls may be displaced or broken and sometimes roads may be moved out of alignment. The rate of movement depends not only on climatic conditions and the angle of slope, but also on the soil type and parent material.

Talus (scree) creep occurs wherever a steep talus exists. Its movement is quickest and slowest in cold and arid regions respectively.

Solifluction is a form of creep which occurs in cold climates or high altitudes where masses of saturated rock waste move downslope. Generally, the bulk of the moving mass consists of fine debris but blocks of appreciable size may also be moved. Saturation may be due to either water from rain or melting snow. Moreover in periglacial regions water cannot drain into the ground since it is frozen permanently (see Chapter 8). Solifluction differs from mudflow in that it moves much more slowly, the movement is continuous and it occurs over the whole slope. Solifluction processes vary at different altitudes owing to the progressive comminution of materials in their downward migration and to differences in growth of vegetation. At higher elevations surfaces are irregular and terraces are common, but at lower elevations the most common solifluction phenomena are continuous aprons of detritus which skirt the bases of all the more prominent relief features.

Valley bulges consist of anticlinal folds formed by mass movement of argillaceous material in valley bottoms, the argillaceous material being overlain by thick competent strata. The amplitude of the fold can reach 30 m in those instances where a single anticline occurs along the line of the valley. Alternatively the valley floor may be bordered by a pair of reverse faults or a belt of small-scale folding.

Valley bulging was noted in the Shipton valley in Northamptonshire by Hollingworth et al. (1944). There the Lias Clay exposed in the valley bottom bulges up to form anticlines with dips towards the valley sides. The clays are overlain by ferruginous sandstones and limestones. The authors explained these features as a stress relief phenomenon, that is, as stream erosion proceeded in the valley the excess loading on the sides caused the clay to squeeze out towards the area of minimum loading. This caused the rocks in the valley to bulge upwards.

However, other factors may also be involved in the development of valley bulging such as high piezometric pressures, swelling clays or shales and rebound adjustments of the stress field due to valley loading and excavation by ice. Notable examples of valley bulges occur in some of the valleys carved in Millstone Grit country in the Pennine area (see Chapter 13).

The valleyward movement of argillaceous material results in cambering of the overlying competent strata, blocks of which become detached, and move down the hillside. Fracturing of cambered strata produces deep debris-filled cracks or 'gulls' which run parallel to the trend of the valley. Some gulls may be several metres wide. Small gulls are sometimes found in relatively flat areas away from the slopes with which they are associated.

7.7 Landslides

7.7.1 Factors causing landslides

Landsliding is one of the most effective and widespread mechanisms by which landscape is developed. It is of great interest to the engineer since an understanding of the causes of landsliding should help provide answers relating to the control of slopes, either natural or man-made. An engineer faced with a landslide is primarily interested in curing the harmful effects of the slide. In many instances the principal cause cannot be removed so that it may be more economical to alleviate the effects continually. Indeed in most landslides a number of causes contribute towards movement and any attempt to decide which one finally produced the failure is not only difficult but pointless. Often the final factor is nothing more than a trigger mechanism that set in motion a mass which was already on the verge of failure.

Varnes (1958) defined landslides as the downward and outward movement of slope-forming materials composed of rocks, soils, or artificial fills. Creep, solifluction and avalanching were excluded from his definition. Movement may take place by falling, sliding or flowing, or some combination of these factors. Obviously this movement involves the development of a slip surface between the separating and remaining masses. The majority of stresses found in most slopes are the sum of the gravitational stress from the self-weight of the material plus the residual stress.

Landslides occur because the forces creating movement, the disturbing forces (M_D), exceed those resisting it, the resisting forces (M_R) that is, the shear strength of the material concerned. In general terms the stability of a slope may be defined by a factor of safety (F) where

$$F = M_R/M_D$$

If the factor of safety exceeds one, then the slope is stable, whereas if it is less than one, the slope is unstable.

The common force tending to generate movements on slopes is, of course, gravity. Over and above this a number of causes of landslides can be recognised. These were grouped into two categories by Terzaghi (1950), namely, internal causes and external causes (Table 7.4). The former included those mechanisms within the mass which brought about a reduction of its shear strength to a point below the external forces imposed on the mass by its environment, thus inducing failure. External mechanisms were those outside the mass involved, which were responsible for overcoming its internal shear strength, thereby causing it to fail.

An increase in the weight of slope material means that shearing stresses are increased, leading to a decrease in the stability of a slope, which may ultimately give rise to a slide. This can be brought about by natural or artificial (man-made) activity. For instance, removal of support from the toe of a slope, either by erosion or excavation, is a frequent cause of slides, as is overloading the top of a slope. Such slides are external slides in that an external factor causes failure.

Submerged and most partly submerged slopes are comparatively stable as the water pressure acting on the surface of the slope reduces the shearing stresses. If, however, the water level falls (as it does due to rapid drainage of a reservoir or due to tidal effects) the stabilising influence of the water disappears. If the slopes consist of cohesive soils, the water table will not be lowered at the same rate as the body of water. This means that the slope is temporarily overloaded with excess pore water, which may lead to failure. Thus rapid drawdown may be critical for slope stability (see Morgenstern (1963)).

Other external mechanisms include earthquakes or other shocks and vibrations. Earthquake shocks and vibrations in granular soils not only increase the external stresses on slope material but they can cause a reduction in the pore space which effectively increases pore pressures.

Internal slides are generally caused by an increase of pore pressures within the slope material, which causes a reduction in the effective shear strength. Indeed it is generally agreed that in most landslides groundwater constitutes the most important single contributory cause. An increase in water content also means an increase in the weight of the slope material or its bulk density, which can induce slope failure. Significant volume changes may occur in some materials, notably clays, on wetting and drying out. Not only does this weaken the clay by developing desiccation cracks within it, but the enclosing strata may also be adversely affected. Seepage forces within a granular soil can produce a reduction in strength by reducing the number of contacts between grains. Water can also weaken slope material by causing minerals to alter or by bringing about their solution. However, according to Terzaghi (1950) in humid climates, at least, there is always enough moisture in the ground to act as a lubricant for movement. He therefore dismissed the idea that slides which developed after rainfall were principally due to the lubricating effect of excess water. In such instances slides occur due to the consequent increase in pore pressures.

Weathering can effect a reduction in strength of slope material, leading to sliding. The necessary breakdown of equilibrium to initiate sliding may take decades. For example, Chandler (1974) quoted a case of slope failure in Lias Clay in Northamptonshire, which was primarily due to swelling in the clay. This took 43 years to reduce the strength of the clay below the critical level at which sliding occurred. Indeed, in relatively impermeable cohesive soils the swelling process is probably the most important factor leading to a loss of strength and therefore to delayed failure.

Table 7.4. Processes leading to landslides (after Terzaghi (1950))

Name of agent	Event or process which brings agent into action	Mode of action of agent	Slope materials most sensitive to action	Physical nature of significant actions of agent	Effects on equilibrium conditions of slope
Transporting agent	Construction operations or erosion	(1) Increase of height or rise of slope	Every material	Changes state of stress in slope-forming material	Increases shearing stresses
			Stiff, fissured clay, shale	Changes state of stress and causes opening of joints	Increases shearing stresses and initiates process 8
Tectonic stresses	Tectonic movements	(2) Large-scale deformations of Earth crust	Every material	Increases slope angle	Increase of shearing stresses
Tectonic stresses or explosives	Earthquakes or blasting	(3) High-frequency vibrations	Every material	Produces transitory change of stress	
			Loess, slightly cemented sands, and gravel	Damages intergranular bonds	Decrease of cohesion and increase of shearing stresses
			Medium or fine loose sand in saturated state	Initiates rearrangement of grains	Spontaneous liquefaction
Weight of slope-forming material	Process which created the slope	(4) Creep on slope	Stiff, fissured clay, shale remnants of old slides	Opens up closed joints, produces new ones	Reduces cohesion, accelerates process 8
		(5) Creep in weak stratum below foot of slope	Rigid materials resting on plastic ones		
Water	Rains or melting snow	(6) Displacement of air in voids	Moist sand	Increases pore water pressure	
		(7) Displacement of air in open joints	Jointed rock, shale		Decrease of frictional resistance

Table 7.4—cont.

Water—*Cont.*

Event	Physical or chemical process	Material	Mode of action	Effect
Rains or melting snow	(8) Reduction of capillary pressure associated with swelling	Stiff, fissured clay and some shales	Causes swelling	Decrease of cohesion
	(9) Chemical weathering	Rock of any kind	Weakens intergranular bonds (chemical weathering)	Decrease of cohesion
Frost	(10) Expansion of water due to freezing	Jointed rock	Widens existing joints, produces new ones	
	(11) Formation and subsequent melting of ice layers	Silt and silty sand	Increases water content of soil in frozen top-layer	Decrease of frictional resistance
Dry spell	(12) Shrinkage	Clay	Produces shrinkage cracks	Decrease of cohesion
Rapid drawdown	(13) Produces seepage toward foot of slope	Fine sand, silt, previously drained	Produces excess pore water pressure	Decrease of frictional resistance
Rapid change of elevation of water table	(14) Initiates rearrangement of grains	Medium or fine loose sand in saturated state	Spontaneous increase of pore water pressure	Spontaneous liquefaction
Rise of water table in distant aquifer	(15) Causes a rise of piezometric surface in slope-forming material	Silt or sand layers between or below clay layers	Increases pore water pressure	Decrease of frictional resistance
Seepage from artificial source of water (reservoir or canal)	(16) Seepage toward slope	Saturated silt	Increases pore water pressure	Decrease of frictional resistance
	(17) Displaces air in the voids	Moist, fine sand	Eliminates surface tension	Decrease of frictional resistance
	(18) Removes soluble binder	Loess	Destroys intergranular bond	Decrease of cohesion
	(19) Subsurface erosion	Fine sand or silt	Undermines the slope	Increase of shearing stress

7.7.2. Landslides in soils

Displacement in soil, usually along a well-defined plane of failure, occurs when shear stress rises to the value of shear strength. The shear strength of the material along the slip surface is reduced to its residual value so that subsequent movement can take place at a lower level of stress. The residual strength of a soil is of fundamental importance as far as the behaviour of landslides is concerned (see Skempton (1964); Skempton and Early (1972); and in the problem of progressive failure (see Bishop (1971)).

A slope in dry frictional soil should be stable provided its inclination is less than the angle of repose. Slope failure tends to be caused by the influence of water. For instance, seepage of groundwater through a deposit of sand in which slopes exist can cause them to fail. Failure on a slope composed of granular soil involves the translational movement of a shallow surface layer. The slip is often appreciably longer than it is in depth. This is because the strength of granular soils increases rapidly with depth. If, as is generally the case, there is a reduction in the density of the granular soil along the slip surface, then the peak strength is reduced ultimately to the residual strength. The soil will continue shearing without further change in volume once it has reached its residual strength. Although shallow slips are common, deep-seated shear slides can occur in granular soils. They are usually either due to rapid drawdown or to the placement of heavy loads at the top of the slope.

In cohesive soils slope and height are interdependent and can be determined when the shear characteristics of the material are known. Because of their water-retaining capacity, due to their low permeability, pore pressures are developed in cohesive soils. These pore pressures reduce the strength of the soil (see Chapter 6). Thus, in order to derive the strength of an element of the failure surface within a slope in cohesive soil, the pore pressure at that point needs to be determined to obtain the total and effective stress. This effective stress is then used as the normal stress in a shear box or triaxial test to assess the shear strength of the clay concerned. Skempton (1964) showed that on a stable slope in clay the resistance offered along a slip surface, that is, its shear strength (s), is given by

$$s = \bar{c}' + (\sigma - u)\tan\bar{\phi}'$$

where $\bar{c}'$ = cohesion intercept, $\bar{\phi}'$ = angle of shearing resistance (these are average values around the slip surface and are expressed in terms of effective stress), σ = total overburden pressure, u = pore water pressure. In a stable slope only part of the total available shear resistance along a potential slip surface will be mobilised to balance the total shear force (τ), hence

$$\sum\tau = \sum\bar{c}'/F + \sum(\sigma - u)\tan\bar{\phi}'/F$$

If the total shear force equals the total shear strength a slip will occur (that is, $F = 1.0$).

Cohesive soils, especially in the short-term conditions, may exhibit relatively uniform strength with increasing depth. As a result slope failures, particularly short-term failures, may be comparatively deep-seated, with roughly circular slip surfaces. This type of failure is typical of relatively small slopes. Landslides on larger slopes are often non-circular failure surfaces following bedding planes or other weak horizons.

Shearing in cohesive soils, if relatively slow, gives rise to the reorientation of platy minerals, as well as to an increase in volume, along the shear surface. These two factors account for the reduction in strength in dense or over-consolidated deposits of clay, along the shear surface, from the peak to residual value. Indeed, after a survey of natural slopes in the London Clay, Skempton and DeLory (1957) concluded that there is a general tendency for slope inclination to decrease with increasing orders of magnitude of time and the angles of ultimate stability could be correlated with residual strength.

Chandler (1977) showed that where drainage is impeded in cohesive soils, the threshold slope for landsliding approximates to half the angle of residual shearing resistance (ϕ_r'). However, he went on to point out that as the normal stresses increase, the value of ϕ_r' frequently falls. This means that where there is a potential for deep-seated landsliding on clay slopes, the range of normal stress will be large and there will not be a unique value of ϕ_r'. This means that larger landslides will move on flatter slopes than small slides, other factors being equal.

The long-term stability of a clay slope in overconsolidated sensitive clay was considered by Silvestri (1980). He attributed the almost non-existent cohesion intercept at failure to the fissured nature of the clay, and to the spreading of discontinuities caused by stress relief following exposure and subsequent desiccation.

Bjerrum (1967) maintained that failures in overconsolidated plastic clays and clay shales are preceded by the development of a surface on which sliding is continuous. This mechanism of progressive failure gradually reduces the strength of the soil from the peak to the residual value. He went on to state that for this sliding surface to develop, then it is necessary that

(1) the internal lateral stresses are high enough to cause stress concentrations in front of an advancing sliding surface, where the shear stresses exceed peak shear strength,
(2) the clay should contain enough recoverable strain energy to produce the required amount of expansion in the direction of sliding to strain the clay in the zone of failure,
(3) the residual shear strength should be relatively low compared with the peak shear strength.

Bjerrum (1967) noted that the magnitude of lateral stresses in overconsolidated clays is largely dependent on strain energy. In this context he recognised two types of clay, namely, those in which the strain energy is recovered simultaneously with a change of stress and those in which it is not. In the latter type of clay, strain energy is not immediately available because diagenetic bonds were formed when the deposit was carrying maximum overburden. These bonds 'welded' the contacts between particles and prevented bent particles from straightening when the load was reduced by subsequent erosion.

Weathering, however, does lead to the breakdown of these bonds, when the clay is exposed at the surface. The strain energy is thereby gradually liberated. Consequently the behaviour of overconsolidated clays depends on whether the diagenetic bonds are weak or strong. In the former case most of the strain energy is recovered on unloading. Swelling in the clay is more or less unrestricted and the ratio of horizontal to vertical effective stress increases during rebound. Weathering has little effect on the upper layer of such clay. By contrast swelling is restricted in overconsolidated clays with strong diagenetic bonds because recoverable strain energy is not released on unloading. The horizontal effective stress is relatively small. Weathering, however, does give rise to substantial swelling and to an increase in effective stress parallel to the surface, with expansion occurring in the same direction.

Hence the development of progressive failure in overconsolidated clays depends very much upon the time at which the stored recoverable strain energy is liberated. Those deposits with strong diagenetic bonds that are subjected to weathering are the most dangerous since this leads to a large amount of stored strain energy being released. This in turn gives rise to high lateral stresses and, as remarked, to a notable expansion parallel to the surface. The behaviour of overconsolidated clay with weak bonds is more or less the same in the unweathered or weathered condition and can prove suspect. If overconsolidated clay is unweathered, it is least dangerous.

Bjerrum (1967) illustrated this order of susceptibility by an analysis of some 60 slides in overconsolidated clays. He found that 55% had occurred in the weathered material of those clays with strong diagenetic bonds, 35% in clays with weak bonds and the remainder in unweathered clays with strong diagenetic bonds. He also suggested that progressive failure becomes more likely as the plasticity index increases.

Notable landsliding has periodically occurred at Folkestone Warren on the Kent coast and is aided by the continual removal of material by marine erosion. The great slip of 1915 was due to renewal of movement on pre-existing shear surfaces in the Gault Clay, which underlies the Chalk. Movement commenced at the western end of the Warren and spread eastwards, triggering the various falls in the Chalk scarp. Landslip in the Chalk is influenced by the dominant joint trends. A recent investigation of the Gault Clay has shown that it varies in character. Of particular importance as far as landsliding is concerned is the fact that at average normal effective stress levels in the Warren, ϕ'_r for low and high liquid limit clay are about 12 and 7° respectively (see Hutchinson et al. (1980)).

7.7.3 Landslides in rock masses

The factors which determine the degree of stability of steep slopes in hard unweathered rock (defined as rock with an unconfined compressive strength of 35 MPa and over) have been examined by Terzaghi (1962). More recent accounts have been provided by Hoek and Bray (1977) and by Richards et al. (1978). Terzaghi contended that landsliding in such rock is largely dependent on the incidence, orientation and nature of the discontinuities present. As in soil, the shearing resistance of rock with a random pattern of jointing can be obtained from the Coulomb equation. The value of the angle of shearing resistance (ϕ) depends on the type and degree of interlock between the blocks on either side of the surface of sliding, but in such rock masses interlocking is independent of the orientation of the surface of sliding. Hence ϕ has the same value for every surface.

From the results of various workers Terzaghi concluded that the critical slope angle for slopes underlain by hard massive rocks with a random joint pattern is about 70°, provided the walls of the joints are not acted on by seepage pressures.

In a bedded and jointed rock mass, if the bedding planes are inclined, the critical slope angle depends upon their orientation in relation to the slope and the orientation of the joints. The relation between the angle of shearing resistance (ϕ) along a discontinuity, at which sliding would occur under gravity, and the inclination of the discontinuity (α) is important. If $\alpha < \phi$ the slope is stable at any angle, whilst if $\phi < \alpha$ then gravity would induce movement along the discontinuity surface and the slope would not exceed the critical angle, which would have a maximum value equal to the inclination of the discontinuities. It must be borne in mind, however, that rock masses are generally interrupted by more than one set of discontinuities (see Chapter 5).

Terzaghi (1962) observed that little or nothing was known about the mechanics of deep-seated, large-scale rock slides, such as those which have occurred in narrow sections of deep valleys in the Austrian Alps. For example, it is not known whether these slides took place slowly or rapidly, and it is doubtful whether they were preceded by important creep deformation of rocks located within the shear zone. The rock in the immediate proximity of the surface of sliding is completely broken or crushed, and it would appear that the rock above the surface of sliding has also been broken, at least to a moderate extent. Furthermore existing joints in the rock material concerned have opened and new joints have been formed. Hence the compressibility and secondary permeability of the rocks has increased.

7.7.4 Classification of landslides

Varnes (1958) classified landslides according to the type of movement undergone on the one hand and the type of materials involved on the other (Figure 7.6). Types of movement were grouped into falls, slides and flows; one can, of course, merge into another. The materials concerned were simply grouped as rocks and soils.

Table 7.5. Classification of mass movements on slopes (after Hutchinson (1968))

Water content	Stress level in slope	
	Greater than soil strength	Less than soil strength
Infinite	Stream flow	
Very high	Mudflow	
High	Mudslide ⎫ shear	Continuous creep
Low	Landslide ⎬ surfaces	Climatic creep
Very low	Rockslide ⎭ formed	(a) Fluctuating groundwater
	Rockfall	(b) Frost creep

A classification of mass movement on slopes has been proposed by Hutchinson (1968); (Table 7.5). It is based, first, on the water content of the material involved and, second, on whether the movement occurs with shear stresses greater or less than the strength of the material. Shearing or sliding occurs in the former case and the resultant movement is a form of landsliding or flowage; in the latter case the movement is some form of creep.

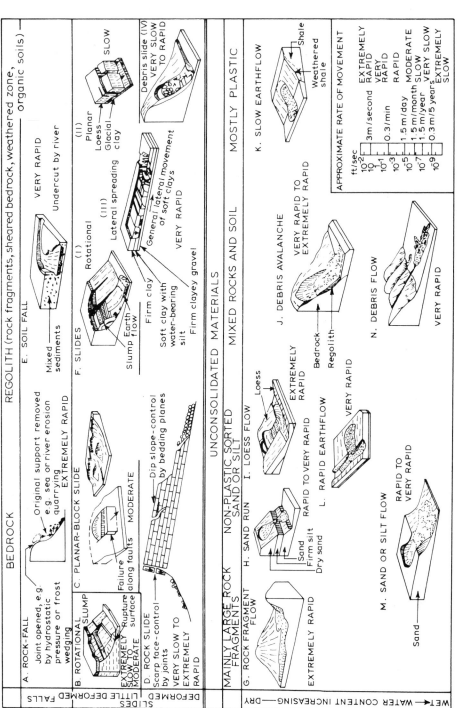

Figure 7.6 A classification of landslides (after Varnes (1958))

7.7.5 Falls

Falls are very common. The moving mass travels mostly through the air by free fall, saltation or rolling, with little or no interaction between the moving fragments. Movements are very rapid and may not be preceded by minor movements. In rockfalls the fragments are of various sizes and are generally broken in the fall. They accumulate at the bottom of a slope as scree deposit. If a rockfall is active or very recent then the slope from which it was derived is scarped. Frost–thaw action is one of the major causes of rockfall.

Toppling failure is a special type of rockfall which can involve considerable volumes of rock. According to Hoek (1971) the condition for toppling is defined by the position of the weight vector in relation to the base of the block. If the weight vector, which passes through the centre of gravity of the block, falls outside the base of the block, toppling will occur (Figure 7.7). The danger of a slope toppling increases with increasing discontinuity angle and steep slopes in vertically jointed rocks frequently exhibit signs of toppling failure (Figure 5.14).

Skempton and Hutchinson (1969) noted that falls in clay soils generally represented short-term failures, which originated from a tension crack, in newly exposed slopes.

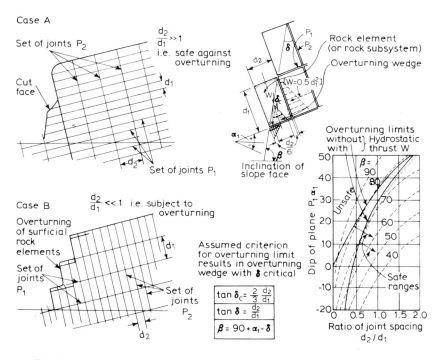

The diagram summarizes the criteria required for stability against overturning for two-dimensional conditions in terms of the mean spacing and orientation of the joint sets, the slope angle and the presence of hydrostatic forces.

Figure 7.7 Criteria for toppling failure in two dimensions (after Richards *et al.* (1978))

7.7.6 Slides

In true slides the movement results from shear failure along one or several surfaces, such surfaces offering the least resistance to movement. The mass involved may or may not experience considerable deformation. One of the most common types of slide occurs in clay soils where the slip surface is approximately spoon-shaped. Such slides are referred to as rotational slides. They are commonly deep-seated (0.15 depth/length < 0.33).

Although the slip surface is concave upwards it seldom approximates to a circular arc of uniform curvature. For instance, if the shearing strength of the soil is less in the horizontal than vertical direction, the arc may flatten out; if the soil conditions are reversed then the converse may apply. What is more the shape of the slip surface is very much influenced by the existing discontinuity pattern.

Rotational slides usually develop from tension scars in the upper part of a slope, the movement being more or less rotational about an axis located above the slope (Figure 7.8). The tension cracks at the head of a rotational slide are generally concentric and parallel to the main scar. Undrained depressions and perimeter lakes, bounded upward by the main scar, characterise the head regions of many rotational slides.

Skempton and Hutchinson (1969) recognised circular slips in relatively uniform overconsolidated deposits of fissured clay such as the London Clay. Non-circular slips occur in overconsolidated clays in which weathering has led to the development of quasi-planar slide surfaces, or in unweathered structurally anisotropic clays. Both circular and non-circular shallow rotational slips tend to form on moderately inclined slopes in weathered or colluvial clays.

When the scar at the head of a rotational slide is almost vertical and unsupported, then further failure is usually just a matter of time. As a consequence successive rotational slides occur until the slope is stabilised. These are retrogressive slides and they develop in a headward direction. All multiple retrogressive slides have a common basal shear surface in which the individual planes of failure are combined.

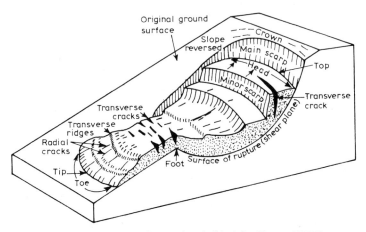

Figure 7.8 The main features of a rotational slide (after Varnes (1958))

Bottle neck slides, according to Skempton and Hutchinson (1969), are peculiar to quick clays, being a form of retrogressive failure. They start as a rotational failure of the stiffer weathered crust at the banks of an incised stream. This is very rapidly followed by further failures which involve the weaker, less weathered material. Failed, remoulded clay is forced through the opening in the bank and is removed by the stream. The failure spreads out away from the bank, hence the name.

Translational slides occur in inclined stratified deposits, the movement occurring along a planar surface, frequently a bedding plane (Figure 7.9). The mass involved in the movement becomes dislodged because the force of gravity overcomes the frictional resistance along the potential slip surface, the mass having been detached from the parent rock by a prominent discontinuity such as a major joint. Slab slides, in which the slip surface is roughly parallel to the ground surface, are a common type of translational slide. Such a slide may progress almost indefinitely if the slip surface is sufficiently inclined and the resistance along it is less than the driving force, whereas rotational sliding usually brings equilibrium to an unstable mass. Slab slides can occur on gentler surfaces than rotational slides and may be more extensive.

According to Skempton and Hutchinson (1969), compound and translational slides develop in clay deposits when rotation is inhibited by an underlying planar feature, such as a bedding plane or the base of a weathered boundary layer.

Translational slides tend to be more superficial than compound slides, being governed by more shallow inhomogeneities. Clay which is subjected to part-rotational, part-translational sliding is often distorted and broken. Block slides may develop in the more lithified, jointed deposits of clay, blocks of clay first separating and then sliding on well-defined bedding, joint or fault planes. Slab slides are characteristic of more weathered clay slopes of low inclination. Material moves *en masse* with little internal distortion.

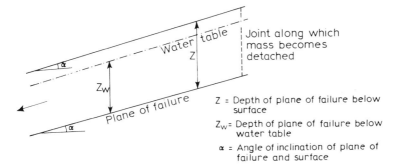

Figure 7.9 A translational slide. In a translational slide it is assumed that the potential plane of failure lies near to and parallel to the surface. The water table also is inclined parallel to the surface. If the water table creates a hydrostatic component of pressure on the slip surface with flow out of the slope then it has been shown that the factor of safety (F) is as follows (see Skempton and DeLory (1957): $F = c' + (\gamma Z \cos^2 \alpha - \gamma_w Z_w) Z \tan \phi' / \gamma Z \sin \alpha \cos \alpha$ where γ_w is the unit weight of water and c' and ϕ' are the effective cohesion and angle of shearing resistance respectively

Weathered mantle and colluvial materials are particularly prone to slab failure, which rarely occurs with depth:length ratios greater than 0.1. If a sufficient number of overlapping slips develop they may form a shallow translational retrogressive slide.

Failures which involve lateral spreading may develop in clays; quick clays and varved clays appear particularly susceptible. This type of failure is due to high pore water pressure in a more permeable zone at relatively shallow depth, dissipation of pore pressure leading to the mobilisation of the clay above. The movement is usually complex being dominantly translational, although rotation and liquefaction and consequent flow may also be involved. Such masses, however, generally move over a planar surface and may split into a number of semi-independent units. Like other landslides these are generally sudden failures, although sometimes movement can take place slowly.

Rock slides and debris slides are usually the result of a gradual weakening of the bonds within a rock mass and are generally translational in character. Most rock slides are controlled by the discontinuity patterns within the parent rock. Water is seldom an important direct factor in causing rock slides although it may weaken bonding along joints and bedding planes. Freeze–thaw action, however, is an important cause. Rock slides commonly occur on steep slopes and most of them are of single rather than multiple occurrence. They are composed of rock boulders. Individual fragments may be very large and may move great distances from their source. Debris slides are usually restricted to the weathered zone or to surficial talus. With increasing water content debris slides grade into mudflows. These slides are often limited by the contact between loose material and the underlying firm bedrock. Skempton and Hutchinson (1969) defined colluvium slides as those involving the movement of weathered material which has collected at the base of steep slopes or cliffs. Much colluvial material in the United Kingdom is Pleistocene in age and colluvial slides can involve the regeneration of older slide surfaces.

7.7.7 Flows

In a flow the movement resembles that of a viscous fluid (see Bishop (1973)). Slip surfaces are usually not visible or are short lived and the boundary between the flow and the material over which it moves may be sharp or may be represented by a zone of plastic flow. Some content of water is necessary for most types of flow movement but dry flows can and do occur (bear in mind Terzaghi's remark relating to water content and lubrication referred to above). Consequently the range of water content in flows must be regarded as ranging from dry at one extreme to saturated at the other. Dry flows, which consist predominantly of rock fragments, are simply referred to as rock fragment flows or rock avalanches and generally result from a rock slide or rockfall turning into a flow. They are generally very rapid and short lived, and frequently are composed mainly of silt or sand. As would be expected they are of frequent occurrence in rugged mountainous regions where they usually involve the movement of many millions of tonnes of material. Wet flows occur when fine-grained soils, with or without coarse debris, become mobilised by an excess of water. They may be of great length.

Progressive failure is rapid in debris avalanches and the whole mass, either because it is quite wet or is on a steep slope, moves downwards, often along a stream channel, and advances well beyond the foot of a slope. Lumb (1975) reported speeds of 30 m/s for debris avalanches in Hong Kong. The main characteristics of many slips which occur in the residual soils (mainly decomposed granite) of Hong Kong are the rapid fall of debris (once movement starts the whole mass separates from the main slope within minutes) and the shallow depth of the slide, usually less than 3 m. The ratio of thickness to length of the scar is usually less than 1:5. There is rarely any prior warning that a slip is imminent. The prime cause of failures is direct infiltration of rain water into the surface zones of the slopes, leading to soil saturation and its loss of effective cohesion. Debris avalanches are generally long and narrow and frequently leave V-shaped scars tapering headwards. These gulleys often become the sites of further movement.

Debris flows are distinguished from mudflows on a basis of particle size, the former containing a high percentage of coarse fragments, whilst the latter consist of at least 50% sand-size or less. Almost invariably debris flows follow unusually heavy rainfall or sudden thaw of frozen ground. These flows are of high density, perhaps 60–70% solids by weight, and are capable of carrying large boulders. Like debris avalanches they commonly cut V-shaped channels, at the sides of which coarser material may accumulate as the more fluid central area moves down-channel. Both debris flows and mudflows may move over many kilometres.

Mudflows may develop when a rapidly moving stream of storm water mixes with a sufficient quantity of debris to form a pasty mass (Figure 7.10). Because such mudflows frequently occur along the same courses, they should be kept under observation when significant damage is likely to result. Mudflows frequently move at rates ranging between 10 and 100 m/min and can travel over slopes inclined at one degree or less. Indeed they usually develop on slopes with shallow inclinations, that is, between 5 and 15°. Skempton and Hutchinson (1969) observed that mudflows also develop along discretely sheared boundaries in fissured clays and varved or laminated fluvio-glacial deposits where the ingress of water has led to softening at the shear zone. Movement involves the development of forward thrusts due to undrained loading of the rear part of the mudflow where the basal shear surface is inclined steeply downwards. A mudflow continues to move down shallow slopes due to this undrained loading, which is implemented by frequent small falls or slips of material from a steep rear scarp on to the head of the moving mass. This not only aids instability by loading, but it also raises the water pressures along the back part of the slip surface (see Hutchinson and Bhandari (1971); Bromhead (1978)).

Vallego (1980) has advanced an alternative explanation for the mobilisation of mudflows on low-angled slopes. He suggested that in a mixture of solid particles and fluid, the force on the particles in the direction of movement is mainly due to the effective weight of the particles themselves. Movement is not brought about by the stress exerted by the interstitial fluid. Numerous notable mudflows have occurred in the quick clays of Canada and Norway.

An earthflow involves mostly cohesive or fine-grained material which may move slowly or rapidly. The speed of movement is to some extent dependent on water content in that the higher the content, the faster the movement.

Figure 7.10 Aerial photograph taken a few hours after the 1966 slip showing the extent and nature of the flowslide and subsequent mudflow Aberfan

Slowly moving earthflows may continue to move for several years. These flows generally develop as a result of a build-up of pore pressure, so that part of the weight of the material is supported by interstitial water with consequent decrease in shearing resistance. If the material is saturated a bulging frontal lobe is formed and this may split into a number of tongues which advance with a steady rolling motion. Earthflows frequently form the spreading toes of rotational slides due to the material being softened by the ingress of water. Skempton and Hutchinson (1969) restricted the term earthflow to slow movements of softened weathered debris, as forms at the toe of a slide. They maintained that movement was transitional between a slide and a flow and that earthflows accommodated less breakdown than mudflows.

Solifluction lobes and sheets are flows which develop under periglacial conditions.

7.7.8 Landslide investigation and mapping

Because most landslides occur in areas previously affected by instability and because few occur without prior warning, Cotecchia (1978) emphasised the importance of carrying out careful surveys of areas which appear potentially unstable and of making systematic records of the relevant phenomena. He provided a review of the techniques involved in mapping mass movements, as well as itemising which data should be included on such maps. He maintained that the ultimate aim should be the production of maps of landslide hazard zoning.

A useful starting point in landslide investigation is a checklist, as suggested by Cooke and Doornkamp (1974). This enables each separate slope unit to be classified according to a stability rating (Figure 7.11). The checklist can be used either during an investigation of aerial photographs or during a field survey and provides a systematic examination of the main factors influencing mass movement. The more boxes which are ticked on the right-hand side of the checklist, the more the slope concerned is approaching an unstable state. Yague (1978) and Novosad (1978) have discussed the value of aerial photographs, precise surveying and geophysical methods in landslide investigation.

Huma and Radulescu (1978) described the use of a computer to produce maps of slope stability. The necessary data were obtained from aerial photographs as well as field and laboratory investigations. They included a survey of the lithology and mechanical characteristics of the rocks concerned, their structural and hydrogeological conditions, the slope angle, the amount of vegetation cover and exposure. A computer was then used to assess the data in terms of slope stability and to plot a map of land stability zones (Figure 7.12).

Moser (1978) suggested that various geotechnical maps could be produced in order to make an assessment of the necessary treatment of mass movements in upland areas. These included maps showing the classification of mass movements, the type and thickness of soils, and the type of bedrock (in both latter cases the geotechnical properties of the materials are investigated).

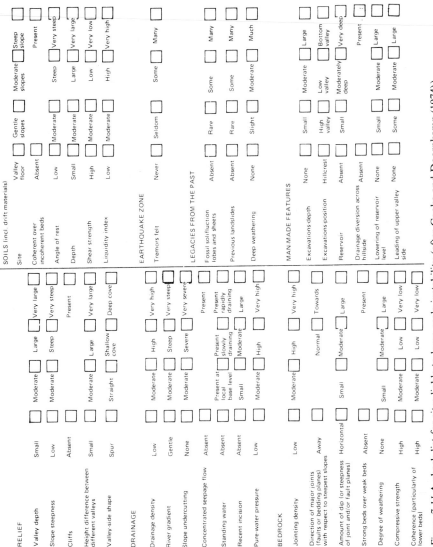

Figure 7.11 A check-list for sites liable to large-scale instability (after Cooke and Doornkamp (1974))

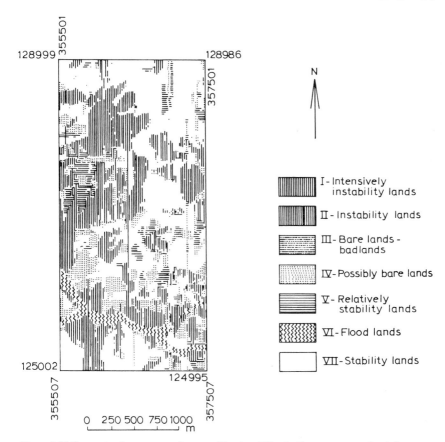

128999   355501   128986
357501

I - Intensively
instability lands

II - Instability lands

III - Bare lands -
badlands

IV - Possibly bare lands

V - Relatively
stability lands

VI - Flood lands

VII - Stability lands

125002
124995
355507   357507

0   250 500 750 1000
m

Figure 7.12 Computer-drawn map of zones of land stability in Vrancea mountains (after Huma and Radulescu (1978))

7.8 Preventive and corrective measures

As the same preventive or corrective work cannot always be applied to different types of slides it is important to identify the type of slide which is likely to take, or which has taken place (see Hutchinson (1977)). In this context, however, it is important to bear in mind that landslides may change in character and that they are usually complex, frequently changing their physical characteristics as time proceeds. When it comes to the correction of a landslide, as opposed to its prevention, since the limits and extent of the slide are generally well defined, the seriousness of the problem can be assessed. Nevertheless in such instances consideration must be given to the stability of the area immediately adjoining the slide. Obviously any corrective treatment must not adversely affect the stability of the area about the slide.

If landslides are to be prevented then areas of potential landsliding must first be identified, as must their type and possible amount of movement. Then if the hazard is sufficiently real the engineer can devise a method of preventive treatment. Preventive measures in areas undergoing excavation entail proper slope design and drainage. Economic considerations, however, cannot be disregarded. In this respect it is seldom economical to design cut slopes sufficiently flat to preclude the possibility of landslides and indeed many roads in rough terrain could not be constructed with the finance available without accepting some risk of landslides. All the same this is no justification for lack of thorough investigation and adoption of all economical means of slide prevention.

Landslide prevention may be brought about by reducing the activating forces, by increasing the forces resisting movement or by avoiding or eliminating the slide. Reduction of the activating forces can be accomplished by removing material from that part of the slide which provides the force which will give rise to movement. Complete excavation of potentially unstable material from a slope may be feasible and such treatment is applicable to all types of mass movement. However, Baker and Marshall (1958) placed an upper limit of 50 000 m³ on the amount of material which can be removed economically. Hence the use of this form of treatment is limited.

Although partial removal is suitable for dealing with most types of mass movement, for some types it is inappropriate. For example, removal of head has little influence on flows or slab slides. On the other hand, this treatment is eminently suitable for rotational slips. Slope flattening, however, is rarely applicable to rotational or slab slides. Slope reduction may be necessary in order to stabilise the toe of a slope and so prevent successive undermining with consequent spread of failure upslope. Benching can be used on steeper slopes. It brings about stability by dividing a slope into segments.

Drainage is the most generally applicable preventive and corrective treatment for slides, regardless of type. Indeed drainage is the only economic way of dealing with landslides involving the movement of several million cubic metres of material. The surface of a landslide is generally uneven, hummocky and traversed by deep fissures. This is particularly the case when the slipped area consists of a number of slices. Water collects in depressions and fissures, and pools and boggy areas are thus formed. In such cases the first remedial measure to be carried out is surface drainage.

Surface run-off or water flowing from springs should not be allowed to drain across an unsuitable area. This is usually accomplished by a drainage ditch at the top of a slope. Herringbone ditch drainage is usually employed to convey water from the surface of a slope into a ditch at its base. Ditches, especially in soils, should be lined to prevent their erosion and may be filled with cobble aggregate. Infiltration can be lowered by sealing the cracks in a slope with cement, bitumen or clay, or by regrading. A surface covering serves similar purposes and function.

Successful use of subsurface drainage depends on tapping the source of water and on the location of the drains on relatively unyielding material to ensure continuous operation (flexible PVC drains are frequently used). Filters should be installed to minimise silting in the drainage channels. According to Zaruba and Mencl (1969) where the drainage of large slips has to be carried out over lengths of 200 m or more, galleries are indispensable. The drainage galleries should be backfilled with stone to ensure their drainage capacity, even if they are partially deformed by subsequent movements. Borings may be made from the perimeter of a gallery to enhance drainage.

Drainage holes with perforated pipes are much cheaper than galleries and are satisfactory over short lengths.

Restraining structures control landslides by increasing the resistance to movement. They include retaining walls, cribs, gabions, buttresses, piling and rock bolts. The following minimum information is required to determine the type and size of a restraining structure

(1) the boundaries and depth of the unstable area, its moisture content and its relative stability,
(2) the type of slide which is likely to develop or has occurred,
(3) the foundation conditions since restraining structures require a satisfactory anchorage.

There are certain limitations which must be considered before retaining walls, cribs or piles are uses as a method of landslide control. These involve the ability of the structure to resist shearing action, overturning and sliding on or below the base of the structure. Retaining walls are often used where there is a lack of space for the full development of a slope, such as along many roads. As retaining walls are subjected to unfavourable loading, a large wall width is necessary to increase slope stability. Retaining walls should be designed for a predetermined load which they are to transmit to the foundation of known bearing capacity, and should include adequate provision for drainage.

FLUVIAL PROCESSES

7.9 The development of drainage systems

It is assumed that the initial drainage pattern which develops on a new surface consists of a series of sub-parallel rills flowing down the steepest slopes. The drainage pattern then becomes integrated by micropiracy (the beheading of the drainage system of a small rill by that of a larger) and cross-grading. According to Horton (1945) micropiracy occurs when the ridges which separate the initial rills are overtopped and broken down. When the divides are overtopped the water tends to move towards those rills at a slightly lower elevation, and in the process the divides are eroded. Eventually water drains from rills of higher elevation into adjacent ones of lower elevation (Figure 7.13). The flow towards the master rill steadily increases and its development across the main gradient was termed cross-grading by Horton. The tributaries which flow into the master stream are subsequently subjected to cross-grading and so a dendritic system is developed.

There is an area around the perimeter of every river basin where all flow is overland flow before it is concentrated into channels. Horton (1945) defined this as the critical length of overland flow (x_c) from the hillcrest needed to give sufficient run-off to initiate erosion and therefore to change from overland to confined flow. This distance depends on gradient, runoff intensity, soil infiltration capacity, and the roughness and resistance to sheet erosion of the surface. It can be expressed as follows

$$x_c = \frac{65}{Q_s n} \times \left(\frac{R_i}{f(s)}\right)^{5/3}$$

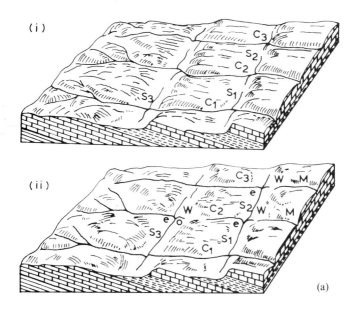

Figure 7.13 (a) (i) Trellised drainage pattern of consequent streams (C) and their subsequents (S) showing the desiccation of a gently dipping series of hard and soft beds into escarpments and inner lowlands (ii) later development illustrating river capture by the headward growth of the more vigorous subsequent streams. e = elbow of capture; W = wind gap; M = misfit stream; o = obsequent stream. (b) River capture Brecon Beacons, Wales. The streams in the foreground will capture the headwaters of the river Taff in the none too distant future

where Q_s = runoff intensity (in/h), n = surface roughness factor, R_i = threshold value of resistance of surface, f = eroding force (lb/ft^2), s = tangent of slope.

The texture of the drainage system is influenced by rock type and structure, the nature of the vegetation cover and the type of climate. The drainage density affords a measure of comparison between the development of one drainage system and another. It is calculated by dividing the total length of a stream by the area it drains, and is generally expressed in kilometres per square kilometre.

Horton (1945) classified streams into orders. First-order streams are unbranched, and when two such streams become confluent they form a second-order stream. It is only when streams of the same order meet that they produce one of higher rank, for example, a second-order stream flowing into a third-order stream does not alter its rank. The frequency with which streams of a certain order flow into those of the next order above them is referred to as the bifurcation ratio. The bifurcation ratio for any consecutive pair of orders is obtained by dividing the total number of streams of the lower order by the total number in the next higher order. Similarly the stream length ratio is found by dividing the total length of streams of the lower order by the total length of those in the next higher order. Values of stream length ratio depend mainly on drainage density and stream entrance angles, and increase somewhat with increasing order. A river system is also assigned an order, which is defined numerically by the highest stream order it contains.

The dominant action of master streams is vertical downcutting which is accomplished by the formation of pot-holes, which ultimately coalesce, and by the abrasive action of the load. Hence in the early stages of river development the cross profile of the valley is sharply V-shaped. As time proceeds valley widening due to soil creep, slippage, rainwash and gullying becomes progressively more important and eventually lateral corrasion replaces vertical erosion as the dominant process. A river possesses few tributaries in the early stages, but as the valley widens their numbers increase, which affords a growing increment of rock waste to the master stream thereby enhancing its corrasive power.

In the early stages of development in particular, rivers tend to accommodate themselves to the local geology. For instance, tributaries may develop along fault zones (see Howard (1967)). What is more, rock type has a strong influence on the drainage texture or channel spacing. In other words a low drainage density tends to form on resistant or permeable rocks whereas weak highly erodible rocks are characterised by a high drainage density.

During valley widening the stream erodes the valley sides by causing undermining and slumping to occur on the outer concave curves of meanders where steep cliffs or bluffs are formed (see Langbein (1947)). These are most marked on the upstream side of each spur. Deposition usually takes place on the convex side of a meander. They migrate both laterally and downstream, and their amplitude is progressively increased. In this manner spurs are continually eroded, first becoming more asymmetrical until they are eventually truncated (Figure 7.14a). The slow deposition which occurs on the convex side of a meander will, as lateral migration proceeds, produce a gently sloping area of alluvial ground called the flood plain (Figure 7.14b). The flood plain gradually grows wider as the river bluffs recede until it is

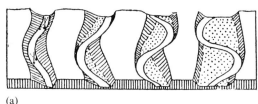

(a)

(b)

Figure 7.14 (a) Widening of valley floor by lateral corrasion. (b) Ebro Valley, Spain (courtesy of Popperfoto)

as broad as the amplitude of the meanders. It was at this period in river development that Davis (1909) regarded maturity as having been reached (Figure 7.15). From now onwards the continual migration of meanders slowly reduces the valley floor to an almost flat plain which slopes gently downstream and is bounded by shallow valley sides.

Throughout its length a river channel has to adjust to several factors which change independently of the channel itself. These include the different rock types and structures across which it flows. The tributaries and inflow of water from underground sources affect the long profile but are independent of the channel. Other factors which bring about adjustment of a river channel are, flow resistance which is a function of particle size and the shape of transistory deposits such as bars, the method of load transport, and the channel pattern, including meanders and islands. Lastly, the river channel must also adjust itself to the river slope, width, depth and velocity. Summarising, the

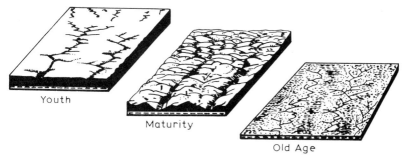

Figure 7.15 Diagram illustrating the three main periods in the denudation of
an uplifted land surface according to the Davisian interpretation of the
'normal' cycle of erosion; in youth, parts of the initial surface survive;
in maturity, most or all of the initial surface has vanished and the landscape
is mainly slopes, apart from valley floors; in old age or senility, the
landscape becomes subdued and gently undulating, rising only to residual
hills representing the divides between adjoining drainage basins. Eventually
such hills are worn down and the region becomes a peneplain.

longitudinal profile of a river may be regarded as a function of discharge,
load (contributed to the channel), size of the particles, flow resistance, slope,
depth, width and flow velocity.

As the longitudinal profile or thalweg of a river is developed, the differences
between the upper, middle and lower sections of its course become more
clearly defined until three distinctive tracts are observed. These are the upper
or torrent, the middle or valley, and the lower or plain stage. The torrent
stage includes the headstreams of a river where small fast-flowing streams
are engaged principally in active downwards and headwards erosion. They
possess steep-sided cross profiles and irregular thalwegs. The initial longitudinal profile of a river reflects the irregularities which occur in its path.
For instance, it may exhibit waterfalls or rapids where it flows across resistant
rocks. However, such features are transient in the life of a river. In the valley
tract the predominant activity is lateral corrasion. The shape of the valley
sides depends upon the nature of the rocks being excavated, the type of
climate, the rate of rock wastage and meander development. Some reaches
in the valley tract may approximate to grade and there the meanders may
have developed alluvial flats, whilst other stretches may be steep-sided with
irregular longitudinal profiles. The plain tract is formed by the migration
of meanders and deposition is the principal river activity.

Meanders, although not confined to, are characteristic of flood plains. The
consolidated veneer of alluvium, spread over a flood plain, offers little resistance to continual meander development, so the loops become more and more
accentuated. As time proceeds the swelling loops approach one another.
During flood the river may cut through the neck separating two adjacent
loops and thereby straightens its course. As it is much easier for the river
to flow through this new course, the meander loop is silted off and abandoned
as an oxbow lake (Figure 7.16).

Deformed and compressed meanders commonly develop when a river
migrates freely back and forth across a valley floor. Such features reflect
the influence of more resistant alluvium or bedrock. In other words a meander

Figure 7.16 Meanders and cut-offs in Mudjalik river, Saskatchewan. (This aerial photograph A1814–27 © 1929 Her Majesty the Queen in Right of Canada, reproduced with permission of Energy Mines and Resources, Canada)

deforms when its downstream limb is fixed in place by resistant materials. The continuing downstream movement of the upstream limb leads to the formation of a compressed meander.

Meander lengths vary from 7–10 times the width of the channel whilst cross-overs occur at about every 5–7 channel widths. The amplitude of a meander bears little relation to its length but is largely determined by the erosion characteristics of the river bed and local factors. For instance, Friedkin (1945) maintained that in uniform material the amplitude of meanders does not increase progressively nor do meanders form oxbow lakes during the downstream migration of bends. Subsequently Schumm (1960) concluded that higher sinuosity was associated with small width relative to depth and a larger percentage of silt and clay, which afforded greater cohesiveness, in the river banks. Dury (1964) showed that the dimensions of meanders are related to discharge. Relatively sinuous channels with a low width-depth ratio are developed by rivers transporting large quantities of suspended sediment. By contrast, the channel tends to be wide and shallow, and less sinuous, when the amount of bedload discharge is high.

A river is described as being braided if it splits into a number of separate channels or anabranches to adjust to a broad valley. The areas between the anabranches are occupied by islands built of gravel and sand. For the islands to remain stable the river banks must be more erodible so that they give way rather than the islands. Braided channels occur on steeper slopes than do meanders.

Climatic changes and earth movements alter the base level to which a river grades. When a land surface is elevated the downcutting activity of rivers

flowing over it is accelerated. The rivers begin to regrade their courses from
their base level and as time proceeds their newly graded profiles are extended
upstream until they are fully adjusted to the new conditions. Until this time
the old longitudinal profile intersects with the new to form a knick point.
The upstream migration of knick points tends to be retarded by outcrops
of resistant rock, consequently after an interval of time they are usually
located at hard rock exposures. The acceleration of downcutting consequent
upon uplift frequently produces a new valley within the old, the new valley
extending upstream to the knick point.

River terraces are also developed by rejuvenation. In the lower course of
a river uplift leads to the river cutting into its alluvial plain. The lateral and
downstream migration of meanders means that a new flood plain is formed
but very often paired alluvial terraces, representing the remnants of the former
flood plain, are left at its sides (Figure 7.17).

Incised meanders are also associated with rejuvenation and are often found
together with river terraces. When uplift occurs the downcutting action of
meanders is accelerated and they carve themselves into the terrain over which
they flow. The landforms which are then produced depend upon the character
of the terrain and the relative rates of downcutting and meander migration.
If vertical erosion is rapid, meander shift has little opportunity to develop
and consequently the loops are not greatly enlarged. The resulting incised
meanders are described as entrenched (Figure 7.18a). However, when time
is afforded for meander migration, they incise themselves by oblique erosion
and the loops are enlarged; they are then referred to as ingrown meanders
(Figure 7.18b).

When incision occurs in the alluvium of a river plain the meanders migrate
back and forth across the floor. On each successive occasion that a meander
swings back to the same side, it does so at a lower level. As a consequence

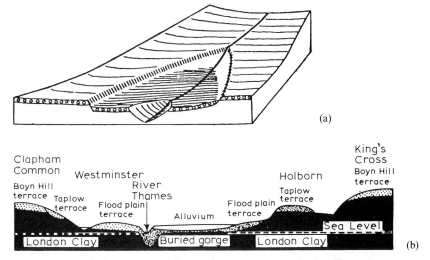

Figure 7.17 (a) Paired river terraces due to rejuvenation, note valley in valley and
knick-points. (b) Section across London to show the paired terraces and one of the
buried 'gorges' of the Thames valley

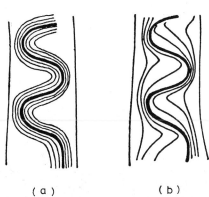

(a)   (b)

Figure 7.18 Incised meanders. (a) Entrenched. (b) Ingrown. (c) Entrenched meanders of the San Juan river, Monument Valley, Utah. (d) Incised meanders in the Rheidol Valley, Pont Erwyd, Cardigan. Note the upper valley with broad flat floor below which the meanders are incised (courtesy the Institute of Geological Sciences)

(c)

(d)

Figure 7.19 Alluvial terraces of the river Findhorn, looking north from Daless, Nairn, Scotland, cut through glacial deposits and marking successive stages in the erosion of the valley (courtesy the Institute of Geological Sciences)

small remnant terraces may be left above the newly formed plain. These terraces are not paired across the valley and their position and preservation depends on the swing of meanders over the valley (Figure 7.19). If downcutting is very slow, then erosion terraces are unlikely to be preserved.

7.10 The work of rivers

7.10.1 River flow

The velocity of a river depends upon channel gradient, volume and configuration. One of the most commonly used equations applicable to open channel hydraulics is the Chezy formula

$$v = C\sqrt{(Rs)}$$

where v is the mean velocity, C is a coefficient which varies with the characteristics of the channel, R is the hydraulic radius and s is the slope. Numerous attempts have been made to find a generally acceptable expression for C.

The Manning formula, based on field and experimental determinations of the value of the resistance coefficient (n) is also widely used

$$v = \frac{1}{n}R^{2/3}s^{1/2}$$

The experimental values of n vary from approximately 0.01 for smooth metal surfaces to 0.06 for natural, irregular channels containing large stones. Some values of n used in the Manning formula are as follows

(1) clean straight channel with no pools   0.025–0.033
(2) channel containing weeds and stones   0.030–0.040
(3) channel containing large stones   0.045–0.060

The velocity to initiate movement, that is, erosional velocity, is appreciably higher than that required to maintain movement. Figure 7.20 shows the ranges for current velocity for erosion, transportation and deposition of well sorted sediments.

The quantity of flow can be estimated from measurements of cross-sectional areas and current speed of a river. Generally channels become wider relative to their depth and adjusted to larger flows with increasing distance downstream. Bankfull discharges also increase downstream in proportion to the square of the width of the channel or of the length of individual meanders, and in proportion to the 0.75 power of the total drainage area focused at the point in question.

Statistical methods are used to predict river flow and assume that recurrence intervals of extreme events bear a consistent relationship to their magnitudes. A recurrence interval, generally of 50 or 100 years, is chosen in accordance with given hydrological requirements. Sherman's concept (1932) of unit hydrograph postulates that the most important hydrological characteristics of any basin can be seen from the direct run-off hydrograph resulting from 25 mm of rainfall evenly distributed over 24 h. This is produced by

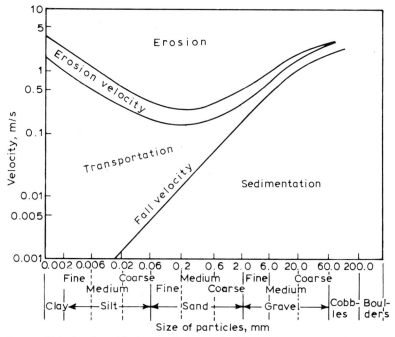

Figure 7.20 Ratio of particle size to velocity required for erosion, transportation and deposition (after Hjulstrum, F. (1935). 'Studies of the morphological activity of rivers, as illustrated by the river Fynis', Uppsala Univ. Geol. Inst., *Bull. 25*.)

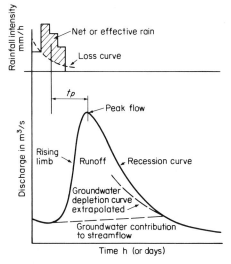

Figure 7.21 Component parts of a hydrograph.
When rainfall commences there is an initial
period of interception and infiltration before
any measurable runoff reaches the stream
channels. During the period of rainfall these
losses continue in a reduced form so that the
rainfall graph has to be adjusted to show
effective rain. When the initial losses are met,
surface runoff begins and continues to a peak
value which occurs at time *tp*, measured from
the centre of gravity of the effective rain on the
graph. Thereafter surface runoff declines along
the recession limb until it disappears. Baseflow
represents the groundwater contribution along
the banks of the river

drawing a graph of the total stream flow at a chosen point as it changes
with time after such a storm, from which the normal base-flow caused by
groundwater is subtracted (Figure 7.21).

Carlston (1963) found a highly significant relationship between mean
annual flood discharge per unit area and drainage density. Peak discharge
and the lag time of discharge (the time which elapses between maximum
precipitation and maximum run-off) are also influenced by drainage density,
as well as by the shape and slope of the drainage basin (see Gregory and
Walling (1973)). Stream flow is generally most variable and flood discharges
at a maximum per unit area in small basins. This is because storms tend
to be local in occurrence. Carlston also showed that an inverse relationship
existed between drainage density and base-flow or groundwater discharge.
This is related to the permeability of the rocks present in a drainage basin.
In other words the greater the quantity of water which moves on the surface
of the drainage system, the higher the drainage density, which in turn means
that the base-flow is lower. As pointed out previously, in areas of high
drainage density the soils and rocks are relatively impermeable and water
runs off rapidly. The amount of infiltration is reduced accordingly.

Statistically significant relationships have been demonstrated between channel width, depth and cross-sectional area (see Leopold and Maddock (1953)); Brown (1971). Schumm (1977) has developed expressions relating channel hydrology and other morphological characteristics to channel width (an index of discharge) and channel width–depth ratio (an index of the type of sediment load). The consensus of opinion believes that the discharge necessary for the formation of channels is that which approaches bankfull stage. Hence mean annual flood should be closely related to channel dimensions. Schumm indicated that if a flood event is chiefly responsible for the dimensions of a channel, then mean annual discharge is not so closely related to channel morphology as the mean flood. Water discharge (Q) influences the channel width (b), depth (Z), meander wavelength (λ) and gradient of a river as follows

$$Q = \frac{BZ\lambda}{s}$$

where s is the slope. In stable rivers with sand-beds the relation between bedload (L) and channel morphology is given by

$$L \simeq \frac{B\lambda s}{Zp}$$

where p is sinuosity (the ratio of channel length to valley length). Lane's (1955) expression

$$Ld \simeq Qs$$

indicates that as bedload (L) and sediment size (d) increase, either water discharge (Q) or slope (s) or both, increase to compensate. This provides a view of the way a river slope changes in response to the changes in water-sediment discharge.

7.10.2 Erosion

The work undertaken by a river is threefold: it erodes rocks and transports the products thereof, which it eventually deposits. Erosion occurs when the force provided by the river flow exceeds the resistance of the material over which it runs. Four types of fluvial erosion have been distinguished, namely, hydraulic action, attrition, corrasion and corrosion. Hydraulic action is the force of the water itself. Attrition is the disintegration which occurs when two or more particles which are suspended in water collide. Corrasion is the abrasive action of the load carried by a river on its channel. Most of the erosion done by a river is attributable to corrasive action. Hence a river carrying coarse, resistant, angular rock debris possesses a greater ability to erode than does one transporting fine particles in suspension. Corrosion is the solvent action of river water.

In the early stages of river development erosion tends to be greatest in the lower part of the drainage basin. However, as the basin develops the zone of maximum erosion moves upstream and accordingly it is concentrated along the divides in the later stages. If the zone of maximum erosion cannot be located from a survey of aerial photographs or by field reconnaisance, then average slope curves or hypsometric curves can be used to locate it (see Strahler (1964)). Gully enlargement and the destruction of flood plains can be measured and the rates of erosion estimated if surveys, either of aerial photographs or in the field, with resultant maps, are frequently made.

The amount of erosion accomplished by a river in a given time depends upon its volume and velocity of flow, the character and size of its load, the rock type and geological structure over which it flows, the infiltration capacity of the area it drains and the vegetation which directly affects the stability and permeability of the soil. The volume and velocity of a river influences the quantity of energy it possesses. When flooding occurs the volume of a river is greatly increased which leads to an increase in its velocity and competence. However, much energy is spent in overcoming the friction between the river and its channel so that energy losses increase with any increase in channel roughness. Obstructions, changing forms on a river bed such as sand-bars and vegetation offer added resistance to flow. Bends in a river also increase friction. Each of these factors causes deflection of the flow which dissipates energy by creating eddies, secondary circulation and increased shear rate.

There is no means of measuring the state of balance between a river's energy and its load, nor is there any evidence for assuming that only by virtue of possessing energy in excess of that required to carry its load can a river erode. If that were so, then, by reducing the load, more and more energy would become available for erosion. Pursuing this reasoning to its logical conclusion would mean that the maximum amount of river erosion would occur when the river transported no debris. This is absurd because river erosion is overwhelmingly due to corrasion.

The ratio between the cross-sectional area of a river channel and the length of its wetted perimeter determines the efficiency of the channel. This ratio is termed the hydraulic radius, and the higher its value, the more efficient is the river. The most efficient forms of channel are those with approximately circular or rectangular sections with widths approaching twice their depths. On the other hand the most inefficient channel forms are very broad and shallow with wide wetted perimeters.

The effects of scour and fill during flood can be illustrated by examining the passage of snow melt water in the San Juan near Bluff, Utah, between September and December 1941 (see Leopold et al. (1964). The flood started on September 9 when the discharge was 18 m³/s and the bed was at an elevation of 1.2 m on the gauge datum. As the discharge increased to 186 m³/s on September 15 the bed rose approximately 0.6 m. However, between then and October 14, when the maximum discharge occurred, the bed was scoured about 1.8 m. Subsequently the bed filled about 1.5 m as the discharge fell from 1688 m³/s on October 14 to 513 m³/s twelve days later. It is interesting to note that the bed filled during the first stages of flood and then was scoured as discharge further increased. Although the scour did not occur in proportion to the discharge, the maximum scour did coincide with the highest discharge.

7.10.3 Transportation

The load which a river carries is transported in four different ways. First, there is traction, that is, rolling of the coarsest fragments along the river bed. Second, smaller material, when it is caught in turbulent upward moving eddies, proceeds downstream in a jumping motion referred to as saltation. Third, fine sand, silt and mud are transported in suspension. Fourth, soluble material is carried in solution.

Sediment yield may be determined by sampling both the suspended load and the bedload. It can also be derived from the amount of deposition which takes place when a river enters a relatively still body of water such as a lake or a reservoir.

The competence of a river to transport its load is demonstrated by the largest boulder it is capable of moving; it varies according to a river's velocity and volume, being at a maximum during flood. It has been calculated that the competence of a river varies as the sixth power of its velocity. The capacity of a river refers to the total amount of sediment which it carries. It varies according to the size of the rock fragments which form the load, and the velocity of the river. When the load consists of fine particles, the capacity is greater than when it is comprised of coarse material. Usually the capacity of a river varies as the third power of its velocity.

Both the competence and capacity of a river are influenced by changes in the weather, and the lithology and structure of the rocks over which it flows, as well as by vegetative cover and land-use. Because the discharge of a river varies, sediments are not transported continuously, for instance, boulders may be moved only a few metres during a single flood. Sediments which are deposited over a flood plain may be regarded as being stored there temporarily.

The character of a valley floor influences flooding (see Wolman (1971)). For example, water spreads more widely over a valley with a flat floor than where the floor is concave.

7.10.4 Deposition

Deposition occurs where turbulence is at a minimum or where the region of turbulence is near the surface of a river. For example, lateral accretion occurs, with deposition of a point bar, on the inside of a meander bend. The settling velocity for small grains in water is roughly proportional to the square of the grain diameter, whereas for larger particles settling velocity is proportional to the square root of the grain diameter.

An individual point bar grows as a meander migrates downstream and new ones are formed as a river changes its course during flood. Indeed old meander scars are a common feature of flood plains. The combination of point bar and filled slough or oxbow lake gives rise to ridge and swale topography. The ridges consist of sandbars and the swales are sloughs filled with silt and clay.

An alluvial flood plain is the most common depositional feature of a river. The alluvium is made up of many kinds of deposits, laid down both in the channel and outside it. Vertical accretion of a flood plain is accomplished by in-channel filling and the growth of overbank deposits during and immediately after floods. Gravel and coarse sands are moved chiefly at flood stages and deposited in the deeper parts of a river. As the river overtops its banks, its ability to transport material is lessened so that coarser particles are deposited near the banks to form levees. Levees stand above the general level of the adjoining plain so that the latter is usually poorly drained and marshy. This is particularly the case when levees have formed across the confluences of minor tributaries, so forcing them to wander over the flood plain until they find another entrance to the main river. Finer material is carried farther and laid down as backswamp deposits (Figure 7.22). At this point a river sometimes aggrades its bed, eventually raising it above the level of the

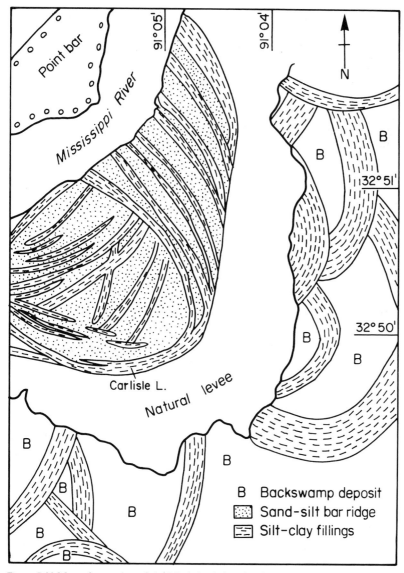

Figure 7.22 Map of a portion of the Mississippi river flood plain, showing various kinds of deposits. (From Fisk, H. N. (1944). *Geological Investigations of the Alluvial Valley of the Lower Mississippi*, US Corps of Engineers, Mississippi river Commission, Vicksburg.)

surrounding plain. Consequently when levees are breached by flood water, hundreds of square kilometres may be inundated.

7.11 A note on river control

River control refers to projects designed to hasten the run-off of flood waters or confine them within restricted limits, to improve drainage of adjacent lands, to check stream bank erosion or to provide deeper water for navigation. What the engineer has to bear in mind, however, is that a river in an alluvial channel is continually changing its position due to the hydraulic forces acting on its banks and bed. For instance, as meanders migrate downstream they may increase their amplitude or the bend may rotate. As a consequence any major modifications to the river regime which are imposed without consideration of the channel will give rise to a prolonged and costly struggle to maintain the change.

Flooding is a serious problem in the plain stage of a river. A study of flood frequency is important since it provides the engineer with data relating to the probability of future flood occurrences.

Artificial strengthening and heightening of levees or constructing artificial levees are frequent measures employed as protection against flooding. Because this confines a river to its channel its efficiency is increased. The efficiency of a river is also increased by cutting through the constricted loops of meanders. River water can be run off into the cut-off channels when high floods occur. Canalisation or straightening of a river can help to regulate flood flow, and improves the river for navigation. However, canalisation in some parts of the world has only proved a temporary expedient. This is because canalisation steepens the channel, as sinuosity is reduced. This, in turn, increases the velocity of flow and therefore the potential for erosion. Moreover the base level of the upper reaches of the river has, in effect, been lowered, which means that channel incision will begin there. Although the flood and drainage problems are temporarily solved, the increase in erosion upstream not only creates problems of lost land, but increases sediment load. This is then deposited in the canalised sections which can lead to a return of the original flood and drainage problems.

Diversion is another method used to control flooding, this involves opening a new exit for part of the river water. But any diversion must be designed in such a way that it does not cause excessive deposition in the main channel otherwise it is defeating its purpose. If, in some localities little damage will be done by flooding, then these areas can be inundated during times of high flood and so act as safety valves.

Channel regulation can be brought about by training dykes, jetties or wing dams which are used to deflect channels into a more desirable alignment or to confine them to lesser widths. Dykes and dams can be used to close secondary channels and thus divert or concentrate the river into a preferred course. In some cases ground sills or weirs need to be constructed to prevent undesirable deepening of the bed by erosion. Bank revetment by pavement, rip-rap or protective mattresses to retard erosion is usually carried out along with channel regulation or it may be undertaken independently for the protection of the lands bordering the river.

Control of the higher stretches of a river in those regions prone to soil removal by gullying and sheet erosion is very important. The removal of the soil mantle means that run-off becomes increasingly more rapid and consequently the problem of flooding is aggravated. These problems were tackled in the Tennessee valley by establishing a well planned system of agriculture so that soil fertility was maintained, and the valley slopes were re-afforested. The control of surface water thereby was made more effective. Gullies were filled and small dams were constructed across the headstreams of valleys to regulate run-off. Larger dams were erected across tributary streams to form catchment basins for flood-waters. Finally, large dams were built across the main river to smooth flood flow.

Nonetheless the construction of a dam across a river can lead to other problems. It decreases the peak discharge and reduces the quantity of bedload through the river channel. The width of a river may be reduced downstream of a dam as a response to the decrease in flood peaks (see Schumm (1969)). Moreover scour normally occurs immediately downstream of a dam. Removal of the finer fractions of the bed material by scouring action may cause armouring of the channel.

River channels may be improved for navigation purposes by dredging. When a river is dredged its floor should not be lowered so much that the water level is appreciably lowered. In addition the nature of the materials occupying the floor and their stability should be investigated. This provides data from which the stability of the slopes of the projected new channel can be estimated and indicates whether blasting is necessary. The rate at which sedimentation takes place provides some indication of the regularity at which dredging should be carried out.

7.12 Deltas and estuaries

A delta develops when a river enters a quiet body of water such as a lake, or at its mouth when the tidal and current action of the sea is incapable of removing most of the sediment it carries. For example, the river Mississippi deposits sediment in its delta at a rate of $17\,000$ m^3/s. The speed at which a delta grows depends upon the rate of sediment supply, the size of the basin of accumulation, the rate at which the basin is subsiding or being uplifted, and the rate at which sediments are being removed by marine action.

Bates (1954) advanced a theory of delta formation based upon the relationship between the density of the river and that of the body of water into which it flows. He distinguished three types of flow, hypopycnal flow, when the river water is denser than that into which it flows, homopycnal flow, when the river water is of similar density and hyperpycnal flow, when the river water is less dense. Bates likened the flow of river water into the sea to a jet stream moving into a comparatively quiet body of water. The three types of flow produced three different types of deltas. When river water is denser than the water into which it flows, by virtue of the amount of load it carries, it sinks through the lighter water to form a turbidity current. Only the coarser fraction of the load is deposited at the river mouth where it forms a visible delta. This kind of hydrodynamic relationship is usually found where heavily laden rivers discharge into clear-water lakes. Under homopycnal flow river water spreads out in fanlike fashion and mixes with the water into which it is flowing, and its velocity is therefore quickly reduced. Sediment is then deposited to form an arcuate delta. However, these sediments tend to obstruct the flow of the river through the main channel and distributaries develop. The latter further accentuate the arcuate shape. During floods levees form alongside these distributaries and shallow lagoons and swamps develop outside the levees. Lastly, when a river is less dense than the body of water into which it runs a bird's foot delta is constructed. The less dense river water flows over the denser sea water. The latter offers little resistance to the frontal movement of the river water, but does retard its velocity at the side. Hence sedimentation occurs along the sides of the river and levees are built up. These levees are breached in places by the river and in this way new channels develop to give the bird's foot appearance (Figure 7.23).

An estuary is that part of a river mouth which is subjected to tidal action. Unlike a delta, most sediment is removed from an estuary. This is accomplished when the tidal waters have sufficient strength to flush the river mouth clean. The most effective flushing action takes place where the tidal range is large and it flows straight in and out of a river's mouth. Funnel shaped estuaries commonly occur where river mouths have been recently drowned. Sands and silts which are deposited in an estuary form bars, which are usually oriented parallel to the direction of tidal movement thereby offering least resistance to this force. These bars may move their positions.

7.13 Limestone topography and underground drainage

Limestones are well jointed and bedded carbonate rocks which are susceptible to chemical attack by weakly acidified water. The solution of limestone etches

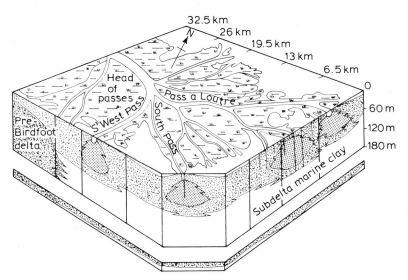

Figure 7.23 Block diagram of the birdfoot' subdelta of the Mississippi Linear sand bodies made by main distributaries shown in open stipple (vertical exaggeration about 30) (courtesy John Wiley & Sons Ltd)

(a)

(b)

Figure 7.24 (a) Limestone pavement (courtesy Institute of Geological Sciences).
(b) Gaping Ghyll pothole which descends about 150 m in the Carboniferous Limestone,
near Clapham, Yorkshire

out and gradually enlarges joints to form clints, the blocks of limestone between being termed grykes. Limestone is frequently removed along bedding planes to form limestone pavements (Figure 7.24a). The continued enlargement of joints, particularly where they intersect, eventually leads to the formation of funnel-shaped hollows known as swallow holes (Figure 7.24b). Surface streams disappear underground via swallow holes. Some of the larger swallow holes are connected near the surface by irregular, inclined shafts known as ponors, to underground integrated systems of caverns and galleries. Larger surface depressions form when enlarged swallow holes coalesce to form uvalas. These features may range up to a kilometre in diameter. Some of these depressions are formed essentially by solution but some at least are partly created by the collapse of underground caverns. Still larger depressions which occur, for instance, in the limestone region of western Yugoslavia, are called poljes. However, some authors have suggested that these depressions are of tectonic origin. Any residual masses of limestone which, after a lengthy period of continuous erosion, remain as isolated hills are known as hums. These are invariably honeycombed with galleries, shafts and caverns. An extensive account of karstic landforms has been provided by Sweeting (1972).

Underground river systems in limestone deepen and widen their courses by mechanical erosion as well as by solution. During flood they usually erode the roofs as well as the sides and floors of the caverns and galleries through which they flow. In this way caverns are enlarged and their roofs are gradually thinned until they become so unstable that they wholly or partly collapse to form gorges or natural arches respectively (see Smith (1977)).

Surface drainage is usually sparse in areas of thick limestone. Dry valleys are common although they may be occupied by streams during periods of heavy rainfall. Underground streams may appear as vaclusian springs where the water table meets the surface. Occasionally streams which rise on impervious strata may traverse a broad limestone outcrop without disappearing.

Chalk is a very pure form of limestone but, unlike most other forms, it is soft and possesses an irregular pattern of joints. Although chalk is permeable and large quantities are dissolved by acidified waters, caverns occur in it only with exceptional rarity. This is because chalk lacks the strength to support them. Dry valleys are characteristic of chalklands. Only the main valleys are occupied by perennial streams and even some of these fail to traverse chalk outcrops after periods of protracted drought. The form and pattern of the dry valleys developed on chalklands leave little doubt that they have been excavated by rivers in the normal manner at some previous time, although they may have been modified by solution effects. They are usually graded to the main rivers with which they are associated.

References

BADGER, C. W., CUMMINGS, A. D. & WHITMORE, R. L. (1956). 'The disintegration of shale', *J. Inst. Fuel*, **29**, 417–23.

BAKER, R. F. & MARSHALL, H. E. (1958). 'Control and correcting', (in *Landslides in Engineering Practice*), Eckel, E. B. (ed.), Committee on Landslide Investigations, *Highway Res. Board, Special Report*, **29**, Washington, 150–88.

BATES, C. C. (1954), 'Rational theory of delta formation', *Bull. Am. Ass. Petrol. Geol.*, **37**, 2119–62.

BISHOP, A. W. (1971). 'The influence of progressive failure on the choice of method of stability analysis', *Geotechnique*, **21**, 168–72.

BISHOP, A. W. (1973). 'The stability of tips and spoil heaps', *Q. J. Engg Geol.*, **6**, 335–76.

BJERRUM, L. (1967). 'Progressive failure in slopes of overconsolidated plastic clay and clay shales', *Proc. ASCE. J. Soil. Mech. Foundation Engg Div.*, **93**, 3–49.

BROMHEAD, E. N. (1978). 'Large landslides in London Clay, Herne Bay, Kent', *Q. J. Engg Geol.*, **11**, 291–304.

BROWN, D. A. (1971). 'Stream channels and flow relations', *Water Resources Res.*, **7**, 304–10.

CARLSTON, C. W. (1963). 'Drainage density and streamflow', *US Geol. Surv.*, Prof. Paper 422 C, 8pp.

CHANDLER, R. J. (1969). 'The effect of weathering on the shear strength properties of the Keuper Marl', *Geotechnique*, **19**, 321–34.

CHANDLER, R. J. (1974). 'Lias Clay: the long-term stability of cutting slopes', *Geotechnique*, **24**, 21–38.

CHANDLER, R. J. (1977). 'The application of soil mechanics methods to the study of slopes'. (in *Applied Geomorphology*), Hails, J. R. (ed.), Elsevier, Amsterdam, 157–82.

COOKE, R. U. & DOORNKAMP, J. C. (1974). *Geomorphology in Environmental Management*, Clarendon Press, Oxford.

COTECCHIA, V. (1978). 'Systematic reconnaissance mapping and registration of slope movements', *Bull. Int. Ass. Engg Geol.*, No. 17, 5–37.

DAVIS, W. M. (1909). *Geographical Essays*, Dover, New York.

DEARMAN, W. B. (1974). 'Weathering classification in the characterisation of rock for engineering purposes in British practice', *Bull. Int. Ass. Engg Geol.*, No. 9, 33–42.

DUNCAN, N., DUNNE, M. H. & PETTY, S. (1968). 'Swelling characteristics of rocks', *Water Power*, May, 185–92.

DURY, G. H. (1964). 'Principles of underfit streams', *US Geol. Surv.*, Prof. Paper, 452 A, 67pp.

FOOKES, P. G., DEARMAN, W. R. & FRANKLIN, J. A. (1972). 'Some engineering aspects of weathering with field examples from Dartmoor and elsewhere', *Q. J. Engg Geol.*, **3**, 1–24.

FRANKLIN, J. A. & CHANDRA, R. (1972). 'The slake durability test', *Int. J. Rock Mech. Min. Sci.*, **9**, 325–41.

FRIEDKIN, J. F. (1945). 'A laboratory study of the meandering of alluvial rivers', *US Waterways Engr. Exper. Stn. Report*, 40pp.

GOLDICH, S. S. (1938). 'A study of rock weathering', *J. Geol.*, **46**, 17–58.

GREGORY, K. J. & WALLING, D. E. (1973). *Drainage, Basin Forms and Process, a Geomorphological Approach*, Edward Arnold, London.

HAMROL, A. (1961). 'A quantitative classification of weathering and weatherability of rocks', *Proc. 5th Int. Conf. Soil Mech. Foundation Engg*, Paris, **2**, 771–3.

HOBBS, N. B. (1975). *Foundations on Rocks*, Soil Mechanics, Bracknell.

HOEK, E. (1971). 'The influence of structure on the stability of rock slopes', *Proc. 1st Symp on Stability in Open Pit Mining, Vancouver*, AIME, 49–63.

HOEK, E. & BRAY, T. (1977). *Rock Slope Engineering*, Inst. Min. Metall., London.

HOLLINGWORTH, S. E., TAYLOR, J. H. & KELLAWAY, G. A. (1944). 'Large-scale superficial structures in the Northampton Ironstone field', *Q. J. Geol. Soc.*, **100**, 1–44.

HORTON, R. E. (1945). 'Erosional development of streams and their drainage basins: hydrophysical approach to quantitative morphology', *Bull. Geol. Soc. Am.*, **56**, 275–370.

HOWARD, A. D. (1967). 'Drainage analysis in geologic interpretation. A summation', *Bull. Am. Ass. Petrol. Geol.*, **51**, 2246–59.

HUMA, I. & RADULESCU, D. (1978). 'Automatic production of thematic maps of slope stability', *Bull. Int. Ass. Engg Geol.*, No. 17, 95–9.

HUTCHINSON, J. N. (1968). 'Mass movement', (in *The Encyclopedia of Geomorphology*), Fairbridge, R. W. (ed.), Van Nostrand Reinhold, New York, 668–95.

HUTCHINSON, J. N. (1977). 'Assessment of the effectiveness of corrective measures in relation to geological conditions and types of slope movement', *Bull. Int. Ass. Engg Geol.*, No. 16, 131–55.

HUTCHINSON, J. N. & BHANDARI, R. K. (1971). 'Undrained loading—a fundamental mechanism of mudflows and other mass movements', *Geotechnique*, **21**, 353–8.

HUTCHINSON, J. N., BROMHEAD, E. N. & LUPINI, J. F. (1980). 'Additional observations on the Folkstone Warren landslides', *Q. J. Engg Geol.*, **13**, 1–32.

ILIEV, I. G. (1967). 'An attempt to estimate the degree of weathering of intrusive rocks from their physico-mechanical properties', *Proc. 1st Cong. Int. Soc. Rock Mech., Lisbon*, 109–14.

IRFAN, T. Y. & DEARMAN, W. R. (1978a). 'Engineering classification and index properties of weathered granite', *Bull. Int. Ass. Engg Geol.*, No. 17, 79–90.

IRFAN, T. Y. & DEARMAN, W. R. (1978b). 'The engineering petrography of a weathered granite in Cornwall, England', *Q. J. Engg Geol.*, **11**, 233–44.

KIERSCH, G. A. & TREACHER, R. C. (1955). 'Investigations, areal and engineering geology—Folsam Dam project, central California', *Econ. Geol.*, **50**, 271–310.

KNILL, J. L. & JONES, K. S. (1965). 'The recording and interpretation of geological conditions in the foundations of the Rosieres, Kariba and Latiyan Dams', *Geotechnique*, **15**, 94–124.

LANE, E. W. (1955). 'Importance of fluvial morphology in hydraulic engineering', *Proc. ASCE, J. Hyd. Div.*, **81**, 1–17.

LANGBEIN, W. B. (1947). 'Topographic characteristic of drainage basins', *US Geol. Surv. Water Supply Paper*, 68C, 99–114.

LEOPOLD, L. B. & MADDOCK, T. (1953). 'The hydraulic geometry of stream channels and some physiographic applications', *US. Geol. Surv.*, Prof. Paper 252, 57pp.

LEOPOLD, L. B., WOLMAN, M. G. & MILLER, L. P. (1964). *Fluvial Processes in Geomorphology*, Freeman & Co., San Francisco.

LITTLE, A. L. (1969). 'The engineering classification of residual tropical soils', *Proc. 7th Int. Conf. Soil Mech. Foundation Engg, Mexico*, **1**, 1–10.

LOVEGROVE, C. W. & FOOKES, P. G. (1972). 'The planning and implementation of a site investigation for a highway in tropical conditions in Fiji, *Q. J. Engg Geol.*, **5**, 43–68.

LUMB, P. (1962). 'The properties of decomposed granite', *Geotechnique*, **12**, 226–43.

LUMB, P. (1975). 'Slope failures in Hong Kong', *Q. J. Engg Geol.*, **8**, 31–66.

MENDES, F. M., AIRES-BARROS, L. & RODRIGUES, F. P. (1966). 'The use of modal analysis in the mechanical characterisation of rock masses', *Proc. 1st Int. Cong. Rock Mech.*, Lisbon, **1**, 217–23.

MORGENSTERN, N. R. (1963). 'Stability charts for earth slopes during rapid drawdown', *Geotechnique*, **13**, 121–31.

MOSER, M. (1978). 'Proposals for geotechnical maps concerning slope stability potential in mountain watersheds', *Bull. Int. Ass. Engg Geol.*, No. 17, 100–108.

MOYE, D. G. (1955). 'Engineering geology for the Snowy Mountain scheme', *J. Inst. Engrs*, Aust., **27**, 287–98.

NAKANO, R. (1967). 'On weathering and change of properties of Tertiary mudstone related to landslide', *Soil & Found.*, **7**, 1–14.

NOVOSAD, S. (1978). 'The use of modern methods in investigating slope deformations', *Bull. Int. Ass. Engg Geol.*, No. 17, 71–3.

OLIVIER, H. J. (1976). 'Importance of rock durability in the engineering classification of Karroo rock masses for tunnelling', (in *Exploration for Rock Engineering*), Bieniawski, Z. T. (ed.), A. A. Balkema, Cape Town, **1**, 137–44.

OLIVIER, H. J. (1979). 'A new engineering–geological rock durability classification', *Engg Geol.*, **14**, 255–79.

OLLIER, C. D. (1969). *Weathering*, Oliver & Boyd, Edinburgh.

ONODERA, T. F. YOSHINAKA, R. & ODA, M. (1974). 'Weathering and its relation to mechanical properties of granite', *Proc. 3rd Cong. Int. Soc. Rock Mech. Denver*, **2A**, 71–98.

RICHARDS, L. R., LEG, G. M. M. & WHITTLE, R. A. (1978). 'Appraisal of stability conditions in rock slopes, (in *Foundation Engineering in Difficult Ground*) Bell, F. G. (ed.), Butterworths, London, 449–513.

SCHUMM, S. A. (1960). 'The shape of alluvial channels, in relation to sediment type', *US Geol. Surv.*, Prof. Paper 352B, 17–30.

SCHUMM, S. A. (1969). 'River metamorphosis', *Proc. ASCE, J. Hyd. Div.*, **95**, 255–73.

SCHUMM. S. A. (1977). 'Applied fluvial geomorphology', (in *Applied Geomorphology*), Hails, J. R. (ed.), Elsevier, Amsterdam, 119–56.

SHARPE, C. F. S. (1938). *Landslides and Related Phenomena*, Columbia University Press, New York.

SHERMAN, L. K. (1932). 'Streamflow from rainfall by unit graph method', *Engg News Record*, **108**, 501–5.

SILVESTRI, V. (1980). 'The long-term stability of a cutting slope in an overconsolidated sensitive clay', *Canadian Geot. J.*, **17**, 337–51.

SKEMPTON, A. W. (1964). 'Long-term stability of clay slopes' (4th Rankine Lecture), *Geotechnique*, **14**, 77–101.

SKEMPTON, A. W. & DELORY, F. A. (1957). 'Stability of natural slopes in London Clay', *Proc. 4th Int. Conf. Soil Mech. Foundation Engg*, London, **2**, 378–81.

SKEMPTON, A. W. & EARLY, K. R. (1972). 'Investigations of the landslide at Walton's Wood, Staffordshire', *Q.J. Engg Geol.*, **5**, 19–42.

SKEMPTON, A. W. & HUTCHINSON, J. N. (1969). 'Stability of natural slopes and embankment foundations', *Proc. 7th Int. Conf. Soil Mech. Foundation Engg*, State-of-the-Art-Volume, Mexico, 221–42.

SMITH, D. I. (1977). 'Applied geomorphology and hydrology of karst regions', (in *Applied Geomorphology*), Hails, J. R. (ed.), Elsevier, Amsterdam, 85–118.

STRAHLER, A. N. (1964). 'Quantitative geomorphology of drainage basins and channel networks', (in *Handbook of Applied Hydrology*), Chow, V. T. (ed.), McGraw-Hill, New York, 39–76.

SWEETING, M. M. (1964). 'Weathering of limestones', (in *Essays in Geomorphology*), Dury, G. H. (ed.), Heinemann, London, 177–210.

SWEETING, M. M. (1972). *Karst Landforms*, Macmillan, London.

TAYLOR, R. K. & SPEARS, D. A. (1970). 'The breakdown of British Coal Measures rocks', *Int. J. Rock Mech. Min. Sci.*, **7**, 481–501.

TERZAGHI, K. (1950). 'Mechanisms of landslides', (in *Applications of Geology to Engineering Practice*), Paige, S. (ed.), Berkey Volume, *Am. Geol. Soc.*, 83–124.

TERZAGHI, K. (1962). 'Stability of steep slopes on hard unweathered rock', *Geotechnique*, **12**, 251–70.

VALLEGO, L. E. (1980). 'Mechanics of mudflow mobilization in low-angled clay slopes,' *Engg Geol.*, **16**, 63–70.

VARNES, D. J. (1958). 'Landslide types and processes', (in *Landslides in Engineering Practice*), Eckel, E. B. (ed.), Committee on Landslide Investigations, *Highway Res. Board, Spec. Report 29*, Washington, 20–47.

WARD, W. H., BURLAND, J. B. & GALLOIS, R. W. (1968). 'Geotechnical assessment of a site at Mundford, Norfolk, for a large proton accelerator', *Geotechnique*, **18**, 399–431.

WINKLER, E. M. (1973). *Stone Properties, Durability in Man's Environment*, Springer-Verlag, New York.

WOLMAN, M. G. (1971). 'Evaluating alternative techniques of flood plain mapping', *Water Resources Res.*, **7**, 1383–92.

YAGUE, A. G. (1978). 'Modern methods used in the study of mass movements', *Bull. Int. Ass. Engg Geol.*, No. 17, 65–71.

ZARUBA, Q. & MENCL, V. (1969). *Landslides and Their Control*, Elsevier, Prague.

Chapter 8

Geomorphological processes II

GLACIATION

At the present day glaciation is of minor importance in shaping landscapes. However, in Pleistocene times ice-masses were much more extensive and they have left their imprint on over 12 million square kilometres of the Earth's surface. Recent work suggests that there probably were about 20 major events during Pleistocene times, which lasted for some two million years. Each glacial episode produced its own effects and its own suite of deposits. Deposits of earlier glacial episodes were reshaped by later glacial advances. In this way complex glacial sequences can develop due to glaciers advancing and retreating over given areas.

A glacier may be defined as a mass of ice which is formed from recrystallised snow and refrozen melt water and which moves under the influence of gravity. Glaciers develop above the snow-line, that is, in regions of the world which are cold enough to allow snow to remain on the surface throughout the year. The snow-line varies in altitude from sea level in polar regions to above 5000 m in equatorial regions. As the area of a glacier which is exposed to wastage is small compared with its volume, this accounts for the fact that glaciers penetrate into the warmer zones below the snow-line.

Glaciers can be grouped into three types, namely, valley glaciers, piedmont glaciers and ice sheets and ice caps (see Flint (1967)). Valley glaciers flow down pre-existing valleys from mountains where snow has collected and formed into ice. They disappear where the rate of melting exceeds the rate of supply of ice. When a number of valley glaciers emerge from a mountain region onto a plain, where they coalesce, they then form a piedmont glacier. At the present day piedmont glaciers are found in Alaska and Antarctica. Ice sheets are huge masses of ice which extend over areas which may be of continental size, ice caps are of smaller dimensions. At the present day there are two ice sheets in the world, one extends over the Antarctic continent whilst the other covers most of Greenland.

8.1 Glacial erosion

Although pure ice is a comparatively ineffective agent as far as eroding massive rocks is concerned, it does acquire rock debris which enhances its abrasive power. The larger fragments of rock embedded in the sole of a glacier tend to carve grooves in the path over which it travels whilst the finer material smooths and polishes rock surfaces. Ice also erodes by a quarrying process, whereby fragments are plucked from rock surfaces. Generally 'quarrying' is a more effective form of glacial erosion than abrasion.

The rate of glacial erosion is extremely variable and depends upon the velocity of the glacier, the weight of the ice, the abundance and physical character of the rock debris carried at the bottom of the glacier and the resistance offered by the rocks of the glacier channel (see Nye (1952); Von Engeln (1937)). The erodibility of the surface over which a glacier travels will vary with depth and hence with time. Once the weathered over-burden and open jointed bedrock have been removed, the rate of glacial erosion slackens. This is because 'quarrying' becomes less effective and hence the quantity of rock fragments contributed to abrasive action is gradually reduced.

In the case of continental ice sheets, these move very slowly and may be effective agents of erosion only temporarily, removing the weathered mantle from, and smoothing off the irregularities of a landscape. The pre-glacial relief features are consequently afforded protection by the overlying ice against denudation, although the surface is somewhat modified by the formation of hollows and hummocks.

The commonest features produced by glacial abrasion are striations on

Figure 8.1 Ice-sculptured surface of *roche moutonée* type east of Sanna Bay, Ardnamurchan, Scotland (courtesy the Institute of Geological Sciences)

rock surfaces which were formed by rock fragments embedded in the base of the glacier.

Many glaciated slopes formed of resistant rocks which are well jointed display evidence of erosion in the form of ice-moulded hummocks (see Carol (1947)). These hummocks, which are known as *roche moutonnées*, vary in size but are usually asymmetrical in outline (Figure 8.1).

Large, highly resistant obstructions, like volcanic plugs, which lie in the path of advancing ice give rise to features individually called *crag* and *tail* (Figure 8.2). The resistant obstruction forms the crag and offers protection to the softer rocks which occur on its lee-side (see Linton (1963)). They there-

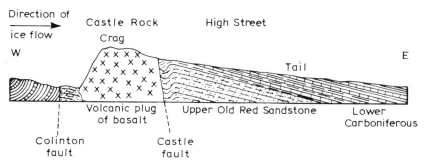

Figure 8.2 Crag and Tail, Edinburgh. The castle rock probably represents an early phase of volcanic activity associated with the ancient volcano, Arthur's Seat

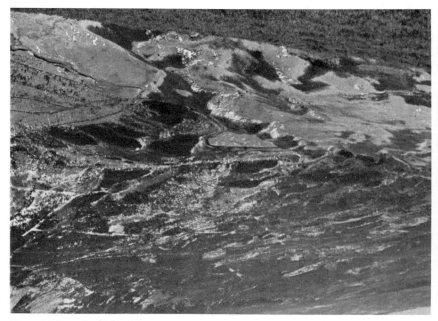

Figure 8.3 Drumlins, Isle of Arran

fore form a tail which slopes gently away from the crag. The tail may or may not possess a covering of till.

Drumlins are mounds which are rather similar in shape to the inverted bowl of a spoon (Figure 8.3). Their long axis is orientated in the direction of ice movement and their stoss end (this faces the direction from which the ice came) is frequently blunted. Drumlins vary in composition ranging from 100% bedrock to 100% glacial deposit. It has been suggested that they were developed under thick ice some distance from the snout of a glacier (see Gravenor (1953)). Obviously those types formed of bedrock originated as a consequence of glacial erosion, however, even those composed of glacial debris were, at least in part, moulded by glacial action. Drumlins range up to a kilometre in length and some may be over 70 m in height. Usually they do not occur singly but in scores or even hundreds in drumlin fields. Some drumlins may be indistinctly separated from each other; they may form double or treble ridges which are united at their stoss ends but which possess distinct tails. The tail of one may rise from the flank of another or small drumlins may arise from the flanks of larger ones. A hummocky drumlin landscape, with its irregular drainage, is commonly referred to as basket of eggs topography.

Corries are located at the head of glaciated valleys, being the features in which ice accumulated. Hence they formed at or close to the snow line. Corries are frequently arranged in tiers and in such instances give rise to a corrie stairway up a mountain side. Because of their shape, corries are often likened to amphitheatres in that they are characterised by steep backwalls and steep sides (Figure 8.4a). Their floors are generally rock basins. Corries vary in size, some of the largest being about 1 km across (see Thompson (1950)). The dominant factor influencing their size is the nature of the rock in which they were excavated. Corries begin life as small nivation hollows. As erosion proceeds they first adopt a circular outline but when mature, their outline becomes rectangular. As the corries about a mountain mass grow, the area between them is progressively reduced until they are separated from each other by sharp ridges termed *arêtes*. Furthermore the headward extension of the backwalls of each corrie into the mountain side eventually produces a pyramidal peak (Figure 8.4b).

The cross profile of a glaciated valley is typically steep sided with a comparatively broad, flat bottom and it is commonly referred to as U-shaped (Figure 8.5). Most glaciated valleys are straighter than those of rivers since their spurs have been truncated by ice. In some glaciated valleys a pronounced bench or shoulder occurs above the steep walls of the trough (see Cotton (1947)). Tributary streams of ice flow across the shoulders to the main glacier. When the ice disappears the tributary valleys are left hanging above the level of the trough floor. The valleys are then occupied by streams. Those in the hanging valleys cascade down the slopes of the main trough as waterfalls. An alluvial cone may be deposited at the base of the waterfall.

Generally glaciated valleys have a scalloped or stepped long profile and sometimes the head of the valley is terminated by a major rock-step known as a trough's end. Such rock-steps develop where a number of tributary glaciers, descending from corries at the head of the valley, converge and thereby effectively increase erosive power. A simple explanation of a scalloped valley floor can be found in the character of the rock type. Not only is a

Figure 8.4 (a)

glaciated valley stepped but reversed gradients are also encountered within its path. The reversed gradients are located in rock floored basins which occur along the valley. Rock basins appear to be formed by localised ice action.

Fiords are found along the coasts of glaciated highland regions which have suffered recent submergence, they represent the drowned part of a glaciated valley (see Holtedahl (1967)). Frequently a terminal rock barrier, the threshold, occurs near the entrance of a fiord. Some thresholds rise very close to sea level, indeed some may be uncovered at low tide. However, water landward of the thresholds very often is deeper than the known post-glacial rise in sea level. For example, depths in excess of 1200 m have been recorded in some Norwegian fiords. Many fiords occur along belts of structural weakness, such as areas shattered by faults or heavily dissected by joints.

8.2 Glacial deposits: unstratified drift

Glacial deposits form a more significant element of the landscape in lowland areas than they do in highlands. Two kinds of glacial deposits are distinguished, namely, unstratified drift or till and stratified drift (see Flint (1967)). However, one type commonly grades into the other. Till is usually regarded as being synonymous with boulder clay and is deposited directly

(b)

Figure 8.4 (a) Snowdon and Glaslyn, a tarn occupying a corrie viewed from the long arête of Crib Goch, North Wales (courtesy the Institute of Geological Sciences). (b) The Matterhorn, Switzerland, an excellent example of a pyramidal peak (courtesy the Swiss National Tourist Office)

Figure 8.5 The Lauterbrunnen Valley, Switzerland, showing the influence of glaciation with its steep sides and truncated spurs (courtesy the Swiss National Tourist Office)

by ice whilst stratified drift is deposited by melt waters issuing from ice (see Boulton (1975)).

Till consists of a variable assortment of rock debris which ranges in size from fine rock flour to boulder clay (see Chapter 10, Section 10.4). Till deposits are characteristically unsorted (see Krumbein (1936); Elson (1961)). The compactness of a till varies according to the degree of consolidation undergone, the amount of cementation and size of the grains. Tills which contain less than 10% clay fraction are usually friable whilst those with over 10% clay tend to be massive and compact. Frequently there is a concentration of boulders near the surface of a till. Such concentrations may have been brought about by post-glacial erosion, either by sheet water action or by deflation winnowing out the finer particles, prior to the establishment of vegetation.

Distinction has been made between tills derived from rock debris carried along at the base of a glacier and those deposits which were transported within or at the terminus of the ice (see Fookes *et al.* (1975)). The former is sometimes referred to as lodgement till whilst the latter is termed ablation till. Lodgement till is commonly compact and fissile, and the fragments of rock it contains are frequently orientated in the path of ice movement. Ablation till accumulates as the ice, in which the parent material is entombed, melts. Hence it is usually uncompacted and non-fissile, and the boulders present display no particular orientation. Since ablation till consists only of the load carried at the time of ablation it usually forms a thinner deposit than does lodgement till.

A moraine is an accumulation of drift deposited directly from a glacier (see Embleton and King (1968)). There are six types of moraines deposited by valley glaciers. Rock debris which a glacier wears from its valley sides and which is supplemented by material that falls from the valley slopes above the ice forms the lateral moraine. When two glaciers become confluent a medial moraine develops from the merger of the two inner lateral moraines. Material which falls onto the surface of a glacier and then makes its way via crevasses into the centre, where it becomes entombed, is termed englacial moraine. Some of this debris, however, eventually reaches the base of the glacier and there enhances the material eroded from the valley floor. This constitutes the subglacial moraine. The ground moraine is often irregularly distributed since it is formed when basal ice becomes overloaded with rock debris and is forced to deposit some of it. That material which is deposited at the snout of a glacier when the rate of wastage is balanced by the rate of outward flow of ice is known as a terminal moraine (Figure 8.6). Terminal moraines possess a curved outline impressed upon them by the lobate nature of the snout of the ice. They are usually discontinuous being interrupted where streams of melt water issue from the glacier. Frequently a series of terminal moraines may be found traversing a valley, the farthest down-valley marking the point of maximum extension of the ice, the others indicating pauses in glacial retreat. Sometimes the latter types are called recessional moraines.

Ground moraines and terminal moraines are the two principal types of moraines deposited by ice sheets which spread over lowland areas (see Charlesworth (1957)). In lowland areas the terminal moraines of ice sheets may rise to a height of some 60 m. In plan they commonly form a series of crescents, each crescent corresponding to a lobe at the snout of the ice. If copious amounts of melt water drained from the ice front then morainic material was washed away, hence a terminal moraine either did not develop or if it did, was of inconspicuous dimension.

8.3 Fluvio-glacial deposits; stratified drift

Stratified deposits of drift are often subdivided into two categories, namely, those deposits which accumulate beyond the limits of the ice, forming in

(a)

(b)

Figure 8.6 (a) Terminal moraine, Isle of Arran. (b) close-up of material in terminal moraine

the streams, lakes or seas and those deposits which develop in contact with the ice. The former type are referred to as pro-glacial deposits whilst the latter are termed ice-contact deposits.

Most melt water streams which deposit outwash fans do not originate at the snout of a glacier but from within or upon the ice. Many of the streams which flow through a glacier have steep gradients and are therefore efficient transporting agents, but when they emerge at the snout, they do so on to a shallower incline and deposition results. Outwash deposits are typically cross bedded and range in size from boulders to coarse sand. When first deposited the porosity of these sediments varies from 25–50%. They are therefore very permeable and hence can resist erosion by local run-off. The finer silt-clay fraction is transported further downstream. Also in this direction an increasing amount of stream alluvium is contributed by tributaries so that eventually the fluvio-glacial deposits cannot be distinguished. Most outwash masses are terraced.

Valley trains are outwash deposits which are confined within long narrow valleys. Since deposition occurs more rapidly at the centres of valleys than at the sides, they are thickest there. If outwash quickly accumulates in a valley it may eventually dam tributary streams so that small lakes form along the sides of the main valley. Such deposition may bury small watersheds and divert pro-glacial streams.

Five different types of stratified drift deposited in glacial lakes have been recognised, namely, terminal moraines, deltas, bottom deposits, ice-rafted erratics and beach deposits. Terminal moraines that formed in glacial lakes differ from those which arose on land in that lacustrine deposits are interstratified with drift. Glacial lake deltas are usually composed of sands and gravels which are typically cross bedded. By contrast those sediments which accumulated on the floors of glacial lakes are fine grained, consisting of silts and clays (see Kuenen (1951)). These fine grained sediments are sometimes composed of alternating laminae of finer and coarser grain size. Each couplet has been termed a varve and sediments so stratified are consequently described as varved (see Chapter 10). Large boulders which occur on the floors of glacial lakes were transported on rafts of ice and were deposited when the ice melted. Usually the larger the glacial lake, the larger were the beach deposits which developed about it. If changes in lake level took place then these may be represented by a terraced series of beach deposits.

Deposition which takes place at the contact of a body of ice is frequently sporadic and irregular. Locally the sediments possess a wide range of grain size, shape and sorting. Most are granular and variations in their engineering properties reflect differences in particle size distribution and shape. Deposits often display abrupt changes in lithology and consequently in density index. They are characteristically deformed since they sag, slump or collapse as the ice supporting them melts.

Kame terraces are deposited by melt water streams which flow along the contact between the ice and the valley side (Figure 8.7). The drift is principally derived from the glacier although some is supplied by tributary streams. They occur in pairs, one each side of the valley. If a series of kame terraces occurs on the valley slopes then each pair represents a pause in the process of glacier thinning. The surfaces of these terraces are often pitted

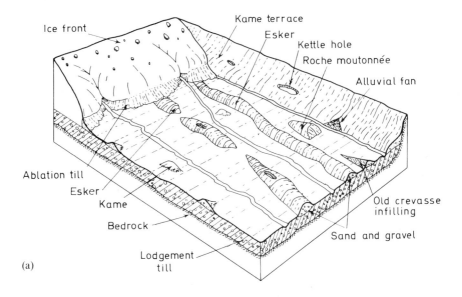

Ice front
Kame terrace
Esker
Kettle hole
Roche moutonnée
Alluvial fan
Ablation till
Esker
Kame
Bedrock
Lodgement till
Old crevasse infilling
Sand and gravel

(a)

(b)

Figure 8.7 (a) Block diagram of a glaciated valley showing some typical glacial deposits. (b) Esker composed mainly of material derived from the Chalk, which has been partially cemented, near Hunstanton, Norfolk

with kettle holes (these are depressions where large blocks of ice remained unmelted whilst material accumulated around them). Narrow kame terraces are usually discontinuous, spurs having impeded deposition.

Kames are mounds of stratified drift which originate as small deltas or fans built against the snout of a glacier where a tunnel in the ice, along which melt water travels, emerges (see Holmes (1947); Figure 8.7). Other small ridge-like kames accumulate in crevasses in stagnant or near-stagnant ice. Many kames do not survive deglaciation for any appreciable period of time.

Eskers are long, narrow, sinuous, ridge-like masses of stratified drift which are unrelated to surface topography (see Figure 8.7). For example, eskers may climb up valley sides and cross low watersheds. They represent sediments deposited by streams which flowed within channels in a glacier (see Lewis (1947)). Although eskers may be interrupted their general continuity is easily discernible and indeed some may extend lengthwise for several hundred kilometres. Eskers may reach up to 50 m in height, whilst they range up to 200 m wide. Their sides are often steep. Eskers are composed principally of sands and gravels, although silts and boulders are found within them. These deposits are generally cross bedded.

Lenses of openwork gravel may mean that the permeability of a fluvio-glacial deposit is higher than expected. The bearing capacity of openwork gravel is generally less than surrounding sandy gravel because of the higher void space. On loading this may give rise to differential settlement.

Lacustrine clays and silts generally occur as layers interbedded with fluvio-glacial sand and gravel deposits, but may also occur as pockets and lenses. Such inclusions may lead to differential settlement or even bearing capacity failure. These interbedded deposits may reduce vertical permeability significantly. Inclusions of till may have similar effects.

8.4 Other glacial effects

Ice sheets have caused diversions of drainage in areas of low relief. In some areas which were completely covered with glacial deposits the post-glacial drainage pattern may bear no relationship to the surface beneath the drift, indeed moraines and eskers may form minor water divides (see Charlesworth (1957)). As would be expected notable changes occurred at or near the margin of the ice. There lakes were formed which were drained by streams whose paths disregarded pre-glacial relief. Evidence of the existence of pro-glacial lakes is to be found in the lacustrine deposits, strandlines and over-flow channels which they leave behind.

Where valley glaciers extend below the snow-line they frequently pond back streams which flow down the valley sides and thereby give rise to lakes. If any col between two valleys is lower than the surface of the glacier occupying one of them, then the water from an adjacent lake dammed by this glacier eventually spills into the adjoining valley and in so doing erodes an overflow channel. Marginal spillways may develop along the side of a valley at the contact with the ice (see Linton (1949); Sissons (1960; 1961).

Lakes of glacial origin form significant features of the post-glacial land-scape. They have been formed either by the scouring action of ice, by being dammed by morainic debris, or occur in depressions in the surface of glacial drift.

The enormous weight of an overlying ice sheet causes the Earth's crust beneath it to sag. Once the ice sheet disappears the land slowly rises to recover its former position and thereby restores isostatic equilibrium. Con-sequently the areas of northern Europe and North America presently affected by isostatic uplift more or less correspond with those areas which were formerly covered with ice. At the present day the rate of isostatic recovery in, for example, the centre of Scandinavia is approximately one metre per century (Figure 8.8). Evidence concerning crustal recovery is demonstrated by river and lake terraces, and raised beaches. Isostatic uplift is neither regular nor continuous. Consequently the rise in the land surface has at times been overtaken by a rise in sea level. The latter was caused by melt water from the retreating ice sheets.

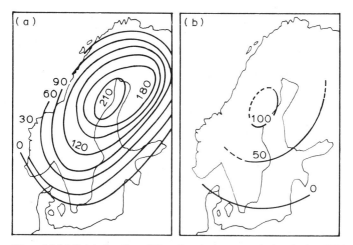

Figure 8.8 (a) Total elevation of Scandinavia, in metres during the last 9700 years. (b) Present rate or rise, in millimetres per 10 years

With the advance and retreat of ice sheets in Pleistocene times the level of the sea fluctuated. Marine terraces (strandlines) were produced during interglacial periods when the sea was at a much higher level. The post-glacial rise in sea level has given rise to drowned coastlines such as rias and fiords, young developing cliff lines, aggraded lower stretches of river valleys, buried channels, submerged forests, marshlands, shelf seas, straits and the reformation of numerous islands.

Buried channels represent abandoned erosional features occupied by coarse stream bedload deposits or, in the case of channels formed by glacial erosion, by coarse granular deposits derived from sub-glacial drainage or a pre-glacial river.

8.5 Frozen ground phenomena in periglacial environments

Frozen ground phenomena are found in regions which experience a tundra climate, that is, in those regions where the winter temperatures rarely rise above freezing point and the summer temperatures are only warm enough to cause thawing in the upper metre or so of the soil (see Derbyshire (1977)). Beneath the upper or active zone the subsoil is permanently frozen and so is known as the permafrost level (see Black (1954)). Because of this layer summer melt water cannot seep into the ground, the active zone then becomes waterlogged and the soils on gentle slopes are liable to flow. Layers or lenses of unfrozen ground termed taliks may occur, often temporarily, in the permafrost (Figure 8.9).

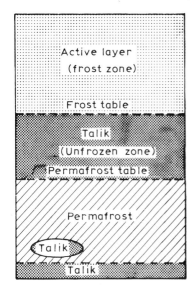

Figure 8.9 Terminology of some features associated with permafrost

Permafrost is an important characteristic, although it is not essential to the definition of periglacial conditions, the latter referring to conditions under which frost action is the predominant weathering agent (see Washburn (1973)). Permafrost covers 20% of the Earth's land surface and during Pleistocene times it was developed over an even larger area. Ground cover, surface water, topography and surface materials all influence the distribution of permafrost. The temperature of perennially frozen ground below the depth of seasonal change ranges from slightly less than 0°C to −12°C. Generally the depth of thaw is less, the higher the latitude. It is at a minimum in peat or highly organic sediments and increases in clay, silt and sand to a maximum in gravel where it may extend to 2 m in depth.

The development of ice means that expansive forces are set up which shatter frozen material and cause ground heave. In the United Kingdom chalk proved especially suspect to frost heave during Pleistocene times. Most chalk possesses a surface zone of very closely jointed material in which the fractures tend to be crazed, with curvi-planar surfaces typical of frost shattering.

Continued frost churning frequently led to the obliteration of the macro-structure in the upper metre or so of chalk and to the formation of a mass of pasty remoulded chalk (putty chalk) enclosing angular fragments of un-altered material, which increase in size with depth as the undisturbed parent chalk is approached (Figure 8.10).

Figure 8.10 Upper surface of the Chalk showing high degree of fracturing due to periglacial action and solution pipes, near Wells, Norfolk

Prolonged freezing gives rise to shattering in the frozen layer, fracturing taking place along joints and cracks. Frost shattering, due to ice action in Pleistocene times, has been found to extend to depths of 30 m in the Chalk (see Higginbottom and Fookes (1970)) and to 12 m in the Borrowdale Volcanic Series (see Knill (1968)). In this way the rock concerned suffers a reduction in bulk density and an increase in deformability and perme-ability. Fretting and spalling are particularly rapid where the rock is closely fractured. Frost shattering may be concentrated along certain preferred planes, if joint patterns are suitably oriented. Preferential opening takes place most frequently in those joints which run more or less parallel with the ground surface. Silt and clay frequently occupy the cracks in frost shattered ground, down to appreciable depth, having been deposited by melt water. Their presence may cause stability problems. If the material possesses a certain range of grain size and permeability, the freezing of intergranular water causes expansion and disruption of previously intact material and frost shattering may be very pronounced.

Knill (1968) described flat-lying discontinuities within a zone of more general shattering. These discontinuities ran more or less sub-parallel to the bedrock surface and possessed an anastomising curvi-planar form. They enclosed lenticular units of frost-shattered rock. Knill noted that such discontinuities commonly are associated with slightly cleaved rocks of Lower Palaeozoic age such as the Borrowdale Volcanic Series, and that they have only been reported from glaciated areas. They are best developed where the strike of the cleavage runs at an appreciable angle to the ground surface, and in flat terrain. Silt and clay infill is usually present in the discontinuities. Knill maintained that these discontinuities are shear surfaces developed initially by glacial drag. Such shear surfaces can penetrate to depths of 30 m. Frost heave under subsequent periglacial conditions is thought to have caused the general frost shattering and fracture enlargement.

Stress relief following the disappearance of ice on melting may cause the enlargement of joints. This may aid failure on those slopes which were over-steepened by glaciation.

Stone polygons are common frozen ground phenomena and fossil forms are found in Pleistocene strata (Figure 8.11). They consist of marginal rings of stone which embrace mounds of finer material. Their diameters range up to 12 m. The stones are gradually raised by frost heaving through the active zone to the surface. Once lifted to the surface the stones are slowly moved to the peripheries of individual mounds where they help construct polygons. The polygons are regular in pattern on slopes up to 2° but become elongated on slopes of 3–7° and form alternating stripes of stones and finer debris where the gradient exceeds 7°. Ice wedges frequently surround polygons, indeed their expansion may be responsible for producing the mound-like appearance of polygons.

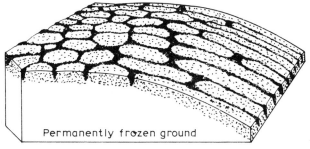

Figure 8.11 Diagram illustrating the merging of stone polygons on a flat surface into stone-stripes on a slope

Frost wedging is one of the chief factors of mechanical weathering in tundra regimes (Figure 8.12). A review of the manner in which frost wedges are formed has been given by Lachenbruch (1966). Ice crystals form in the active zone but patches of ice develop more rapidly below stones in the soil, since they have a greater thermal conductivity than the soil. Such action raises the stones very slightly through the surrounding material and the alternate freeze–thaw action in the active zone eventually carries the stones to the surface. Frozen soils often display a polygonal pattern of cracks (see Shotton (1960)). Individual cracks may be 1.2 m wide at their top, may penetrate to depths of 10 m and may be some 12 m apart. They form when, because of exceptionally low temperatures, shrinkage of the ground occurs. Ice wedges occupy these cracks and cause them to expand. When the ice disappears an ice wedge pseudomorph is formed by sediment, frequently sand, filling the crack.

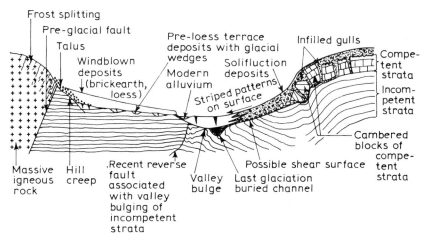

Figure 8.12 Idealised section of valley showing important periglacial features and deposits (after Higginbottom and Fookes (1970))

Ground may undergo notable disturbance as a result of mutual interference of growing bodies of ice or from excess pore pressures developed in confined water-bearing lenses. Involutions are plugs, pockets or tongues of highly disturbed material, generally possessing inferior geotechnical properties, which have been intruded into overlying layers. They are formed as a result of hydrostatic uplift in water trapped under a refreezing surface layer (see West (1968)). They usually are confined to the active layer.

Ice wedges and involutions usually mean that one material suddenly replaces another. This can cause problems in shallow excavations (see Morgan (1971)).

Segregation of lenticular masses of ground ice produces uplift and fissuring in the surface layers and leads to the formation of ice mounds. These measure several metres across and a few metres thick. Poor drainage facilitates the development of ice mounds. On the other hand, solifluction leads to wastage of ice mounds. Consequently water-filled depressions are formed where the ice melts. These depressions are eventually occupied by peat, silt and clay.

Fortunately, however, these deposits are not usually thick enough to cause serious difficulties in ground engineering.

Pingos or hydrolaccoliths are found in Canada, Greenland and Siberia. They are large mounds, being anything up to 300 m in diameter and 50 m in height, which have been upheaved by intrusions of ground ice. Such cone-like features are usually deeply fissured and a crater may occur at their summit. Boreholes sunk into these craters have penetrated several metres of soil and then drilled into ice.

The movement downslope as a viscous flow of saturated rock waste is referred to as solifluction. It is probably the most significant process of mass wastage in tundra regions. Such movement can take place down slopes with gradients as low as 2°. The movement is extremely slow, most measurements showing rates ranging between 10 and 300 mm per year. Solifluction deposits commonly consist of gravels, which are characteristically poorly sorted, sometimes gap-graded, and poorly bedded. These gravels consist of fresh, poorly worn, locally derived material. Individual deposits are rarely more than 3 m thick and frequently display flow structures.

Periglacial action accelerates hill creep, the latter being particularly well developed on thinly bedded or cleaved rocks. Creep material may give way to solifluction deposits on approaching the surface. These deposits consist mainly of flat rock fragments oriented parallel with the hillside and are interrupted by numerous shallow slips.

Sheets and lobes of solifluction debris, transported by mudflow activity, are commonly found at the foot of slopes. These materials may be reactivated by changes in drainage, by stream erosion, by sediment overloading or during construction operations. Solifluction sheets may be underlain by slip surfaces, the residual strength of which controls their stability.

Such material in the United Kingdom is commonly referred to as head, rubble drift or coombe rock, the latter occurring at the base of the Chalk escarpment in the South Downs. Head deposits are usually of late Pleistocene age and have not been consolidated by ice loading. They tend to be relatively permeable, weak and compressible materials. However, their strength is affected by the proportion of rock fragments they contain. Those deposits derived from calcareous rocks may be weakly cemented by calcium carbonate.

The clay-with-flints which caps much of the high ground on the chalk of southern England is regarded as a type of solifluction deposit. It was presumably formed by continued frost-churning of chalk, and sometimes of residual Tertiary material, on relatively flat surfaces where lateral transportation was not significant. It generally is strongly leached and decalcified, and some solution of the underlying chalk may have occurred. The engineering properties of clay-with-flints are usually fairly good because of the high proportion of flint debris and the excellent drainage into the underlying chalk.

Oversteepening of glaciated valleys and melt water channels occurs when ground is stabilised by deep permafrost or supported by ice masses. Frost sapping at the bottom of scarp features also causes oversteepening. When the support disappears, the oversteepened slopes become potentially unstable. Melt water in a shattered rock mass gives rise to an increase in pore pressures which in turn leads to movement or instability along bedding planes and joints. Increase in the moisture content of cohesive material brings about a reduction in its strength and may cause it to swell, thereby aggravating the instability, due to oversteepening, in the near-surface zone. As a result landsliding on a large scale is associated with such oversteepened slopes.

Higginbottom and Fookes (1970) suggested that cambering and valley bulging may have been initiated due to the preferential development of ground ice in valley bottoms where water would have been more readily available to supply growing ice. However, other factors could also have been responsible for the development of these features (see Chapter 7). In the area around Bath, Avon, limestones overlie deposits of clay, both of which are of Jurassic age. In Pleistocene times valley bulging caused thrusting and contortion in the clays of the valley floors and oversteepening of valley slopes (see Chandler *et al.* (1976)). Mass movement by sliding took place under periglacial conditions. Slopes exceeding 15° remain unstable.

The solubility of carbon dioxide in water varies inversely with temperature, for example, it is 1.7 times greater at 0°C than at 15°C. Accordingly cold melt waters frequently have had a strong leaching effect on calcareous rocks. Some pipes and swallow holes in chalk may have been produced by such melt waters. The problem of pipes and swallow holes in chalk is aggravated by their localised character and the frequent absence of surface evidence. They are often undetected by a conventional site investigation.

According to Muller (1947) there are two methods of construction in permafrost, namely, the passive and the active methods. In the former the frozen ground is not disturbed and heat from a structure is prevented from thawing the ground below, thereby reducing its stability. Certainly settlement of notable proportions can occur if the permafrost is melted (see Thomson (1980)). Prevention of heat flow to permafrost can be accomplished by providing an air space beneath the structure. Placing an insulating layer between the structure and the frozen soil delays, but does not stop, thawing.

By contrast the ground is thawed prior to construction in the active method. It is either kept thawed or removed and replaced by materials not affected by frost action. The latter method is used where permafrost is thin, sporadic or discontinuous, and where thawed ground has an acceptable bearing capacity. On the other hand, if permafrost is well developed, then the removal of frozen ground will probably prove impracticable, hence the passive method will be employed.

WIND ACTION AND DESERT LANDSCAPES

Although winds are frequently violent in the United Kingdom, the part they play in sculpting landscapes is negligible compared with that of weathering and rivers. This is largely because vegetation acts as a protective cover against wind erosion. However, in desert regions, because there is little vegetation, wind action is much more significant and sediment yield may be high.

In temperate humid latitudes the work of wind is most significant along low-lying stretches of sandy coasts. In such localities, if the prevailing winds are onshore, sand is heaped into dunes behind the beach. Successive waves of dunes migrate landwards and spoil land which lies in their path.

By itself wind can only remove uncemented rock debris, which it can perform more effectively if the debris is dry rather than wet. But once armed with rock particles, the wind becomes a noteworthy agent of abrasion. The size of individual rock particles which the wind can transport depends on the strength of the wind, and particle shape and weight. The distance which the wind, given that its velocity remains constant, can carry rock particles depends principally upon their size.

8.6 Wind action

Wind erosion takes place when air pressure overcomes the force of gravity on surface particles. At first particles are moved by saltation. The impact of saltating particles on others may cause them to move by creep, saltation or suspension. Saltation accounts for three-quarters of the grains transported by wind, most of the remainder being carried in suspension, the rest are moved by creep or traction. Saltating grains may rise to a height of up to 2 m, their trajectory then being flattened by faster moving air and tailing off as the grains fall to the ground. It has been found that the saltation height generally is inversely related to particle size and directly related to roughness. The length of the trajectory is roughly ten times the height.

One of the most important factors in wind erosion is its velocity. Its turbulence, frequency, duration and direction are also important. As far as the mobility of particles is concerned the important factors are their size, shape and density. It would appear that particles less than 0.1 mm in diameter are usually transported in suspension, those between 0.1 and 0.5 mm are normally transported in saltation and those larger than 0.5 mm tend to be moved by traction or creep. Grains with a relative density of 2.65, such as quartz sand, are most suspect to wind erosion in the size range 0.1–0.15 mm. A wind blowing at 12 km/h will move grains 0.2 mm diameter—a lesser velocity will keep the grains moving.

Because wind can only remove particles of a limited size range, if erosion is to proceed beyond the removal of existing loose particles, then remaining material must be sufficiently broken down by other agents of erosion or weathering. Material which is not sufficiently reduced in size will seriously inhibit further wind erosion. Obviously removal of fine material leads to a proportionate increase in that of larger size which cannot be removed. The latter affords increasing protection against continuing erosion and eventually a wind-stable surface is created. Binding agents, such as silt, clay and organic matter, hold particles together and so make wind erosion more difficult. Soil moisture also contributes to cohesion between particles. Indeed according to Woodruff and Siddoway (1965) the rate of soil movement varies approximately inversely as the square of effective surface soil moisture.

Generally a rough surface tends to reduce the velocity of the wind immediately above it. Consequently particles of a certain size are not as likely to be blown away as they would on a smooth surface. Even so Bagnold (1941) found that grains of sand less than 0.03 mm diameter were not lifted by the wind if the surface on which they lay was smooth. On the other hand, particles of this size can easily remain suspended by the wind. Vegetation affords surface cover and increases surface roughness. The longer the surface distance over which a wind can blow without being interrupted, the more likely it is to attain optimum efficiency.

There are three types of wind erosion, namely, deflation, attrition and abrasion (see Chepil and Woodruff (1963)). Deflation results in the lowering of land surfaces by loose unconsolidated rock waste being blown away by the wind. The effects of deflation are most acutely seen in arid and semi-arid regions. For instance, deflation produced serious soil erosion in the 'Dust Bowl' of the United States. Basin-like depressions are formed by deflation in the Sahara and Kalahari deserts. However, downward lowering is almost invariably arrested when the water table is reached since the wind cannot readily remove moist rock particles. What is more, deflation of sedimentary material, particularly alluvium, creates a protective covering if the material contains pebbles. The fine particles are removed by the wind, leaving a surface formed of pebbles which are too large to be blown away. The suspended load carried by the wind is further comminuted by attrition, turbulence causing the particles to collide vigorously with one another.

When the wind is armed with grains of sand it possesses great erosive force, the effects of which are best displayed in rock deserts. Accordingly any surface subjected to prolonged attack by wind-blown sand is polished, etched or fluted. Abrasion has a selective action, picking out the weaknesses in rocks (see Cooke and Warren (1973)). For example, discontinuities are opened and rock pinnacles developed. Since the heaviest rock particles are transported near to the ground, abrasion is there at its maximum and rock pedestals are formed. In deserts, flat, smoothed surfaces produced by wind erosion are termed desert pavements.

The differential effects of wind erosion are well illustrated in areas where alternating beds of hard and soft rock are exposed. If strata are steeply tilted then, because soft rocks are more readily worn away than hard, a ridge and furrow relief develops. Such ridges are called vardangs. Conversely when an alternating series of hard and soft rocks are more or less horizontally bedded, features known as zeugens are formed. In such cases the beds of hard rock act as resistant caps affording protection to the soft rocks beneath. Nevertheless any weaknesses in the hard caps are picked out by weathering and the caps are eventually breached exposing the underlying soft rocks. Wind erosion rapidly eats into the latter and in the process the hard cap is undermined. As the action continues tubular masses, known as mesas and buttes, are left in isolation.

A brief review of the methods used to control and prevent wind erosion, including windbreaks, and agricultural practices, has been provided by Cooke and Doornkamp (1974).

8.7 Desert dunes

About one-fifth of the land surface of the Earth is desert. Approximately four-fifths of this desert area consists of exposed bedrock or weathered rock waste. The rest is mainly covered with deposits of sand (see Glennie (1970)). Bagnold (1941) recognised five main types of sand accumulations, namely, sand drifts and sand shadows, whalebacks, low-scale undulations, sand sheets and true dunes. He further distinguished two kinds of true dunes, the barkhan and the seif.

Several factors control the form which an accumulation of sand adopts. First, there is the rate at which sand is supplied; second, there is wind speed, frequency and constancy of direction; third, there is the size and shape of the sand grains and fourth, there is the nature of the surface across which the sand is moved. Sand drifts accumulate at the exits of the gaps through which wind is channelled and are extended down-wind. However, such drifts, unlike true dunes, are dispersed if they are moved down-wind. Whalebacks are large mounds of comparatively coarse sand which are thought to represent the relics of seif dunes. Presumably the coarse sand is derived from the lower parts of seifs, where accumulations of coarse sand are known to exist. These features develop in regions devoid of vegetation. By contrast undulating mounds are found in the peripheral areas of deserts where the patchy cover of vegetation slows the wind and creates sand traps. Large undulating mounds are composed of fine sand. Sand sheets are also developed in the marginal areas of deserts. These sheets consist of fine sand which is well sorted; indeed they often present a smooth surface which is capable of resisting wind erosion. A barkhan is crescentic in outline and is orientated at right angles to the prevailing wind direction, whilst a seif is a long ridge-shaped dune running parallel to the direction of the wind. Seif dunes are much larger than barkhans, they may extend lengthwise for up to 90 km and reach heights up to 100 m. Barkhans are rarely more than 30 m in height and their width is usually about twelve times their height. Generally seifs occur in great numbers running approximately equidistant from each other, the individual crests being separated from one another by anything from 30–500 m.

It is commonly believed that sand dunes come into being where some obstacle prevents the free flow of sand, sand piling up on the windward side of the obstacle to form a dune. But, in areas where there is an exceptionally low rainfall and therefore little vegetation to impede the movement of sand, observation has revealed that dunes develop most readily on flat surfaces, devoid of large obstacles. It would seem that where the size of the sand grains varies or where a rocky surface is covered with pebbles, dunes grow over areas of greater width than 5 m. Such patches exert a frictional drag on the wind causing eddies to blow sand towards them. Sand is trapped between the larger grains or pebbles and an accumulation results. If a surface is strewn with patches of sand and pebbles, deposition takes place over the pebbles. However, patches of sand exert a greater frictional drag on strong winds than do patches of pebbles and so deposition under such conditions takes place over the former. When strong winds sweep over a rough surface they become transversely unstable and barkhans may develop.

It would appear that barkhans do not form unless a pile of sand exceeds 0.3 m in height. At this critical height the heap of sand develops a shallow windward and a steep leeward slope. Sand is driven up the former slope and emptied down the latter, but eddying of the wind occurs along the leeward slope, impeding the fall of sand and imparting a concave outline to it. With a smaller height than 0.3 m the sand deposit cannot maintain its form since individual grains, when moved by the wind, fall beyond its boundaries. It is then dispersed to form sand ripples. Wind-blown sand is commonly moved by saltation, that is, in a series of jumps. The distance which an individual particle can leap depends on its size, shape and weight, the wind velocity and the angle of lift which in turn is influenced by the nature of the surface. If a deposit of sand reaches a height of 0.3 m then the grains which leap from its windward face usually land on the leeward slope. As the infant barkhan grows, an increasing amount of sand is piled at the top of the leeward slope since the trajectories of sand grains cannot carry them further. The leeward slope is increasingly steepened until the angle of rest is reached, which is approximately 35°. Subsequent deposition leads to sand slumping down the face. The dune advances in this manner.

The question which now has to be answered is, how does the sand become piled to the critical height? It has been asserted that if the wind increases in velocity it does not immediately become fully loaded with sand. Hence until this happens the wind is capable of moving a greater quantity of sand than normal. Although down-wind the sand-flow is increasing this also means that frictional drag and turbulence are also increasing. The velocity of the wind is therefore reduced and so more sand is deposited than is removed.

As there is more sand to move in the centre of a barkhan than at its tails the latter advance at a faster rate and the deposit gradually assumes an arcuate shape. The tails are drawn out until they reach a length where their obstructive power is the same as that of the centre of the dune. At this point the dune adopts a stable form which is maintained as long as the factors involved in dune development do not radically alter. Barkhans tend to migrate down-wind in waves, their advance varying from about 6–16 m/year, depending on their size (the rate of advance decreases rapidly with increasing size). If there is a steady supply of material the dune tends to advance at a constant speed and to maintain the same shape, whereas when the supply is increased the dune grows and its advance decelerates. The converse happens if the supply of sand decreases (Figure 8.13).

Longitudinal dunes may develop from barkhans. Suppose that the tails of a barkhan for some reason become fixed, for example, by vegetation or

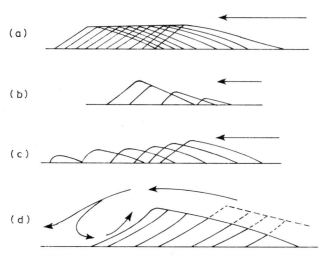

Figure 8.13 Travel and growth of dunes. (a) Regular advance with constant supply. (b) Advance decelerating, with increasing height and sand supply. (c) Advance accelerating, with decreasing height and sand supply. (d) Influence of wind-eddy on sand-fall face

by the water table rising to the surface, then the wind continues to move the central part until the barkhan eventually loses its convex shape becoming concave towards the prevailing wind. As the central area becomes further extended, the barkhan may split. The two separated halves are gradually rotated by the eddying action of the wind until they run parallel to one another, in line with the prevailing wind direction. Dunes that develop in this manner are often referred to as blow-outs.

Seif dunes appear to form where winds blow in two directions, that is, where the prevailing winds are gentle and carry sand into an area, the sand then being driven by occasional strong winds into seif-like forms. Seifs may also develop along the bisectrix between two diagonally opposed winds of roughly equal strength.

Because of their size, seif dunes can trap coarse sand much more easily than can barkhans. This material collects along the lower flanks of the dune. Indeed barkhans sometimes occur in the troughs of seif dunes. On the other hand, the trough may be floored by bare rock. The wind is strongest along the centre of the troughs as the flanks of the seifs slow the wind by frictional

Table 8.1. Objectives and methods of aeolian sand control (from Kerr and Nigra (1952))

Objectives

(1) The destruction or stabilisation of sand accumulations in order to prevent their further migration and encroachment.

(2) The diversion of wind-blown sand around features requiring protection.

(3) The direct and permanent stoppage or impounding of sand before the location or object to be protected.

(4) The rendition of deliberate aid to sand movement in order to avoid deposition over a specific location, especially by augmenting the saltation coefficient through surface smoothing and obstacle removal.

Methods

The above objectives are achieved by the use of one or more types of surface modification.

(1) *Transposing*. Removal of material (using anything from shovels to bucket cranes)—rarely economical or successful, and does not normally feature in long-term plans.

(2) *Trenching*. Cutting of transverse or longitudinal trenches across dunes destroys their symmetry and may lead to dune destruction. Excavation of pits in the lee of sand mounds or on the windward side of features to be protected will provide temporary loci for accumulation.

(3) *Planting* of appropriate vegetation is designed to stop or reduce sand movement, bind surface sand and provide surface protection. Early stages of control may require planting of sand-stilling plants (e.g. *Ammophila arenaria*, beach grass), protection of surface (e.g. by mulching), seeding, and systematic creation of surface organic matter. Planting is permanent and attractive, but expensive to install and maintain.

(4) *Paving* is designed to increase the saltation coefficient of wind-transported material by smoothing or hard-surfacing a relatively level area, thus promoting sand migration and preventing its accumulation at undesirable sites. Often used to leeward of fencing, where wind is unladen of sediment, and paving prevents its re-charge. Paving may be with concrete, asphalt or wind-stable aggregates (e.g. crushed rock).

(5) *Panelling* in which solid barriers are erected to the windward of areas to be protected, is designed either to stop or to deflect sand movement (depending largely on the angle of the barrier to wind direction). In general, this method is inadequate, unsatisfactory and expensive, although it may be suitable for short-term emergency action.

(6) *Fencing*. The use of relatively porous barriers to stop or divert sand movement, or destroy or stabilise dunes. Cheap, portable and expendable structures are desirable (using, for example, palm fronds or chicken wire).

(7) *Oiling* involves the covering of aeolian material with a suitable oil product (e.g. high-gravity oil) which stabilises the treated surface and may destroy dune forms. It is, in many deserts, a quick, cheap and effective method.

drag so creating eddies. The wind blowing along the flanks of seifs is accordingly diverted up their slopes.

The movement of wind-borne material, not always in dune form, often gives rise to problems in many arid regions as far as settlements and agricultural land are concerned. Kerr and Nigra (1952) presented a review of the objectives and methods of control of wind-blown sand (Table 8.1).

8.8 Stream action in arid and semi-arid regions

It must not be imagined that stream activity plays an insignificant role in the evolution of landscape in arid and semi-arid regions. Admittedly the amount of rainfall occurring in arid regions is small and falls irregularly whilst that of semi-arid regions is markedly seasonal. Nevertheless in both instances it falls as heavy and often violent showers. The result is that the river channels cannot cope with the amount of rain water and extensive flooding takes place. These floods develop with remarkable suddenness and either form raging torrents, which tear their way down slopes excavating gullies as they go, or they may assume the form of sheet floods. Dry wadis are rapidly filled with swirling water and are thereby enlarged. However, such floods are short-lived since the water soon becomes choked with sediment and the consistency of the resultant mudflow eventually reaches a point when further movement is checked. Much water is lost by percolation, and mudflows are also checked where there is an appreciable slackening in gradient.

Fookes (1978) gave a general description of the ground conditions met with in alluvial plains and base-level plains in the arid regions of the Middle East. The alluvial plain mainly consists of fine gravels and sands, which are normally spread in layers. Two types have been distinguished, namely, sandy-stony and silty-stony desert, which Fookes interpreted as deposition from stream flow and overbank sheet flow respectively. The terrain can be subject to hazardous sheet flood or stream flood. Desert pavements composed of a single layer of single-sized stone can occur extensively.

The base-level plain is composed of the finest and furthest travelled sediments, that is, wind-blown silts and sands. These have been frequently reworked by flooding or marine inundation. Young granular deposits which are not bound by clay material or cement are often subject to erosion by wind or water. Aggressive salty conditions frequently occur in areas where the water table is high.

Some of the most notable features produced by stream action in arid and semi-arid regions are found in intermontane basins, that is, where mountains circumscribe a basin of inland drainage. The rain which falls on the encircling mountains causes flooding and active erosion. Mechanical weathering plays a significant role in the mountain zone (see Fookes (1978)). Boulders, 2 m or more in diameter, are found in gullies which cut the mountain slopes whilst finer gravels, sands and muds are washed downstream.

Alluvial cones or fans, which consist of irregularly assorted sediment, are found along the foot of the mountain belt where it borders the pediment—the marked change in gradient accounting for the rapid deposition. The particles composing the cones are almost all angular in shape, boulders and cobbles being more frequent upslope, grading downslope into fine gravels. The cones have a fairly high permeability. When these alluvial cones merge into one another they form a bahada. The streams which descend from the mountains rarely reach the centre of the basin since they become choked by their own deposits and split into numerous distributaries which spread a thin veneer of gravels over the pediment.

Although stream flow on alluvial cones is ephemeral, flooding nevertheless can constitute a serious problem, occurring along the margins of the main channels and in the zone of deposition beyond the ends of supply channels. The flood waters are problematic because of their high velocities, their variable sediment content and their tendency to change locations with successive floods, abandoning and creating channels in a relatively short time.

Hydrocompaction may occur on alluvial cones, particularly if they are irrigated. The dried surface layer of these cones may contain many voids. Percolating water frequently reduces the strength of this material which, in turn, reduces the void space. This gives rise to settlement or hydrocompaction.

Pediments in arid regions are graded plains cut by the lateral erosion of ephemeral streams. King (1963) described the pediments found in the semi-arid areas of South Africa as smooth, flat zones across which a maximum of water is discharged (that from the mountains together with that of the pediments). As a consequence pediments are adjusted to dispose of water in the most efficient way and the heavy rainfall characteristic of semi-arid regions means that this is often in the form of sheet wash. Although true laminar flow occurs during sheet wash, as the flowing water deepens laminar flow yields to turbulent flow. The latter possesses much greater erosive power and occurs during and immediately after heavy rainfall. This, it is argued, is why these pediments carry only a thin veneer of rock debris. With a lesser amount of rainfall there is insufficient water to form sheets and it is confined to rills and gullies. According to King the rock waste transported across the pediment is relatively fine and is deposited in hollows thereby smoothing the slope. The abrupt change in the slope at the top of the pediment is caused by a change in the principal processes of earth sculpture, the nature of the pediment being governed by sheet erosion whilst that of the steep hillsides is controlled by the downward movement of rock debris.

Aeolian and fluvial deposits, notably sand, also may be laid down in the intermediate zone between the pediment and the central depression or playa. However, if deflation is active this zone may be barren of sediments. Sands are commonly swept into dunes and the resultant deposits are cross bedded.

The central area of a basin is referred to as the playa and it sometimes contains a lake (Figure 8.14). This area is covered with deposits of sand, silt, clay and evaporites. The silts and clays often contain crystals of salt whose development further comminutes their host. Silts usually exhibit ripple marks whilst clays are frequently laminated. Desiccation structures such as mudcracks are developed on an extensive scale in these fine grained sediments. If the playa lake has contracted to leave a highly saline tract then this area is termed a salina. The capillary rise generally extends to the surface, leading to the formation of a salt crust. Where the capillary rise is near to, but does not normally reach the surface, desiccation ground patterns provide an indication of its closeness.

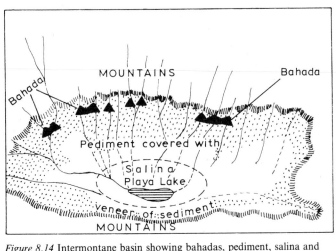

Figure 8.14 Intermontane basin showing bahadas, pediment, salina and playa lake

Duricrust is a surface or near-surface hardened accumulation or encrusting layer, formed by precipitation of salts on evaporation of saline ground-water. It may be composed of calcium or magnesium carbonate, gypsum, silica, alumina or iron oxide, or even halite, in varying proportions (see Goudie (1973)). It may occur in a variety of forms, ranging from a few milli-metres in thickness to over a metre. A leached cavernous, porous or friable zone is frequently found beneath the duricrust. When describing duricrusts those terms ending in *crete* (for example, calcrete, see Chapter 10) refer to hardened surfaces usually occurring on hard rock; those ending in *crust* (for example, gypcrust) represent softer accumulations which usually are found in salt playas, salinas or sabkhas (coastal salt marshes). Locally, especially near the coast, sands may be cemented with calcrete to form cap-rock or miliolite. Desert fill often consists of mixtures of nodular calcrete, calcrete fragments and drifted sand (see Fookes (1978)).

Aggressive salty ground occurs, for instance, in sabkhas, salinas, salt playas and some duricrusts. Salt-weathering brings about the disintegration of rock through crystallisation, hydration and thermal expansion (see Evans (1970)).

The extension of pediments on opposing sides of a mountain mass means that the mountains are slowly reduced until a pediplain is formed. The pediments are first connected through the mountain mass by way of pediment passes. The latter become progressively enlarged forming pediment gaps. Finally opposing pediments meet to form a pediplain on which there are residual hills. Such isolated, steep-sided residual hills have been termed inselbergs or bornhardts. They are characteristically developed in the semi-arid regions of Africa, where they are usually composed of granite or gneiss, that is, of more resistant rock than that which forms the surrounding pedi-plain.

Hazard maps of arid regions, such as maps of unstable sand dunes and their migration, of areas susceptible to flooding, and of the degree of salt-weathering, are frequently produced for development purposes (see Cooke *et al.* (1978); Figure 8.15).

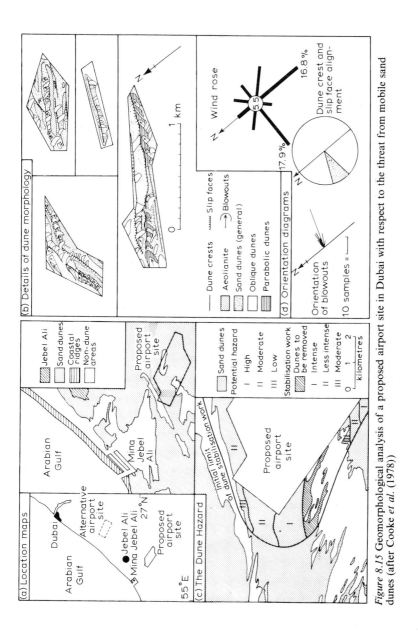

Figure 8.15 Geomorphological analysis of a proposed airport site in Dubai with respect to the threat from mobile sand dunes (after Cooke *et al.* (1978))

8.9 Loess

Loess is an unconsolidated deposit which is believed to be of aeolian origin. It is composed principally of particles of silt size. Although loess is usually unstratified it does occasionally contain laminated units which may have been deposited in temporary small lakes. Loess commonly occurs as sheets less than 100 m thick. Some deposits may extend over thousands of square kilometres.

Loess is largely composed of quartz particles and small amounts of feldspars, micas and clay minerals. Carbonate minerals are usually present, indeed in some instances they may form as much as 40%, by weight, of the deposit. Irregularly shaped calcareous concretions generally occur at specific horizons. Unweathered loess is usually yellow but it may also be various shades of red or brown. These colours are due to the presence of ferric oxides, however, if the iron has been reduced then the loess is a shade of grey.

Crude vertical parting planes are often exhibited, particularly by the finer grained type of loess. These may give a subcolumnar appearance to some outcrops.

Almost without exception, loess is confined to Pleistocene times. Nevertheless two kinds of loess can be recognised, namely, that of non-glacial and that of glacial derivation. Loess derived from deserts can be distinguished from that from a glacial source by the wider range of grain size it displays; it is also less well sorted. A further account of loess and its engineering properties is given in Chapter 10.

COASTS AND SHORELINES

Johnson (1919) distinguished three elements in the shore zone—the coast, the shore and the offshore. The coast was defined as the land immediately behind the cliffs, whilst the shore was regarded as that area between the base of the cliffs and low-water mark. That area which extended seawards from the low-water mark was termed the offshore. The shore was further divided into foreshore and backshore, the former embracing the inter-tidal zone whilst the latter extended from the foreshore to the cliffs.

8.10 Wave action

When wind blows across the surface of deep water it causes an orbital motion of the water particles in the plane normal to the wind direction. Because adjacent particles are at different stages of their circular course a wave is produced. The motion is transmitted to the water beneath the surface but the orbitals are rapidly reduced in size with increasing depth and the motion dies out at a depth equal to that of the wave length (Figure 8.16). There is no progressive forward motion of the water particles in such a wave although the form of the wave profile moves rapidly in the direction in which the wind is blowing. Such waves are described as oscillatory waves.

The parameters of a wave are defined as follows: the wave length (L) is

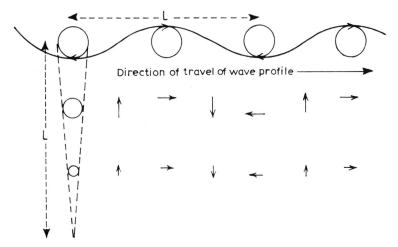

Figure 8.16 Orbital motion of waves in deep water

the horizontal distance between each crest; the wave height (H) is the vertical distance between the crest and the trough; the wave period (T) is the time interval between the passage of successive wave crests. The rate of propagation of the wave form is the wave length divided by the wave period. The height and period of waves are functions of the wind velocity, the fetch (the distance over which the wind blows), and the length of time for which the wind blows (Figure 8.17).

Fetch is the most important factor determining wave size and efficiency of transport. Winds of moderate force which blow over a wide stretch of water generate larger waves than do strong winds which blow across a short reach. Where the fetch is less than 32 km the wave height increases directly as, and the wave period increases as the square root of, the wind velocity. Long waves only develop where the fetch is large, for instance, the largest waves are generated in the southern oceans where their lengths may exceed 600 m and their periods be greater than 20 s. Usually wave lengths in the open sea are less than 100 m and the speed of propagation is approximately 50 km/h.

Those waves which are developed in storm centres in the centre of an ocean may journey to its limits. This explains why large waves may occur along a coast during fine weather.

Waves frequently approach a coastline from different areas of generation. If they are in opposition then their height is decreased whilst their height is increased if they are in phase.

Four types of waves have been distinguished—forced, swell, surf and forerunners. Forced waves are those formed by the wind in the generating area, they are usually irregular. On moving out of the area of generation the waves become long and regular. They are then referred to as swell or free waves. As these waves approach a shoreline they feel bottom which disrupts their pattern of motion, changing them from oscillation to translation waves, in other words, they break into surf. The longest and lowest waves are commonly termed forerunners of swell.

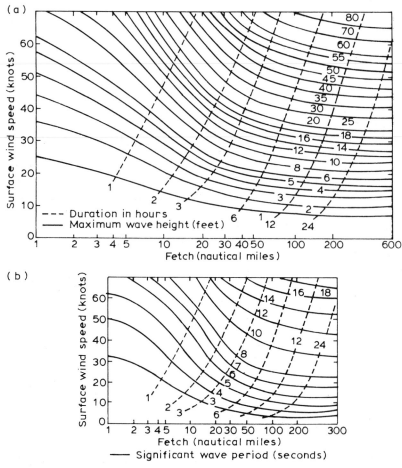

Figure 8.17 Graphs relating wave height (a) and wave period (b) to wind speed, wind duration and fetch in oceanic waters (after Darbyshire, J. and Draper, L. (1963). 'Forecasting wind-generated sea waves', *Engg.*, **195**, 482–4)

Waves, acting on beach material, are a varying force. They vary with time and place due to, first, changes in wind force and direction over a wide area of sea, and second, changes in coastal aspect and offshore relief. This variability means that the beach is rarely in equilibrium with the waves, in spite of the fact that it may only take a few hours for equilibrium to be attained under new conditions. Such a more or less constant state of disequilibrium occurs most frequently where the tidal range is considerable, as waves are continually acting at a different level on the beach.

The breaking of a wave is influenced by its steepness, the slope of the sea floor and the presence of an opposing or supplementary wind. When waves enter water equal in depth to their wave length they begin to feel bottom, and their length decreases whilst their height increases. Their velocity of travel or celerity (*c*) is also reduced in accordance with the expression

$$c = \left[\frac{gL}{2\pi} \tanh \left(\frac{2\pi Z}{L} \right) \right]^{1/2}$$

where L is the wave length, Z is the depth of the water and g is acceleration due to gravity. As a result, the wave steepens until the wave train consists of peaked crests separated by relatively flat troughs. The wave period, however, remains constant. Steepening accelerates towards the breaker zone and the wave height grows to several times what it was in deep water. Three types of breaking waves are distinguished, plunging, spilling and surging or swash. Plunging breakers collapse when their wave height is approximately equal to the depth of the water. They topple suddenly and fall with a crash. They are usually a consequence of long, low swell and their formation is favoured by opposing winds. Spilling breakers begin to break when the wave height is just over one-half of the water depth and they do so gradually over some distance. Generally they result from steep wind waves and they commonly occur when the wind is blowing in the direction of wave propagation. Surging breakers or swash rush up the beach and are usually encountered on beaches with steep profiles. The term backwash is used to describe the water which subsequently descends the beach slope.

Four dynamic zones have been recognised within the nearshore current system of the beach environment, they are, the breaker zone, the surf zone, the transition zone and the swash zone (Figure 8.18). The breaker zone is that in which waves break. The surf zone refers to that region between the breaker zone and the effective seaward limit of the backwash. The presence and width of a surf zone is primarily a function of the beach slope and tidal phase. Those beaches which have gentle foreshore slopes are often characterised by wide surf zones during all tidal phases whereas steep beaches seldom possess this zone. The transition zone includes that region where backwash interferes with the water at the leading edge of the surf zone and it is characterised by high turbulence. That region where water moves up and down the beach is termed the swash zone.

Swash tends to pile water against the shore and thereby gives rise to currents which move along the shore termed longshore currents. After flowing parallel

Water motion	Oscillatory waves	Wave collapse	Waves of translation(bores); longshore currents, seaward return flow, Rip currents	Collision	Swash, backwash	Wind
Dynamic zone	Offshore	Breaker	Surf	Transition	Swash	Berm crest
Profile						MWLW
Sediment size trends	Coarser→	Coarsest grains	←Coarser	Bi-modal Lag deposit	←Coarser	Wind-winnowed Lag deposit
Predominant action	Accretion	Erosion	Transportation	Erosion	Accretion and erosion	
Sorting	←Better—	Poor	Mixed	Poor	Better→	
Energy	—Increase→	High	—Gradient→	High	← →	

Figure 8.18 Summary diagram schematically illustrating the effect of the four major dynamic zones in the beach environment. Hatched areas represent zones of high concentrations of suspended grains. Dispersion of fluorescent sand and electro-mechanical measurements indicate that the surf zone is bounded by two high-energy zones; the breaker zone and the transition zone. MWLW = mean water low water (after Ingle (1966))

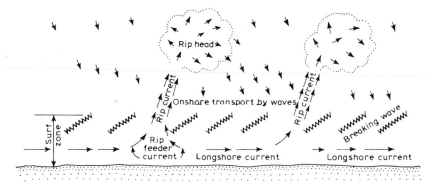

Figure 8.19 Nearshore circulation system and related terms

to the beach the water runs back to the sea in narrow flows called rip currents (Figure 8.19). In the neighbourhood of the breaker zone rip currents extend from the surface to the floor whilst in the deeper reaches they override the bottom water which still maintains an overall onshore motion. The positions of rip currents are governed by submarine topography, coastal configuration and the height and period of the waves. They frequently occur on the up-current sides of points and on either side of convergences where the water moves away from the centre of convergence and turns seawards. The onshore movement of water by wave action in the breaker zone; the lateral transport by longshore currents in the breaker zone; the seaward return of flow as rip currents through the surface zone and the longshore movement in the expanding head of a rip current, all form part of the nearshore circulation system.

Harlow (1979) described an investigation, which involved offshore sampling, to ascertain the sediment distribution pattern between Selsey Bill and Gilkicker Point, Hampshire. He was able to distinguish different sub-zones from analysis of the particle size distribution. A shingle storm beach sub-zone occurs at Hayling Island, which is composed of very well sorted coarse sediments, representing a high energy environment. The sand foreshore sub-zone represents a low energy environment with a gentle slope, the median grain size of the sand ranging between 0.21 and 0.14 mm. A mixed mid-beach sub-zone occurs between the two, where sand is packed in the voids between poorly sorted shingle. A backshore sub-zone occurs behind the storm beach and is made up of very uniformly sorted wind-blown sands.

Tides may play an important part in beach processes. In particular the tidal range is responsible for the area of the foreshore over which waves are active. Tidal streams are especially important where a residual movement resulting from differences between ebb and flood occurs, and where there is abundant loose sediment for the tidal streams to transport. They are frequently fast enough to carry coarse sediment but the forms, notably bars, normally associated with tidal streams in the offshore zone or in tidal estuaries usually consist of sand. These features, therefore, only occur where sufficient sand is available. In quieter areas tidal mud flats and salt marshes are developed, where the tide ebbs and floods over large flat expanses depositing muddy material. Mud also accumulates in runnels landward of high ridges, in lagoons and on the lower foreshore where shelter is provided by offshore banks.

8.11 Marine erosion

Coasts undergoing erosion display two basic elements of the coastal profile, namely, the cliff and the bench or platform. In any theoretical consideration of the evolution of a coastal profile it is assumed that the coast is newly uplifted above sea-level. After a time a wave-cut notch may be excavated and its formation intensifies marine erosion in this narrow zone. The development of a notch varies according to the nature of the rock in which excavation is proceeding, indeed it may be absent if the rocks are unconsolidated or if the bedding planes dip seawards. Where a notch develops it gives rise to a bench and the rock above is undermined and collapses to form a cliff face.

Pot-holes are common features on most wave-cut benches, they are excavated by pebbles and boulders being swilled around in depressions. As they increase in diameter, they coalesce and so lower the surface of the bench. The debris produced by cliff recession gives rise to a rudimentary beach. In tidal seas the base of a cliff is generally at high-tide level whilst in non-tidal seas it is usually above still-water level.

As erosion continues the cliff increases in height and the bench widens. The slope adopted by the bench below sea-level is determined by the ratio of the rate of erosion of the slope to the recession of the cliff. A submarine accumulation terrace forms in front of the bench and is extended out to sea. Because of the decline in wave energy, consequent upon the formation of a wide flat bench, the submarine terrace deposits may be spread over the lower part of the bench in the final stages of its development. The rate of cliff recession is correspondingly retarded and the cliff becomes gently sloping and moribund.

If the relationship between the land and sea remains constant, then erosion and consequently the recession of land beneath the sea is limited. Although sand can be transported at a depth of half the length of storm waves, bedrock is abraded at only half this depth or less. As soon as a submarine bench slope of 0.01–0.05 is attained, bottom abrasion generally ceases and any further deepening is brought about by organisms or chemical solution. Such rock destruction can occur at any depth but the floor can only be lowered where currents remove the altered material.

Like river erosion, marine erosion can be divided into four types, namely hydraulic action, corrosion, corrasion and attrition (see Chapter 7). Studies of quartz sand abrasion in the swash-surf zones have indicated that the loss of sand from a beach due to abrasion is insignificant. Abrasion is, however, an important factor as far as material of cobble size is concerned (see Komar (1976)).

The nature of the impact of a wave upon a coastline depends to some extent on the depth of the water and partly on the size of the wave. The vigour of marine action drops sharply with increasing depth from the water surface, in fact, at approximately the same rate as the decline in the intensity of wave motion. Erosion is unlikely to take place at a depth of more than 60 m along the coast of an open sea and at less than that in closed seas.

If deep water occurs alongside cliffs then waves may be reflected without breaking and in so doing they may interfere with incoming waves. In this way clapotis (standing waves which do not migrate) are formed. It is claimed that the oscillation of standing waves causes an alternate increase and decrease of pressure along discontinuities in rocks which occur in that part of the cliff face below the water line. Also, when waves break, a jet of water is thrown against the cliff at approximately twice the velocity of the wave and, for a few seconds, this increases the pressure within the discontinuities. Such action gradually dislodges blocks of rock.

The force of a wave (P) can be calculated indirectly by observing the maximum height to which the water is thrown when it breaks. According to Zenkovich (1967) this can be expressed by the formula

$$P = 10.8H \text{ kPa}$$

where H is the height to which the water is forced upwards. The energy and the force of translation waves is greater than that of oscillation waves. For instance, it has been estimated that translation waves reflected from a vertical wall exerted six times as much pressure on the wall as oscillation waves of equal dimension. Zhadnov (see Zenkovich (1967)) discovered that the maximum impact recorded at the base of cliffs was not produced by the largest waves, which were partially destroyed on their approach to the coast, but by the waves of moderate storms. He concluded that it was the constant wearing away of rock by waves of any size, rather than by catastrophic ones, that was the most significant aspect of marine erosion.

Those waves with a period of approximately 4 s are usually destructive whilst those with a lower frequency, that is, a period of about 7 s or over are constructive. When high frequency waves collapse they form plunging breakers and the mass of water is accordingly directed downwards at the

(a)

beach. In such instances swash action is weak and, because of the high frequency of the waves, is impeded by backwash. As a consequence material is removed from the top of the beach. The motion within waves which have a lower frequency is more elliptical and produces a strong swash which drives material up the beach. In this case the backwash is reduced in strength because water percolates into the beach deposits and therefore little material is returned down the beach. Although large waves may throw material above the high-water level and thus act as constructive agents, they nevertheless have an overall tendency to erode the beach whilst small waves are constructive.

Swash is relatively ineffective compared with backwash on some shingle beaches. This action frequently leads to very rapid removal of the shingle from the foreshore into the deeper water beyond the break-point. Storm waves on such beaches, however, may throw some pebbles to considerable

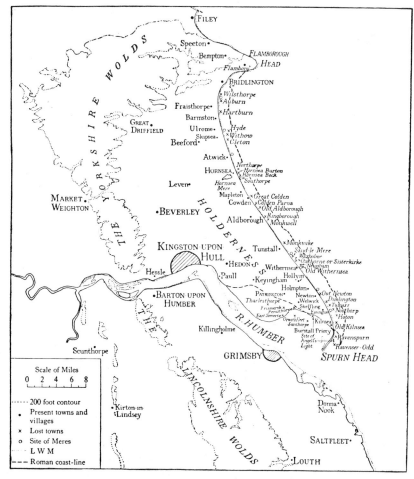

Figure 8.20 (a) Erosion of glacial deposits which form the cliffs north of (b)
Bridlington, Yorkshire (courtesy of *Hull Daily Mail*). (b) The lost towns of
East Yorkshire and Holderness 'Bay' (from Steer (1964))

elevations above mean sea-level creating a storm-beach ridge and, because of rapid percolation of water through the shingle, backwash does not remove these pebbles. On the other hand, when steep storm waves attack a sand beach they are usually entirely destructive and the coarser the sand, the greater the quantity which is removed. Some of this sand may form a submarine bar at the break-point, whilst some is carried into deeper water offshore. It is by no means a rarity for the whole beach to be removed by storm waves, witness the disappearance of sand from some of the beaches along the Lincolnshire coast after the storm-flood of January–February 1953, which exposed their clay base. In some cases a vertical scarp is left on the beach at the high-tide limit.

The rate at which coastal erosion proceeds is influenced by the nature of the coast itself. Around the shores of Britain marine erosion is most rapid where the sea attacks soft unconsolidated sediments of Pliocene and Pleistocene age (Figure 8.20a). For example, some stretches of the coast of Holderness, Yorkshire, disappear at a rate of a metre or so annually (Figure 8.20b). An even more vivid illustration of the ferocity of marine erosion is provided by the fact that the sea at Sheringham, Norfolk, reached a depth of 6 m in 1829 at a point where only 48 years earlier there had been a 15 m cliff (see Steers (1964)). When soft deposits are being actively eroded the cliff displays signs of landsliding together with evidence of scouring at its base (see Hutchinson (1965)). For erosion to continue the debris produced must be removed by the sea. This is usually accomplished by longshore drift. If on the other hand material is deposited to form extensive beaches and the detritus is reduced to a minimum size, then the submarine slope becomes very wide. Wave energy is dissipated as the water moves over such beaches and cliff erosion ceases.

Waves usually leave little trace on massive smooth rocks except to polish them. However, where there are irregularities or projections on a cliff face the upward spray of breaking waves quickly removes them (it has been noted that the force of upward spray along a seawall can be as much as twelve times that of the horizontal impact of the wave).

The degree to which rocks are traversed by discontinuities affects the rate at which they are removed by marine erosion. In particular the attitude of joints and bedding planes is important. Where the bedding planes are vertical or dip inland, then the cliff recedes vertically under marine attack. But if beds dip seawards blocks of rock are more readily dislodged since the removal of material from the base of the cliff means that the rock above lacks support and tends to slide into the sea. Joints may be enlarged into deep narrow inlets such as the geos developed in the Caithness Flagstones, Scotland. Marine erosion is also concentrated along fault planes.

The height of a cliff is another factor which influences the rate at which coastal erosion takes place. The higher the cliff the more material falls when its base is undermined. This in turn means that a greater amount of debris has to be broken down and removed before the cliff is once more attacked with the same vigour.

Erosive forms of local relief include such features as wave-cut notches, caves, blowholes, marine arches and stacks (Figure 8.21). Marine erosion is concentrated in areas along a coast where the rocks offer less resistance. Caves and small bays or coves are excavated where the rocks are softer or

(a)

(b)

Figure 8.21 (a) Blow-hole, Housel Bay, Cornwall. (b) Stacks and marine
arches in the Chalk Falaise d'Etretat, Côte de Caux, France

strongly jointed. At the landward end of large caves there is often an opening to the surface, through which spray issues, which is known as a blowhole. Blowholes are formed by the collapse of jointed blocks loosened by wave compressed air. A marine arch is developed when two caves on opposite sides of a headland unite. When the arch falls the isolated remnant of the headland is referred to as a stack.

Wave refraction is the process whereby the direction of wave travel changes because of changes in the topography of the nearshore sea floor. When waves approach a straight beach at an angle they tend to swing parallel to the shore due to the retarding effect of the shallowing water. At the break-point such waves seldom approach the coast at an angle exceeding 20° irre-spective of the offshore angle to the beach. As the waves break they develop a longshore current, indeed wave refraction is often the major factor in dic-tating the magnitude and direction of longshore drift (see Henderson and Webber (1979)). Wave refraction is also responsible for the concentration of erosion on headlands which leads to a coast being gradually smoothed in outline. As waves approach an irregular shoreline refraction causes them to turn into shallower water so that the wave crests run roughly parallel to the depth contours. The pattern of wave behaviour in shallow water can be determined from refraction diagrams and they provide a qualitative assess-ment of longshore wave power (Figure 8.22). Along an indented coast shallower water is first met with off headlands. This results in wave con-vergence and an increase in wave height, with wave crests becoming concave towards headlands. Conversely where waves move towards a depression in the sea floor they diverge, are decreased in height and become convex towards the shoreline.

Dunes are formed along low-lying stretches of coast, where there is an abundance of sand on the foreshore, by onshore winds carrying sand-sized material landward from the beach. The sand is trapped by obstructions, notably vegetation. Leatherman (1979) maintained that dunes act as barriers, energy dissipators and sand reservoirs during storm conditions. For example, the broad sandy beaches and high dunes along the coast of the Netherlands present a natural defence against inundation during storm surges (see Edel-man (1966)). Because dunes provide a natural defence against erosion, once they are breached the ensuing coastal changes may be long-lasting. During severe storms, a dune face is eroded to give a more or less vertical face. Such faces then act as barriers to storm waves. Dunes, particularly large dunes, subjected to erosion during storm conditions supply much of the sand required to fill the offshore profile and to build up an outer bar, thereby providing an increasing width of beach which is more able to dissipate incident wave energy.

A dune may reflect waves during a high storm surge. Chesnutt and Galvin (1974) showed that reflection values decreased as the inshore zone widened and greatly increased as the offshore profile steepened. The super-position of reflected waves on incident waves produces a new wave system, the velocity components of which equal the sum of the velocity components of the incident and reflected waves. From the point of view of sediment budget the effect of increased beach erosion due to wave reflection from dunes may be masked by the sand contributed to the beach by dune erosion.

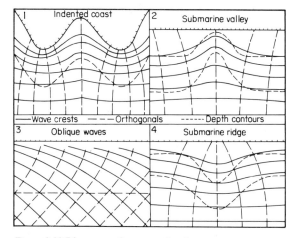

Figure 8.22 Diagrammatic wave refraction patterns

Breaching of dunes by overwash can occur when the supply of sand in the subaerial environment is not large, or when sediment is removed from the beach–inshore system to form back-barrier washover fans. Nevertheless, an appreciable amount of energy is lost when overwash surge flows through a dune field.

Although large amounts of material can be lost by overwash during storms the sediment which is deposited landward of dunes in fact acts as a temporary sediment reservoir. As erosion continues and dunes retreat landward, this overwash sand is reintroduced into the beach environment.

In spite of the fact that dunes inhibit erosion, Leatherman (1979) stressed that they cannot be relied upon to provide protection over the long term along rapidly eroding shorelines, without beach nourishment. The latter widens the beach and maintains the proper functioning of the beach-dune system during normal and storm conditions.

8.12 Constructive action of the sea

Ingle (1966) used fluorescent dyed grains to trace the movement of sand on five beaches along the coast of southern California. He found that the pattern of tracer dispersion on planar beaches showed that a significant proportion of the dyed grains were transported obliquely offshore during all wave conditions. Fewer grains moved along the shore beneath the surf zone and landward into the swash zone. Grains continually moved into the breaker zone where they travelled along the shore. There turbulence kept most grains smaller than 0.15 mm in suspension, whilst larger grains saltated over the floor. The grains followed numerous paths during their transport along the beach, each grain searching for a position of equilibrium. The persistent offshore motion of many grains may reflect this process in that such grains were unable to find equilibrium positions. It was observed that longshore currents could not initiate grain motion although they played a notable role in transporting grains disturbed by bores. If the longshore current velocity was less than 0.3 m/s then grain motion was usually at right angles to the coast but above this grain movement swung shoreward until eventually, at speeds in excess of 0.6 m/s it paralleled the shoreline.

The process whereby grains adjust to the hydrodynamic environment is termed sorting. Material is sorted according to its size, shape and density, and the variations in the energy of the transporting agent.

During swash material of various sizes may be swept along in traction or suspension whereas during backwash the lower degree of turbulence results in a lessened lifting effect so that most grains roll along the bottom. Grains of larger diameter may roll downslope farther than the smaller particles. Tests carried out by Ingle (1966) indicated that coarser grains tended to move obliquely offshore towards the breaker zone whereas the finer material was transported along the shore beneath the surf zone. The median diameter of the grains was less than 0.2 mm except beneath the breaker zone, showing that the coarser particles were not in equilibrium on the beach surface and so were moved offshore to a position of equilibrium beneath the breaker zone. Grains smaller than 0.25 mm found positions of oscillating equilibrium beneath the surf zone and were carried along the shore by longshore currents. Those grains which were not in equilibrium beneath the surf zone usually moved at right angles to the shore.

Beaches may be supplied with sand which is almost entirely derived from the adjacent sea floor although in some areas a larger proportion is produced by cliff erosion. During periods of low waves the differential velocity between onshore and offshore motion is sufficient to move sand onshore except where rip-currents are operational. Onshore movement is particularly notable when long period waves approach a coast whereas sand is removed from the foreshore during high waves of short period.

The beach slope is produced by the interaction of swash and backwash. It is also related to the grain size (Table 8.2) and permeability of the beach. For example, the loss of swash due to percolation into beaches composed of grains of 4 mm in median diameter is ten times greater than into those where the grains average 1 mm. As a result there is almost as much water in the backwash on fine beaches as there is in the swash, so the beach profile is gentle and the sand is hard packed.

Table 8.2. Average beach face slopes compared to sediment diameters (after Shepard (1963))

Type of beach sediment	Size	Average slope of the beach face
Very fine sand	0.0625–0.125 mm	1°
Fine sand	0.125–0.25 mm	3°
Medium sand	0.25–0.5 mm	5°
Coarse sand	0.5–1 mm	7°
Very coarse sand	1–2 mm	9°
Granules	2–4 mm	11°
Pebbles	4–64 mm	17°
Cobbles	64–256 mm	24°

According to King (1972) some wave processes affect sand and shingle in a different manner. Those waves which produce the most conspicuous constructional features on a shingle beach are storm waves, which remove material from a sandy beach. A small foreshore ridge develops on a shingle beach at the limit of the swash when constructional waves are operative. Similar ridges or berms may form on a beach composed of coarse sand. Berms represent a marked change in slope and usually occur a small distance above the high water mark. However, they may be overtopped by high spring tides. Berms are not such conspicuous features on beaches of fine sand. Greater accumulation occurs on coarse sandy beaches because their steeper gradient means that the wave energy is dissipated over a relatively narrow width of beach. King asserted that accretion frequently does not extend to low-water level.

Beach cusps and sandbars are constructional features of small size. The former are commonly found on shingle beaches. They consist of a series of ridges composed of shingle which are separated by troughs in which finer material occurs. Sandbars are characteristic of tideless seas. Their location is related to the break-point which in turn is related to wave size. Consequently more than one bar may form, the outermost being attributable to storm waves, the inner to normal waves. On tidal beaches the break-point migrates over a wide zone hence sandbars do not form and they may be replaced by ripple marks.

Bay-head beaches are one of the commonest types of coastal deposits and they tend to straighten a coastline. Wave refraction causes longshore drift to move from headlands to bays where sediments are deposited. Marine deposition also helps straighten coastlines by building beach plains.

When waves move parallel to the coast they simply move sand and shingle up and down the beach. On the other hand when they approach the coast at an angle material is moved up the beach by the swash in the direction normal to that of wave approach and then it is rolled down the steepest slope of the beach by the backwash. Consequently material is moved in a zig-zag path along the beach. This is known as longshore drift. Such action can transport pebbles appreciable distances along a coast. The duration of movement along a coastline is dependent upon the direction of the dominant winds. An indication of the direction of longshore drift is provided by the orientation of spits along a coast.

The amount of longshore drift is also influenced by coastal outline and wave length. Short waves can approach the shore at a considerable angle, and generate consistent downwave currents. This is particularly the case on straight or gently curving shores and can result in serious erosion where the supply of beach material reaching the coast from updrift is inadequate, such as along the Holderness coast, Yorkshire. Conversely long waves suffer appreciable refraction before they reach the coast. Currents usually flow from zones of convergence, where waves are higher, to zones of divergence where they are lower. Where longshore currents converge they often flow offshore through the surf zone as rip currents.

Material may be supplied to the littoral sediment budget by coastal erosion, by feed from offshore or by contributions from rivers. After sediment has been distributed along the coast by longshore drift it may be deposited in a sediment reservoir and therefore lost from the active environment. Sediment reservoirs formed offshore take the form of bars where the material is in a state of dynamic equilibrium, but from which it may easily re-enter the system. Dunes are the commonest type of onshore reservoir, from which

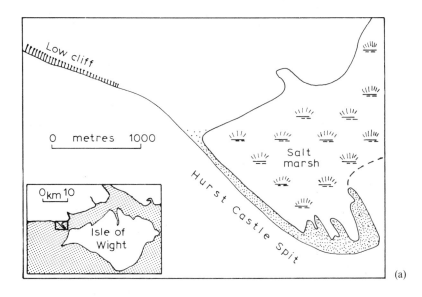

Figure 8.23 (a) The form of Hurst Castle Spit. (b) Fire Island inlet, Long Island, USA. This offshore bar resembles a spit in that it consists of an embankment fronting the shore and has recurred laterals, but it is not tied to the land

sediment is less likely to re-enter the system. According to Harlow (1979), estimation of the littoral sediment budget for a coastal unit could be based on measurement of coastal changes from historical surveys and calculation of the minimum rate of unidirectional drift that could be responsible for the observed changes. It can also be based on surveying sediment distribution patterns by diving in the offshore zones.

Spits are deposits which grow out from the coast. They are supplied with material chiefly by longshore drift. Their growth is spasmodic and alternates with episodes of retreat. The stages in the development of many complex spits and forelands are marked by beach ridges which are frequently continuous over long distances. While longshore drift provides the material for construction their building results from spasmodic progradation, by frontal wave accretion during major storms. The distal end of a spit is frequently curved (Figure 8.23). Those spits which extend from the mainland to link up with an island are known as tombolas (Figure 8.24).

Bay-bars are constructed across the entrance to bays by the growth of a spit being continued from one headland to another. Bays may also be sealed

Figure 8.24 Chesil Beach, Dorset. This tombolo joins the Isle of Portland to the mainland. It is over 25 km long and the shingle composing it is remarkably graded, being about the size of peas near Bridport and potatoes near Portland. The beach rises 5.8 m above high water at Bridport and 13.1 m above at Portland

Figure 8.25 A cuspate foreland, Dungeness. This is one of the largest shingle forelands in the world and its construction has added about 250 km² to the area of Kent

off if spits, which grow from both headlands, merge. If two spits, extending from an island, meet then they form a looped bar.

A cuspate bar arises where a change in the direction of spit growth takes place so that it eventually joins the mainland again, or where two spits coalesce. If progradation occurs then cuspate bars give rise to cuspate forelands, of which the outstanding example in Britain is found at Dungeness (Figure 8.25).

Offshore bars or barriers consist of ridges of shingle or sand. They usually project above sea level, extend for several kilometres and are located a few kilometres offshore. The most frequently quoted explanation of offshore bars is that advanced by Johnson (1919). In the first instance breaking waves erode material from the sea floor which is subsequently deposited to form an offshore bar (Figure 8.26). Material also may be contributed by longshore drift. As time proceeds the bar builds above sea level and in so doing cuts off a lagoon on its landward side. The lagoon is slowly filled with sediment and organic material brought from the land, to give rise to a marshy area. Concurrently storm waves throw material from the seaward to the landward side of the bar. In this manner the bar slowly migrates over the marsh with the result that the marsh bed eventually appears on the seaward side of the bar. With further landward movement of the bar the marsh bed is exposed to marine erosion and finally it disappears when the bar moves onto the land to form a sand-dune.

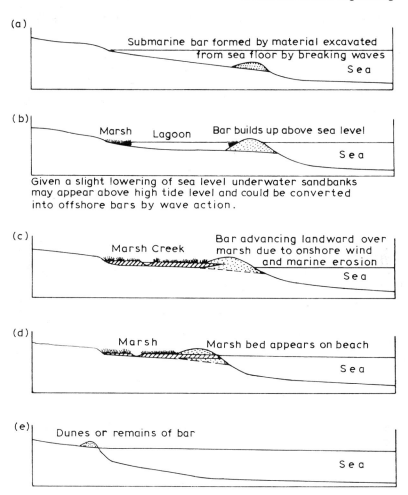

Figure 8.26 Formation of an offshore bar (after Johnson (1919))

8.13 Beach and coastal engineering

The best protection to any coast is afforded by a wide, high beach. The ideal scheme of protection would be one which created zones in which the natural processes could operate without hindrance.

Before any project can be started a complete study of the beach must be made. According to Krumbein (1950) the preliminary investigation of the area concerned should first consider the landforms and rock formations along the beach and adjacent rivers, giving particular attention to their durability and stability. Estimates of the rates of erosion and the proportion of eroded material contributed to the beach must be made. The engineer needs to know how the rocks subjected to erosion behave under wave attack. Not only must the character of the rocks be investigated but the presence of joints and fractures must also be studied. The latter afford some estimate of the average size of fragments which may result from erosion. The rapidity of the break-down of rock fragments into individual particles must also be noted. The behaviour of unconsolidated materials when weathered, particularly when subjected to alternate wetting and drying and freeze–thaw action, together with their slope stability and likelihood of sliding has to be taken into account. Consideration must be given to the width, slope, composition and state of erosion or accretion of the beach, the presence of bluffs, dunes, marshy areas or vegetation in the backshore area, the rock formations which compose the headlands and the presence of beach structures such as groynes. Samples of the beach and underwater material have to be collected and analysed for such factors as their particle size distribution and mineral content. Mechanical analyses may prove useful in helping to determine the amount of material which is likely to remain on the beach, for beach sand is seldom finer than 0.1 mm in diameter. The amount of material moving along the shore must be investigated inasmuch as the effectiveness of the structures erected may depend upon the quantity of drift available.

Topographic and hydrographic surveys of an area allow the compilation of maps and charts from which a study of the changes along the coast may be made. Observations are taken of winds, waves and currents and information is gathered on streams which enter the sea in or near the area concerned. Inlets across a beach need particular evaluation. During normal times there may be relatively little longshore drift but if upbeach breakthroughs occur in a bar off the inlet mouth, then sand is moved downbeach and is subjected to longshore drift. The contributions made by large streams may vary. For example, material brought down by large floods may cause a temporary, but nevertheless appreciable, increase in the beach width around the mouths of rivers.

Selection of the measures to be taken necessitates consideration of whether the beach conditions represent a long-term trend or whether they are cyclical phenomena which may recur at some time in the future. The recent history of a beach and the marine processes operating on it may be evaluated from the following criteria suggested by the American Beach Erosion Board* (1938). There is usually a well developed berm where sand has recently been deposited on a beach but if the beach has suffered erosion then there is usually an almost vertical scarp along the berm front. Moreover, if erosion is active along the beach the foreshore slope tends to be concave upwards from the low-water mark to the cliff or other shoreward limit of erosion. A prograding beach is often softer and has a gentler slope than one which is being eroded. Observations must be extended over at least one complete storm cycle because beach slopes and other features may change rapidly during such times.

The groyne is the most important structure used to stabilise beaches (Figure 8.27). They stabilise or increase the width of the beach by arresting longshore drift (see Silvester (1974)). Consequently they are constructed at right angles to the shore. Groynes should be approximately 50% longer than the beach on which they are erected. Standard types usually slope at about the same angle as the beach. Permeable groynes have openings which increase in size seawards and thereby allow some drift material to pass through

* Superseded by the Coastal Engineering Research Center (CERC) in 1963.

Figure 8.27 Beach drift impeded by groynes at Eastbourne, north east of Beachy Head, Sussex. The direction of drifting is to the north east, i.e. up-Channel (courtesy of Aerofilms Ltd)

them (see Thorn and Simmons (1971)). The common spacing rule for groynes is to arrange them at intervals of one to three groyne-lengths. The selection of the type of groyne and its spacing depends upon the direction and strength of the prevailing or storm waves, the amount and direction of longshore drift and the relative exposure of the shore. With abundant longshore drift and relatively mild storm conditions almost any type of groyne appears satisfactory whilst when the longshore drift is lean, the choice is much more difficult. Groynes, however, reduce the amount of material passing downdrift and therefore can prove detrimental to those areas of the coastline. Indeed Hails (1977) pointed out that groynes may simply transfer the problem of erosion from one section of a coast to another. Their effect on the whole coastal system should therefore be considered.

Artificial replenishment of the shore by building beach fills is used either if it is economically preferable or if artificial barriers fail to defend the shore adequately from erosion (see Newman (1976)). In fact beach nourishment is becoming more common since it represents the only form of coastal protection which does not adversely affect other sectors of the coast. Unfortunately, because of the number of factors involved, it is often difficult to predict how frequently a beach should be renourished. Ideally the beach fill used for renourishment should have a similar particle size distribution to the natural beach material, it should be at least as coarse grained (see Krumbein and James (1965)).

The dynamic sediment budget along the intertidal area of sandy surf beaches is influenced by the interaction of swash–backwash flows with the beach water table. The height at which the water table outcrops on the beach is affected by the previous tide. Chappell et al. (1979) carried out an investigation which indicated that beach aggradation can be induced by maintaining the beach water table at a low level. This leads to increased infiltration in the swash zone and sediment entrainment during backwash is reduced. They recommended that pumping from the water table should take place during storms, when beach face slumping due to liquefaction of sand can present a problem and during prolonged periods of moderate swell, when sand accretion may be accelerated.

Seawalls and bulkheads are protective waterfront structures. The former range from a simple rip-rap deposit to a regular masonry retaining wall. Bulkheads are vertical walls either of timber boards or of steel sheet piling. Foundation ground conditions for these retaining structures must be given careful attention and due consideration must be given to the likelihood of scour occurring at the foot of the wall and to changes in beach conditions (see Herbich and Ko (1968)). Cut-off walls of steel sheet piling below reinforced concrete superstructures may provide an effective method of construction, ensuring protection against scouring.

Provided the wall structure does not encroach far below high tide mark, its existence should not affect the stability of the beach. Nevertheless, because walls are impermeable they can increase the backwash and therefore its erosive capability. Walls provide some protection against marine inundation.

Harris and Ralph (1980) described coastal protection measures at Clacton-on-Sea, Essex. The cliffs at Clacton consist of Pleistocene gravels overlying London Clay and reach a height of 20 m AOD. Buried channels occur in the upper surface of the clay and they conduct water to the front of the cliffs, thereby weakening the clay. The situation was further complicated by the fact that the stability of the existing seawalls protecting the base of the cliffs was endangered by erosion on the downdrift side of two large groynes. Remedial measures included drainage of perched water tables and the cliffs by vertical and inclined bored drains, protection of the base of the cliffs with stepped aprons and seawalls and the installation of a new groyne system, together with beach replenishment. The scheme has proved successful so far. This illustrates that a coastal protection scheme may have to include more than one measure if it is going to be effective.

Offshore breakwaters and jetties are designed to protect inlets and harbours. Both structures are designed to prevent serious wave action in the area they protect. Breakwaters disperse the waves of heavy seas and provide shelter for harbours and ships whilst jetties impound longshore drift material upbeach of the inlet and thereby prevent sanding of the channel they protect. Offshore breakwaters commonly run parallel to the shore or at slight angles to it, chosen with respect to the direction which storm waves approach the coast. Although long offshore breakwaters shelter their leeside they also cause wave refraction and so may generate currents in opposite directions along the shore towards the centre of the sheltered area with resultant impounding of sand. Jetties are usually built at right angles to the shore although their outer segments may be set at an angle. Two parallel jetties may extend from each side of a river for some distance out to sea and because of its confinement, the velocity of river flow is increased, which in turn lessens the amount of deposition which takes place. Like groynes, such structures inhibit the downdrift movement of material, downdrift beaches being deprived of sediment and serious erosion may result (see Johnson (1974)).

Reclamation of land from the sea together with schemes for protection against marine inundation can be illustrated by a brief outline of the schemes initiated in The Netherlands. In 1927 the Dutch began work on the Zuider Zee scheme which set out to reclaim 550 000 hectares and convert the remainder of the bay into a fresh water lake. This project will increase the total area of The Netherlands by about 7%. The enclosing dam across the Zuider Zee is some 32 km long and 90 m wide at its crest and was completed in 1932. Pumping began immediately after its completion and the salt water which it enclosed was converted into a fresh water lake, namely, Yssel Meer. The Wieringer Meer Polder was the first to be drained (1930) and has an area of 49 000 hectares. The North Eastern Polder, 119 000 hectares, was drained by 1942 and Eastern Flevoland, 133 000 hectares, by 1957. The Markerwaard and Southern Flevoland add a further 150 000 and 100 000 hectares respectively.

The Delta scheme aims to seal the four major channels of the Schelde estuary by permanent barrages (see Ferguson (1972)). Two main channels will be left to serve Rotterdam and Antwerp but these will be isolated by structures from the impounded area, the water in which will gradually become fresh. The project does not set out to reclaim land but to offer protection against future marine inundation. Another proposed scheme seeks to impound the northern part of the Zuider Zee by using the Frisian Islands as the backbone of an immense enclosing dam.

References

AMERICAN BEACH EROSION BOARD (1938). *Manual of Procedure in Beach Erosion Studies*, Beach Erosion Board, Washington.

BAGNOLD, R. A. (1941). *The Physics of Blown Sand and Desert Dunes*, Methuen, London.

BLACK, R. F. (1954). 'Permafrost—a review', *Bull. Geol. Soc. Am.*, **65**, 839–56.

BOULTON, G. S. (1975). 'The genesis of glacial tills, a framework for geotechnical interpretation', (in *The Engineering Behavior of Glacial Materials*), *Proc. Symp. Midland Soil Mech. Foundation Engg Soc.*, Birmingham University, 52–69.

CAROL, H. (1947). 'The formation of roche moutonnées', *J. Glacial*, **1**, 57–9.

CHANDLER, R. J., KELLAWAY, G. A., SKEMPTON, A. W. & WYATT, R. J. (1976). 'Valley slope sections in Jurassic strata near Bath, Somerset', *Phil. Trans. Roy. Soc. London*, **A283**, 527–56.

CHAPPELL, J., ELIOT, I. G., BRADSHAW, M. P. & LONSDALE, E. (1979). 'Experimental control of beach face dynamics by water table pumping', *Engg Geol.*, **14**, 29–41.

CHARLESWORTH, J. K. (1957). *The Quarternary Era*, Edward Arnold, London.

CHEPIL, W. S. & WOODRUFF, N. P. (1963). 'The physics of wind erosion and its control', *Adv. in Agronom.*, **15**, 211–302.

CHESNUTT, C. B. & GALVIN, C. J. (1974). 'Laboratory profile and reflection changes for $H_0/L_0 = 0.02$', *Proc. 14th Int. Conf., Coastal Engg Copenhagen*, 958–77.

COOKE, R. U. & DOORNKAMP, J. C. (1974). *Geomorphology in Environmental Management*, Clarendon Press, Oxford.

COOKE, R. U., GOUDIE, A. S. & DOORNKAMP, J. C. (1978). 'Middle East—review and bibliography of geomorphological contributions', *Q. J. Engg Geol.*, **11**, 9–18.

COOKE, R. M. & WARREN, A. (1973). *Geomorphology in Deserts*, Batsford, London.

COTTON, C. A. (1947). *Climatic Accidents in Landscape Making*, Whitcombe-Tombs, Christchurch, New Zealand.

DERBYSHIRE, E. (1977). 'Periglacial environments' (in *Applied Geomorphology*), Hails, J. R. (ed.), Elsevier, Amsterdam, 227–76.

EDELMAN, T. (1966). 'Systematic measurements along the Dutch coast', *Proc. 10th Int. Conf., Coastal Engg, Tokyo*, **1**, 489–501.

ELSON, J. A. (1961). 'Geology of tills', *Proc. 14th Canadian Conf. Soil Mech.*, Sect. 3, 5–17.

EMBLETON, C. & KING, C. A. M. (1968). *Glacial and Periglacial Geomorphology*, Edward Arnold, London.

EVANS, I. S. (1970). 'Salt crystallization and rock weathering: a review', *Revue Geomorph. Dyn.*, **19**, 153–77.

FERGUSON, H. A. (1972). 'The Netherlands Delta project: problems and lessons', *Proc. Inst. Civil Engrs*, **47**, Paper 7449, 465–80.

FLINT, R. F. (1967). *Glacial and Pleistocene Geology*, Wiley, New York.

FOOKES, P. G. (1978). 'Middle East—inherent ground problems', *Q. J. Engg Geol.*, **11**, 33–50.

FOOKES, P. G., GORDON, D. L. & HIGGINBOTTOM, I. E. (1975). 'Glacial landforms, their deposits and engineering characteristics' (in *The Engineering Behaviour of Glacial Materials*), *Proc. Symp. Midland Soil Mech. Foundation Engg Soc.*, Birmingham University, 18–51.

GLENNIE, K. W. (1970). *Desert Sedimentary Environments*, Elsevier, Amsterdam.

GOUDIE, A. S. (1973). *Duricrusts in Tropical and Subtropical Landscapes*, Clarendon Press, Oxford.

GRAVENOR, C. P. (1953). 'The origin of drumlins', *Am. J. Sci.*, **251**, 624–81.

HAILS, J. R. (1977). 'Applied geomorphology in coastal zone planning and management' (in *Applied Geomorphology*), Hails, J. R. (ed.), Elsevier, Amsterdam, 317–62.

HARLOW, D. A. (1979). 'The littoral sediment budget between Selsey Bill and Gilkicker Point, and its relevance to coastal protection works on Hayling Island', *Q. J. Engg Geol.*, **12**, 257–66.

HARRIS, W. B. & RALPH, K. J. (1980). 'Coastal engineering problems at Clacton-on-Sea, Essex', *Q.J. Engg Geol.*, **13**, 97–104.

HENDERSON, G. & WEBBER, N. B. (1979). 'The application of wave refraction diagrams to shoreline protection problems: with particular reference to Poole and Christchurch', *Q.J. Engg Geol.*, **12**, 319–27.

HERBICH, J. B. & KO, S. C. (1968). 'Scour on sand beaches in front of seawalls', *Proc. 11th Int. Conf. Coastal Engg, London*, 622–43.

HIGGINBOTTOM, I. E. & FOOKES, P. G. (1970). 'Engineering aspects of periglacial features in Britain', *Q. J. Engg Geol.*, **3**, 85–118.

HOLMES, C. D. (1947). 'Kames', *Am. J. Sci.*, **245**, 240–49.

HOLTEDAHL, H. (1967). 'Notes on the formation of fjords and fjord valleys', *Geogr. Annlr.*, **49**, 188–203.

HUTCHINSON, J. N. (1965). 'A survey of the coastal landslides of Kent', *Build. Res. Stn Watford*, Note EN/35/65.

INGLE, J. G. (1966). *Movement of Beach Sand*, Elsevier, Amsterdam.

JOHNSON, D. W. (1919). *Shoreline Processes and Shoreline Development*, Wiley, New York.

JOHNSON, J. W. (1974). 'History of some aspects of modern coastal engineering', *Proc. 14th Int. Conf. Coastal Engg, Copenhagen*, **1**, 21–44.

KERR, R. C. & NIGRA, J. O. (1952). 'Eolian sand control', *Bull. Am. Ass. Petrol. Geologists*, **36**, 1541–73.

KING, C. A. M. (1972). *Beaches and Coasts*, Edward Arnold, London.

KING, L. C. (1963). *South African Scenery—a Textbook of Geomorphology*, Oliver & Boyd, Edinburgh.

KNILL, J. L. (1968). 'Geotechnical significance of some glacially induced rock discontinuities', *Bull. Ass. Engg Geologists*, **5**, 49–62.

KOMAR, P. D. (1976). *Beach Processes and Sedimentation*, Prentice-Hall, Englewood Cliffs, New Jersey.

KRUMBEIN, W. C. (1936). 'Textural and lithologic variations in glacial till', *J. Geol.*, **41**, 382–408.

KRUMBEIN, W. C. (1950). 'Geological aspects of beach engineering' (in *Application of Geology to Engineering Practice*), Berkey Volume, Geol. Soc. Am., 195–223.

KRUMBEIN, W. C. & JAMES, W. R. (1965). 'A lognormal size distribution model for estimating stability of beach fill material', *U.S. Army Coastal Engg Res. Center, Tech. Mem.*, **16**, Washington DC.

KUENEN, PH. H. (1951). 'The mechanics of varve formation and the action of turbidity currents', *Geol. För. Stockh. Forh.*, **6**, 149–62.

LACHENBRUCH (1966). 'Contraction theory of ice wedge polygons', *Proc. Int. Conf. Permafrost, Nat. Acad. Sci., Nat. Res. Coun., Publ. 1287*, 65–71.

LEATHERMAN, S. P. (1979). 'Beach and dune interactions during storm conditions', *Q.J. Engg Geol.*, **12**, 281–90.

LEWIS, W. V. (1947). 'An esker in process of formation, Boverbreen Jotunheim', *J. Glacial*, **1**, 314–19.

LINTON, D. L. (1963). 'The forms of glacial erosion', *Trans. Inst. Br. Geogr.*, **33**, 1–28.

LINTON, D. R. (1949). 'Watershed breaching by ice in Scotland', *Trans. Inst. Br. Geogr.*, **15**, 1–16.

MORGAN, A. V. (1971). 'Engineering problems caused by fossil permafrost features in the English Midlands', *Q. J. Engg Geol.*, **4**, 111–14.

MULLER, S. W. (1947). *Permafrost or Permanently Frozen Ground and Related Engineering Problems*, Edwards Bros., Ann Arbor, Mich.

NEWMAN, D. E. (1976). 'Beach replenishment: sea defences and a review of the role of artificial beach replenishment', *Proc. Inst. Civil Engrs*, **60**, 445–60.

NYE, J. F. (1952). 'Mechanics of glacier flow', *J. Glacial*, **2**, 82–93.

SHEPARD, F. P. (1963). *Submarine Geology*, Harper & Row, New York.

SHOTTON, F. W. (1960). 'Large-scale patterned ground in the valley of the Worcestershire Avon', *Geol. Mag.*, **97**, 404–408.

SILVESTER, R. (1974). *Coastal Engineering*, Elsevier, Amsterdam.

SISSONS, J. B. (1960, 1961). 'Some aspects of glacial drainage channels in Britain Parts I & II', *Scott. Geogr. Mag.*, **76**, 131–46; **77**, 15–36.

STEERS, J. A. (1964). *The Coastline of England and Wales*, Cambridge University Press, Cambridge.

THOMPSON, H. R. (1950). 'Some corries of north west Sutherland', *Proc. Geol. Ass.*, **61**, 145–154.

THOMSON, S. (1980). 'A brief review of foundation construction in the western Canadian Arctic', *Q.J. Engg Geol.*, **13**, 67–76.

THORN, R. B. & SIMMONS, J. C. F. (1971). *Sea Defence Works*, Butterworths, London.

VON ENGELN, A. D. (1937). 'Glacial geomorphology and glacier motion', *Am. J. Sci.*, **235**, 426–40.

WASHBURN, A. L. (1973). *Periglacial Processes and Environments*, Edward Arnold, London.

WEST, R. G. (1968). *Pleistocene Geology and Biology*, Longmans, London.

WOODRUFF, N. P. & SIDDOWAY, F. H. (1965). 'A wind erosion equation', *Proc. Soil. Sci. Soc. Am.*, **29**, 602–608.

ZENKOVICH, V. P. (1967). *Processes of Coastal Development*, Steers, J. A. (ed.), Fry, D. G. (trans.), Oliver & Boyd, Edinburgh.

Chapter 9

The mechanics of soil

9.1 Origin of soil

Soil is an unconsolidated assemblage of solid particles between which are voids. These may contain water or air, or both. Soil is derived from the breakdown of rock material by weathering and/or erosion and it may have suffered some amount of transportation prior to deposition. It may also contain organic matter. The type of breakdown process(es) and the amount of transport undergone by sediments influence the nature of the macro- and microstructure of the soil, which in turn influence its engineering behaviour (Table 9.1).

Probably the most important methods of soil formation are mechanical and chemical weathering. The agents of weathering, however, are not capable of transporting material. Transport is brought about by gravity, water, wind or moving ice. If sedimentary particles are transported then this affects their character, particularly their grain size distribution, sorting and shape. For

Table 9.1. Effects of transportation on sediments

	Gravity	*Ice*	*Water*	*Air*
Size	Various	Varies from clay to boulders	Various sizes from boulder gravel to muds	Sand size and less
Sorting	Unsorted	Generally unsorted	Sorting takes place both laterally and vertically. Marine deposits often uniformly sorted. River deposits may be well sorted	Uniformly sorted
Shape	Angular	Angular	From angular to well rounded	Well rounded
Surface texture	Striated surfaces	Striated surfaces	Gravel: rugose surfaces. Sand: smooth, polished surfaces. Silt: little effect	Impact produces frosted surfaces

instance, stream channel deposits are commonly well graded, although the grain size characteristics may vary erratically with location. On the other hand, wind-blown deposits are usually uniformly sorted with well rounded grains.

Changes occur in soils after they have accumulated. In particular, seasonal changes take place in the moisture content of sediments above the water table. Volume changes associated with alternative wetting and drying occur in cohesive soils with high plasticity indices (see below). Exposure of a soil to dry conditions means that its surface dries out and that water is drawn from deeper zones by capillary action. The capillary rise is associated with a decrease in pore pressure in the layer beneath the surface and a corresponding increase in effective pressure. This supplementary pressure is known as capillary pressure and it has the same mechanical effect as a heavy surcharge. Therefore surface evaporation from very compressible soils produces a conspicuous decrease in the void ratio of the layer undergoing desiccation. If the moisture content in this layer reaches the shrinkage level then air begins to invade the voids and the soil structure begins to break down. Moreover if the plasticity index of the soil exceeds 20% then the seasonal variations in moisture content of the upper layers are accompanied by ground movement. The decrease in void ratio consequent upon desiccation of a cohesive sediment leads to an increase in its shearing strength. Thus if a dry crust is located at or near the surface above softer material it acts as a raft. The thickness of dry crusts often varies erratically.

Chemical changes which take place in the soil due, for example, to the action of weathering, may bring about an increase in its clay mineral content, the latter developing from the breakdown of less stable minerals. In such instances the plasticity of the soil increases whilst its permeability decreases. Leaching, whereby soluble constituents are removed from the upper, to be precipitated in a lower horizon, occurs where rainfall exceeds evaporation. The porosity may be increased in the zone undergoing leaching, which in a clay soil may mean that it becomes more compressible. Profound changes are thought to have occurred in many marine clays of Pleistocene age as a result of base exchange, which has involved a reduction in the salt content in the pore water (see Chapter 10).

During sediment accumulation the stress at any given elevation continues to build up as the thickness of the overburden increases. As a result the properties of the sediment are continually changing, the void space, in particular, being reduced. If, subsequently, overburden is removed by erosion, or for that matter by extensive excavation, the void ratio tends to increase. A cohesive deposit which is in equilibrium under the maximum stress it has ever experienced is referred to as normally consolidated whereas one experiencing less stress than it did previously because of removal of overburden, is described as overconsolidated.

With continuing exposure soil develops a characteristic profile from the surface downwards. This development involves the accumulation and decay of organic matter, leaching, precipitation, oxidation or reduction and further mechanical and biological breakdown. The profile which forms is influenced by the character of the parental material but climatic conditions, vegetative cover, groundwater level and relief also play their part, and the time factor allows the distinction between immature and mature soils. The profile forms the basis of pedological classifications of soil. Although such classifications are used mainly for agricultural purposes, individual groups of soil tend to have particular engineering characteristics, especially as far as structure and drainage are concerned.

9.2 Basic properties of soil

As remarked a soil consists of an assemblage of particles between which are voids, and as such contains three phases—solids, water and air. The interrelationships of the weights and volumes of these three phases are important since they help define the character of a soil. The weights and volumes of the different soil phases can be represented diagrammatically (Figure 9.1).

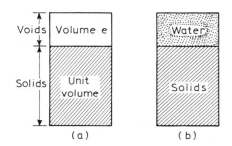
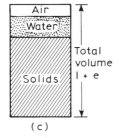

Figure 9.1 Unit soil

One of the most fundamental properties of a soil is the void ratio (e), which is the ratio of the volume of the voids (V_v) to that of the volume of the solids (V_s) (the volume of the solids is usually expressed as unity)

$$e = V_v/V_s$$

In the case of a fully saturated soil

$$e = mG_s$$

where m is the moisture content and G_s is the relative density (or specific gravity). The porosity (n) is a similar property, it being the ratio of the volume of the voids (V_v) to the total volume (V) of the soil, expressed as a percentage

$$n = (V_v/V) \times 100$$

Both void ratio and porosity indicate the relative proportion of void volume in a soil sample and the relationship between the two can be expressed as follows

$$n = e/(1 + e)$$

and

$$e = n/(1 - n)$$

As water plays a fundamental part in determining the engineering behaviour of any soil the moisture content (m) is a very relevant property to ascertain. This can be done by weighing a sample, drying it in an oven at 105–110°C and then weighing it again. The sample is assumed dry when there is no further reduction in weight on successive weighings, these being taken at four-hour intervals. Drying for 24 h is usually adequate. The moisture content is expressed as a percentage of the weight of the solid material in the soil sample (see BS 1377 (1975))

$$m = (W_w/W_s) \times 100$$

where W_w is the weight of water and W_s the weight of solid. Nuclear methods have also been used to determine the moisture content of soils, as well as their density (see Meigh and Skipp (1960)). The degree of saturation (S_r) expresses the relative volume percentage of water (V_w) in the voids

$$S_r = (V_w/V_v) \times 100$$

or

$$S_r = mG_s/e \times 100$$

The air content (A) is the ratio of the volume of air (V_a) to the total volume of the soil

$$A = V_a/V$$

It can also be derived as follows

$$A = (e - mG_s)/(1 + e)$$

or

$$A = n(1 - S_r)$$

The range of values of phase relationships for cohesive soils is much larger than for granular soils. For instance, saturated sodium montmorillonite at low confining pressure can exist at a void ratio of more than 25, its moisture content being some 900%. On the other hand saturated clays under high stress that exist at great depth may have void ratios of less than 0.2, with about 7% moisture content.

The unit weight (γ) of a soil is its weight (W) per unit volume (V), whilst its relative density (G_s) is the ratio of its weight to that of an equal volume of water

$$\gamma = W/V$$

The unit weight can also be derived from

$$\gamma = \gamma_w \frac{G_s(1 + m)}{1 + e}$$

or

$$\gamma = \gamma_w \frac{G_s + S_r e}{1 + e}$$

where γ_w is the unit weight of water, 9.8 kN/m³. The unit weight is expressed in kN/m³.

When the soil *in situ* is fully saturated the solid particles (volume 1 unit, weight $G_s\gamma_w$) are subjected to upthrust (γ_w). Hence the buoyant or submerged unit weight (γ_{sub}) is given by

$$\gamma_{sub} = \frac{G_s\gamma_w - \gamma_w}{1+e}$$
$$= \gamma_w\frac{G_s - 1}{1+e}$$

In soil mechanics the relative density or specific gravity (G_s) is that of the actual soil particles, which is expressed as follows

$$G_s = \frac{W_s}{V_s\gamma_w}$$

or

$$G_s = \frac{M_s}{V_s\rho_w}$$

where W_s is the weight of the solids, γ_w is the unit weight of water, ρ_w is the density of water and M_s is the mass of the solids. The relative density of soil particles is determined by the density bottle method or the pycnometer method (see Vickers (1978); BS1377 (1975)).

The dry density (ρ_d) is the weight of the solid particles divided by the total volume

$$\rho_d = \frac{W_s}{V}$$

$$= \frac{G_s\gamma_w}{1+e}$$

$$= \frac{G_s\rho_w}{1+e}$$

The International Association of Engineering Geology (Anon (1979)) proposed that dry density, porosity, void ratio and degree of saturation could each be divided into five classes as follows

Class	Dry density (t/m³)	Description	Porosity (%)	Description
1	Less than 1.4	Very low	Over 50	Very high
2	1.4–1.7	Low	50–45	High
3	1.7–1.9	Moderate	45–35	Medium
4	1.9–2.2	High	35–30	Low
5	Over 2.2	Very high	Less than 30	Very low

Class	Void ratio	Description	Degree of saturation (%)	Description
1	Over 1	Very high	Less than 25	Naturally dry
2	1.0–0.8	High	25–50	Wet
3	0.8–0.55	Medium	50–80	Very wet
4	0.55–0.43	Low	80–95	Highly saturated
5	Less than 0.43	Very low	Over 95	Saturated

The bulk density (ρ_b) or the natural *in situ* weight of a soil is expressed as

$$\rho_b = \frac{W}{V}$$

$$= \frac{W_s + W_w}{V_s + V_v}$$

$$= \frac{G_s V_s \rho_w + V_v \rho_w S_r}{V_s + V_v}$$

$$= \rho_w \frac{G_s + e S_r}{1 + e}$$

$$= \rho_w \frac{G_s(1 + m)}{1 + e}$$

When a soil is saturated its saturated density (ρ_{sat}) is given by

$$\rho_{sat} = \frac{W_{sat}}{V}$$

$$= \rho_w \frac{G_s + e}{1 + e}$$

If a soil is below the water table then part of its weight is offset by the buoyant effect of the water; therefore the submerged density (ρ_{sub}) is derived as follows

Submerged density = Saturated density − Density of water

$$\rho_{sub} = \rho_w \frac{G_s + e}{1 + e} - \rho_w$$

$$= \rho_w \frac{G_s - 1}{1 + e}$$

The density of a soil is governed by the manner in which its solid particles are packed. For example, granular soils may be densely or loosely packed. The void ratio of a densely packed soil is increased on excavation, that is, it is bulked; whilst it may be reduced in loosely packed soils by mechanical means such as by vibration. Indeed a maximum and minimum density can be distinguished. The smaller the range of particle sizes present, the smaller the particles; the more angular the particles, the smaller the minimum density. Conversely, if a wide range of particle sizes is present the void space is reduced accordingly, hence the maximum density is higher. A useful way to characterise the density of a granular soil is by its density index (I_D) which is defined as

$$I_D = \frac{e_{max} - e}{e_{max} - e_{min}}$$

$$= \frac{\rho_{d\,max}}{\rho_d} \times \frac{\rho_d - \rho_{d\,min}}{\rho_{d\,max} - \rho_{d\,min}}$$

where e is the naturally occurring void ratio.

Five degrees of density index have been distinguished.

Class	Density index	Description
1	Less than 0.2	Very loose
2	0.2–0.4	Loose
3	0.4–0.6	Medium dense
4	0.6–0.8	Dense
5	Over 0.8	Very dense

9.3 Particle size distribution

The particle size distribution expresses the size of particles in a soil in terms of percentages by weight of boulders, cobbles, gravel, sand, silt and clay. The United Soil Classification (see Wagner (1957)) and the British Standard *Code of Practice* (BS5930 (1981)) give the following limits for these size grades (*cf* Table 3.1).

Boulders		Over 200 mm
Cobbles		60–200 mm
Gravel	Coase	20–60 mm
	Medium	6–20 mm
	Fine	2–6 mm
Sand	Coarse	0.6–2 mm
	Medium	0.2–0.6 mm
	Fine	0.06–0.2 mm
Silt	Coarse	0.02–0.06 mm
	Medium	0.006–0.02 mm
	Fine	0.002–0.006 mm
Clay		Less than 0.002 mm

In nature there is a deficiency of soil particles in the fine gravel and silt ranges, and boulders and cobbles are quantitatively speaking not significant (see Glossop and Skempton (1945)). Sands and clays are therefore the most important soil types.

Two methods are usually employed to determine the particle size distribution in a soil, that is, sieving and sedimentation (see BS1377(1975); Akroyd (1964); Vickers (1978); Head (1980)). Sieving can be done either dry or wet and is used for the particle size analysis of sands and gravels. The sample is placed on a nest of standard sieves, of decreasing size from top to bottom, and shaken by a ro-tap vibrator. At the end of the test the soil fractions retained on each sieve are weighed. However, if a significant amount of fine-grained material is present in the sample, it may have to be washed through the sieves, after treating with a deflocculating agent. The dry fractions retained on each sieve are then weighed.

There are two principal sedimentation techniques, the pipette and the hydrometer methods, and these are used for size analysis of cohesive soils. Sedimentation depends on Stokes' Law which relates the velocity (v) of settlement of a spherical particle in suspension to its size

$$v = \frac{2}{9} \frac{(\rho_1 - \rho_2)gr^2}{\eta}$$

where ρ_1 is the density of the particle, ρ_2 is the density of the liquid, g is the acceleration due to gravity, r is the radius of the particle and η is the viscosity of the liquid. The size of a particle is assumed to be the same as the diameter of a sphere which would settle at the same velocity. The soil sample is pretreated to remove organic material, calcium compounds and soluble salts (see Head (1980)). It is then made up as a suspension in distilled water, to which a deflocculating agent (sodium hexametaphosphate) is added. The suspension is placed in a suspension tube. Stokes' Law permits the time (t) taken for settlement of particles of a certain size (D) to be calculated. In the pipette method a sample of the suspension is drawn off from a depth of 100 mm at given times, which correspond to certain particles sizes

Time (min)	Size (mm)
0.5	0.06
4.6	0.02
51.6	0.006
464.0	0.002
1856.0	0.001

The weights of the residues, after evaporating off the water, provide an indication of particle size distribution. Alternatively in the hydrometer method the relative density of the suspension is measured at the same given times. The size of the soil particles in the suspension, and consequently its density, decrease with time as the larger particles settle out. A detailed account of particle size analysis is given by Krumbein and Pettijohn (1938) and Milner (1962).

The results of particle size analysis are given in the form of a series of fractions, by weight, of different size grades. These fractions are expressed as a percentage of the whole sample and are generally summed to obtain a cumulative percentage. Cumulative curves can then be plotted on semilogarithmic paper to give a graphical representation of the particle size distribution. The slope of the curve provides an indication of the degree of sorting. If, for example, the curve is steep as in curve A in Figure 9.2, then the soil is uniformly sorted, whilst curve B represents a well-sorted soil. The sorting or uniformity of a particle size distribution has been expressed in a great many ways but one simple statistical measure which has been used by engineers is the coefficient of uniformity (U). This makes use of the effective size of the grains (D_{10}), that is, the size on the cumulative curve where 10% of the particles are passing and is defined as

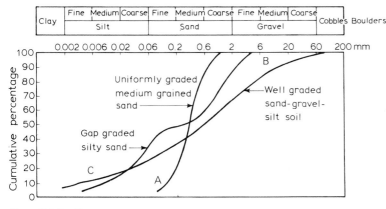

Figure 9.2 Grading curves

$$U = \frac{D_{60}}{D_{10}}$$

Similarly D_{60} is the size on the curve at which 60% of the particles are passing. A soil having a coefficient of uniformity of less than 2 is considered uniform whilst one with a value of 10 is described as well graded. In other words the higher the coefficient of uniformity, the larger is the range of particle sizes. The coefficient of curvature (C_c) is obtained from the following expression

$$C_c = \frac{D_{30}^2}{D_{60} D_{10}}$$

A well-graded soil has a coefficient of curvature between 1 and 3.

9.4 Consistency limits

The Atterberg or consistency limits of cohesive soils are founded on the concept that they can exist in any of four states depending on their water content. These limits are also influenced by the amount and character of the clay mineral content. In other words a cohesive soil is solid when dry but as water is added, it first turns to a semi-solid, then to a plastic, and finally to a liquid state. The water content at the boundaries between these states are referred to as the shrinkage limit (SL), the plastic limit (PL) and the liquid limit (LL) respectively.

Schofield and Wroth (1968) indicated that the liquid and plastic limits for saturated remoulded soils represent critical states such that the plasticity index (that is, the difference between these two limits) corresponds to an arbitrary increase in strength. Hence these limits can be regarded as being defined by water contents associated with particular strengths. Indeed the plasticity index has been redefined by Wroth and Wood (1978) as the change in water content giving rise to a one hundred-fold change in the strength of the soil. More recently Whyte (1982) proposed arbitrary strengths of 1.6 kPa and

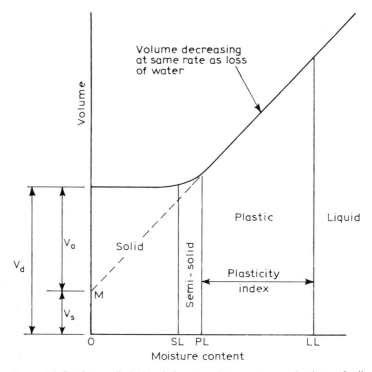

Figure 9.3 Consistency limits in relation to moisture content and volume of soil

110 kPa for liquid and plastic limits respectively. Whyte's suggested values of strength for liquid and plastic limits, however, involve a strength ratio of 70.

The shrinkage limit is defined as the percentage moisture content of a soil at the point where it suffers no further decrease in volume on drying. In practice, the shrinkage limit is taken as the moisture content which corresponds to the point of intersection of the tangents to the volume–moisture content curve (Figure 9.3).

The plastic limit is the percentage moisture content at which a soil can be rolled, without breaking, into a thread 3 mm in diameter, any further rolling causing it to crumble. The sample to be tested is prepared in the same way as for the liquid limit tests (see below). Unfortunately the inadequacy of control involved in this simple test means that the results obtained are not consistent for any particular clay. Various alternative tests therefore have been proposed, of which it would appear that cone penetrometers and extrusion techniques look the most promising since these are based on more reliable determinations of shear strength. For example, Whyte (1982) suggested that an extension technique could be used to establish the strength–moisture content relationship for a remoulded soil and thereby obtain the moisture content at a strength of 110 kPa, that is, the plastic limit.

Turning to the liquid limit this is defined as the minimum moisture content at which a soil will flow under its own weight. Clays may be classified according to their liquid limit as follows

Description	Plasticity	Range of liquid limit
Lean or silty	Low plasticity	Less than 35
Intermediate	Intermediate plasticity	35–50
Fat	High plasticity	50–70
Very fat	Very high plasticity	70–90
Extra fat	Extra high plasticity	Over 90

The liquid limit is determined by using either a liquid limit apparatus (see Casagrande (1958)) or a cone penetrometer. In both the tests the soil sample is dried and then disaggregated in a mortar with a rubber pestle (see Head (1980)). Only that material passing the No.425 BS sieve is used for the tests. The sample is mixed with distilled water to form a paste and stored for 24 h. Some of the paste is then placed in the liquid limit apparatus, levelled off and grooved with a grooving tool (Figure 9.4). The handle of the apparatus is then rotated which causes the bowl to be jarred against the base. The number of blows required to close the groove over 13 mm is recorded, along with the moisture content. The test is repeated on the same sample but with added moisture contents, hence successively fewer blows are required to close the groove. Moisture content is plotted against the number of blows (plotted on the logarithmic scale) on semi-logarithmic paper and the best straight line drawn between the points. The moisture content at 25 blows defines the liquid limit. It is now generally accepted that the percussion cup test generally does not yield reliable results and Wroth (1979) has shown that the liquid limit strength derived from such testing varies with soil type.

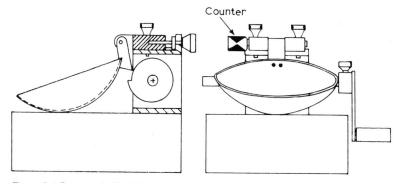

Figure 9.4 Casagrande liquid limit apparatus

When determining the liquid limit with the cone penetrometer (Figure 9.5) the soil sample is again mixed with distilled water to form a paste (BS 1377 (1975)). The paste is placed in a cylindrical container and levelled off at the top. Then the cone is lowered onto the surface of the soil and released for five seconds so that it penetrates the soil. The difference between the dial reading before and after cone penetration is recorded. The test is repeated several times with increasing moisture contents, the latter also being recorded for each test. Cone penetration is then plotted against moisture content and the best straight line drawn between the points. The liquid limit is taken as the moisture content at which the cone penetrates 20 mm into the soil.

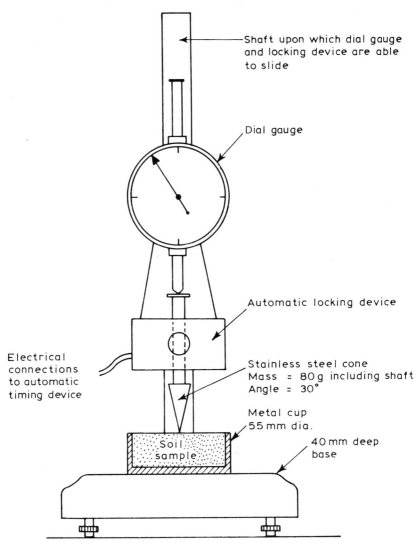

Figure 9.5 Cone penetrometer for liquid limit test

Wroth and Wood (1978) maintained that for a particular cone the un-drained cohesive strength (c_u) of a saturated remoulded clay is

$$c_u = CWk/z^2$$

where C is a constant, W is the weight of the cone, $k \approx 0.8$ for $30°$ cones and 0.27 for $60°$ cones and z is the depth of cone penetration.

Littlejohn and Farmilos (1977) investigated the relationship between the liquid limit obtained by the cone penetrometer and by the Casagrande apparatus. They showed that for values of liquid limit between 20% and 100% there was a reasonable correlation, the following linear equation being an acceptable approximation

$$LL_c = 1.6 + 0.97LL$$

where LL_c in the liquid limit obtained by the cone penetrometer method. Once above 100% the correlation between the values obtained from both tests is unsatisfactory, the values derived by the cone penetrometer test tending to decrease in relation to those found by the Casagrande method.

The consistency of cohesive soils depends on the interaction between the clay particles. Any decrease in water content results in a decrease in cation layer thickness and an increase in the net attractive forces between particles. For a soil to exist in the plastic state the magnitudes of the net interparticle forces must be such that the particles are free to slide relative to each other with cohesion between them being maintained. The plasticity of fine grained soils refers to their ability to undergo irrecoverable deformation at constant volume without cracking or crumbling.

The numerical difference between the liquid and plastic limits is referred to as the plasticity index (PI). This indicates the range of moisture content over which the material exists in a plastic condition. The plasticity index has been divided into five classes which are as follows (see Anon (1979)).

Class	Plasticity index (%)	Description
1	Less than 1	Non-plastic
2	1–7	Slightly plastic
3	7–17	Moderately plastic
4	17–35	Highly plastic
5	Over 35	Extremely plastic

As noted above, the plasticity index has been redefined by Wroth and Wood (1978) as the change in water content giving rise to a 100 fold change in the strength of the soil.

The determination of the liquid and plastic limit may prove difficult for soils with a low proportion of clay minerals. The degree of plasticity of such soils can be indicated by the linear shrinkage test. A sample of soil is mixed with distilled water at a water content approximating to the liquid limit. The sample is then placed in a brass mould, which has a semi-circular cross-section of 13 mm radius, and is 140 mm long. The sample is dried out completely and the linear shrinkage (LS) is calculated as

$$LS = 1 - \frac{\text{Length after drying}}{\text{Initial length}}$$

If the linear shrinkage is expressed as a percentage, an approximate value of the plasticity index is

$$PI = 2.13 \, LS$$

The liquidity index (LI) of a soil is defined as its moisture content in excess of the plastic limit, expressed as a percentage of the plasticity index

$$LI = \frac{m - PL}{LL - PL} \times 100$$

It describes the moisture content of a soil with respect to its index limits and indicates in which part of its plastic range a soil lies, that is, its nearness to the liquid limit.

The consistency index (IC) is the ratio of the difference between the liquid limit and natural moisture content to the plasticity index

$$IC = \frac{LL - m}{LL - PL}$$

It can be used to classify the different types of consistency of cohesive soils, as is shown in Table 9.2.

Table 9.2. Consistency of cohesive soils

Description	Consistency index	Approximate unconfined compressive strength kPa	Field identification
Hard		Over 300	Indented with difficulty by thumbnail, brittle
Very stiff	Above 1	150–300	Readily indented by thumbnail, still very tough
Stiff	0.75–1	75–150	Readily indented by thumb but penetrated only with difficulty. Cannot be moulded in the fingers
Firm	0.5–0.75	40–75	Can be penetrated several centimetres by thumb with moderate effort, and moulded in the fingers by strong pressure
Soft	Less than 0.5	20–40	Easily penetrated several centimetres by thumb, easily moulded
Very soft		Less than 20	Easily penetrated several centimetres by fist, exudes between fingers when squeezed in fist

The plasticity of a soil is influenced by the amount of its clay fraction, since clay minerals greatly influence the amount of attracted water held in the soil. With this in mind Skempton (1953) defined the activity of a clay as

$$Activity = \frac{Plasticity\ index}{\%\ by\ weight\ finer\ than\ 0.002\ mm}$$

He suggested three classes of activity, namely, active, normal and inactive which he further subdivided into five groups as follows

(1) inactive with activity less than 0.5,
(2) inactive with activity range 0.5–0.75,
(3) normal with activity range 0.75–1.25,
(4) active with activity range 1.25–2,
(5) active with activity greater than 2.

As can be seen from Figure 9.6, the activity of many British soils varies between 0.75 and 1.25, the exceptions being lacustrine and estuarine clays which are lower and higher respectively.

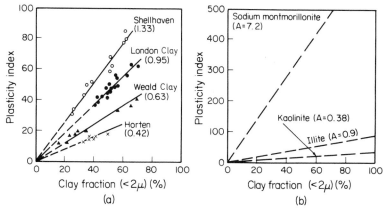

Figure 9.6 Relation between plasticity index and clay fraction. Figures in brackets represent the activities of the clays (after Skempton (1953))

It would appear that there is only a general correlation between the clay mineral content of a deposit and its activity, that is, kaolinitic and illitic clays are usually inactive whilst montmorillonitic clays range from inactive to active. Usually active clays have a relatively high water holding capacity and a high cation exchange capacity. They are also highly thixotropic, have a low permeability and a low resistance to shear.

9.5 Soil classification

Any system of soil classification involves grouping the different soil types into categories which possess similar properties and in so doing providing the engineer with a systematic method of soil description. Such a classification should provide some guide to the engineering performance of the soil type and should provide a means by which soils can be identified quickly.

Although soils include materials of various origins, for purposes of engineering classification it is sufficient to consider their simple index properties, which can be assessed easily, such as their particle size distribution, consistency limits or density.

Casagrande (1948) advanced one of the first comprehensive engineering classifications of soil. In the Casagrande system the coarse grained soils are distinguished from the fine on a basis of particle size. Gravels and sands are the two principal types of coarse grained soils and in this classification both are subdivided into five subgroups on a basis of grading (Table 9.3). Well graded soils are those in which the particle size distribution extends over a wide range without excess or deficiency in any particular sizes, whereas in uniformly graded soils the distribution extends over a very limited range of particle sizes. In poorly graded soils the distribution contains an excess of some particle sizes and a deficiency of others.

Each of the main soil types and subgroups are given a letter, a pair of which are combined in the group symbol, the former being the prefix, the latter the suffix. A plasticity chart (Table 9.4a) is also used when classifying fine grained soils. On this chart the plasticity index is plotted against liquid

Table 9.3. Symbols used in the Casagrande soil classification

Main soil type		Prefix
Coarse-grained soils	Gravel	G
	Sand	S
Fine-grained soils	Silt	M
	Clay	C
	Organic silts and clays	O
Fibrous soils	Peat	Pt

Subdivisions		Suffix
For coarse-grained soils	Well graded, with little or no fines	W
	Well graded with suitable clay binder	C
	Uniformly graded with little or no fines	U
	Poorly graded with little or no fines	P
	Poorly graded with appreciable fines or well graded with excess fines	F
For fine-grained soils	Low compressibility (plasticity)	L
	Medium compressibility (plasticity)	I
	High compressibility (plasticity)	H

limit. The A line is taken as the boundary between organic and inorganic soils, the latter lying above the line. Subsequently the Unified Soil Classification (Table 9.4a) was developed from the Casagrande system. The engineering uses of these various soils are given in Table 9.4b.

The British Soil Classification for engineering purposes (Anon (1981)) also uses particle size as a fundamental parameter and is very much influenced by the Casagrande system. Boulders, cobbles, gravels, sands, silts and clays are distinguished as individual groups, each group being given the following symbol and size range

(1) boulders (B), over 200 mm,
(2) cobbles (Cb), 60–200 mm,
(3) gravel (G), 2–60 mm,
(4) sand (S), 0.06–2 mm,
(5) silt (M), 0.002–0.06 mm,
(6) clay (C), less than 0.002 mm.

The gravel, sand and silt ranges may be further divided into coarse, medium and fine categories (see above). Sands and gravels are granular materials, ideally possessing no cohesion, whereas silts and clays are cohesive materials. Mixed soil types can be indicated as follows

Term	Composition of the coarse fraction
Slightly sandy GRAVEL	Up to 5% sand
Sandy GRAVEL	5–20% sand
Very sandy GRAVEL	Over 20% sand
GRAVEL/SAND	About equal proportions of gravel and sand
Very gravelly SAND	Over 20% gravel
Gravelly SAND	20–5% gravel
Slightly gravelly SAND	Up to 5% gravel

These major soil groups are again divided into subgroups on a basis of grading in the case of cohesionless soils, and on a basis of plasticity in

Table 9.4a. Unified Soil Classification (after Wagner (1957))

Field identification procedures (excluding particles larger than 76 mm and basing fractions on estimated weights)			Group symbols[a]	Typical names	Information required for describing soils	Laboratory classification criteria
Coarse-grained soils — More than half of material is *larger* than No. 200 sieve size[b]	Gravels — More than half of coarse fraction is larger than No. 7 sieve*	Clean gravels (little or no fines)	GW	Well graded gravels, gravel-sand mixtures, little or no fine	Give typical name; indicate approximate percentages of sand and gravel; maximum size; angularity, surface condition, and hardness of the coarse grains; local or geologic name and other pertinent descriptive information: and symbols in parentheses	Determine percentages of gravel and sand from grain size curve. Depending on fines (fraction smaller than No. 200 sieve size) coarse-grained soils are classified as follows: Less than 5%: *GW, GP, SW, SP.* More than 12%: *GM, GC, SM, SC.* 5% to 12%: *Borderline* cases require use of dual symbols — $C_u = \dfrac{D_{60}}{D_{10}}$ Greater than 4; $C_c = \dfrac{(D_{30})^2}{D_{10} \times D_{60}}$ Between 1 & 3
			GP	Poorly graded gravels, gravel-sand mixtures, little or no fine		Not meeting all gradation requirements for *GW*
		Gravels with fines (appreciable amount of fines)	GM	Silty gravels, poorly graded gravel-sand-silt mixtures	For undisturbed soils add information on stratification, degree of compactness, cementation, moisture conditions and drainage characteristics	Atterberg limits below 'A' line, or *PI* less than 4 — Above 'A' line with *PI* between 4 and 7 are *borderline* cases requiring use of dual symbols
			GC	Clayey gravels, poorly graded gravel-sand-clay mixtures		Atterberg limits above 'A' line with *PI* greater than 7
	Sands — More than half of coarse fraction is smaller than No. 7 sieve size*	Clean sands (little or no fines)	SW	Well graded sands, gravelly sands, little or no fines	Example: *Silty sand, gravelly;* about 20% hard, angular gravel particles 12.5 mm maximum size; rounded and subangular sand grains coarse to fine, about 15% nonplastic fines with low dry strength; well compacted and moist in place; alluvial sand; *(SM)*	$C_u = \dfrac{D_{60}}{D_{10}}$ Greater than 6; $C_c = \dfrac{(D_{30})^2}{D_{10} \times D_{60}}$ Between 1 & 3
			SP	Poorly graded sands, gravelly sands, little or no fines		Not meeting all gradation requirement for *SW*
		Sands with fines (appreciable amount of fines)	SM	Silty sands, poorly graded sand-silt mixtures		Atterberg limits below 'A' line of *PI* less than 5 — Above 'A' line with *PI* between 4 and 7 are *borderline* cases requiring use of dual symbols
			SC	Clayey sands, poorly graded sand-clay mixtures		Atterberg limits above 'A' line with *PI* greater than 7

Use grain size curve in identifying the fractions as given under field identification

Wide range in grain size and substantial amounts of all intermediate particle sizes

Predominantly one size or a range of sizes with some intermediate sizes missing

Nonplastic fines (for identification procedures see *ML* below)

Plastic fines (for identification procedures, see *CL* below)

Wide range in grain sizes and substantial amounts of all intermediate particle sizes

Predominantly one size or a range of sizes with some intermediate sizes missing

Nonplastic fines (for identification procedures, see *ML* below)

Plastic fines (for identification procedures, see *CL* below)

Table 9.4a—cont.

Plasticity chart for laboratory classification of fine-grained soils

Fine-grained soils	Identification procedures on fraction smaller than No. 40 sieve size				Symbol	Information required for describing soils
		DRY STRENGTH (crushing characteristics)	DILATANCY (reaction to shaking)	TOUGHNESS (consistency near plastic limit)		
More than half of material is smaller than No. 200 sieve size[b]	Silts and clays liquid limit less than 50	None to slight	Quick to slow	None	Inorganic silts and very fine sands, rock flour, silty or clayey fine sands with slight plasticity	ML	Give typical name: indicate degree and character of plasticity, amount and maximum size of coarse grains; colour in wet condition, odour if any, local or geologic name, and other pertinent descriptive information, and symbol in parentheses
		Medium to high	None to very slow	Medium	Inorganic clays of low to medium plasticity, gravelly clays, sandy clays, silty clays, lean clays	CL	For undisturbed soils add information on structure, stratification, consistency in undisturbed and remoulded states, moisture and drainage conditions
		Slight to medium	Slow	Slight	Organic silts & organic silt-clays of low plasticity	OL	Example
	Silts and clays liquid limit greater than 50	Slight to medium	Slow to none	Slight to medium	Inorganic silts micaceous or diatomaceous fine sandy or silty soils, elastic silts	MH	Clayey silt, brown: slightly plastic; small percentage of fine sand, numerous vertical root holes; firm and dry in place; loess; (ML)
		High to very high	None	High	Inorganic clays of high plasticity, fat clays	CH	
		Medium to high	None to very slow	Slight to medium	Organic clays of medium to high plasticity	OH	
Highly organic soils	Readily identified by colour, odour, spongy feel and frequently by fibrous texture				Peat and other highly organic soils	Pt	

Use grain size curve in identifying the functions as given under field identification

Footnotes to Table 9.4a

^a*Boundary classifications.* Soils possessing characteristics of two groups are designated by combinations of group symbols. For example *GW-GC*, well graded gravel-sand mixture with clay binder.

^bAll sieve sizes on this chart are US standard. The No. 200 sieve size is about the smallest particle visible to the naked eye.

*For visual classification, the 6.3 mm size may be used as equivalent to the No. 7 sieve size.

Field identification procedure for fine-grained soils or fractions
These procedures are to be performed on the minus No. 40 sieve size particles, approximately 0.4 mm. For field classification purposes, screening is not intended, simply remove by hand the coarse particles that interfere with the tests.

Dilatancy (reacting to shaking):
After removing particles larger than No. 40 sieve size, prepare a pat of moist soil with a volume of about one cubic centimetre. Add enough water if necessary to make the soil soft but not sticky.
Place the pat in the open palm of one hand and shake horizontally, striking vigorously against the other hand several times. A positive reaction consists of the appearance of water on the surface of the pat which changes to a livery consistency and becomes glossy. When the sample is squeezed between the fingers, the water and gloss disappear from the surface, the pat stiffens and finally it cracks and crumbles. The rapidity of appearance of water during shaking and of its disappearance during squeezing assist in identifying the character of the fines in a soil.
Very fine clean sands give the quickest and most distinct reaction whereas a plastic clay has no reaction. Inorganic silts, such as a typical rock flour, show a moderately quick reaction.

Dry strength (crushing characteristics):
After removing particles larger than No. 40 sieve size, mould a pat of soil to the consistency of putty, adding water if necessary. Allow the pat to dry completely by oven, sun or air drying, and then test its strength by breaking and crumbling between the fingers. This strength is a measure of the character and quantity of the colloidal fraction contained in the soil. The dry strength increases with increasing plasticity.
High dry strength is characteristic for clays of the CH group. A typical inorganic silt possesses only very slight dry strength. Silty fine sands and silts have about the same slight dry strength, but can be distinguished by the feel when powdering the dried specimen. Fine sand feels gritty whereas a typical silt has the smooth feel of flour.

Toughness (consistency near plastic limit):
After removing particles larger than the No. 40 sieve size, a specimen of soil about one cubic centimetre in size, is moulded to the consistency of putty. If too dry, water must be added and if sticky, the specimen should be spread out in a thin layer and allowed to lose some moisture by evaporation. Then the specimen is rolled out by hand on a smooth surface or between the palms into a thread about 3 mm in diameter. The thread is then folded and re-rolled repeatedly. During this manipulation the moisture content is gradually reduced and the specimen stiffens, finally loses its plasticity, and crumbles when the plastic limit is reached.
After the thread crumbles, the pieces should be lumped together and a slight kneading action continued until the lump crumbles.
The tougher the thread near the plastic limit and the stiffer the lump when it finally crumbles, the more potent is the colloidal clay fraction in the soil. Weakness of the thread at the plastic limit and quick loss of coherence of the lump below the plastic limit indicate either inorganic clay of low plasticity, or materials such as kaolin-type clays and organic clays which occur below the A-line.
Highly organic clays have a very weak and spongy feel at the plastic limit.

Table 9.4b. Engineering use chart (after Wagner (1957))

Typical names of soil groups	Group symbols	Important properties				Relative desirability for various uses (Graded from 1 (highest) to 14 (lowest))									
		Permeability when compacted	Shearing strength when compacted and saturated	Compressibility when compacted and saturated	Workability as a construction material	Rolled earth dams			Canal sections		Foundations		Roadways		
						Homogeneous embankment	Core	Shell	Erosion resistance	Compacted earth lining	Seepage important	Seepage not important	Fills		Surfacing
													Frost heave not possible	Frost heave possible	
Well-graded gravels, gravel-sand mixtures, little or no fines	GW	Pervious	Excellent	Negligible	Excellent	—	—	1	1	—	—	1	1	1	1
Poorly graded gravels, gravel-sand mixtures, little or no fines	GP	Very pervious	Good	Negligible	Good	—	—	2	2	—	—	3	3	3	—
Silty gravels, poorly graded gravel-sand-silt mixtures	GM	Semi-pervious to impervious	Good	Negligible	Good	2	4	—	4	4	1	4	4	9	5
Clayey gravels, poorly graded gravel-sand-clay mixtures	GC	Impervious	Good to fair	Very low	Good	1	1	—	3	1	2	6	5	5	1
Well-graded sands, gravelly sands, little or no fines	SW	Pervious	Excellent	Negligible	Excellent	—	—	3 if gravelly	6	—	—	2	2	2	4
Poorly graded sands, gravelly sands, little or no fines	SP	Pervious	Good	Very low	Fair	—	—	4 if gravelly	7 if gravelly	—	—	5	6	4	—

Table 9.4b—cont.

Important properties						Relative desirability for various uses (Graded from 1 (highest) to 14 (lowest))									
						Rolled earth dams			Canal sections		Foundations		Roadways		
													Fills		
Typical names of soil groups	Group symbols	Permeability when compacted	Shearing strength when compacted and saturated	Compressibility when compacted and saturated	Workability as a construction material	Homogeneous embankment	Core	Shell	Erosion resistance	Compacted earth lining	Seepage important	Seepage not important	Frost heave not possible	Frost heave possible	Surfacing
Silty sands, poorly graded sand-silt mixtures	SM	Semi-pervious to impervious	Good	Low	Fair	4	5	—	8 if gravelly	5 erosion critical	3	7	8	10	6
Clayey sands, poorly graded sand-clay mixtures	SC	Impervious	Good to fair	Low	Good	3	2	—	5	2	4	8	7	6	2
Inorganic silts and very fine sands, rock flour, silty or clayey fine sands with slight plasticity	ML	Semi-pervious to impervious	Fair	Medium	Fair	6	6	—	—	6 erosion critical	6	9	10	11	—
Inorganic clays of low to medium plasticity, gravelly clays, sandy clays, silty clays, lean clays	CL	Impervious	Fair	Medium	Good to fair	5	3	—	9	3	5	10	9	7	7
Organic silts and organic silt-clays of low plasticity	OL	Semi-pervious to impervious	Poor	Medium	Fair	8	8	—	—	7 erosion critical	7	11	11	12	—
Inorganic silts, micaceous or diatomaceous fine sandy or silty soils, elastic silts	MH	Semi-pervious to impervious	Fair to poor	High	Poor	9	9	—	—	—	8	12	12	13	—
Inorganic clays of high plasticity, fat clays	CH	Impervious	Poor	High	Poor	7	7	—	10	8 volume change critical	9	13	13	8	—
Organic clays of medium to high plasticity	OH	Impervious	Poor	High	Poor	10	10	—	—	—	10	14	14	14	—
Peat and other highly organic soils	Pt	—	—	—	—	—	—	—	—	—	—	—	—	—	—

the case of fine material. Granular soils are described as well graded (W) or poorly graded (P). Two further types of poorly graded granular soils are recognised, namely, uniformly graded (Pu) and gap-graded (Pg). Silts and clays are generally subdivided according to their liquid limits (LL) into low (under 35%), intermediate (35–50%) and high (50–70%) subgroups. Very high (70–90%) and extremely high (over 90%) categories have also been recognised. As in the Casagrande classification each subgroup is given a combined symbol in which the letter describing the predominant size fraction is written first (e.g. GW = well graded gravels; CH = clay with high liquid limit).

Any group may be referred to as organic if it contains a significant proportion of organic matter, in which case the letter O is suffixed to the group symbol (e.g. CVSO = organic clay of very high liquid limit with sand). The symbol Pt is given to peat.

In many soil classifications boulders and cobbles are removed before an attempt is made at classification, for example, their proportions are recorded separately in the British Soil Classification. Their presence should be recorded in the soil description, a plus sign being used in symbols for soil mixtures, for example, G + Cb for gravel with cobbles. The British Soil Classification has proposed that very coarse deposits should be classified as follows

(1) BOULDERS   Over half of the very coarse material is of boulder size (over 200 mm). May be described as cobbly boulders if cobbles are an important second constituent in the very coarse fraction.

(2) COBBLES   Over half of the very coarse material is of cobble size (200 mm–60 mm). May be described as bouldery cobbles if boulders are an important second constituent in the very coarse fraction.

Mixtures of very coarse material and soil can be described by combining the terms for the very coarse constituent and the soil constituent as follows

Term	Composition
BOULDERS (or COBBLES) with a little finer material*	Up to 5% finer material
BOULDERS (or COBBLES) with some finer material*	5–20% finer material
BOULDERS (or COBBLES) with much finer material*	20–50% finer material
FINER MATERIAL* with many BOULDERS (or COBBLES)	50–20% boulders (or cobbles)
FINER MATERIAL* with some BOULDERS (or COBBLES)	20–5% boulders (or cobbles)
FINER MATERIAL* with occasional BOULDERS (or COBBLES)	Up to 5% boulders (or cobbles)

* Give the name of the finer material (in parentheses when it is the minor constituent), e.g. sandy GRAVEL with occasional boulders; cobbly BOULDERS with some finer material (sand with some fines).

The British Soil Classification can be made either by rapid assessment in the field or by full laboratory procedure (see Tables 9.5a and 9.5b respectively). The classification is made, like that of the Unified System, on a basis of grading the soil according to particle size distribution and in the case of silts and clays by determining plasticity.

As far as soil description is concerned BS5930 (1981) recommends that the principal features to be included should be

(1) *Mass characteristics*
  (a) Field strength or compactness (Table 9.2) and indication of moisture condition.
  (b) Bedding (see Chapter 3)
  (c) Discontinuities (see Chapter 5)
  (d) Degree of weathering (see Chapter 7)
(2) *Material characteristics*
  (a) Colour (Table 9.5a)
  (b) Particle shape (see below) and composition.
  (c) Soil name (in capitals, e.g. SAND), grading and plasticity.
(3) *Geological formation, age and type of deposit* (see Chapters 1, 2 and 3)
(4) *Classification* (optional).
  Soil group symbol.

In particular instances it may be necessary to describe the shape of soil particles. The following terms have been suggested by BS5930 (Anon. (1981)).

Angularity	angular
	subangular
	subrounded
	rounded
Form	equidimensional
	flat
	elongated
	flat and elongated
	irregular
Surface texture	rough
	smooth

Since it is rarely possible to carry out significant soil tests on made ground, good descriptions are, highly important. According to BS5930 (1981) such descriptions should include information on the following, as well as on the soil constituents.

(a) Mode of origin of the material.
(b) Presence of large objects such as concrete, masonry or old motor cars.
(c) Presence of voids or collapsible hollow objects.
(d) Chemical waste, and dangerous or poisonous substances.
(e) Organic matter, with a note on the degree of decomposition.
(f) Odourous smell.
(g) Striking colour tints.
(h) Any dates readable on buried newspapers.
(i) Signs of heat or internal combustion under ground, e.g. steam emerging from boreholes.

Table 9.5a. Field identification and description of soils (after Anon (1981))

	Basic soil type	Particle size (mm)	Visual identification	Particle nature and plasticity	Composite soil types (mixtures of basic soil types)
Very coarse soils	BOULDERS		Only seen complete in pits or exposures.	Particle shape:	*Scale of secondary constituents with coarse soils*
		— 200			
	COBBLES		Often difficult to recover from boreholes	Angular Subangular	Term — % of or silt
		— 60		Subrounded Rounded	
Coarse soils (over 65% sand and gravel sizes)	GRAVELS	coarse	Easily visible to naked eye; particle shape can be described; grading can be described.	Flat Elongate	slightly clayey ⎫ GRAVEL ⎬ or under slightly silty ⎭ SAND
		— 20	Well graded: wide range of grain sizes, well distributed. Poorly graded: not well graded. (May be uniform: size of most particles lies between narrow limits; or gap graded: an intermediate size of particle is markedly under-represented.)		— clayey ⎫ GRAVEL ⎬ or 5 to 15 — silty ⎭ SAND
		medium			
		— 6			
		fine		Texture:	very clayey ⎫ GRAVEL ⎬ or 15 to 3 very silty ⎭ SAND
		— 2			
	SANDS	coarse	Visible to naked eye; very little or no cohesion when dry; grading can be described.	Rough Smooth Polished	Sandy GRAVEL ⎫ Sand or gravel a ⎬ important secor Gravelly SAND ⎭ constituent of t coarse fraction
		— 0.6			
		medium	Well graded: wide range of grain sizes, well distributed. Poorly graded: not well graded. (May be uniform: size of most particles lies between narrow limits; or gap graded: an intermediate size of particle is markedly under-represented.)		For composite types described as: clayey: fines are plastic, cohesive; silty: fines non-plastic or of low plasticity
		— 0.2			
		fine			
		— 0.06			
Fine soils (over 35% silt and clay sizes)	SILTS	coarse	Only coarse silt barely visible to naked eye; exhibits little plasticity and marked dilatancy; slightly granular or silky to the touch. Disintegrates in water; lumps dry quickly; possess cohesion but can be powdered easily between fingers.	Non-plastic or low plasticity	*Scale of secondary constituents with fine soils*
		— 0.02			
		medium			Term — % of sa gravel
		— 0.006			
		fine			sandy ⎫ CLAY ⎬ or 35 to gravelly ⎭ SILT
		— 0.002			
	CLAYS		Dry lumps can be broken but not powdered between the fingers; they also disintegrate under water but more slowly than silt; smooth to the touch; exhibits plasticity but no dilatancy; sticks to the fingers and dries slowly; shrinks appreciably on drying usually showing cracks. Intermediate and high plasticity clays show these properties to a moderate and high degree, respectively.	Intermediate plasticity (Lean clay)	— CLAY:SILT under 3
					Examples of composite types
					(Indicating preferred order for description)
				High plasticity (Fat clay)	Loose, brown, subangular very san fine to coarse GRAVEL with smal pockets of soft grey clay
Organic soils	ORGANIC CLAY, SILT or SAND	Varies	Contains substantial amounts of organic vegetable matter.		Medium dense, light brown, clayey, fine and medium SAND
					Stiff, orange brown, fissured sandy CLAY
	PEATS	Varies	Predominantly plant remains usually dark brown or black in colour, often with distinctive smell; low bulk density.		Firm, brown, thinly laminated SIL and CLAY
					Plastic, brown, amorphous PEAT

...pactness/strength		Structure			Colour
	Field test	Term	Field identification	Interval scales	
...se	By inspection of voids and particle packing.	Homogeneous	Deposit consists essentially of one type.	Scale of bedding spacing	Red Pink Yellow Brown Olive Green Blue White Grey Black, etc.
...se		Interstratified	Alternating layers of varying types or with bands or lenses of other materials. Interval scale for bedding spacing may be used.	**Term** — **Mean spacing (mm)**; Very thickly bedded — Over 2000; Thickly bedded — 2000–600	
...se	Can be excavated with a spade; 50 mm wooden peg can be easily driven.	Heterogeneous	A mixture of types.	Medium bedded — 600–200; Thinly bedded — 200–60	
...se	Requires pick for excavation; 50 mm wooden peg hard to drive.	Weathered	Particles may be weakened and may show concentric layering.	Very thinly bedded — 60–20; Thickly laminated — 20–6	Supplemented as necessary with:
...tly ...nted	Visual examination; pick removes soil in lumps which can be abraded.			Thinly laminated — under 6	Light Dark Mottled, etc.
...or ...e	Easily moulded or crushed in the fingers.				and
...or ...e	Can be moulded or crushed by strong pressure in the fingers.	Fissured	Break into polyhedral fragments along fissures. Interval scale for spacing of discontinuities may be used.		Pinkish Reddish Yellowish Brownish, etc.
...soft	Exudes between fingers when squeezed in hand.	Intact	No fissures.		
	Moulded by light finger pressure.	Homogeneous	Deposit consists essentially of one type.	Scale of spacing of other discontinuities	
	Can be moulded by strong finger pressure.	Interstratified	Alternating layers of varying types. Interval scale for thickness of layers may be used. Usually has crumb or columnar structure.	**Term** — **Mean spacing (mm)**	
	Cannot be moulded by fingers. Can be indented by thumb.	Weathered		Very widely spaced — Over 2000	
...stiff	Can be indented by thumb nail.			Widely spaced — 2000–600	
	Fibres already compressed together			Medium spaced — 600–200; Closely spaced — 200–60	
...ngy	Very compressible and open structure.	Fibrous	Plant remains recognisable and retain some strength.	Very closely spaced — 60–20	
...ic	Can be moulded in hand, and smears fingers.	Amorphous	Recognisable plant remains absent.	Extremely closely spaced — Under 30	

Table 9.5b. British Soil Classification System for Engineering Purposes (after Anon (1981))

Soil groups (see note 1)

Subgroups and laboratory identification

GRAVEL and SAND may be qualified Sandy GRAVEL and Gravelly SAND. etc. where appropriate

Soil groups				Group symbol (see notes 2 & 3)	Subgroup symbol (see note 2)	Fines (% less than 0.06 mm)	Liquid limit (%)	Name
COARSE SOILS — less than 35% of the material is finer than 0.06 mm	GRAVELS — More than 50% of coarse material is of gravel size (coarser than 2 mm)	Slightly silty or clayey GRAVEL	G	GW	GW	0 to 5		Well graded GRAVEL
				GP	GPu GPg			Poorly graded/Uniform/Gap graded GRAVEL
		Silty GRAVEL	G-F	G-M	GWM GPM	5 to 15		Well graded/Poorly graded silty GRAVEL
		Clayey GRAVEL		G-C	GWC GPC			Well graded/Poorly graded clayey GRAVEL
		Very silty GRAVEL	GF	GM	GML, etc.	15 to 35		Very silty GRAVEL; subdivide as for GC
		Very clayey GRAVEL		GC	GCL GCI GCH GCV GCE			Very clayey GRAVEL (clay of low, intermediate, high, very high, extremely high plasticity)
	SANDS — More than 50% of coarse material is of sand size (finer than 2 mm)	Slightly silty or clayey SAND	S	SW	SW	0 to 5		Well graded SAND
				SP	SPu SPg			Poorly graded/Uniform/Gap graded SAND
		Silty SAND	S-F	S-M	SWM SPM	5 to 15		Well graded/Poorly graded silty SAND
		Clayey SAND		S-C	SWC SPC			Well graded/Poorly graded clayey SAND
		Very silty SAND	SF	SM	SML, etc.	15 to 35		Very silty SAND; subdivided as for SC
		Very clayey SAND		SC	SCL SCI SCH SCV SCE			Very clayey SAND (clay of low, intermediate, high, very high, extremely high plasticity)

Table 9.5b—cont.

		FG	MG			
FINE SOILS more than 35% of the material is finer than 0.06 mm — **35% to 65% fines** Gravelly or sandy SILTS and CLAYS	Gravelly SILT	MLG, etc.				Gravelly SILT; subdivide as for CG
	Gravelly CLAY (see note 4)		CG	CLG CIG CHG CVG CEG	< 35 35 to 50 50 to 70 70 to 90 > 90	Gravelly CLAY of low plasticity of intermediate plasticity of high plasticity of very high plasticity of extremely high plasticity
		FS	MS			
	Sandy SILT (see note 4)	MLS, etc.				Sandy SILT; subdivide as for CG
	Sandy CLAY		CS	CLS, etc.		Sandy CLAY; subdivide as for C
65% to 100% fines SILTS and CLAYS		F	M			
	SILT (M-SOIL)	ML, etc.				SILT; subdivide as for C
	CLAY (see notes 5 & 6)		C	CL CI CH CV CE	< 35 35 to 50 50 to 70 70 to 90 > 90	CLAY of low plasticity of intermediate plasticity of high plasticity of very high plasticity of extremely high plasticity
ORGANIC SOILS	Descriptive letter 'O' suffixed to any group or subgroup symbol.			Organic matter suspected to be a significant constituent. Example MHO: Organic SILT of high plasticity.		
PEAT	Pt			Peat soils consist predominantly of plant remains which may be fibrous or amorphous.		

NOTE 1. The name of the soil group should always be given when describing soils, supplemented, if required, by the group symbol, although for some additional applications (e.g. longitudinal sections) it may be convenient to use the group symbol alone.

NOTE 2. The group symbol or subgroup symbol should be placed in brackets if laboratory methods have not been used for identification, e.g. (GC).

NOTE 3. The designation FINE SOIL, or FINES, F, may be used in place of SILT, M, or CLAY, C, when it is not possible or not required to distinguish between them.

NOTE 4. GRAVELLY if more than 50% of coarse material is of gravel size. SANDY if more than 50% of coarse material is of sand size.

NOTE 5. SILT (M-SOIL), M, is material plotting below the 'A' line, and has a restricted plastic range in relation to its liquid limit, and relatively low cohesion. Fine soils of this type include clean silt-sized materials and rock flour, micaceous and diatomaceous soils, pumice, and volcanic soils, and soils containing halloysite. The alternative term 'M-soil' avoids confusion with materials of predominantly silt size, which form only a part of the group.

Organic soils also usually plot below the 'A' line on the plasticity chart, when they are designated ORGANIC SILT, MO.

NOTE 6. CLAY, C, is material plotting above the 'A' line, and is fully plastic in relation to its liquid limit.

9.6 Shear strength of soil

9.6.1 Concept of shear strength

The shear strength of a soil is the maximum resistance which it can offer to shear stress. When this maximum has been reached the soil is regarded as having failed, its strength having been fully mobilised (see Menzies (1976)). However, the shear strength value determined experimentally is not a unique constant which is characteristic of the material but varies with the method of testing. Shear displacement also continues to take place after the shear strength is exceeded. Shear displacements usually occur across a well defined single plane of rupture or across a shear zone.

The stress on any plane surface can be resolved into the normal stress, σ_n, which acts perpendicular to the surface, and the shearing stress, τ, which acts along the surface, the magnitude of the resistance being given by Coulomb's equation

$$\tau = c + \sigma_n \tan\phi$$

where ϕ is the angle of shearing resistance or internal friction and c is the cohesion.

Clays are often described as cohesive soils in which the shear strength or cohesion is independent of applied stresses, and sands and gravels are described as non-cohesive or frictional soils in which the shearing stress (τ) along any plane is directly proportional to the normal stress (σ_n) across that plane

$$\tau = \sigma_n \tan\phi$$

In a truly cohesive soil it is assumed that $\phi = 0$ therefore

$$\tau = c$$

The stress that controls changes in the volume and the strength of a soil is known as the effective stress (σ'). When a load is applied to a saturated soil it is either carried by the pore water, which gives rise to an increase in the pore pressure (u), or the soil skeleton, or both. The effect that a load has on a soil is therefore affected by the drainage conditions but it has been shown that for most practical cases the effective stress is equal to the intergranular stress and can be determined from the equation

$$\sigma' = \sigma - u$$

where σ is the total stress. Hence shear strength depends upon effective stress and not total stress (see Skempton (1961)). Accordingly Coulomb's equation must be modified in terms of effective stress and becomes

$$\tau = c' + \sigma' \tan\phi'$$

The internal frictional resistance of a soil, for example, as is developed in a sand, is generated by friction when the grains in the zone of shearing are caused to slide, roll and rotate against each other. Local crushing may occur at the points of contacts which suffer the highest stress. The total resistance to rolling is the sum of the behaviour of all the particles and is influenced by the confining stress, the coefficient of friction and angles of contact between the minerals as well as their surface roughness. However, the angle of internal friction does not depend solely on the internal friction between grains because a proportion of the shearing stress on the plane of failure is utilised in overcoming interlocking, that is, it is also dependent upon the initial void ratio or density of a given soil. It is also influenced by the size and shape of the grains. The larger the grains, the wider is the zone which is affected. The more angular the grains, then the greater is the frictional resistance to their relative movement. Electrical forces of attraction and repulsion may also be involved in shearing resistance.

It is commonly believed that interparticle friction is influenced by the composition of the particles involved and the state of their surface chemistry. However, after an investigation of quartz (siliceous) and calcareous grains, Frossard (1979) proposed that interparticle friction appears to increase with angularity, independently of grain composition. He also noted that sphericity has a notable influence on volumetric strain, particularly on the maximum rate of dilatancy. A review of the effects of stress history on the deformation of sand has been given by Lambrechts and Leonards (1978). They also referred to how minor differences in packing can influence stress–strain behaviour.

In clay soils the cohesion which is developed by the molecular attractive forces between the minute soil particles is mainly responsible for the resistance offered to shearing. Because molecular attractive forces depend to a large extent on the mineralogical composition of the particles and on the type and concentration of electrolytes present in the pore water, the magnitude of the true cohesion also depends on these factors.

Lambe's concept of the shear strength in clay postulated the existence of forces of attraction and repulsion between the particles with a net repulsive force in accordance with physico-chemical principles (see Lambe (1960)). Hence the equilibrium of internal stresses in a clay soil can be expressed as

$$\sigma' = \sigma - u = R - A$$

where σ' is the effective stress, σ is the total stress, u is the pore water pressure, R is the total force of repulsion and A is the total force of attraction per unit area between the particles. Unfortunately the R and A forces cannot be measured. The relationship is demonstrated by the fact that clay behaviour differs when immersed in solutions with different electrolyte concentrations. For a discussion of the shear strength of saturated clays see Sridharan and Rao (1979).

Mineral agglomerations often result from the deposition of cementitious compounds which adhere to the surfaces of the minerals and bond them together. The magnitude of the cohesion in bulk is then affected by the nature and number of bonded interparticle contacts per unit volume, which in turn is influenced by the stress history, the soil fabric and the mineralogical composition.

9.6.2 The triaxial test

Because of the complex nature of the shearing resistance of soils several methods of testing are employed in its assessment. The principal tests in use today are the triaxial test and the shear box test. The triaxial test is

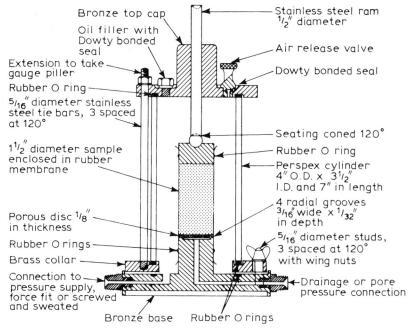

Figure 9.7 Cross section of a typical triaxial cell (from Bishop and Henkel (1962))

the more reliable of the two and so is much more frequently used. Both the cohesion and angle of shearing resistance can be derived from the triaxial test. There is a relatively uniform distribution of stress on the failure plane in the triaxial test and the soil is free to fail along the weakest surface. Furthermore water can either be drained from or forced through the soil during the test to simulate ground conditions.

In the triaxial test a cylindrical specimen of soil, generally 38 mm in diameter by 76 mm long, but sometimes 100 mm diameter by 200 mm in length, is enclosed in a rubber membrane and placed within the triaxial cell (Figure 9.7). The specimen is subjected to an all-round fluid pressure in the cell, then the axial load is gradually applied, at a constant rate of strain, by a plunger until the specimen fails. The stress applied by the plunger is referred to as the deviator stress $(\sigma_1 - \sigma_3)$* and the cell pressure is the minor principal stress σ_3 (and the intermediate principal stress $\sigma_2 = \sigma_3$). Connections to the top and bottom of the specimen control the drainage of the specimen and can be connected to the pore pressure measurement apparatus.

The deviator stress at failure is used to plot the diameter of the Mohr circle for the test in question (Figure 9.8a). At least three tests at different cell pressures should be performed on the material from one sample. Hence three Mohr circles are plotted and the common tangent, known as the Mohr envelope, is drawn to the circles. The Mohr envelope allows the determination of the angle of shearing resistance (ϕ) and the cohesion (c). In a truly cohesive

---

* σ_1 = Load at failure which consists of load applied by plunger plus cell pressure.

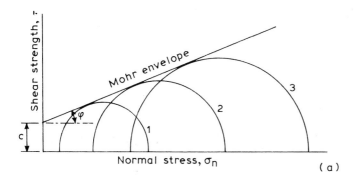

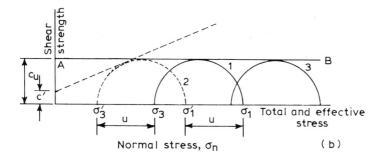

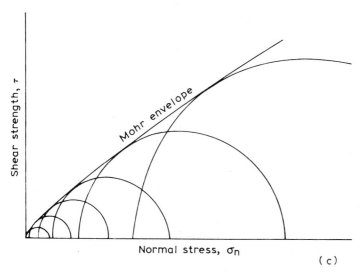

Figure 9.8 (a) Mohr circles and envelope for determination of ϕ and c for a mixed soil. (b) Mohr circles and envelope for undrained triaxial test or clay. (c) Mohr circles and envelope for sand

soil $\phi = 0$ (see Skempton (1948)) and as a consequence all the Mohr's circles have the same radius (Figure 9.8b), that is, $\tau = c$. By contrast in cohesionless soils $\tau = \sigma_n \tan\phi$ and the Mohr envelope is focused on the origin (Figure 9.8c).

There are three principal types of triaxial test (see Bishop and Henkel (1962)).

(1) *Undrained tests.* The specimen is subjected to the cell pressure and the axial load is applied immediately, with no drainage throughout the test. The results are plotted in terms of total stresses and the strength parameters are denoted as ϕ_u and c_u.

(2) *Consolidated undrained tests.* The cell pressure is applied and drainage is permitted until the excess pore pressures, due to the application of the cell pressure, have dissipated. Then the deviator stress is applied but with no further drainage. The unconsolidated undrained shear strength parameters are obtained if the results are plotted in terms of total stresses. These values have limited practical use. If pore pressures are measured during the second stage of the test the results can be plotted in terms of effective stress, the effective stress shear strength parameters ϕ', c' being obtained. This test is often used in preference to the drained test since it generally requires less time and the results are, for many practical purposes, the same. Repeated cyclic loading on an undrained specimen causes a net increase in both pore pressure and strain (see Castro and Christian (1976)).

(3) *Drained tests.* Drainage of the specimen is allowed until consolidation is complete and it is continued throughout the test. The rate of test is sufficiently slow to ensure that full dissipation of excess pore pressure is maintained. The results are plotted in terms of effective stresses (in this case equal to total stresses); the terms ϕ_d and c_d are sometimes used in place of ϕ' and c' respectively.

These tests attempt to simulate conditions in the field where the strength of a soil is different under undrained conditions from drained conditions. It is assumed that undrained conditions exist immediately after loading in soils of low permeability, prior to any appreciable consolidation. However, the shear strength of such a soil will gradually change as consolidation takes place, ultimately reaching the drained strength when it is complete. The drained condition in highly permeable soils will apply at any time since excess pore pressures can be dissipated rapidly.

A value of Young's modulus can also be obtained from the triaxial test. Hanna and Adams (1968) showed that the modulus of deformation of clay as determined in the laboratory is sensitive to the test load system used and the stress level applied, as well as the magnitude of the undrained shear strength. The values are generally less than those determined by plate load tests.

9.6.3 Other strength tests

The shear strength of granular soils can be determined from a shear box test. The sample is contained in a box split horizontally in which the bottom of the box can be moved relative to the top, thus shearing the sample along a horizontal plane (Figure 9.9a). A vertical load is applied to the sample

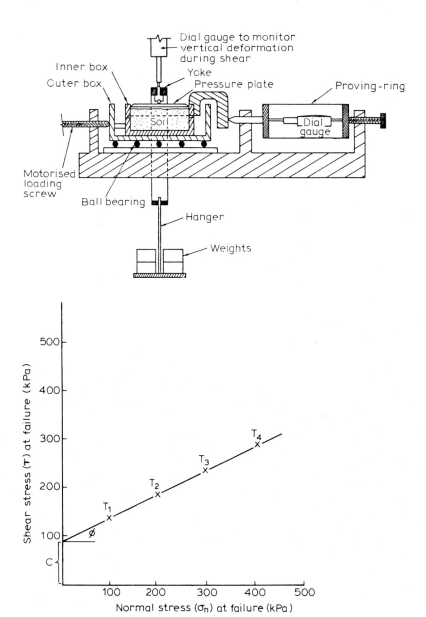

Figure 9.9 (a) Shear box apparatus. (b) Graph of results from a shear box test carried out on sand with some fines

by means of a weighted hanger and a lever arm system. Sufficient tests are carried out under different vertical loads and the results are plotted as shear stress against normal pressure. The cohesion is taken as the intercept of the shear stress axis and the angle of shearing resistance is obtained from the slope of the plotted line (Figure 9.9b). The shear box is also used to determine the residual strength of clays under drained conditions, extending the strain by reversing the movement a number of times (see Skempton (1964)). However, a value of the residual strength of a clay soil is more readily obtained by using the ring shear apparatus (see Bishop *et al.* (1971)).

The shear strength of soft clays can be determined by using a shear vane test (see Vickers (1978)). A cruciform vane is inserted into the clay to a depth equal to the height of the vane and then rotated until the soil fails. The shear strength is obtained from

$$\tau = \frac{T}{\pi D^2 (H/2 + D/6)}$$

where H is the height of the vane, D is the diameter of the vane and T is the torque. This laboratory test is used infrequently.

A value of the shear strength of clay can be obtained quickly from the unconfined compression test (see Cooling and Golder (1940)). The specimen is loaded through a calibrated spring by a manually operated screwjack (Figure 9.10). The shear strength is taken as half the unconfined compressive strength.

The fall cone test provides a quick method of measuring the strength of a soil (Figure 9.11). Hansbo (1957) noted the relationship between cone penetration and strength for different weights and cone angles. He demonstrated that the relationship is governed by the sensitivity of the clay and the manner in which it is sampled.

Wroth and Wood (1978) have established correlations between the remoulded undrained shear strength (c_u) of a clay and its liquidity index on the one hand, and between the compression index of remoulded clay and its plasticity index on the other. The liquid and plastic limits involved must be determined by the cone penetrometer, the plasticity index being redefined as the change of water content giving rise to a 100-fold change in strength of the soil. In other words the shear strength at the plastic limit is 100 times greater than that at the liquid limit. The remoulded strength of normally consolidated clay frequently is an underestimate of the *in situ* strength, whilst for heavily overconsolidated clays it is usually close to the *in situ* value.

9.6.4 Sensitivity of clays

As noted the shear strength of an undisturbed clay is generally found to be greater than that obtained when it is remoulded and tested under the same conditions and at the same water content. The ratio of the undisturbed to the remoulded strength at the same moisture content is defined as the sensitivity of a clay. Skempton and Northey (1952) proposed the following grades of sensitivity

(1) insensitive clays, under 1,
(2) low sensitive clays, 1–2,

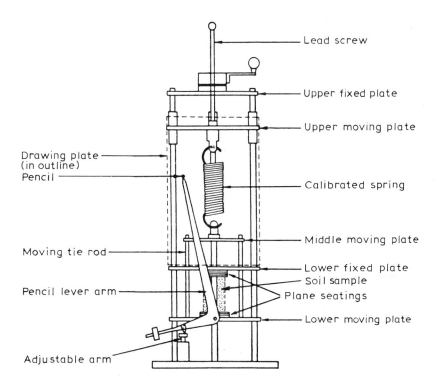

Lead screw

Upper fixed plate

Upper moving plate

Drawing plate
(in outline)
Pencil

Calibrated spring

Moving tie rod

Middle moving plate

Lower fixed plate
Soil sample
Plane seatings

Pencil lever arm

Lower moving plate

Adjustable arm

Figure 9.10 (a) Unconfined compression apparatus. (b) Samples of clay failed in unconfined compression

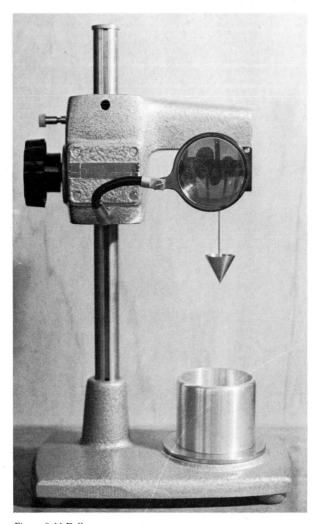

Figure 9.11 Fall cone apparatus

(3) medium sensitive clays, 2–4,
(4) sensitive clays, 4–8,
(5) extra-sensitive clays, 8–16,
(6) quick clay, over 16.

Clays with high sensitivity values have little or no strength after being disturbed. Indeed if they suffer slight disturbance this may cause an initially fairly strong material to behave as a viscous fluid. High sensitivity seems to result from the metastable arrangement of equidimensional particles. The strength of the undisturbed clay is chiefly due to the strength of the framework developed by these particles and the bonds between their points of contact. If the framework is destroyed by remoulding the clay loses most of its strength. Sensitive clays generally possess high moisture contents,

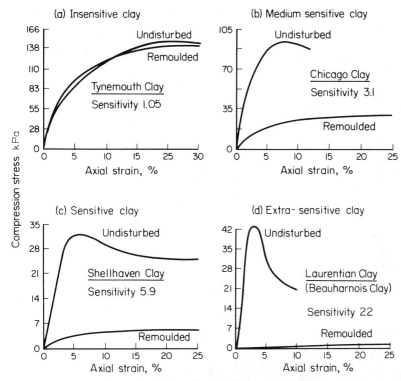

Figure 9.12 Stress–strain curves of clay soils with different sensitivities
(after Skempton, A. W. (1953). 'Soil mechanics in relation to geology',
Proc. Yorks Geol. Soc., **29**)

frequently with liquidity indices well in excess of unity. A sharp increase
in moisture content may cause a great increase in sensitivity, sometimes with
disastrous results. Heavily overconsolidated clays are insensitive. The effect
of remoulding on clays of various sensitivities is illustrated in Figure 9.12.

Some clays with moderate to high sensitivity show a regain in strength
when, after remoulding, they are allowed to rest under unaltered external
conditions. Such soils are thixotropic. Thixotropy is the property of a material
which allows it to undergo an isothermal gel-to-sol-to-gel transformation
upon agitation and subsequent rest. This transformation can be repeated
indefinitely without fatigue and the gelation time under similar conditions
remains the same. The softening and subsequent recovery of thixotropic
soils appears to be due, first, to the destruction, and then, secondly, to
the rehabilitation of the molecular structure of the adsorbed layers of the
clay particles. For example, the loss of consistency in soils containing Na
montmorillonite occurs since large volumes of water are adsorbed upon
and held between the colloidal clay particles. Furthermore the ionic forces
attracting the colloidal clay particles together have a definite arrangement
which is an easily destroyable microstructure when subjected to agitation.
When the material is at rest the ions and water molecules tend to reorientate
themselves and strength is thereby recovered.

9.7 Consolidation

The theory of consolidation, as advanced by Terzaghi (1925) has enabled engineers to determine the amount and rate of settlement which is likely to occur when structures are erected on cohesive soils. When a layer of soil is loaded, some of the pore water is expelled from its voids, moving slowly away from the region of high stress as a result of the hydrostatic gradient created by the load. The void ratio accordingly decreases and settlement occurs. This is termed primary consolidation. Further settlement, usually of minor degree, may occur due to the rearrangement of the soil particles under stress, this being referred to as secondary consolidation. However, in reality primary and secondary consolidation are not distinguishable.

Terzaghi showed that the relationship between unit load and the void ratio for a sediment can be represented by plotting the void ratio, e, against the logarithm of the unit load, p (Figure 9.13). In the laboratory the relationship between e and p is investigated by using either a standard oedometer or a Rowe cell oedometer (see Vickers (1978)). In the standard oedometer test a disc-shaped specimen is contained in a metal ring, usually 76 mm

Figure 9.13 Results of consolidation test, e log p curve

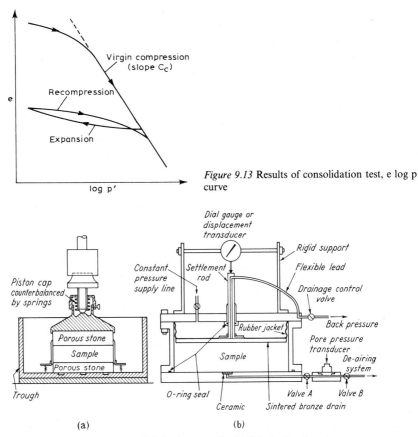

Figure 9.14 Oedometer (consolidation) press cells. (a) Standard oedometer cell. (b) Rowe hydraulic oedometer cell

diameter and lies between two porous stones (Figure 9.14a). The apparatus sits in an open cell of water, to which the pore water of the specimen has free access. The specimen is loaded axially and the initial load applied depends on the type of soil. Subsequently a sequence of loads is applied to the specimen, each being double the previous load. Each load increment is usually imposed for 24 h, and is applied when compression due to the previous load has ceased, that is, when excess pore pressure is dissipated. The compression with time is recorded for each load increment. The ultimate loading should be greater than any likely to be imposed on the soil concerned. The moisture content of the specimen is determined at the end of the test, its initial moisture content having been measured previously.

In the Rowe cell test (Figure 9.14b) the specimen, either 150 mm or 250 mm diameter, with either vertical or radial drainage, is loaded hydraulically with pore pressure measurement at the bottom surface (see Rowe and Barden (1966)). Pore pressure dissipation is also observed for each pressure increment in the Rowe cell test.

The void ratio at the end of the test is obtained from the moisture content (m_1)

$$e_1 = m_1 G_s$$

The other values of e are obtained by working backwards from this value of e_1 by finding the change in void ratio for each load increment

$$\Delta e = \Delta H \frac{(1 + e_o)}{H_o}$$

where H is the thickness of the specimen and the suffix o indicates original, $e_0 = e_1 + \Delta e$. The pressure/voids ratio curve ($e/\log p$ curve) is then plotted.

The shape of the $e/\log p$ curve is related to the stress history of a clay (Figure 9.13). In other words the $e/\log p$ curve for a normally consolidated

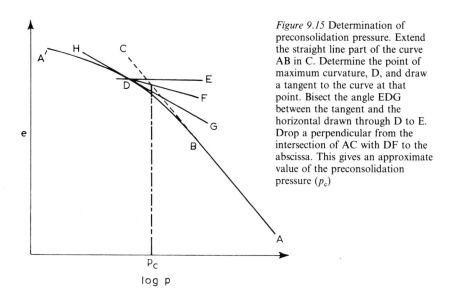

Figure 9.15 Determination of preconsolidation pressure. Extend the straight line part of the curve AB in C. Determine the point of maximum curvature, D, and draw a tangent to the curve at that point. Bisect the angle EDG between the tangent and the horizontal drawn through D to E. Drop a perpendicular from the intersection of AC with DF to the abscissa. This gives an approximate value of the preconsolidation pressure (p_c)

clay is linear and is referred to as the virgin compression curve (see Terzaghi and Peck (1967)). On the other hand if the clay is overconsolidated the $e/\log p$ curve is not straight and the preconsolidation pressure can be derived from the curve (Figure 9.15). The preconsolidation pressure refers to the maximum overburden pressure to which a deposit has been subjected. Overconsolidated clay is appreciably less compressible than normally consolidated clay.

The compressibility of a clay can be expressed in terms of the compression index (C_c) or the coefficient of volume compressibility (m_v). The compression index is the slope of the linear section of the $e/\log p$ curve and is dimensionless. It can be determined from any two points on this part of the curve as follows

$$C_c = \frac{e_1 - e_2}{\log p_2/p_1}$$

The value of C_c for cohesive soils ranges between about 0.15 for lean sandy clays to more than 1.0 for highly colloidal bentonic clays. Hence the compressibility index increases with increasing clay content and so with increasing liquid limit. Indeed Skempton (1944) found that C_c for normally consolidated clays is closely related to their liquid limit, the relationship between the two being expressed as

$$C_c = 0.009(\text{LL} - 10)$$

The coefficient of volume compressibility is defined as the volume change per unit volume per unit increase in load, its units are the inverse of pressure (m^2/kN). The volume change can be expressed according to specimen thickness (H) or void ratio (e)

$$m_v = \frac{1}{H_1} \left(\frac{H_1 - H_2}{p_2 - p_1} \right)$$

or

$$m_v = \frac{1}{1 + e_1} \left(\frac{e_1 - e_2}{p_2 - p_1} \right)$$

The value of m_v for a given soil depends upon the stress range over which it is determined; BS 1377 (1975) recommends that it should be calculated for a pressure increment of 100 kPa in excess of the effective overburden pressure on the soil at the depth in question. The stress–strain behaviour of a clay undergoing confined compression can be described in terms of m_v as follows

$$m_v = \frac{\Delta \varepsilon}{\Delta p}$$

where ε is the vertical strain.

Crawford (1966) showed that the compressibility of a soil is dependent on the average rate of compression and that the soil structure has a substantial time-dependent resistance to compression. At the instant when a load, p, on a layer of clay is suddenly increased by Δp, the thickness of the layer remains unchanged. Hence the application of the load Δp, produces an equal increase, Δu, in the hydrostatic pressure of the pore water. As time proceeds, the excess pore pressure is gradually dissipated and finally disappears, whilst the grain to grain pressure simultaneously increases from an initial value p to $p + \Delta p$. The ratio between the decrease of the void ratio, Δe, at time, t, and the ultimate decrease, Δe_1 represents the degree of consolidation, U, at time, t

$$U = 100 \frac{\Delta e}{\Delta e_1}$$

With a given thickness, H, of a layer of clay the degree of consolidation at time, t, depends exclusively on the coefficient of consolidation, c_v

$$c_v = \frac{k}{\gamma_w m_v}$$

where k is the hydraulic conductivity and γ_w is the unit weight of water. The coefficient of consolidation which determines the rate at which settlement takes place is calculated for each load increment and either a mean value is used or that value appropriate to the pressure range in question (c_v is measured in m^2/year). With increasing values of p both k and m_v decrease, therefore c_v is fairly independent of p. The coefficient of consolidation decreases for normally consolidated clays from about 31.5 m/year for very lean clays to about 0.03 m/year for highly colloidal clays. At any value of c_v, the time (t) at which a given degree of consolidation, U, is reached, increases in simple proportion to the square of the thickness, H, of the layer.

Mesri and Rokhsar (1974) pointed out that the Terzaghi theory of one-dimensional consolidation made several unrealistic assumptions regarding the consolidation characteristics of soils. For instance, the theory assumes a linear relationship between void ratio and effective stress independent of time and stress history. Also a constant hydraulic conductivity during the consolidation process is assumed. Consequently, Mesri and Rohksar proposed a theory of consolidation which considered finite strain, variations in compressibility and permeability during the consolidation process, and the effects of a critical pressure and secondary compression.

9.8 Settlement

Settlement presents a problem in clayey soils so that the amount which is likely to take place when they are loaded needs to be determined. It invariably continues after the construction period, indeed settlement in clays may continue for several years.

9.8.1 Initial and primary settlement

Initial or immediate settlement is that which occurs under constant volume (undrained) conditions when clay undergoes elastic deformation to accommodate the imposed shear stresses. The immediate settlement below the corner

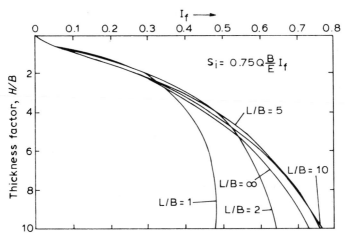

Figure 9.16 Steinbrenner's influence factors for loaded area $L \times B$ on compressible stratum of thickness H

of a uniformly loaded rectangular foundation structure can be obtained from the Steinbrenner expression (see Terzaghi (1943))

$$S_i = 0.75 \, Q\frac{B}{E} \, I_f$$

where Q is the load, B is the width of the foundation, E is Young's modulus for the clay, and I_f is an influence factor related to the length and width of the foundation and the thickness of the compressible layer. Values of I_f for a value of Poisson's ratio of 0.5 are given in Figure 9.16. Settlements beneath any point within a rectangular foundation can be calculated by splitting the area into a number of rectangles, with one of their corners centred on the point concerned, and then summing the individual settlements for each rectangle. Young's modulus may be determined in the laboratory by compression tests or in the field by plate load tests. The Menard pressuremeter may also be used to determine the undrained deformation modulus of clay.

The average immediate settlement can be found by using the method advanced by Janbu *et al.* (1956)

$$S_i = \mu_1 \, \mu_0 \, Q \, B/E$$

The factors μ_0 and μ_1 are related to the depth of the foundation, the thickness of the compressible layer and the length (L), breadth (B) ratio of the foundation (Figure 9.17). The value of S_i and S_c (consolidation settlement) are added to obtain the final amount of settlement. Lambe (1973) reviewed the various procedures used to determine initial settlement.

Primary consolidation in a clay soil takes place due to the void space being gradually reduced as the pore water is expelled therefrom on loading (see Wahls (1962)). The rate at which it occurs depends on the rate at which the excess pore pressure induced by a structural load is dissipated, thereby allowing the structure to be supported entirely by the soil skeleton. Consequently the permeability of the clay is all important. The increase in effective

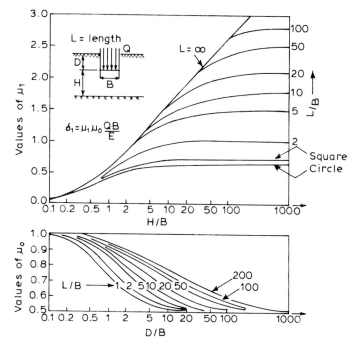

Figure 9.17 Determination of initial settlement (after Janbu *et al.* (1956))

stress in the soil corresponds to a decrease in volume which again is controlled by the rate at which the pore water can escape from the voids.

9.8.2 Stress distribution

In order to derive the amount of settlement likely to occur beneath a foundation structure, the differing amounts of vertical stress, imposed by the load, in the layers of soil beneath the foundation have to be found. A reasonable approximation of how stress is distributed in soil upon loading can be obtained by assuming that the soil behaves in an elastic manner as if it was a homogeneous material. In such an instance the vertical stress (σ_z), produced at any point (N) in the soil by a load (Q) on the surface (Figure 9.18a) may be derived from the Boussinesq (1885) expression

$$\sigma_z = \frac{3Q}{2\pi \, z^2} \left[\frac{1}{1 + (r/z)^2} \right]^{5/2}$$

The expression has been simplified as follows

$$\sigma_z = I_f \frac{Q}{z^2}$$

where I_f is an influence factor depending on the depth and position of the point at which the stress is required in relation to the point load (Figure 9.18b).

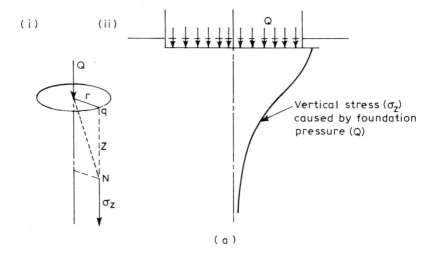

(i)   (ii)

Vertical stress (σ_z)
caused by foundation
pressure (Q)

(a)

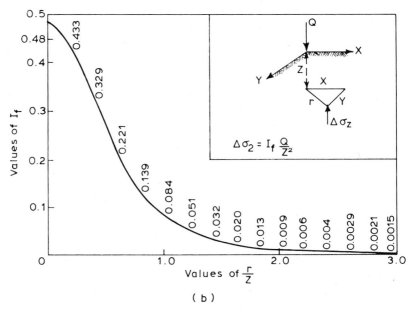

$$\Delta \sigma_2 = I_f \frac{Q}{Z^2}$$

Values of I_f

0.5
0.48
0.4
0.3
0.2
0.1
0

0.433
0.329
0.221
0.139
0.084
0.051
0.032
0.020
0.013
0.009
0.006
0.004
0.0029
0.0021
0.0015

1.0   2.0   3.0

Values of $\frac{r}{Z}$

(b)

Figure 9.18 (a) Vertical stress distibution beneath a surface load.
(i) Intensity of vertical pressure at point N in interior of a semi-infinite solid
acted on by load Q; (ii) vertical stress (σ_2) with depth generated by
foundation pressure (q). (b) Influence coefficients for vertical stress from a
concentrated load (after Boussinesq (1885))

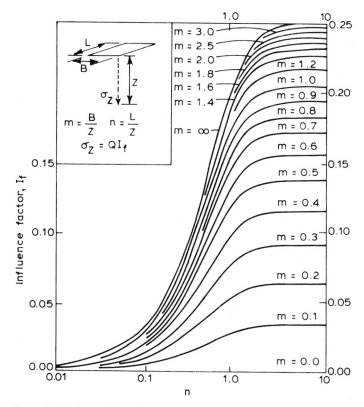

Figure 9.19 Influence factors for the vertical stress beneath the corner of a rectangular foundation (Fadum (1948))

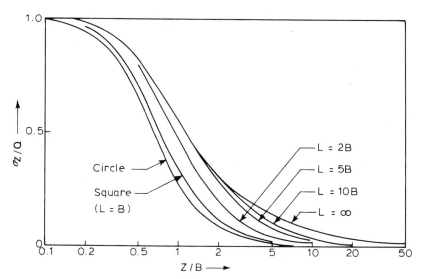

Figure 9.20 Determination of increase in vertical stress under the centre of uniformly loaded flexible footings (after Janbu et al. (1956))

Various design charts have been advanced for estimating stresses beneath a foundation structure. For example, Fadum (1948) provided a set of curves from which the increment of vertical stress (σ_z) beneath the corner of a uniformly loaded rectangular area could be obtained by substituting values in the following expression

$$\sigma_z = QI_f$$

where Q is the load and I_f is an influence factor depending on the dimensions of the rectangle and the depth concerned (Figure 9.19). The method can also be used to determine the increment of vertical stress beneath a point anywhere beneath a structure as long as the area can be subdivided into rectangles and the point occurs beneath the coincident corners of the rectangles. The vertical stress is then simply the sum of the stresses produced by the rectangles, if the point is within the foundation area. If the point lies outside the foundation area, then rectangles are extended to include the point. In this case the areas outside the foundation are subtracted in the computation of vertical stress.

Simons and Menzies (1975) suggested that the most useful design chart, at present, for deriving stresses beneath a foundation is that of Janbu *et al.* (1956). This chart indicates the increase in vertical stress beneath the centre of a uniformly loaded strip, rectangular or circular foundation structure (Figure 9.20).

The Newmark chart (see Newmark (1935)) provides a graphical method of estimating the vertical stress produced at a given point in the interior

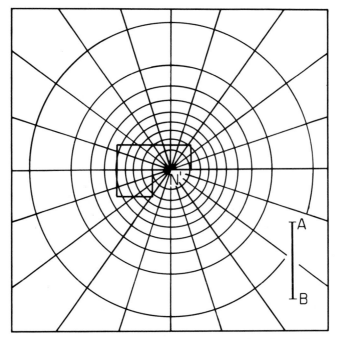

Figure 9.21 The Newmark chart for vertical stress under a foundation (see Newmark (1935))

of the soil mass by a load uniformly distributed over an irregularly shaped foundation (Figure 9.21). A tracing of the foundation is drawn to the same scale as the depth of the position at which the stress is to be evaluated (the line AB on the chart represents the depth, this being used for the scale of the drawing). The scale drawing is then laid over the influence chart with the point, vertically below which the stress is required, being placed at the centre of the chart. Then the number of segments on the chart covered by the drawing are counted and multiplied by the influence factor shown on the chart. This value is then multiplied by the loading to give the increase in pressure at the required depth.

9.8.3 Determination of primary or consolidation settlement

Consolidation settlement (S_c) is determined preferably from the values of the coefficient of volume compressibility as determined from oedometer tests (see Tomlinson (1980)). However, Skempton and Bjerrum (1957) showed that consolidation settlement, indicated from the results of oedometer tests, may be somewhat greater than actually occurs. This is because the amount of settlement is influenced by the type of clay concerned, so that

$$S_c = \mu \, S_o$$

where μ is a coefficient depending on the type of clay and S_o is settlement determined from oedometer tests. Skempton and Bjerrum suggested that the following values of μ can be used in most cases

Type of clay	Value of μ
Very sensitive clays	1.0–1.2
Normally consolidated clays	0.7–1.0
Overconsolidated clays	0.5–0.7
Heavily overconsolidated clays	0.2–0.5

The amount of oedometer settlement (S_o) of a layer of cohesive soil can be obtained from the expression

$$S_o = m_v \, \sigma_z \, H$$

where m_v is the average coefficient of volume compressibility for the effective pressure increment in the layer of soil in question, σ_z is the average imposed vertical stress on the layer due to the net foundation pressure, and H is the thickness of the layer. The settlement for each soil stratum, which is likely to be significantly affected by the structural load, is determined. Similarly, thick deposits of clay should be subdivided into layers and the settlement for each sublayer determined, thereby taking account of variations with depth. The total settlement is equal to the sum of that of the individual layers concerned.

Tomlinson (1980) recommended that if the results of only one or two oedometer tests are available it is more convenient to calculate S_o from the $e/\log p$ curves. In this method reduction in void ratio, brought about by foundation loading, is considered at the centre of the clay stratum. In other words, the depth at which the stress increment is calculated is from the base of the foundation to the centre of the clay layer concerned.

Oedometer settlement is also derived as follows

$$S_o = \frac{H}{1 + e_0}(e_0 - e_i)$$

or

$$S_o = \frac{H}{1 + e_1}C_c \log_{10}\frac{p_0 + \sigma_2}{p_0}$$

where H is the thickness of the layer, e_0 and e_1 are the initial and final void ratios, p_0 is the initial effective overburden pressure and σ_2 is the vertical stress at the centre of the layer due to the net foundation pressure. Individual layers of soil are again considered separately and then summed to give total settlement.

The rate at which a foundation settles is generally determined at the time required for 50% and 90% of final settlement to be completed. The time (t) required is given by the expression $t = (T_v d^2/c_v)$, where T_v is a dimensionless number termed the time factor, and d is the length of the drainage path ($d = H$, i.e. the thickness of the layer of soil concerned, for drainage in one direction only and $d = H/2$ for two-way drainage. In the field drainage above and below a stratum depends on the relative permeability of the beds immediately adjacent; two-way drainage takes place in the oedometer test). The values of the time factor for various degrees of consolidation (U) can be determined from Figure 9.22. The coefficient of consolidation (c_v) for 50% and 90% final settlement is usually determined by the log time method of Casagrande or the root time method of Taylor (1948) respectively. These two methods are outlined respectively in Figures 9.23 and 9.24.

The load imposed on a soil changes during the construction period. Initially it is reduced due to removal of material on excavation, then as the structure is erected the load gradually increases. The instantaneous time-settlement curve therefore needs adjusting to take account of the construction period. An empirical method of correction was proposed by Terzaghi (1943) and is outlined in Figure 9.25.

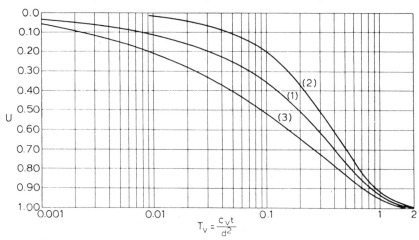

Figure 9.22 Relationships between average degree of consolidation and time factor. Curve (1) two-way drainage, pore pressure equal throughout. Curve (2) one-way drainage, pore pressure increasing downwards. Curve (3) one-way drainage, pore pressure increasing upwards

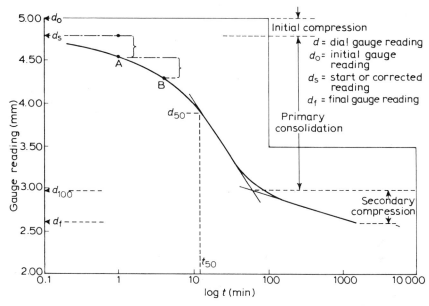

Figure 9.23 The log time method of determination of coefficient of consolidation. To obtain the correct position for zero, two values are chosen of log t on the initial part of the curve, for which the values of t are in the ratio 4:1 (see A and B). The vertical distance between these two points is obtained and this distance is set off vertically above the upper point (A). The new point is the corrected zero, i.e. $U = 0$. The latter point (d_s) will not correspond with the point representing the initial dial gauge reading (d_o), the difference between them is referred to as initial compression. The point $U = 100\%$ is located at the intersection (d_{100}) of the extension from the straight line parts of the curve. Primary consolidation takes place between d_s and d_{100}, beyond which secondary compression occurs. The point corresponding to $U = 50\%$ is midway between d_s and d_{100}. The value of T_v corresponding to $U_2 = 50\%$ is 0.196, the coefficient of consolidation (c_v) at this point is: $c_v = (0.916 \, d^2/t_{50})$, where d = half thickness of specimen for given pressure increment

9.8.4 Secondary settlement

After a sufficient time has elapsed excess pore pressures approach zero but a deposit of clay may continue to undergo a decrease in volume. This is referred to as secondary consolidation and involves compression of the soil fabric. Unfortunately no reliable method is as yet available for determining the amount and rate of secondary consolidation. Consequently if estimates of secondary consolidation are required in practice they are generally based on empirical procedures (see Simons (1975)). It has frequently been suggested that secondary settlement is often of little practical consequence. Nevertheless this does not mean that the possibility of its occurrence should not be investigated, for it has been noted that in certain circumstances a large part of the observed settlement which has occurred beneath structures has developed after the dissipation of excess pore pressure (see Foss (1969)). When considering secondary settlement it should be noted that two different factors may influence the process. The first is reduction in volume at constant effective stress and the second is the vertical strain resulting from lateral movements in the ground beneath the structure. Terzaghi (1943) pointed out that these two factors may be expected to result in completely different types of settle-

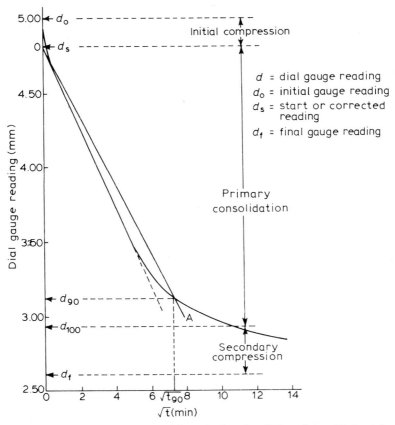

Figure 9.24 The root time method of determination of coefficient of consolidation (after Taylor (1948)). Extend back the initial straight-line part of the curve to intersect the ordinate, to give zero time (0). Multiply the values of $\sqrt{t}$ on the straight part of the curve by 1.15, and redraw a further curve through these points (AO). The intersection of AO with the original curve corresponds to $U = 90\%$ and therefore gives the values of $\sqrt{t_{90}}$. The value of T_v corresponding to $U = 90\%$ is 0.848 and hence the coefficient of consolidation (c_v) is obtained from $c_v = (0.848 \ d^2/t_{90})$, where d = half thickness of specimen for given pressure increment. The point on the original curve corresponding to $U = 100\%$ can be obtained by proportioning

ment. The relative importance of these factors will vary from structure to structure, depending on the stress level, type of clay and the geometry of the problem, and for any given structure will vary with the location of any deforming soil element and with time.

Simons (1975) termed a clay which had been recently deposited and achieved equilibrium under its own weight, but has not undergone significant secondary consolidation, a 'young' normally consolidated clay. Such a clay is only capable of carrying the overburden weight of soil and any additional load results in relatively large settlements. Such clays left under constant effective stresses for thousands of years continue to settle. This produces a more stable soil particle fabric which affords greater strength and reduced compressibility. With time a clay undergoing such secondary consolidation develops a reserve resistance against further compression. It can carry a

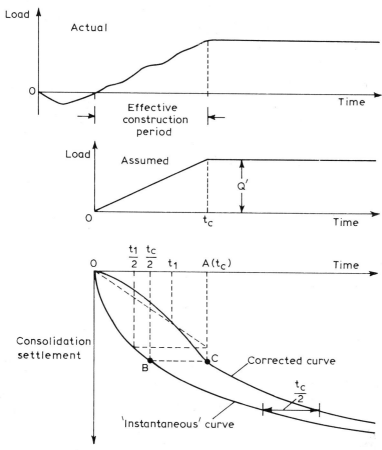

Figure 9.25 Correction for construction period. It is assumed that the net load (gross load less weight of soil removed on excavation) is applied uniformly over the construction period (t_c) but that the degree of consolidation at the end of this period is the same as if the load had been acting for half the period ($t_c/2$). Hence the settlement at any time during the construction period is that which takes place for an instantaneous loading at half the time. Construction: Drop perpendicular from A(t_c) and another from $\frac{1}{2}t_c$, to intersect instantaneous curve in B. Draw BC horizontal to cut AC in C, the first point on the corrected curve. Intermediate points are similarly obtained. Beyond C the settlement curve is assumed to be the instantaneous curve offset to the right by half the loading period, i.e. offset by distance BC. Corrected total settlement can be derived by adding immediate settlement

load in addition to the effective overburden pressure without significant volume change.

An overconsolidated clay is one which has been subjected to a pressure in excess of its present overburden pressure, that is, at some previous time it was buried to a greater depth but subsequently overburden has been removed by erosion. Some heavily overconsolidated clays may have been subjected to several cycles of loading and unloading. If a clay is over-consolidated then the estimated degree of settlement is likely to be in excess of that which actually occurs after loading. The difference decreases as the ratio $\Delta p/(p_o' - p_o)$ approaches unity, p_o' being the maximum pressure under which the clay was previously consolidated. Unfortunately the amount of discrepancy often cannot be predicted reliably. Generally because over-consolidated clays have a more stable fabric than normally consolidated clays they undergo relatively small secondary settlements compared with the latter. Nevertheless exceptions do occur, for example, Bjerrum (1966) maintained that even in overconsolidated soils significant secondary settlements can occur in association with structures with large variations in live load.

9.8.5 Settlement of cohesionless soils

A sudden application of a load on a layer of cohesionless sediment composed of sound, equidimensional particles produces an instantaneous compression followed by a slight additional compression at a decreasing rate. At low pressure both instantaneous and gradual compression are almost exclusively due to slippage at points of contact. As the load increases an increasing proportion of the compression is due to grain crushing.

In granular soils the allowable settlement is usually exceeded before soil rupture considerations become significant. Generally the total settlements of footings on granular soils are small, recorded settlements being of the order of 25 mm or less and rarely exceed 50 mm. The commonly accepted basis of design is that the total settlement of a footing should be restricted to about 25 mm as by so doing the differential settlement between adjacent footings will be confined within limits that can be tolerated by a structure.

The great difficulty in assessing the performance of cohesionless soils is due to the fact that it is almost impossible to obtain an undisturbed sample from them. Hence methods of settlement calculation must be based on the evidence gained from field tests such as the cone penetration test and the plate load test (see Meyerhof (1965); Sutherland (1975); Jordan (1977); Bell (1978)). The density index, as obtained by the standard penetration test, has been used by Arnold (1980) to predict the settlement of footings on sand.

9.9 Bearing capacity of soil

9.9.1 Deformation and failure of soil under loading

Foundation design is primarily concerned with ensuring that movements of a foundation are within limits which can be tolerated by the proposed structure, without adversely affecting its functional requirements. Consequently the design of a foundation structure requires an understanding of the local geological and groundwater conditions and more particularly an appreciation of the various types of ground movement that can occur. Foundation movements may occur when the ground is excavated, when it is loaded or independently of construction operations. Movement of foundations under the influence of loading may be due to the ground being overstressed, which gives rise to plastic deformation in the ground beneath the foundation structure. In extreme cases shear failure may occur. Even when the factor of safety against shear failure is adequate, settlement still takes place.

In order to avoid shear failure or substantial shear deformation the foundation pressures used in design should have an adequate factor of safety when compared with the ultimate bearing capacity of the foundation. The ultimate bearing capacity is the value of the net loading intensity which causes the ground to fail suddenly in shear. If this is to be avoided then a factor of safety must be applied to the ultimate bearing capacity, the value obtained being referred to as the maximum safe bearing capacity. In other words this is the maximum net loading intensity which may be safely carried without the risk of shear failure. But even this value may still mean that there is a risk of excessive or differential settlement. Thus the allowable bearing capacity is the value which is used in design, this taking into account all possibilities of failure, and so its value is frequently less than that of the safe bearing capacity. The value of ultimate bearing capacity depends on the type of foundation structure as well as the soil properties. For example, the dimensions, shape and depth at which a footing is placed all influence the bearing capacity. More specifically the width of the foundation is important in cohesionless soils, the greater the width, the larger the bearing capacity whilst in saturated clays it is of little effect. The density, frictional resistance and cohesion influence the shear strength of a soil, and its permeability, compressibility and consolidation must also be taken into con-

Table 9.6. Presumed bearing values under vertical static loading (from Anon (1972))

(These values are for preliminary design purposes only, and may need alteration upwards or downwards. No addition has been made for the depth of embedment of the foundation)

Group	Class	Types of rocks and soils	Presumed bearing value (kPa)	Remarks
I	1	Hard igneous and gneissic rocks in sound condition	10 000	These values are based on
Rocks	2	Hard limestones and hard sandstones	4000	the assump-
	3	Schists and slates	3000	tion that the
	4	Hard shales, hard mudstones and soft sandstones	2000 —	foundations are carried
	5	Soft shales and soft mudstones	600–1000	down to un-
	6	Hard sound chalk, soft limestone	600	weathered
	7	Thinly bedded limestones, sandstones, shales	} To be assessed after inspection	rock
	8	Heavily shattered rocks		
II	9	Compact gravel, or compact sand and gravel	>600	Width of foundation
Non-cohesive soils	10	Medium dense gravel or medium dense sand and gravel	200–600	(B) not less than 1 m
	11	Loose gravel, or loose sand & gravel	<200	Ground-
	12	Compact sand	>300	water level
	13	Medium dense sand	100–300	assumed to
	14	Loose sand	<100	be at a depth not less than B below the base of the foundation
III	15	Very stiff boulder clays & hard clays	300–600	Group III is
	16	Stiff clays	150–300	susceptible
Cohesive soils	17	Firm clays	75–150	to long-
	18	Soft clays and silts	<75	term consolidation
	19	Very soft clays and silts	Not applicable	settlement

sideration when the allowable bearing pressure, which takes settlement into account, is determined. With uniform soil conditions the ultimate bearing capacity increases with depth of installation of the foundation structure. This increase is associated with the confining effects of the soil, the decreased overburden pressure at foundation level and with the shear forces that can be mobilised between the sides of the foundation and the ground. Where foundations are deep, the contribution to the ultimate bearing capacity by the shear forces on the sides may be large. The presumed bearing values for various types of soil are given in Table 9.6.

When a load is applied to a soil in gradually increasing amounts the soil deforms and a load-settlement curve can be plotted (Figure 9.26). When the failure load is reached the rate of deformation increases and the load-settlement curve goes through a point of maximum curvature which indicates that the soil has failed. The shape of the curve is influenced by the type of soil involved, for example, dense sand and insensitive clay show a more gradual transition, associated with progressive failure.

There are usually three stages in the development of a foundation failure. First, the soil beneath the foundation is forced downwards in a wedge-shaped zone (Figure 9.27). Consequently the soil beneath the wedge is forced downwards and outwards, elastic bulging and distortion taking place within the soil mass. Secondly, the soil around the foundation perimeter pulls away from the foundation and the shear forces propagate outward from the apex

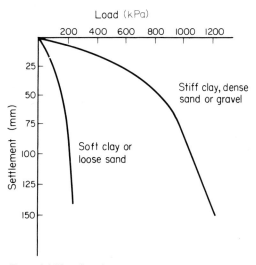

Figure 9.26 Load settlement curve

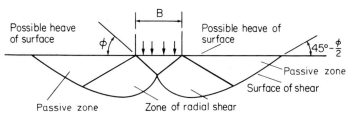

Figure 9.27 Foundation failure

of the wedge. This is the zone of radial shear in which plastic failure by shear occurs. If the soil is very compressible or can endure large strains without plastic flow the failure is confined to fan-shaped zones of local shear. The foundation will displace downwards with little load increase. On the other hand if the soil is more rigid, the shear zone propagates outwards until a continuous surface of failure extends to the ground surface and the surface heaves. This is termed general shear failure. The failure can be symmetrical, particularly if rotation is restricted by a column attached to the foundation, or it can tilt. A symmetrical bearing capacity failure is not common.

The weight of the material in the passive zone resists the lifting force and provides the reaction through the other two zones which counteract downwards motion of the foundation structure. Thus the bearing capacity is a function of the resistance to uplift of the passive zone. This in turn varies with the size of the zone (which is a function of the angle of shearing resistance), with the unit weight of the soil and with the sliding resistance along the lower surface of the zone (which is a function of the cohesion, angle of shearing resistance and unit weight of the soil). A surcharge placed on the passive zone, or increasing the depth of the foundation, therefore increases the bearing capacity.

Where a weak horizon overlies a strong one the shear will be confined to the weaker material and the stronger will not be involved in the failure. The bearing capacity should therefore be determined from the strength of the weaker material. Because the shear zone is restricted the true bearing capacity will exceed that calculated. In the converse situation where a strong layer overlies a weak one, then the former spreads the load thus reducing the pressure on the weaker horizon. Failure occurs by shear in the weaker material as the stronger bends under the load.

9.9.2 Determination of bearing capacity

A number of bearing capacity factors are used to determine the influence of the various characteristics of the soil and foundation structure on the ultimate bearing capacity. The bearing capacity factors suggested by Terzaghi (1943) for shallow footings, that is, at a depth not exceeding the width of the footing, involve cohesion, density and overburden pressure, and the ultimate bearing capacity for general shear is given by the expression

$$q = cN_c + \gamma z N_q + 0.5\gamma B N_\gamma$$

or net ultimate bearing capacity (q_n)

$$q_n = cN_c + \gamma z(N_q - 1) + 0.5\gamma B N_\gamma$$

The bearing capacity factors N_c, N_q and N_γ, are related to cohesion, overburden pressure and density respectively, which are functions of the angle of shearing resistance, c is the cohesion, γ is the unit weight of the soil, z is the depth of foundation emplacement and B is the width of footing. The bearing capacity factors can be obtained from Figure 9.28.

Meyerhof (1951, 1955) showed that Terzaghi's general equations for shallow foundations are conservative. This is because they do not take into account the shearing resistance of the soil along the surface of failure above the level of the base of the foundation. Meyerhof further pointed out

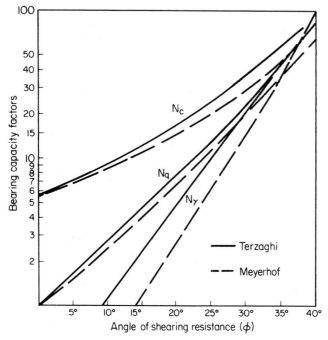

Figure 9.28 Bearing capacity factors for shallow strip loads

that as far as deep foundations are concerned Terzaghi's method suffers in that when the surface of failure does not extend to ground level, the height over which the shearing resistance of the soil is mobilised is very uncertain. As a consequence Meyerhof developed his own bearing capacity factors which depend on the depth and shape of the foundation and the roughness of its base, as well as the angle of shearing resistance (ϕ). However, he continued to use Terzaghi's general bearing capacity equations but with different curves (Figure 9.28). Meyerhof recommended that in the case of rectangular and circular footings the values of the bearing capacity factors should be multiplied by a shape factor (Figure 9.29).

Nonetheless Tomlinson (1980) suggested that the Terzaghi equations should be used for determination of the ultimate bearing capacity of granular soils, arguing that it was better to err on the conservative side. He accepted Meyerhof's curves for shallow and deep foundations on cohesive soils.

Both the Terzaghi and Meyerhof analyses assume the development of the full shear surface and complete shear failure. However, loose sands and highly sensitive clays fail by local or progressive shear when local cracking develops around the foundation or when the wedge forms under the foundation structure. Terzaghi (1943) therefore suggested the following expression for failure in local shear

$$q = \tfrac{2}{3}cN_c + \gamma zN_q + 0.5\gamma BN_\gamma$$

When a footing is rectangular or square the support given by the soil at the ends of the footing must be taken into account and the foundation has a higher ultimate bearing capacity. As has been seen above, when the

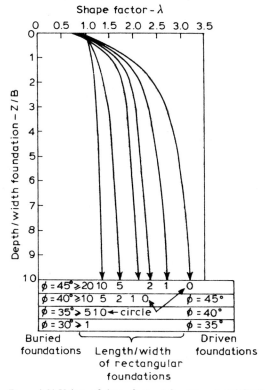

Figure 9.29 Values of shape factor (after Meyerhof (1955))

soil under a strip footing fails in shear the movement is laterally outward. If a square, rectangular or circular footing fails the movement of the soil particles forms a radial pattern. According to Skempton (1951) the ultimate bearing capacity for a rectangular footing is equal to the ultimate bearing capacity for a strip footing multiplied by

$$1 + 0.2\frac{B}{L}$$

where B and L are breadth and length respectively. Terzaghi (1943) suggested the following expressions for square and circular footings

Square $q = 1.3cN_c + \gamma z N_q + 0.4\gamma B N_\gamma$ (1)

Circular $q = 1.3cN_c + \gamma z N_q + 0.3\gamma B N_\gamma$ (2)

where B is the width and diameter of the footing respectively.

In granular soils the cohesion is negligible consequently the general bearing capacity expression is reduced to

$$q = \gamma z N_q + 0.5\gamma B N_\gamma$$

the net bearing capacity being

$$q_n = \gamma z (N_q - 1) + 0.5\gamma B N_\gamma$$

However, obtaining a value of ϕ by laboratory methods, in order to determine the bearing capacity factors, is often virtually impossible. This is due to difficulties in obtaining undisturbed samples of granular soils. Consequently it is common practice to determine the approximate density index of granular materials, and thereby an approximate value of ϕ, by means of penetration tests (see Terzaghi and Peck (1967)).

If the angle of shearing resistance is zero, as in saturated clay in undrained shear, only the cohesion contributes materially to the bearing capacity. Consequently Skempton (1951), in a review of the bearing capacity of clays, concluded that the ultimate bearing capacity of a footing on saturated clays in undrained conditions could be obtained from the following expression

$$q = cN_c + \gamma z$$

the net ultimate bearing capacity being simply

$$q_n = cN_c$$

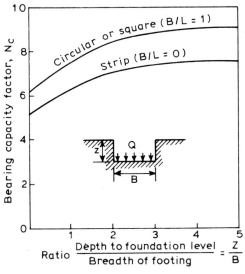

Figure 9.30 Bearing capacity factors for cohesive soils (Crown copyright BRE material reproduced by permission of the Controller, HMSO)

The bearing capacity factor, N_c, also depends upon the shape of the footing (strip, circular or square) and its depth/breadth ratio (Figure 9.30). In order to obtain the bearing capacity for a rectangular foundation the value of N_c for a strip foundation is multiplied by a shape factor (Figure 9.31).

When the water table is located at the base of the foundation the submerged unit weight is used in the bearing capacity expressions, which reduces that part of the bearing

$$q = cN_c + \gamma z N_q + 0.5\gamma_{sub} B N_\gamma$$

or

$$q_n = cN_c + \gamma z(N_q - 1) + 0.5\gamma_{sub} B N_\gamma$$

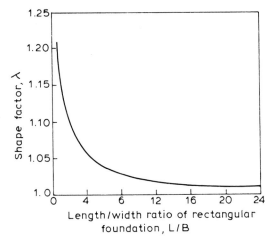

Figure 9.31 Shape factor for rectangular foundations
in clays (after Skempton (1951))

In undrained cohesive soils $\phi = 0$ and the term $0.5\gamma_{sub}BN_\gamma$ is of small
consequence, that is, the bearing capacity is not significantly affected by
groundwater (see Smith (1974)). Groundwater, however, does have an
appreciable influence on the bearing capacity of granular soils. If the water
table is above the base of the foundation the overburden pressure is affected
also.

9.9.3 Contact pressure and pressure bulbs

The pressure acting between the bottom of a foundation structure and the
soil is the contact pressure. The assumption that a uniformly loaded foundation structure will transmit the load uniformly so that the ground will be
uniformly stressed is by no means valid. For example, the intensity of the
stresses at the edges of a rigid foundation structure on hard clay is theoretically
infinite. In fact, the clay yields slightly and so reduces the stress at the edges
(Figure 9.32a). As the load is increased, more and more local yielding of
the ground material takes place until, when the loading is close to that
which would cause failure, the distribution is probably very nearly uniform.
Therefore at working loads a uniformly loaded foundation structure on clay
imposes a widely varying contact pressure. On the other hand, a rigid footing
on the surface of dry sand imposes a parabolic distribution of pressure

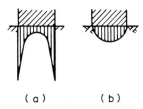

(a)   (b)

Figure 9.32 Contact pressure distribution beneath smooth,
rigid foundations. (a) Cohesive soil, e.g. clay. (b)
Cohesionless soil, e.g. sand

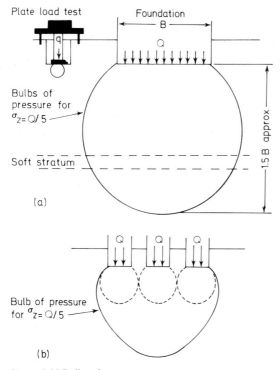

Figure 9.33 Bulbs of pressure

(Figure 9.32b). Since there is no cohesion in such material, no stress can develop at the edges of the footing. If the footing is below the surface of the sand the pressure at the edges is no longer zero but increases with depth. The pressure distribution therefore tends to become more nearly uniform as the depth increases. If a footing is perfectly flexible then it will distribute a uniform load over any type of foundation material. There is an infinite range of types of footing between the perfectly flexible and the perfectly rigid. Also the foundation material may possess both cohesive and frictional properties to varying degrees.

A bulb of pressure is a graphical illustration of the manner in which the loads imposed by a foundation structure are dissipated by the transfer of stress to increasing volumes of soil as the distance from the foundation structure increases (Figure 9.33a). The lines of equal pressure distribution indicate the proportional reduction of stress, exact figures for any set of assumptions can be calculated for any given set of conditions. Closely spaced loads develop a cumulative pressure bulb as shown in Figure 9.33b.

References

AKROYD, T. N. W. (1964). *Laboratory Testing in Soil Engineering*, Soil Mechanics Ltd, London.

ANON. (1972). *Code of Practice on Foundations, CP2004*, British Standards Institution, London.

ANON. (1979). 'Classification of rocks and soils for engineering geological mapping. Part I: Rock and soil materials', *Bull. Int. Ass. Engg. Geol.*, No. 19, 364–71.

ANON. (1981). *Code of Practice on Site Investigation, BS5930*, British Standards Institution, London.

ARNOLD, M. (1980). 'Prediction of footing settlements on sand', *Ground Engg*, **13**, No. 2, 40–49. March.

BELL, F. G. (ed.). (1978). '*In situ* testing and geophysical surveying', (in *Foundation Engineering in Difficult Ground*), Butterworths, London, 33–80.

BISHOP, A. W., GARGA, C. E., ANDERSON, A. & BROWN, J. D. (1971). 'A new ring shear apparatus and its application to the measurement of residual strength', *Geotechnique*, **21**, 273–328.

BISHOP, A. W. & HENKEL, D. J. (1962). *The Measurement of Soil Properties in the Triaxial Cell*, Edward Arnold, London.

BJERRUM, L. (1966). 'Secondary settlements of structures subjected to large variations in live load', *Int. Union Theoretical Appl. Mech. Rheol. Soil Mech. Symp., Grenoble*, 460–71.

BOUSSINESQ, J. (1885). *Application des Potentiels à l'Étude de l'Equilibre et du Mouvement des Solides Elastiques*, Gauthier-Villars, Paris.

BRITISH STANDARDS INSTITUTION (1975). *Methods of Test for Soils for Civil Engineering Purposes, BS1377* British Standards Institution, London.

CASAGRANDE, A. (1936). 'The determination of the pre-consolidation load and its practical significance', *1st Int. Conf. Soil Mech. Foundation Engg, Cambridge, Mass*, **3**, 60–64.

CASAGRANDE, A. (1948). 'Classification and identification of soils,' *Trans ASCE*, **113**, 901–992.

CASAGRANDE, A. (1958). 'Notes on the design of the liquid limit device,' *Geotechnique*, **8**, 84–91.

CASTRO, G. & CHRISTIAN, J. T. (1976). 'Strength of soils and cyclic loading', *Proc. ASCE, J. Geot. Engg Div.*, **102**, GT9, Paper No. 12387, 887–908.

COOLING, L. F. & GOLDER, H. Q. (1940). 'A portable apparatus for compression tests on clay soils', *Engineering*, **149**, 57–60.

CRAWFORD, C. H. (1966). 'The resistance of soil structure to consolidation', *Canadian Geot. J.* **3**, 90–97.

FADUM, R. E. (1948). 'Influence values for estimating stresses in elastic foundations', *Proc. 2nd Int. Conf. Soil Mech. Foundation Engg, Rotterdam*, 77–84.

FOSS, I. (1969). 'Secondary settlements of buildings in Drammen, Norway', *Proc. 7th Int. Conf. Soil Mech. Foundation Engg, Mexico*, **2**, 168–78.

FROSSARD, A. (1979). 'Effect of sand grain shape on interparticle friction: indirect measurements by Rowe's stress dilatancy theory', *Geotechnique*, **29**, 341–50.

GLOSSOP, R. & SKEMPTON, A. W. (1945). 'Particle size in silts and sands', *J. Inst. Civil Engrs*, **25**, Paper No. 5492, 81–105.

HANNA, T. H. & ADAMS, J. I. (1968). 'Comparison of field and laboratory measurements of modulus of deformation of a clay', *Highway Res. Board*, Pub. No. 243, Washington DC.

HANSBO, S. (1957). 'A new approach to the determination of the shear strength of clay by the fall cone test', *Proc. Roy. Swedish Geot. Inst.*, No. 14.

HEAD, K. H. (1980). *Manual of Soil Laboratory Testing*, Volume 1, *Soil Classification and Compaction Tests*, Pentech Press, London.

JANBU, N., BJERRUM, L. & KJAERNSLI, B. (1956). 'Veiledning ved Løsning av Fundamenteringsoppgaver', *Norwegian Geot. Inst.*, Publ. No. 16.

JORDAN, E. E. (Jan. 1977). 'Settlement in sand—methods of calculating and factors affecting', *Ground Engg*, **10**, No. 1, 31–7.

KRUMBEIN, W. C. & PETTIJOHN, F. J. (1938). *Manual of Sedimentary Petrography*, Appleton Century Crofts, New York.

LAMBE, T. W. (1960). 'A mechanistic picture of the shear strength in clay', *ASCE, Proc. Res. Conf., Shear Strength of Cohesive Soils*, Boulder, Colorado, 555–80.

LAMBE, T. W. (1973). 'Predictions in soils engineering', *Geotechnique*, **23**, 151–202.

LAMBRECHTS, J. R. & LEONARDS, G. A. (1978). 'Effects of stress history on the deformation of sand', *Proc. ASCE, J. Geot. Engg Div.*, GT11, 1371–87.

LITTLEJOHN, I. & FARMILOS, M. (May 1977). 'Some observations on liquid limit values with reference to penetration and Casagrande tests', *Ground Engg*, **10**, No. 3, 39–40.

MEIGH, A. C. & SKIPP, B. O. (1960). 'Gamma ray and neutron methods of measuring soil density and moisture', *Geotechnique* **10**, 110–26.

MENZIES, B. K. (1976). 'Strength, stability and similitude', *Ground Engg*, **9**, No. 5, 32–6.

MESRI, G. & ROKHSAR, A. (1974). 'Theory of consolidation for clays', *Proc. ASCE, J. Geot. Engg Div.*, **100**, GT8, 889–94.

MEYERHOF, G. G. (1951). 'The ultimate bearing capacity of foundations', *Geotechnique*, **2**, 301–32.

MEYERHOF, G. G. (1955). 'The influence of roughness of base on the ultimate bearing capacity of foundations', *Geotechnique*, **5**, 227–42.

MEYERHOF, G. G. (1965). 'Shallow foundations', *Proc. ASCE, Div. Soil Mech. Foundation Engg*, **91**, Paper No. 5172, 21–31.

MILNER, H. B. (1962). *Sedimentary Petrography*, Volume 1, *Methods in Sedimentary Petrography*, Allen & Unwin, London.

NEWMARK, N. M. (1935). 'Simplified computation of vertical pressures in elastic foundations', *Engg Exp. Stn, Circ.* 24, University of Illinois.

ROWE, P. W. & BARDEN, L. (1966). 'A new consolidation cell', *Geotechnique*, **16**, 162–70.

SCHOFIELD, A. N. & WROTH, C. P. (1968). *Critical State Soil Mechanics*, McGraw Hill, London.

SIMONS, N. E. (1975). 'Normally consolidated and lightly overconsolidated cohesive materials', (In *Settlement of Structures*), British Geotech. Soc., Pentech Press, London, 500–530.

SIMONS, N. E. & MENZIES, B. K. (1975). *A Short Course in Foundation Engineering*. Newnes-Butterworths, London.

SKEMPTON, A. W. (1944). 'Notes on the compressibility of clays', *Q. J. Geol. Soc. London*, **100**, 119–35.

SKEMPTON A. W. (1948). 'The $\phi = 0$ analysis for stability and its theoretical basis', *Proc. 2nd Int. Conf. Soil Mech. Foundation Engg, Rotterdam*, **1**, 72–5.

SKEMPTON, A. W. (1951). 'The bearing capacity of clays', *Build. Res. Cong.*, Div. 1, 180–89.

SKEMPTON, A. W. (1953). 'The colloidal activity of clays', *Proc. 3rd Int. Conf. Soil Mech. Foundation Engg, Zurich*, **1**, 57–60.

SKEMPTON, A. W. (1961). 'Effective stress in soils', (In *Pore Pressure and Suction in Soils*), Butterworths, London, 4–12.

SKEMPTON, A. W. (1964). 'Long-term stability of clay slopes,' *Geotechnique*, **14** 77–102.

SKEMPTON, A. W. & BJERRUM, L. (1957). 'A contribution to the settlement analysis of foundations on clay,' *Geotechnique*, **7**, 168–78.

SKEMPTON, A. W. & NORTHEY, R. D. (1952). 'The sensitivity of clays', *Geotechnique*, **2**, 30–53.

SMITH, G. N. (1974). *Elements of Soil Mechanics for Civil and Mining Engineers*, Crosby Lockwood Staples, London.

SRIDHARAN, A & RAO, V. G. (1979). 'Shear strength behaviour of saturated clays and the role of the effective stress concept', *Geotechnique*, **29**, 177–93.

SUTHERLAND, H. B. (1975). 'Granular material', (In *Settlements of Structures*), British Geotech. Soc., Pentech Press, London, 473–99.

TAYLOR, D. W. (1948). *Fundamentals of Soil Mechanics*, Wiley, New York.

TERZAGHI, K. (1925). *Erdbaumechanik auf Bodenphysikalischer Grundlage*, Dueticke, Vienna.

TERZAGHI, K. (1943). *Theoretical Soil Mechanics*, Wiley, New York.

TERZAGHI, K. & PECK, R. B. (1967). *Soil Mechanics and Engineering Practice*, Wiley, New York.

TOMLINSON, M. J. (1980). *Foundation Design and Construction*, Pitman, London.

VICKERS, B. (1978). *Laboratory Work in Civil Engineering: Soil Mechanics*, Crosby Lockwood Staples, London.

WAGNER, A A. (1957). 'The use of the Unified Soil Classification System for the Bureau of Reclamation', *Proc. 4th Int. Conf. Soil Mech. Foundation Engg*, London, **1**, 125–34.

WAHLS, H. E. (1962). 'Analysis of primary and secondary consolidation', *Proc. ASCE, J. Soil Mech. Foundation Engg Div.*, **88**, SM6, 207–31.

WHYTE, I. L. (Jan 1982). 'Soil plasticity and strength—a new approach using extrusion', *Ground Engineering*, **15**, No. 1, 16–24.

WROTH, C. P. (1979). 'Correlations of some engineering properties of soils', *Proc. 2nd Int. Conf. BOSS*, London, **1**, 121–32.

WROTH, C. P. & WOOD, D. M. (1978). 'The correlation of index properties with some basic engineering properties of soils', *Canadian Geot. J.*, **15**, 137–45.

Chapter 10

Soil types and their engineering behaviour

10.1 Coarse-grained soils

10.1.1 Fabric of granular soils

The engineering behaviour of a soil is a function of its structure or fabric, which in turn is a result of the geological conditions governing deposition and the subsequent stress history. The macro-structure of a soil includes its bedding, laminations, fissures, joints and tension cracks, all of which can exert a dominant influence on the shear strength and drainage character-istics of a soil mass. The micro-structure of a sand or gravel refers to its particle arrangement which in turn involves the concept of packing (see Chapter 3). The conceptual treatment of packing begins with a consideration of the arrangement of spherical particles of equal size. These can be packed in either a disorderly or a systematic fashion. The closest type of systematic packing is rhombohedral packing whereas the most open type is cubic pack-ing, the porosities approximating to 26 and 48% respectively (Table 10.1). Put another way the void ratio of a well-sorted and perfectly cohesionless aggregate of equidimensional grains can range between values of about 0.35 and 1.00. If the void ratio is more than unity the micro-structure will be collapsible or metastable. If a large number of spheres of equal size are arranged in any systematic packing pattern then there is a certain diameter ratio for smaller spheres which can just pass through the throats between the larger spheres into the interstices. For example, in rhombohedral packing this critical diameter is 0.154 D (D being the diameter of the larger spheres). However, a considerable amount of disorder occurs in most coarse-grained deposits and, according to Graton and Frazer (1935), there are colonies of tighter and looser packing within any deposit.

In a single grain structure individual particles are bulky and pore passages have average diameters of the same order of magnitude as smaller particle diameters. There is virtually no effective combination of particles to form aggregates, each particle functioning individually in the soil framework. Each grain is in contact with the others, in such a manner that the movement of any individual grain is influenced by the position of adjacent grains. For most equilibrium conditions in coarse-grained soil the soil framework serves exclusively as the stressed member.

Table 10.1. Some values of the common properties of soils

A. COHESIONLESS SOILS

	Gravels	Sands
Relative density	2.5 –2.8	2.6–2.7
Bulk density (t/m³)	1.45–2.3	1.4–2.15
Dry density (t/m³)	1.4 –2.1	1.35–1.9
Porosity (%)	20–50	23–35
Shear strength (kPa)	200–600	100–400
Angle of friction (deg)	35–45	32–42

B. COHESIVE SOILS

	Silts	Clays
Relative density	2.64–2.66	2.55–2.75
Bulk density (t/m³)	1.82–2.15	1.5–2.15
Dry density (t/m³)	1.45–1.95	1.2–1.75
Void ratio	0.35–0.85	0.42–0.96
Liquid limit (%)	24–35	Over 25
Plastic limit (%)	14–25	Over 20
Coefficient of consolidation (m²/yr)	12.2	5–20
Effective cohesion (kPa)	75	20–200
Effective angle of friction (deg)	32–36	

C. ORGANIC SOILS AND FILL

	Peat	Coarse discard
Moisture content (%)	650–1100	6–14
Relative density	1.3 –1.7	1.8–2.7
Bulk density (t/m³)	0.91–1.05	1.2–2.4
Dry density (t/m³)	0.07–0.11	1.05–2.0
Void ratio	12.7–14.9	0.35–over 1
Liquid limit (%)		23–45
Plastic limit (%)		Non-plastic–35
Effective angle of friction (deg)	5	28–40
Effective cohesion (kPa)	20	20–50

Size and sorting have a significant influence on the engineering behaviour of granular soils. Generally speaking the larger the particles, the higher the strength, and deposits consisting of a mixture of different sized particles are usually stronger than those which are uniformly graded. For instance, Holtz and Gibbs (1956a) showed that the amount of gravel in a sand–gravel mixture had a very significant effect on shear strength, it increasing considerably as the gravel content is increased up to 50 or 60%. After this point the material becomes less well graded and the actual density does not increase.

However, the mechanical properties of such sediments depend mainly on their density index which depends on packing. Indeed Holtz and Gibbs concluded that the density index of sand or sand-gravel mixtures had an appreciable effect on their shear strength. They quoted densely packed sands as being almost incompressible whereas loosely packed deposits, located above the water table, are relatively compressible but otherwise stable. If the density index of a sand varies erratically this can give rise to differential settlement. Generally settlement is relatively rapid in granular soils; however, when the stresses are large enough to produce appreciable grain fracturing, there is a significant time lag. Greater settlement is likely to be experienced in granular soils when foundation level is below the water table than when above. Additional settlement may occur if the water table fluctuates or the ground is subject to vibrations. Although the density index may decrease in a general manner with decreasing grain size, there is ample evidence to show that, for example, water-deposited sands with similar grain size can vary between wide limits. Hence factors other than grain size, such as rate of deposition and particle shape, influence density index.

10.1.2 Strength and deformation of granular soil

Fundamentally there are two basic mechanisms which contribute towards the deformation of granular soil, namely, distortion of the particles, and the relative motion between them. These two mechanisms are usually interdependent. At any instant during the deformation process different mechanisms may be acting in different parts of the soil and these may change as deformation continues. Interparticle sliding can occur at all stress levels, the stress required for its initiation increasing with initial stress and decreasing void ratio. Crushing and fracturing of particles begins in a minor way at small stresses, becoming increasingly important when some critical stress is reached. This critical stress is smallest when the soil is loosely packed and uniformly graded, and consists of large angular particles with a low strength. Usually fracturing only becomes important when the stress level exceeds 3.5 MPa.

Poulos (1981) defined the steady state of deformation for any mass of particles as that in which deformation is continuous at constant normal effective stress, constant shear stress and constant velocity. Such a state occurs when particle orientation has attained a steady-state condition and when there is no further fracturing of grains. Hence the shear strength required in order for deformation to continue remains constant, as does the velocity of deformation.

The internal shearing resistance of a granular soil is generated by friction when the grains in the zone of shearing are caused to slide, roll and rotate against each other. At the commencement of shearing in a sand some grains are moved into new positions with little difficulty. The normal stress acting in the direction of movement is small but eventually these grains occupy positions in which further sliding is more difficult. By contrast other grains are so arranged in relation to the grains around them that sliding is difficult: they are moved without sliding by the movements of other grains. The frictional resistance of the former is developed as the grains become impeded, whereas in the latter case it is developed immediately. The resistance to rolling represents the sum of the behaviour of all the particles and the resistance to sliding is essentially attributable to friction which is proportional to the confining stress. Frictional resistance is built up gradually and consists of establishing normal stresses in the intergranular structure as the grains push or slide along (see Cornforth (1964)). At the same time sliding allows the structure to loosen in dilatant soils which reduces normal stress. The maximum shearing resistance is a function of these two factors. The packing and external stress conditions govern the amount of sliding by individual grains in mobilising shearing resistance. According to Cornforth the latter factor is the more important and in fact is really a strain condition. He therefore concluded that the strain condition during shear is a major factor contributing to the strength of sand.

Interlocking grains contribute a large proportion of the strength in densely

packed granular soils and shear failure occurs by overcoming the frictional
resistance at the grain contacts. Conversely interlocking has little or no effect
on the strength of very loosely packed coarse-grained soils in which the
mobility of the grains is greater. Interlocking decreases as the confining
stress increases, because the particles become rounded at the points of contact,
sharp corners being crushed and particles may break. Even though this results
in a denser material it is still easier for shear deformations to occur.

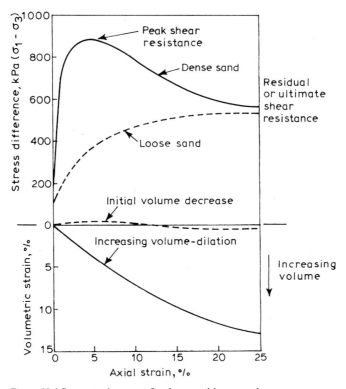

Figure 10.1 Stress–strain curves for dense and loose sand

Figure 10.1 shows that dense sand has a high peak strength and that
when it is subjected to shear stress it expands up to the point of failure,
after which a slight decrease in volume may occur. Conversely loose sand
compacts under shearing stress and its residual strength may be similar to
that of dense sand. Hence a constant void ratio is obtained, that is, the
critical or constant volume condition which has a critical angle of shearing
resistance and a critical void ratio. These are independent of initial density
being a function of the normal effective stress at which shearing occurs.

Both curves in Figure 10.1 exhibit strains which are approximately proportional to stress at low stress levels, suggesting a large component of elastic
distortion. If the stress is reduced the unloading stress–strain curve indicates
that not all the strain is recovered on unloading. The hysteresis loss represents
the energy lost in crushing and repositioning of grains. At higher shear
stresses the strains are proportionally greater indicating greater crushing

and reorientation. Indeed Arnold and Mitchell (1973) showed that as a sample of sand is subjected to cyclic loading the unloading response involves an increasing degree of hysteresis; in other words they found that on unloading, recoverable deformation in sand under triaxial conditions was small. Because irrecoverable strains were larger than the elastic strains this led them to suggest that total strains could in fact be regarded as irrecoverable. As would be expected loose sand with larger voids and fewer points of contact exhibits greater strains and less recovery of strain when unloaded than dense sand.

The angle of shearing resistance is also influenced by the grain size distribution and grain shape. The larger the grains the wider is the zone affected. The more angular the grains, the greater the frictional resistance to their relative movement since they interlock more thoroughly than do rounded ones. They therefore produce a larger angle of shearing resistance (Table 10.2). A well-graded granular soil experiences less breakdown in loading than a uniformly sorted soil of the same mean particle size since in the former type there are more interparticle contacts and hence the load per contact is less than in the latter. In gravels the effect of angularity is less because of particle crushing.

Table 10.2. Effect of grain shape and grading on the peak friction angle of cohesionless soil (after Terzaghi (1955))

Shape and grading	Loose	Dense
(1) Rounded, uniform	30°	37°
(2) Rounded, well graded	34°	40°
(3) Angular, uniform	35°	43°
(4) Angular, well graded	39°	45°

The presence of water in the voids of a granular soil does not usually produce significant changes in the value of the angle of shearing resistance. However, if stresses develop in the pore water they may bring about changes in the effective stresses between the particles whereupon the shear strength and the stress–strain relationships may be radically altered. Whether or not pore pressures develop depends upon the drainage characteristics of the soil mass and its tendency to undergo volume changes when subjected to stress. If the pore water can readily drain from the soil mass during the application of stress then the granular material behaves in a similar manner as it does when dry. On the other hand if loading takes place rapidly, particularly in fine-grained sands which do not drain as easily, then pore pressures are not dissipated. Since the water cannot escape readily from the voids of loosely packed, fine-grained sands, no volume decrease can occur so the pressure increases in the pore water. If the sample is loose enough almost the entire stress difference may be carried by the pore water so that very little increase occurs in the effective stress. In dense sands if the stress and drainage conditions are such that the water cannot flow into the sand as it is stressed then the usual volume increase characteristic of dense dry sand does not occur and a negative pore pressure develops.

Dusseault and Morgenstern (1979) introduced the term *locked sands* to describe certain peculiar sands which were first recognised in the Athabasca Oil Sands in Canada, and are older than Quaternary age. They are characterised by their high quartzose mineralogy, lack of interstitial cement, low porosity, brittle behaviour and high strength, with residual shear strengths (ϕ_r) varying between 30 and 35°.

Locked sands possess very high densities. Indeed the density indices of locked sands exceed 1, that is, their porosities are less than those which can be obtained by laboratory tests for achieving minimum porosity. Dusseault and Morgenstern attributed this to the peculiar fabric of these sands. The latter has been developed by diagenetic processes which reduced the porosity of the sands by solution and by recrystallisation of quartz as crystal overgrowths. Hence locked sands have an interlocked texture with a relatively high incidence of long and interpenetrative grain contacts. Sutured contacts are present and all grains have rugose surfaces.

At low stress levels locked sands undergo high rates of dilation. They have peak frictional strengths considerably in excess of those of dense sands. Dilatancy becomes suppressed as the level of stress increases since the asperities on the surfaces of individual grains are sheared through rather than causing dilation. The failure envelopes of locked sands are steeply curved as a result of the changing energy level required for shearing asperities as the level of stress increases. If undisturbed, locked sands are capable of supporting large loads with only small deformations.

10.2 Silts

The grains in a deposit of silt are often rounded with smooth outlines. This influences their degree of packing. The latter, however, is more dependent on the grain size distribution within a silt deposit, uniformly sorted deposits not being able to achieve such close packing as those in which there is a range of grain size. This in turn influences the porosity and void ratio values, as well as the bulk and dry densities (Table 10.1).

Dilatancy is characteristic of fine sands and silts. The environment is all-important for the development of dilatancy since conditions must be such that expansion can take place. What is more it has been suggested that the soil particles must be well-wetted and it appears that certain electrolytes exercise a dispersing effect thereby aiding dilatancy. The moisture content at which a number of fine sands and silts from British formations become dilatant usually varies between 16 and 35%. According to Boswell (1961) dilatant systems are those in which the anomalous viscosity increases with increase of shear.

Schultze and Kotzias (1961) showed that consolidation of silt was influenced by grain size, particularly the size of the clay fraction, porosity and natural moisture content. Primary consolidation accounted for 76% of the total consolidation exhibited by the Rhine silts tested by Schultze and Kotzias, secondary consolidation contributing the remainder. It was noted that unlike many American silts, which when saturated are often unstable and undergo significant settlements when loaded, the Rhine silts in such a condition were usually stable. The difference no doubt lies in the respective soil structures. Most American silts referred to in the literature are in actual fact loess soils. These have a more open structure then the reworked river silts of the Rhine, which have a void ratio of less than 0.85. Nonetheless in many

silts settlement continues to take place after construction has been completed and may exceed 100 mm. In fact settlement may continue for several months after completion because the rate at which water can drain from the voids under the influence of applied stress is slow.

Schultze and Horn (1961) found that the direct shear test proved unsuitable for the determination of the shear strength of silt, this had to be obtained by triaxial testing. They demonstrated that the true cohesion of silt was a logarithmic function of the water content and that the latter and the effective normal stress determined the shear strength. The angle of friction was dependent upon the plasticity index.

In a series of triaxial tests carried out on silt Penman (1953) showed that in drained tests with increasing strain the volume of the sample first decreased, then increased at a uniform rate and ultimately reached a stage where there was no further change. The magnitude of the dilatancy which occurred when the silt was sheared, and was responsible for these volume changes, was found to increase with increasing density, as it does in sands. The expansion was caused by the grains riding over each other during shearing. The strength of the silt was attributed mainly to the friction between the grains and the force required to cause dilatancy against the applied pressures. These drained tests indicated that the angle of shearing resistance decreased with increasing void ratio and with increasing lateral pressure. Grain interlocking was responsible for the principal increase in the angle of shearing resistance and increased with increasing density. A fall in pore water pressure occurred in the undrained tests during shearing, in fact there was an approximately linear relation between the maximum fall in pore pressure and the void ratio. Provided the applied pressures were sufficiently high, the drop in pore pressure governed the ultimate strength. The fall in the pore pressure was dependent on the density of the silt, the greater the density, the greater the fall in pore pressure. At a given density the ultimate strength was independent of applied pressure (above a critical pressure) and so the silt behaved as a cohesive material ($\phi = 0$). Below this critical pressure, silt behaved as a cohesive and frictional material. An exceptional condition was found to occur when a highly dilatable sample was placed under low cell pressure. When the pore pressure fell below atmospheric pressure, gas was liberated by the pore water and the sample expanded.

Frost heave is commonly associated with silty soils and loosely packed silts can exhibit quick conditions (see Chapter 6).

10.3 Loess

Loess owes its engineering characteristics largely to the way in which it was deposited since this gave it a metastable structure, in that initially the particles were loosely packed. The porosity of the structure is enhanced by the presence of fossil root-holes. The latter subsequently have been lined with carbonate cement, which also helps bind the grains together. This has meant that the initial loosely packed structure has been preserved and the carbonate cement provides the bonding strength of loess. It must be pointed out, however, that the chief binder is usually the clay matrix. On wetting the clay bond in many loess soils becomes soft, which leads to the collapse of the metastable structure. The breakdown of the soil structure occurs under its own weight.

In a detailed examination of the micro-structure of loess soils Larionov (1965) found that the coarser grains were never in contact with each other, being carried in a fine granular dispersed mass. Hence the strength of the soil is largely determined by the character of this fine mass. The ratio of coarser grains to fine dispersed fraction varies not only quantitatively but morphologically. Consequently three micro-structures can be recognised, namely, granular, where a filmy distribution of the fine dispersed fraction predominates; aggregate, consisting mainly of aggregates; and granular-aggregate, having an intermediate character. Larionov suggested that generally loess soils with granular micro-structure have less water resistance than aggregate types, they also have lower cohesion and higher permeability. They are therefore more likely to collapse on wetting than the aggregate types.

Loess deposits generally consist of 50–90% particles of silt size. For example, in the loess deposits of the Missouri Basin, investigated by Clevenger (1958), most of the grains were between 0.019 and 0.074 mm (Figure 10.2a).

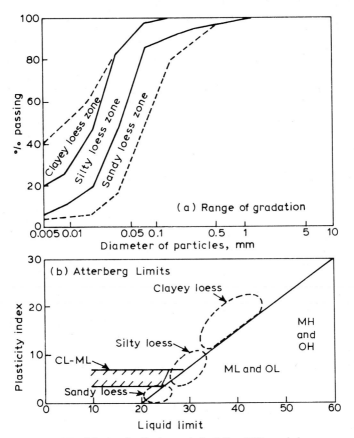

Figure 10.2 Particle size distribution and plasticity of Missouri river basin loess (after Clevenger (1958))

In fact he distinguished sandy, silty and clayey loess. Over three-quarters of the loess examined fell within the silty variety, a fifth being clayey loess, and the rest sandy loess. Clevenger also found that the undisturbed densities of loess in the Missouri basin range from around $1.2-1.36$ t/m³. If this material is wetted or consolidated (or reworked), the density increases, sometimes to as high as 1.6 t/m³. The liquid limit of loess averages about 30% (exceptionally liquid limits as high as 45% have been recorded), and their plasticity index ranges from about $4-9\%$, but averages 6% (Figure 10.2b). As far as their angle of shearing resistance is concerned, this usually varies from $30-34°$. Loess deposits are better drained (their permeability ranges from 10^{-5} to 10^{-7} m/s) than are true silts because of the fossil root-holes. As would be expected their hydraulic conductivity is appreciably higher in the vertical than in the horizontal direction.

Unlike silt, loess does not appear to be frost susceptible, this being due to its more permeable character, but like silt it can exhibit quick conditions and it is difficult, if not impossible, to compact. Because of its porous structure a 'shrinkage' factor must be taken into account when estimating earthwork.

In the unweathered state above the water table the unconfined compressive strength of loess may amount to several hundred kilopascals. On the other hand if loess is permanently submerged the metastable structure breaks down so that loess then becomes a slurry. Several collapse criteria have been proposed which depend upon the void ratios at the liquid limit (e_l) and the plastic limit (e_p). According to Audric and Bouquier (1976) collapse is probable when the natural void ratio is higher than a critical void ratio (e_c) which depends on e_l and e_p. They quoted the Denisov and Feda criteria as providing fairly good estimates of the likelihood of collapse

$$e_c = e_l \text{ (Denisov)}$$
$$e_c = 0.85e_l + 0.15e_p \text{ (Feda)}$$

Audric and Bouquier described a series of consolidated undrained triaxial tests they had carried out, at natural moisture content and after wetting, on loess soil from Roumare in Normandy. They distinguished collapsible and non-collapsible types of loess. The main feature of the collapsible soils was the soil structure, which was formed by soil domains with a low number of grain contacts. The authors noted that when collapsible loess was tested, the deviator stress reached a peak at rather small values of axial strain and then decreased with further strain. The pore pressures continued to increase after the peak deviator stress had been reached. By contrast in non-collapsible soils the deviator stress continued to increase and there was only a small increase of pore pressure. As expected the shear strength of collapsible loess was always less than that of the non-collapsible type.

Normally loess possesses a high shearing resistance and can carry high loadings without significant settlement when natural moisture contents are low. For instance, moisture contents of undisturbed loess are generally around 10% and the supporting capacity of loess at this moisture content is high. However, the density of loess is the most important factor controlling its shear strength and settlement. This is illustrated by the settlements in loess reported by Clevenger (1958). He noted that on wetting large settlements and low shearing resistance are encountered when the density of loess is below 1.28 t/m³, whereas if the density exceeds 1.44 t/m³ then settlement is small and shearing resistance fairly high. Moreover, the results of plate load tests have indicated that the bearing capacity of low-density loess may exceed 540 kPa when dry, and fall to as low as 27 kPa when wetted. The supporting capacity is reduced notably in low density loess when the moisture content exceeds 15%. Indeed laboratory tests have shown that samples of low-density loess consolidate between 15 and 20% when prewetted (little consolidation occurs if there is no prewetting). It is not surprising, therefore, that large settlements have occurred beneath footings in loess after it has been wetted. These have led to foundation failures, some examples of which are quoted by Clevenger (1958). Minkov *et al.* (1977) described settlement problems in loess in Bulgaria.

Clemence and Finbarr (1981) provided a summary of the methods of recognising collapsible soils and predicting their performance. They also noted a number of techniques which could be used hopefully to stabilise collapsible soils (Table 10.3).

Table 10.3. Methods of treating collapsible foundation soils (after Clemence and Finbarr (1981))

Depth of subsoil treatment	Foundation treatment
0–1.4 m	**A.** Current and past methods. Moistening and compaction (conventional extra heavy, impact or vibrating rollers)
1.5–10 m	Over-excavation and recompaction (earth pads with or without stabilisation by additives such as lime or cement). Vibro-flotation (free-draining soils). Vibro-replacement (stone columns). Displacement piles. Injection of lime. Ponding or flooding (if no impervious layers exist)
Over 10 m	Any of the aforementioned, or combinations of the aforementioned, where applicable. Ponding and infiltration wells, or ponding and infiltration wells with the use of explosives
	B. Possible future methods. Heat treatments to solidify the soils in place. Ultrasonics to produce vibrations that will destroy the bonding mechanisms of the soil (possibly electrochemical treatment). Use of grout to fill pore spaces

The presence of vertical root-holes explains why vertical slopes are characteristic of loess landscapes. These slopes may remain stable for long periods and when failure occurs it generally does so in the form of a vertical slice. By contrast an inclined slope is subject to rapid erosion.

10.4 Clay deposits

10.4.1 Clay minerals

Clay deposits are principally composed of fine quartz and clay minerals. The three major clay minerals are kaolinite, illite and montmorillonite. Both kaolinite and illite have non-expansive lattices whilst that of montmorillonite is expansive. In other words montmorillonite is characterised by its ability to swell and by its notable cation exchange properties. The basic reason why montmorillonite can readily absorb water into the interlayer spaces in its sheet structure is that the bonding between them is very weak (see Chapter 3).

The shape, size and specific surface all influence the engineering behaviour of clay minerals. Clay minerals have a plate-like shape and are very small in size. For example, an individual particle of montmorillonite is typically 1000 Å by 10 Å thick, whilst kaolinite is 10 000 Å by 1000 Å thick. The specific surface refers to the surface area in relation to the mass and the smaller the particle, the larger the specific surface (see Table 10.4). The specific surface provides a good indication of the relative influence of electrical forces on the behaviour of a particle.

Table 10.4. Size and specific surface of soil particles

Soil particle	Size (mm)	Specific surface (m^2/g)	Ion exchange/capacity (milli-equivalents/ 100g)
Sand grain	1	0.002	
Kaolinite	$d = 0.3$ to 3^{-3} thickness $= 0.3$ to $0.1d$	10–20	3–15
Illite	$d = 0.1$ to 2^{-3} thickness $= 0.1d$	80–100	20–40
Montmorillonite	$d = 0.1$ to 1^{-3} thickness $= 0.01d$	800	60–100

The surface of a clay particle has a net charge. This means that a clay particle is surrounded by a strongly attracted layer of water. As the diapolar water molecules do not satisfy the electrostatic balance at the surface of the clay particle, some metal cations are also adsorbed. The ions are usually weakly held and therefore can be replaced readily by others. Consequently they are referred to as exchangeable ions. The ion exchange capacity of soils normally ranges up to 40 milli-equivalents per 100 g. However, for some clay soils it may be greater as can be inferred from the ion exchange capacity of kaolinite, illite and montmorillonite (Table 10.4). The type of adsorbed cations influences the behaviour of the soil in that the greater their valency, the better the mechanical properties. For instance, clay soils containing montmorillonite with sodium cations are characterised by high water absorption and considerable swelling. If these are replaced by calcium, a cation with a higher valency, both these properties are appreciably reduced. The thickness of the adsorbed layer influences the soil permeability, that is, the thicker the layer, the lower the permeability since a greater proportion of the pore space is occupied by strongly held adsorbed water. As the ion exchange capacity of a cohesive soil increases so does its plasticity index, the relationship between the two being almost linear.

10.4.2 Micro-structure of clay soils

The micro-structure of cohesive soils is largely governed by the clay minerals present and the forces acting between them. Because of the complex electro-chemistry of clay minerals the spatial arrangement of newly sedimented particles is very much influenced by the composition of the water in which deposition takes place. Single clay mineral platelets may associate in an edge-to-edge (EE), edge-to-face (EF) face-to-face (FF) or random type of arrangement depending on the interparticle balance between the forces of attraction and repulsion, and the amount or absence of turbulence in the water in which deposition occurs. Since sea water represents electrolyte-rich conditions it causes clay particles to flocculate. Flocculation refers to the attraction of particles to one another in a loose, haphazard arrangement (Figure 10.3a). Considerable free water is trapped in the large voids. Flocculent soils are light in weight and very compressible but are relatively strong and insensitive to vibration because the particles are lightly bound by their edge-to-face attraction. They are sensitive to remounding which destroys the bond between the particles so that the free water is released to add to the adsorbed layers at the former points of contact.

Flocculation does not occur amongst clay particles deposited in fresh water. In this case they assume a more or less parallel, close-packed type of orientation. This has been referred to as a dispersed micro-structure (Figure 10.3b). Any bulky grains are distributed throughout the mass and cause localised departures from the pattern. Soils having a dispersed structure are usually dense and watertight. Typical void ratios are often as low as 0.5.

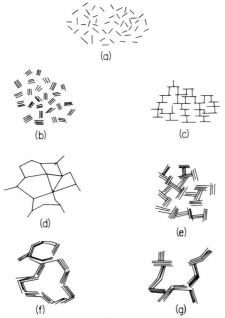

Figure 10.3 Modes of particle association in clay suspensions (after Van Olphen (1963)). (a) Dispersed and flocculated. (b) Aggregated but deflocculated. (c) Edge-to-edge flocculated but dispersed. (d) Edge-to-edge flocculated but dispersed. (e) Edge-to-face flocculated and aggregated. (f) Edge-to-edge flocculated and aggregated. (g) Edge-to-face and edge-to-edge flocculated and aggregated

Estuarine clays, because they have been deposited in marine through brackish to freshwater conditions can contain a mixture of flocculated and dispersed micro-structures. The aggregate micro-structure (Figure 10.3c) was proposed by Van Olphen (1963).

The original micro-structure of a clay deposit is subsequently modified by overburden pressures due to burial, which bring about consolidation. Consolidation tends to produce a preferred orientation with the degree of reorientation of the clay particles being related to both the intensity of stress and the electro-chemical environment, dispersion encouraging and flocculation discouraging clay particle parallelism. For instance, Barden (1972) maintained that lightly consolidated marine clay retains a random open structure, that medium consolidated brackish water clay has a very high degree of orientation, and that extremely heavily consolidated marine clay develops a fair degree of orientation (Figure 10.4).

It is generally accepted that turbostratic groups or domains occur in consolidated clays, although it has been suggested that ill-defined domains are present in unconsolidated clays. These domains consist of aggregations of clay particles which have a preferred orientation but between the aggregates the orientation is random. With increasing overburden pressure it appears that the number of clay particles in each domain increases and that there is an increase in domain orientation. For example, it has been shown that most of the Oxford Clay is made up of small domains, up to 50^{-3}mm across, these showing some alignment with the bedding.

The engineering properties of clay may be influenced by chemical interactions between particles. Olson (1974) sedimented particles in solutions with different pH concentrations and showed that at pH 9 a dispersed structure was developed in which there was a maximum opportunity for diffuse layer forces to influence the failure envelope. At pH 5 edge-to-edge flocculation occurred. A predominant edge-to-edge or edge-to-face particle arrangement minimises double layer repulsions.

Olson (1974) found that the failure envelopes for clays in the calcium form were independent of the electrolyte concentration, indicating that diffuse double layer forces do not develop with sufficient force to influence the shearing strength measurably. For clays in the sodium form, no diffuse double layer effects were found for kaolinite, possibly because of the low level of isomorphous substitution in the kaolinite lattice. Sodium illite had lower strength than calcium illite, that depended on pore water electrolyte concentration. Hence a distinct diffuse double layer effect was noted for sodium illite. The strength of sodium montmorillonite is so low that even the effects of substantial changes in diffuse double layer forces would be masked. It is believed that sodium montmorillonite breaks down to flakes only one unit cell thick (10 Å) and these flakes have little physical strength. The particles are thought to be so flexible that essentially all particle interaction must be of the face-to-face type, which leads to low intergranular stresses and a strong probability that the particles are always separated by at least one layer of adsorbed water, if not a full diffuse double layer. Calcium montmorillonite is believed to aggregate into particles, a number of unit cells thick, which have significant strengths. Olson suggested that location of the effective stress failure envelope was controlled mainly by the size and shape of the individual particles. Particles that are large and more or

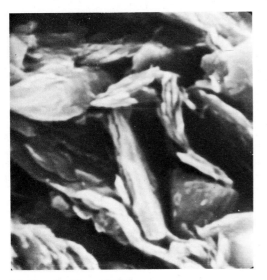

(a) Magnification × 25 500

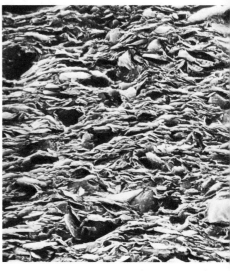

(b) Magnification × 2800

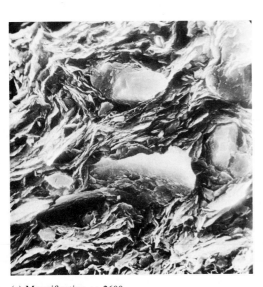

(c) Magnification × 2600

Figure 10.4 (a) Photomicrograph of lightly overconsolidated, post-glacial, marine-brackish water clay from the Clyde estuary. (b) Photomicrograph of lightly overconsolidated brackish water clay from the Clyde estuary. (c) Photomicrograph of heavily overconsolidated marine clay from Luanda, Angola (photographs courtesy Dr K. Collins, Strathclyde University)

less equidimensional, lead to high strengths regardless of chemical effects whereas particles which are very thin with high diameter-to-thickness ratios, such as sodium montmorillonite, have very low strengths.

10.4.3 Engineering properties of clay soils

The engineering performance of clay deposits is very much affected by the total moisture content and by the energy with which this moisture is held. For instance, the moisture content influences their density, consistency and strength. The energy with which moisture is held influences their volume change characteristics since swelling, shrinkage and consolidation are affected by permeability and moisture migration. Furthermore moisture migration may give rise to differential movement in clay soils. The gradients which generate moisture migration in clays may arise from variations in temperature, extent of saturation, and chemical composition or concentration of pore solutions.

The capillary potential or soil water potential is the force required to pull a unit mass of water away from a unit mass of soil, its magnitude indicating the force with which moisture is held. Moisture moves from wet to dry clay and will move upwards under the influence of capillary action against the force of gravity until equilibrium is established. At equilibrium the pore water pressures decrease linearly with height above the water table.

Fully saturated clay soils often behave as incompressible materials when subjected to rapid loading. The amount of elasticity increases continuously as the water content is decreased. Elastic recovery of the original size or shape may be immediate or may take place slowly. The linear relationship between stress and strain only applies to clays at low stresses.

There is no particular value of plastic limit that is characteristic of an individual clay mineral, indeed the range of values for montmorillonite is large. This is due to the inherent variations of structure and composition within the crystal lattice and the variations in exchangeable-cation composition. Generally the plastic limits for the three clay minerals noted decrease in the order montmorillonite, illite and kaolinite. As far as montmorillonite is concerned if the exchangeable ions are Na and Li then these give rise to high plastic limits. In the case of the other two clay minerals the exchangeable cations produce relatively insignificant variation in the plastic limit. On the other hand, poorly crystalline kaolinite of small particle size has a substantially higher plasticity than that of relatively coarse, well-organised particles.

Similarly there is no single liquid limit which is characteristic of a particular clay mineral; indeed the range of limits is much greater than that of the plastic limits. Again the highest liquid limits are obtained with Li and Na montmorillonite; then follow in decreasing order Ca, Mg, K, Al montmorillonite; illite; poorly crystalline kaolinite; and well crystallised kaolinite. Indeed the liquid limit of Li and Na montmorillonite cannot be determined accurately because of their high degree of thixotropy. But the character of the cation is not the sole factor influencing the liquid limit, the structure and composition of the silicate lattice are also important. The liquid limits for illites fall in the range 60–90% whilst those for kaolinites vary from about 30–75%. Again the crystallinity of the lattice and particle size are the controlling factors, for instance, poorly crystallised fine-grained samples may be over 100%. The presence of 10% montmorillonite in an illitic or kaolinitic clay can cause a substantial increase in their liquid limits.

The plasticity indices of Na and Li montmorillonite clays have exceedingly high values, ranging between 300 and 600%. Montmorillonites with other cations have values varying from about 50–300% with most of them in the range of 75–125%. As far as the latter are concerned there is no systematic variation with cation composition. In the case of illitic clays the plasticity indices range from 23–50%. The values for well-crystallised illite are extremely low, indeed they are almost non-plastic. The presence of montmorillonite in these clays substantially increases their indices. The range of plasticity indices for kaolinitic clays varies from about 1–40%, generally being around 25%. As can be inferred from above, the limit values increase with a decrease in particle size and the liquid limit tends to increase somewhat more than the plastic limit.

10.4.4 Volume changes in clay soils

One of the most notable characteristics of clays from the engineering point of view is their susceptibility to slow volume changes which can occur independently of loading due to swelling or shrinkage. The ability of a clay to imbibe water leads to it swelling, and when it dries out it shrinks.

Differences in the period and magnitude of precipitation and evaporation are the major factors influencing the swell–shrink response of an active clay beneath a structure. Poor surface drainage and leakage from underground pipes can produce concentrations of moisture in clay. Trees with high water demand and uninsulated hot-process foundations may dry out a clay causing shrinkage. The density of a clay soil also influences the amount of swelling it is likely to undergo. Expansive clay minerals absorb moisture into their lattice structure, tending to expand into adjacent zones of looser soil before volume increase occurs. In a densely packed soil having small void space, the soil mass has to swell to accommodate the volume change of the expansive clay particles.

Grim (1962) distinguished two modes of swelling in clay soils, namely, intercrystalline and intracrystalline swelling. Interparticle swelling takes place in any type of clay deposit irrespective of its mineralogical composition, and the process is reversible. In relatively dry clays the particles are held together by relict water under tension from capillary forces. On wetting the capillary force is relaxed and the clay expands. In other words intercrystalline swelling takes place when the uptake of moisture is restricted to the external crystal surfaces and the void spaces between the crystals. Intracrystalline swelling, on the other hand, is characteristic of the smectite family of clay minerals, of montmorillonite in particular. The individual molecular layers which make up a crystal of montmorillonite are weakly bonded so that on wetting water enters not only between the crystals but also between the unit layers which comprise the crystals. Swelling in Na montmorillonite is the most notable and can amount to 1000% of the original volume, the clay then having formed a gel. Hydration volume changes are frequently assessed in terms of the free-swell capacity. Mielenz and King

(1955) showed that generally kaolinite has the smallest swelling capacity of the clay minerals and that nearly all of its swelling is of the interparticle type. Illite may swell by up to 15% but intermixed illite and montmorillonite may swell some 60–100%. Swelling in Ca montmorillonite is very much less than in the Na variety, it ranges from about 50–100%. The large swelling capacity of montmorillonites means that they give the most trouble in foundation work.

Cycles of wetting and drying are responsible for slaking in argillaceous sediments which can bring about an increase in their plasticity index and augment their ability to swell. The air pressure in the pore spaces helps the development of the swell potential under cyclic wetting and drying conditions. On wetting the pore air pressure in a dry clay increases and it can become large enough to cause breakdown, which at times can be virtually explosive. The rate of wetting is important, slow wetting allowing the air to diffuse through the soil water so that the pressure does not become large enough to disrupt the soil. In weakly bonded clay soils cyclic wetting and drying brings about a change in the swell potential as a result of the breakdown of the bonds between clay minerals and the alteration of the soil structure.

Bjerrum (1967) suggested that clay soils with high salt contents and a network of cracks undergo an increase in swell potential. This is due to slaking consequent upon osmotic pressures being developed as rain water infiltrates into the cracks.

Freeze-thaw action, like osmotic swelling, also affects the swell potential. Freezing can give rise to large internal pressures at the freezing front in fine-grained soils and if this front advances slowly enough the soil immediately beneath it can become quite desiccated. On melting the desiccation zone is saturated. Hence the swell potential is increased by the freeze pressures.

According to Schmertmann (1969) some clays increase their swell behaviour when they undergo repeated large shear strains due to mechanical remoulding. He introduced the term swell sensitivity for the ratio of the remoulded swelling index to the undisturbed swelling index and suggested that such a phenomenon may occur in unweathered highly overconsolidated clay when the bonds, of various origins, which hold clay particles in bent positions, have not been broken. When these bonds are broken by remoulding the clays exhibit significant swell sensitivity.

An internal swelling pressure will reduce the effective stress in a clay soil and therefore will reduce its shearing strength. As a consequence Hardy (1966) suggested that the Coulomb equation for swelling soils should take the swelling pressure, as well as pore pressure, into account, the expression becoming

$$\tau = c' + (\sigma - u - p_s)\tan \phi'$$

where τ is the shearing strength, c' is the cohesion, σ is the total stress, u is the pore water pressure, p_s is the swelling pressure and ϕ' is the angle of shearing resistance.

Holtz and Gibbs (1956b) showed that expansive clays can be recognised from their plasticity characteristics. However, the most widely used soil property to predict swell potential is the activity of a clay (Figure 10.5).

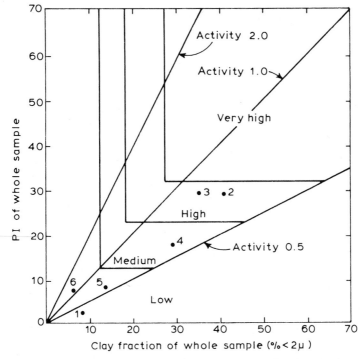

Figure 10.5 Determination of potential expansiveness of soils (after Van der
Merwe, D. H. (1964). 'The prediction of heave from the plasticity index
and percentage clay fraction of soils', *The Civil Engr in S. Afr.*, S. Afr. Inst.
Civil Engrs., **6**, 103–16)

Popescu (1979) reported that the construction damage which occurs in
areas of Rumania where expansive clay forms the surface cover is mainly
caused by seasonal swelling and shrinkage of the foundation subsoil. These
effects are notable in regions which experience alternating wet and dry
seasons. The expansive clays of Rumania are normally residual soils formed
from basic igneous rocks. This accounts for the fact that the clay fraction
of these soils usually contains between 40 and 80% montmorillonite. Popescu
noted that maximum seasonal changes in moisture content in these expansive
clays were around 20% at 0.4 m depth, 10% at 1.2 m, and less than 5%
at 1.8 m. The corresponding cyclic movements of the ground surface are
about 100–120 mm. Desiccation of clays during the dry season leads to
the soil cracking. The cracks can gape up to 150 mm and frequently extend
to 2 m in depth. Popescu stated that the depth of the active zone in expansive clays in Rumania varies from about 2.0–2.5 m. In Australia, India,
South Africa and the United States it may extend to 3 m depth, whereas
in Israel depths of up to 6 m have been recorded.

Popescu recognised three distinct stages in the shrinkage process, the extent
of each depending on the soil structure, particularly the number of interparticle bonds causing resistance. In both the initial and residual stages of
shrinkage the reduction in soil volume is less than the volume of water

lost. Only during the normal shrinkage stage are the two equal. This can give rise to differential movement. The anisotropic character of a clay can also lead to differential movements. Williams and Jennings (1977) found that soil structure has a major influence on the pattern of the shrinkage process, as well as on the total amount of shrinkage. They indicated that in clay soils with the same initial moisture content and density, the more random the particle arrangement, the less the total shrinkage.

Ordemir *et al.* (1977) reported expansion of up to 15%, at a metre or so depth, in the Ankara Clay. They suggested that removal of surface clay and replacement by non-expansive fill was an inexpensive method of treatment.

Volume changes in clays also occur as a result of loading and unloading which bring about consolidation and heave respectively. When clay is first deposited in water, its water content is very high and it may have void ratios exceeding 2. As sedimentation continues overburden pressure increases and the particles are rearranged to produce a new packing mode of greater stability as most of the water is expelled from the deposit. Expulsion of free and adsorbed water occurs until a porosity of about 30% is reached. Any further reduction in pore volume is attributable to mechanical deformation of particles or diagenetic processes. As material is removed from a deposit by erosion the effective overburden stress is reduced and elastic rebound begins. Part of the rebound or heave results from an increase in the water content of the clay. Cyclic deposition and erosion has resulted in multiple loading and unloading of many clay deposits.

The heave potential arising from stress release depends upon the nature of the diagenetic bonds within the soil, that is, the post-depositional changes such as precipitation of cement and recrystallisation which have occurred (see Bjerrum (1967)). It would appear that significant time-dependent vertical swelling may arise, at least in part, from either of two fundamentally different sources, namely, localised shear stress failures or localised tensile stress failures. Localised shear stress failures are associated with long-term deformations of soils having well-developed diagenetic bonds such as clay shales. When an excavation is made in a clay with weak diagenetic bonds elastic rebound will cause immediate dissipation of some stored strain energy in the soil. However, part of the strain energy will be retained due to the restriction on lateral straining in the plane parallel to the ground surface. The lateral effective stresses will either remain constant or decrease as a result of plastic deformation of the clay as time passes (see Bjerrum (1972)). These plastic deformations can result in significant time-dependent vertical heaving. However, creep of weakly bonded soils is not a common cause of heaving in excavations.

The relationships between the stresses, failure mechanisms and time-dependent heaving are complex in clay soils with well-developed diagenetic bonds. According to Obermeier (1974) heaving is in part related to crack development, cracking giving rise to an increase in volume, the rate of crack growth being of particular significance. The initial rate of heaving is probably controlled by cracking due to tensile failures and to plastic deformations arising from shear failures. Furthermore, Obermeier maintained that because of the breakdown of diagenetic bonds there is an increase in the lateral stresses parallel to the ground surface.

10.4.5 Settlement of clay soils

When a load is applied to a clay soil its volume is reduced, this being due principally to a reduction in the void ratio. If such a soil is saturated then the load is initially carried by the pore water which causes a pressure, the hydrostatic excess pressure to develop. The excess pressure of the pore water is dissipated at a rate which depends upon the permeability of the soil mass and the load is eventually transferred to the soil structure. The change in volume during consolidation is equal to the volume of the pore water expelled and corresponds to the change in void ratio of the soil. In clay soils, because of their low permeability, the rate of consolidation is slow. Primary consolidation is brought about by a reduction in the void ratio. Further consolidation may occur due to a rearrangement of the soil particles. This secondary consolidation is usually much less significant. The various factors which influence the compressibility of a clay soil have been reviewed by Wahls (1962).

For all types of foundation structures on clays the factors of safety must be adequate against bearing capacity failures. Generally speaking experience has indicated that it is desirable to use a factor of safety of 3, yet although this will eliminate complete failure, settlement may still be excessive. It is therefore necessary to give consideration to the settlement problem if bearing capacity is to be viewed correctly. More particularly it is important to make a reliable estimate of the amount of differential settlement that may be experienced by the structure. If the estimated differential settlement is excessive it may be necessary to change the layout or type of foundation structure.

During the construction period the net settlement is comprised of immediate settlement due to deformation of the clay without a change in water content and consolidation settlement brought about by pore water being squeezed from the clay. As mentioned above, the rate of consolidation is generally very slow because of the low permeability of clays so that the former type of settlement will have been the greater of the two by the end of the construction period. In the course of time consolidation becomes important, giving rise to long continued settlement, although at a decreasing rate for years or decades after the completion of construction. Accordingly the principal objects of a settlement analysis are, to obtain a reasonable estimate of the net final settlement corresponding to a time when consolidation is virtually complete and, secondly, to estimate the progress of settlement with time. It should be borne in mind that settlement depends primarily on the compressibility of the clay which is in turn intimately related to its geological history, that is, to whether it is normally consolidated or overconsolidated. Clays which have undergone volume increase due to swelling or heave are liable to suffer significantly increased gross settlement when they are subsequently built upon.

10.4.6 Normally consolidated and overconsolidated clays

A normally consolidated clay is that which at no time in its geological history has been subject to pressures greater than its existing overburden pressure; however, an overconsolidated clay is one which has. The major factor in overconsolidation is removal of material that once existed above a clay deposit by erosion. Berre and Bjerrum (1973) carried out a series of triaxial and shear tests on normally consolidated clay, the confining conditions simulating the overburden pressures in the field. They demonstrated that the clay could sustain a shear stress in addition to the *in situ* value, undergoing relatively small deformation as long as the shear did not exceed a given critical value. This critical shear value represents the maximum shear stress which can be mobilised under undrained conditions, and governs the bearing pressure such a clay can carry with limited amount of settlement. Generally this critical shear value varies with plasticity and the rate at which load is applied. An overconsolidated clay is considerably stronger at a given pressure than a normally consolidated clay and it tends to dilate during shear whereas a normally consolidated clay consolidates. Hence when an overconsolidated clay is sheared under undrained conditions negative pore water pressures are induced, the effective strength is increased, and the undrained strength is much higher than the drained strength, the exact opposite to a normally consolidated clay. When the negative pore pressures gradually dissipate the strength falls as much as 60 or 80% to the drained strength.

In both normally consolidated and overconsolidated clays the shear strength reaches a peak value and then, as displacements increase, decreases to the residual strength. The development of residual strength is therefore a continuous process. Put another way, if at any particular point the soil is stressed beyond its peak strength, its strength decreases and additional stress is transmitted to other points in the soil. These in turn become over-stressed and decrease in strength. The failure process continues until the strength at every point along the potential slip surface has been reduced to the residual strength. The long-term strength of an overconsolidated clay is generally lower than its short-term strength. For example, Esu and Grisolia (1977) showed that this was the case as far as the Valdarno Clay is concerned.

10.4.7 Bearing capacity of clay

The ultimate bearing capacity of a foundation in clay soil depends on the shear strength of the soil and the shape and depth at which the foundation structure is placed. In relation to applied stress saturated clays behave as purely cohesive materials provided that no change of moisture content occurs. Thus, when a load is applied to saturated clay it produces excess pore pressures which are not quickly dissipated. In such instances the angle of shearing resistance (ϕ) is equal to zero. The assumption that $\phi = 0$ forms the basis of all normal calculations of ultimate bearing capacity in clays (see Skempton (1951)). The strength may then be taken as the undrained shear strength or one-half the unconfined compressive strength. To the extent that consolidation does occur, the results of analyses based on the premise that $\phi = 0$ are on the safe side. Only in special cases, with prolonged loading periods or with very silty clays, is the assumption sufficiently far from the truth to justify a more elaborate analysis.

The presence of fissures in clay reduces its shear strength since they represent surfaces of weakness along which sliding tends to occur (see Chapter 5). Furthermore fissured clay also tends to soften when subject to strain, which, in turn, gradually reduces the cohesion. Consequently

Skempton and Hutchinson (1969) suggested that this weakening and softening could be taken into account by assuming that the effective cohesion approached zero, while the angle of shearing resistance remained the same as that of intact clay. Subsequently Rivard and Lu (1978), having worked on various Canadian soft fissured clays, supported Skempton and Hutchinson's conclusion, stating that it provides a more reliable means of predicting the *in situ* stability condition.

10.4.8 Estuarine clays

According to Crooks and Graham (1976) the Belfast estuarine deposits, which are post-glacial, generally consist of soft, organic, silty clay. The predominant clay mineral is illite; kaolinite, chlorite and swelling chlorite also being present. The soil is lightly overconsolidated, probably as a result of changes in groundwater level, and is lightly fissured. The soil structure exerts an important controlling effect on the engineering behaviour of the Belfast estuarine deposits. In particular the ability of the structure to sustain stresses greater than the *in situ* stresses is directly related to the degree of overconsolidation.

These deposits show a variation with depth in their grain size distribution and in their sodium/calcium cation ratio. This reflects the changing conditions during sedimentation, that is, the early stages of a marine transgression, its maximum extension, and finally its regression and the return to more littoral conditions. The sodium cation concentration influences the Atterberg limits—they also vary with depth. For instance, the plasticity index decreases from about 70% at 4 m depth to about 45% at 7.5 m depth. The organic content ranges between 1 and 4%.

Crooks and Graham (1976) reported that the material is highly compressible, and has a low undrained shear strength. The latter increases in strength with depth from a minimum of 10 kPa at the top of the clay to about 22 kPa at 8 m. The small-strain undrained strength of these deposits reflects the strength of the soil structure which in turn is influenced by the light overconsolidation. They have a high activity and a medium to high sensitivity.

10.5 Tropical soils

10.5.1 Laterite and lateritic soils

Ferruginous and aluminous clay soils are frequent products of weathering in tropical latitudes. They are characterised by the presence of iron and aluminium oxides and hydroxides. These compounds, especially those of iron, are responsible for the red, brown and yellow colours of the soils. The soils may be fine grained or they may contain nodules or concretions. Concretions occur in the matrix where there are higher concentrations of oxides in the soil. More extensive accumulations of oxides give rise to laterite.

Laterite is a residual ferruginous clay-like deposit which generally occurs below a hardened ferruginous crust or hardpan. Ola (1978a) maintained that in laterites the ratios of silica (SiO_2) to sesquioxides (Fe_2O_3, Al_2O_3) usually are less than 1.33, that those ratios between 1.33 and 2.0 are indicative of lateritic soils, and that those greater than 2.0 are indicative of non-lateritic types.

Laterisation is rapid in tropical regions experiencing periods of heavy rainfall alternating with drier periods (see West and Dumbleton (1970)). Decomposition of the parent rock involves the removal of silica, lime, magnesia, soda and potash, leading to an enrichment in iron and aluminium oxides. Kaolinite is formed and oxides and hydroxides of iron accumulate, since rapid oxidation does not allow the organic compounds in solution to dissolve iron and thereby for it to be carried away.

Laterites tend to occur in areas of gentle topography which are not subject to significant erosion. During drier periods the water table is lowered. The small amount of iron which has been mobilised in the ferrous state by the groundwater is then oxidised, forming hematite, or if hydrated—goethite. The movement of the water table leads to the gradual accumulation of iron oxides at a given horizon in the soil profile. A cemented layer of laterite is formed which may be a continuous or honeycombed mass, or nodules may be formed, as in laterite gravel. Concretionary layers are often developed near the surface in lowland areas because of the high water table.

If laterisation proceeds further, as a result of prolonged leaching, then kaolinite is decomposed, the silica being removed in solution and gibbsite remains. Iron compounds may also be removed so that the soil becomes enriched in alumina, a bauxitic soil being developed. Bauxitic clays often have a pisolitic structure.

Gidigasu (1976) distinguished three major stages in laterisation. The first stage, that of decomposition, is characterised by physico-chemical breakdown of primary minerals and the release of constituent elements. The second stage involves leaching, under appropriate drainage conditions, of combined silica and bases, and the relative accumulation or enrichment from outside sources of oxides and hydroxides of sesquioxides (mainly Fe_2O_3 and Al_2O_3). The third stage involves partial or complete dehydration (sometimes involving hardening) of the sesquioxide-rich materials and secondary minerals.

A typical complete laterite profile is as follows

(5) a hard crust very rich in iron (cuirasse or ironstone caprock),
(4) a zone rich in free sesquioxides, sometimes with kaolinite nodules ((laterite proper),
(3) a zone of kaolinitic clay material, sometimes with small amounts of montmorillonite and micas (lithomarge),
(2) a decomposed zone of bedrock in which occur relicts of the parent material together with decomposing feldspars,
(1) a bedrock usually of igneous material.

The profile indicates how the parent rock is altered, clay minerals developing, with a high proportion of kaolinite. The sesquioxides, especially iron, are concentrated in the upper part of the profile. Aluminium oxides may replace those of iron in the upper part of the profile and the presence of aluminium may take the form of nodular gibbsite at the surface.

Laterites often have a cellular structure and may consist of ironstone pisolites set in ferruginous cement containing remnant quartz. Ola (1978a) found that lateritic soils in Nigeria were composed predominantly of kaolinite with some quartz. Laterite hardens on exposure to air. Hardening may be due to a change in the hydration of iron and aluminium oxides.

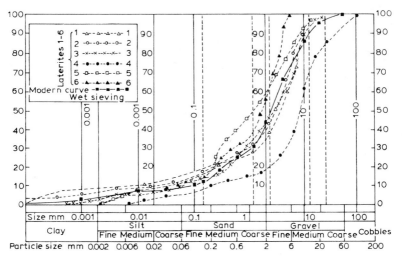

Figure 10.6 Grading curves of laterites (after Madu, R. M. (1977). 'An investigation into the geotechnical properties of some laterites of eastern Nigeria', *Engg. Geol.*, **11**, 101–25)

Laterite commonly contains all size fractions from clay to gravel and sometimes even larger material (Figure 10.6). Nixon and Skipp (1957a) provided the following range of values of common soil properties of laterite

Moisture content,	16–49%
Liquid limit,	33–90%
Plasticity index,	5–59%
Clay fraction,	15–45%

Usually at or near the surface the liquid limits of laterites do not exceed 60% and the plasticity indices are less than 30%. Consequently laterites are of low-to-medium plasticity. The activity of laterites may vary between 0.5 and 1.75.

Lateritic soils, particularly where they are mature, furnish a good bearing stratum. The hardened crust has a low compressibility and therefore settlement is likely to be negligible. In such instances, however, the strength of laterite may decrease with increasing depth. For example, Nixon and Skipp (1957a) quoted values of shear strength of 90 and 25 kPa, derived from undrained triaxial tests, for laterite samples from a site north of Colombo, Sri Lanka, taken from the surface crust and from a depth of 6 m respectively.

Ola (1978a) investigated the effects of leaching on lateritic soils. The cementing agents in lateritic soils help to bond the finer particles together to form larger aggregates. However, as a result of leaching these aggregates break down, which is shown by the increase in liquid limit after leaching. Moreover removal of cement by leaching gives rise to an increase in compressibility of more than 50%. Again this is mainly due to the destruction of the aggregate structure. Conversely there is a decrease in the coefficient of consolidation by some 20% after leaching (Table 10.5). The change in effective angle of shearing resistance and effective cohesion before and after leaching, is similarly explained. Prior to leaching the larger aggregates in the soil cause it to behave as a coarse-grained, weakly bonded particulate material. The strongly curvilinear form of the Mohr failure envelopes can also be explained as a result of the breakdown of the larger aggregates.

Table 10.5. Engineering properties of a lateritic soil before and after leaching (after Ola (1978a))

Property	Before leaching	After leaching
Natural moisture content, %	14	—
Liquid limit, %	42	53
Plastic limit, %	25	21
Relative density	2.7	2.5
Angle of shearing resistance, ϕ'	26.5°	18.4°
Cohesion, c', kPa	24.1	45.5
Coefficient of compressibility, m^2/MN*	12	15
Coefficient of consolidation, $m^2/year$*	599	464

* For a pressure of 215 kPa.

Vertical faces can often be excavated in laterite, up to depths of 6 m, without failing. Deeper cuttings require sloping and drainage at both the top and bottom of the face. Laterite is often a water-bearing stratum. Heavy pumping may be required for excavations.

10.5.2 Red clay and black clay soils

Red earths or latosols are residual ferruginous soils in which oxidation readily occurs. Such soils tend to develop in undulating country and most of them appear to have been derived from the first cycle of weathering of the parent material. They differ from laterite in that they behave as a clay and do not possess strong concretions. They do, however, grade into laterite.

The residual red clays of Kenya have been produced by weathering, leaching having removed the more soluble bases and silica, leaving the soil rich in iron oxide (hematite) and hydroxide (goethite) and in aluminium. The latter usually occurs in the form of kaolinitic clay minerals or sometimes as gibbsite. Dumbleton (1967) found halloysite in these red clays, as did Dixon and Robertson (1970). These soils contain a high percentage of clay size material and have high plastic and liquid limits

Clay content (%)		Liquid limit (%)	Plastic limit (%)
Range	63–88	76–104	34–56
Average	80	86	42

Sherwood (1967) found that the clay particles in Kenyan red clays were cemented together to form aggregates, the cementing agent being free iron oxide.

Tropical red clays may be of alluvial origin. Nixon and Skipp (1957b) found that residual red earths usually plot on or below the 'A' line on the Casagrande plasticity chart whereas alluvial red clays fall above it.

Allam and Sridharan (1981) subjected red earth soil to a number of cycles of wetting and drying in an attempt to simulate the notable changes in natural moisture content which these soils experience as the seasons change. They found that such changes increase the stiffness of the soil fabric which gives rise to a decrease in compressibility and an increase in shear strength.

Black clays are typically developed on poorly drained plains in regions with well-defined wet and dry seasons, where the annual rainfall is not less than 1250 mm. Generally the clay fraction in these soils exceeds 50%, silty material varying between 20 and 40% and sand forming the remainder. The organic content is usually less than 2%. The liquid limits of black clays may range between 50 and 100%, with plasticity indices of between 25 and 70%. The shrinkage limit is frequently around 10–12% (see Clare (1957a)). Montmorillonite is commonly present in the clay fraction and is the chief factor determining the behaviour of these clays. For instance, they undergo appreciable volume changes on wetting and drying due to the montmorillonite content.

Calcium carbonate sometimes occurs in these black clays, frequently taking the form of concretions (kankar), but it is not an essential characteristic. Usually it constitutes less than 1%.

The black cotton soil of Nigeria, described by Ola (1978b), is a highly plastic silty clay which has been formed by the weathering of basaltic rocks, and of shaly and clayey sediments. The soil may contain up to 70% of montmorillonite, kaolinite and quartz comprising the remainder. Shrinkage and swelling of these soils is a problem in many regions of Nigeria which experience alternating wet and dry seasons. These volume changes, however, are confined to an upper critical zone of the soil, which is frequently less than 1.5 m thick. Below this the moisture content remains more or less the same, for instance, around 25%. Ola (1980) noted an average linear shrinkage of 8% for some of these soils, with an average swelling pressure of 120 kPa and a maximum of about 240 kPa. He went on to state that in such situations the dead load of a building should be at least 80 kPa to counteract the swelling pressure.

10.5.3 Soils of arid and semi-arid regions

Calcareous silty clays are important types of soil in arid and semi-arid areas. These silty clays are light to dark brown in colour. They are formed by the deposition of clay minerals in saline or lime-rich waters. These soils possess a stiff to hard desiccated clay crust, referred to as duricrust, which may be up to 2 m thick, and which overlies moist soft silty clay. The crust is usually rich in salts precipitated when saline groundwater evaporates from the ground surface. Enrichment in lime may lead to the formation of small nodules of limestone in the soil, these are caliche deposits. Because of the effect of compacting due to rainfall the upper 25 mm or so of the crust is frequently impervious, beneath which the soil is porous. This impervious surface layer means that run-off is rapid. Uncemented calcareous soils are susceptible to deep erosion, particularly due to sheet flooding. Tomlinson (1957) quoted the following range of values for the consistency limits of these soils

Liquid limit,	40–59%
Plasticity index,	18–35%
Shrinkage limit,	10–15%

Marked swelling and shrinkage is characteristic of calcareous silty clays when alternatively wetted and dried. As a consequence wide, deep cracks may develop in the soil during the dry season.

In arid and semi-arid regions the evaporation of moisture from the surface of the soil may lead to the precipitation of salts in the upper layers. The most commonly precipitated material is calcium carbonate. These caliche deposits are referred to as *calcrete*. Calcrete occurs where soil drainage is reduced due to long and frequent periods of deficient precipitation and high evapotranspiration. Nevertheless, the development of calcrete is inhibited beyond a certain aridity since the low precipitation is unable to dissolve and drain calcium carbonate towards the water table. Consequently in arid climates *gypcrete* may take the place of calcrete. Climatic fluctuations which, for example, took place in North Africa during Quaternary times, therefore led to alternating calcification and gypsification of soils. Certain calcretes were partially gypsified and elsewhere gypsum formations were covered with calcrete hardpans (see Horta (1980)).

The hardened calcrete crust may contain nodules of limestone or be more or less completely cemented (this cement may, of course, have been subjected to differential leaching). In the initial stages of formation calcrete contains less than 40% calcium carbonate and the latter is distributed throughout the soil in a discontinuous manner. At around 40% carbonate content the original colour of the soil is masked by a transition to a whitish colour. As the carbonate content increases it first occurs as scattered concentrations of flaky habit, then as hard concretions. Once it exceeds 60% the concentration becomes continuous (Table 10.6). The calcium carbonate in calcrete profiles decreases from top to base, as generally does the hardness.

Gypcrete is developed in arid zones, that is, where there is less than 100 mm precipitation annually. In the Sahara aeolian sands and gravels often are encrusted with gypsum deposited from selenitic groundwaters. A gypcrete profile may contain three horizons. The upper horizon is rich in gypsified roots and has a banded and/or nodular structure. Beneath this there occurs massive gypcrete–gypsum cemented sands. Massive gypcrete forms above the water table during evaporation from the capillary fringe (newly formed gypcrete is hard but it softens with age). At the water table gypsum develops as aggregates of crystals, this is the sand-rose horizon.

Very occasionally in arid areas enrichment of iron or silica gives rise to ferricrete or silcrete deposits respectively.

Table 10.6. Classification of calcrete formations with continuous concentration of limestone (after Horta (1980))

Designation	Description	CaCO$_3$ (%)	Thickness	Mutual relationships
Non-foliated calcrete		>60	mostly 0.3 to	Under non-foliated calcrete, soft flakes
Massive	Massive or honeycomb structure		2–3 m	with or without hard nodules are always to be found
Nodular	Nodular and honeycomb structure			
Foliated calcrete		>70	some centimetres to	Under crusts, there is nearly always non-
Crusts s. s.	Superimposed and discontinuous sheets of massive or nodular calcrete. The thickness of the sheets is millimetric to centrimetric and increases from base to top		more than 1 m	foliated calcrete. Hardpans exist only at the top of crusts and sometimes totally replace them. Laminated crusts are associated with hardpans and clothe their surface and cracks
Hardpans	Lithified crust, thickness of the sheets from some centimetres to several decimetres			

Clare (1957b) described two types of tropical arid soils rich in sodium salts, namely, kabbas and saltmarsh. Both are characterised by their water retentive properties. Kabbas consists of a mixture of partly decomposed coral with sand, clay, organic matter and salt. The salt content in saltmarsh varies up to 40 or 50%, the soil basically consisting of silt with variable amounts of sand and organic material. It occurs in low lying areas that either have a very high water table or are periodically inundated by the sea. For a review of ground conditions in arid regions such as the Middle East see Fookes (1978) and Epps (1980).

10.6 Tills and other glacial deposits

Till is usually regarded as being synonymous with boulder clay. It is deposited directly by ice whilst stratified drift or tillite is deposited in melt waters associated with glaciers. An extensive review of the various types of glacial deposits and their engineering properties has been provided by Fookes *et al.* (1975).

10.6.1 The character of till

The character of till deposits varies appreciably and depends on the lithology of the material from which it was derived, on the position in which it was transported in the glacier, and on the mode of deposition (see Boulton and Paul (1976)). The underlying bedrock material usually constitutes up to about 80% of basal tills, depending on its resistance to abrasion and plucking. Argillaceous rocks, such as shales and mudstones, are more easily abraded and so produce fine-grained tills which are presumably richer in clay minerals and therefore are more plastic than other tills. Mineral composition also influences the natural moisture content which is slightly higher in tills containing appreciable quantities of clay minerals or mica. Upper tills have a high proportion of far-travelled material and may not contain any of the local bedrock.

Deposits of till consist of a variable assortment of rock debris which ranges in size from fine rock flour to boulders. At one extreme they may consist essentially of sand and gravel with very little binder, at the other they may be more or less stoneless clays. Lenses and pockets of sand, gravel and highly plastic slickensided clay are frequently encountered in some tills. Most tills, however, contain a significant amount of quartz in their silt-clay fractions. Frequently the larger, elongated fragments in till possess a general orientation in the path of ice movement.

The shape of the rock fragments found in till varies but it is largely conditioned by the initial shape of the fragment at the moment of incorporation in the ice. Angular boulders are common, their irregular sharp edges resulting from crushing.

May and Thomson (1978) described lenses of sand in tills of the Edmonton area, Canada. These lenses vary in size and shape from irregularly contorted inclusions, less than 100 mm in size, to more lenticular shaped bodies continuous over distances in excess of 50 m. The sands frequently exhibit cross bedding and normal faulting. They tend to range in size from medium to coarse gravelly sand on the one hand, to fine sandy gravel on the other. Such lenses probably represent meltwater deposits.

10.6.2 Types of till

Distinction has been made between tills derived from rock debris which was carried along at the base of a glacier and those deposits which were transported within and on the ice. The former is referred to as lodgement till whereas the latter is known as ablation till. Lodgement till is plastered onto the ground beneath a moving glacier in small increments as the basal ice melts. Because of the overlying weight of ice such deposits are over-consolidated. Ablation till accumulates on the surface of the ice when englacial debris melts out, and as the glacier decays the ablation till is slowly lowered to the ground. It is therefore normally consolidated. Lodgement till contains fewer, smaller stones (they generally possess a preferred orientation) than ablation till and they are rounded and striated. Clast orientation in ablation till varies from almost random to broadly parallel to the ice flow direction.

Due to abrasion and grinding the proportion of silt and clay size material is relatively high in lodgement till (e.g. the clay fraction varies from 15–40%). Lodgement till is commonly stiff, dense and relatively incompressible. Hence it is practically impermeable. Fissures are frequently present in lodgement till, especially if it is clay matrix dominated. Subhorizontal fissures have been developed as a result of incremental loading and periodic unloading whilst subvertical fissures owe their formation to the overriding effects of ice and stress relief. Oxidation of lodgement till takes place very slowly so that it is usually grey.

Because it has not been subjected to much abrasion ablation till is characterised by abundant large stones that are angular and not striated, the proportion of sand and gravel is high and clay is present only in small amounts (usually less than 10%). Because the texture is loose, ablation till oxidises rapidly and commonly is brown or yellowish brown. It also may have an extremely low *in situ* density. Since ablation till consists of the load carried at the time of ablation it usually forms a thinner deposit than lodgement till.

In fact Elson (1961) went further and divided tills into four categories, namely, superglacial ablation till, subglacial ablation till, comminution till and deformation till. Only superglacial ablation till corresponds with the definition of ablation till given above. It generally consists of thin lenses of sand and gravel of irregular distribution.

Englacial debris occurs mainly in the lower 30–60 m of a glacier where rock detritus may comprise as much as 10 or 20% of its volume. Consequently an appreciable thickness of englacial drift can be melted out from the base of a glacier although this suffers a reduction in volume of anything up to 90%. This subglacial ablation till may be precompressed by the overlying ice and it may be sliced by thrust planes, dipping upstream near the glacial margin. Elongate stones may possess preferred orientations. The average grain size is much smaller than that of superglacial till. At the base of subglacial till deposits striated boulder pavements or thin irregular lenses of sand and pebbles may occur as relics of the last erosive movement of the ice.

Comminution till occurs beneath subglacial ablation till and is formed by the shearing action of the ice at the base of the glacier. Elson (1961) suggested that this action generates enough heat to produce a sufficient quantity of melt water to bring about compaction to maximum density. Particles are oriented in the position of least resistance by the shearing action of moving ice and pebbles show a preferred orientation. Silt-sized tabular particles lie more or less horizontal or parallel to surfaces of larger particles, giving rise to microfoliation. Stones are surrounded by a compact matrix in which there is a high concentration of fines formed by abrasion of their surfaces.

Deformation tills vary from mixtures of local material and erratic debris to disrupted, but little transported, masses of material derived from local bedrock. The materials of which these tills are composed are porous and some initially contain more water than required for compaction to maximum density. Hence they yield soft tills which tend to be deformed rather than crushed.

McGown and Derbyshire (1977) devised a more elaborate system for the classification of tills. Their classification is based upon the mode of formation, transportation and deposition of glacial material and provides a general basis for the prediction of the engineering behaviour of tills (Table 10.7; Figure 10.7).

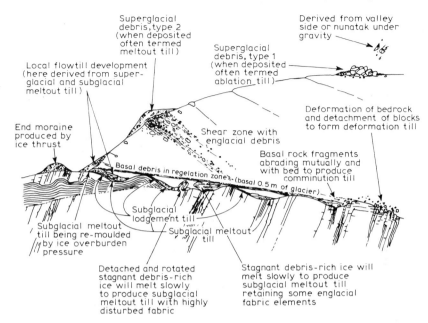

Superglacial debris,type 2 (when deposited often termed meltout till)

Derived from valley side or nunatak under gravity

Superglacial debris, type 1 (when deposited often termed ablation till)

Local flowtill development (here derived from super-glacial and subglacial meltout till)

Deformation of bedrock and detachment of blocks to form deformation till

End moraine produced by ice thrust

Shear zone with englacial debris

Basal rock fragments abrading mutually and with bed to produce comminution till

Basal debris in regelation zone (basal 0.5 m of glacier)

Subglacial lodgement till

Subglacial meltout till being re-moulded by ice overburden pressure

Subglacial meltout till

Detached and rotated stagnant debris-rich ice will melt slowly to produce subglacial meltout till with highly disturbed fabric

Stagnant debris-rich ice will melt slowly to produce subglacial meltout till retaining some englacial fabric elements

Figure 10.7 Acquisition, transportation and deposition of tills by a glacier (after McGown and Derbyshire (1977))

10.6.3 Engineering properties of tills

The particle size distribution and fabric (stone orientation, layering, fissuring and jointing) are among the most significant features as far as the engineering behaviour of a till is concerned. McGown and Derbyshire (1977) therefore used the percentage of fines to distinguish granular, well-graded and matrix dominated tills, the boundaries being placed at 15 and 45% respectively. The fabric of tills includes features of primary and secondary origins such as folds, thrusts, fissures (macrofabric), disposition of clasts (macro- and mesofabric) and the organisation of the matrix. It would seem that distinctive macro- and mesofabric patterns characterise flow till, lodgement till and till deformed by thrusting and loading both in terminal moraines and till flutes.

According to McGown (1971) tills are frequently gap graded, the gap generally occurring in the sand fraction (Figure 10.8). He also noted that large, often very local, variations can occur in the gradings of till which reflect local variations in the formation processes, particularly the comminution processes. The clast size consists principally of rock fragments and composite grains, and presumably was formed by frost action and crushing by ice. Single grains predominate in the matrix. The range in the proportions of coarse and fine fractions in tills dictates the degree to which the properties of the fine fraction influence the properties of the composite soil. The variation in the engineering properties of the fine soil fraction is greater than that of the coarse fraction, and this often tends to dominate the engineering behaviour of the till.

Table 10.7. Till types and processes (from McGown and Derbyshire (1977))

Formative processes	Transportation processes	Depositional processes
Comminution till: produced by abrasion and interaction between particles in the basal zone of a glacier. It is a common element in most tills.	*Superglacial till*:* derived from frost riving of adjacent rocks or by differential melting out of glacier dirt beds. It may or may not become incorporated in the glacier and may suffer frost shattering and washing by meltwaters as it is transported on the top of the glacier.	*Ablation till*:* accumulated by melting out on the surface of a glacier or as a coating on inert ice.
Deformation till: produced by plucking, thrusting, folding and brecciation of the glacier bed.	*Englacial till*: derived from superglacial till subsequently buried by accumulating snow or entrained in shear zones. It is transported within the ice mass	*Meltout till*:* accumulated as the ice of an ice-debris mixture melts out. Meltout tills exhibit a relatively low bulk density but show some variation depending on the particle size distribution. Generally the material is poorly sorted although occasionally thin layers or lenses of washed sediment occur. The micro-fabric is normally rather open. The clast fabrics of meltout tills show a wide variation depending on the disposition and debris concentration of the melting out mass.
	Basal till: derived from comminution products in the ice-rock contact zones particularly the lower-most regions of a glacier. It is generally transported in concentrated bands in the bottom metre or so of a glacier.	*Lodgement till*: accumulated subglacially by accretion from debris-rich basal ice. Lodge-ment tills generally possess a wide range of particle sizes and are frequently anisotropic, fissured or jointed, especially when well-graded or clay matrix dominated. They are usually stiff, dense, relatively incompressible soils. Macro-, meso- and microfabric patterns show high consistency. Sub-horizontal fissuring is due to incremental lodgement and periodic unloading while subvertical fissures are evidence of ice over-riding and stress relief. Low-angle shear failure planes also occur.

Table 10.7—cont.

Flow till: consists essentially of melted out superglacial comminution debris but also occurs due to flow of subglacial meltout tills in subglacial cavities and at the ice margin. It usually contains a wide range of particle sizes. Flow till is frequently interbedded with fluvioglacial material. Orientation of clasts is broadly parallel to the deposition plane and imbricated upflow. The fines also reflect the flow mechanism in the parallelism of clay-silt fractions which produces a locally dense micro-fabric.

Waterlain till: accumulates on subaqueous surface under a variety of depositional processes and may thus show a wide variety of characters. They vary from rather soft lodgement tills to subaqueous mudflows and to crudely stratified lacustrine clay-silts. Stratification, deformation and very diffuse to random clast fabrics are common.

* Note: This distinction between superglacial, meltout and ablation tills is based on the degree of disturbance of their ice-inherited fabric, including loss of fines. The term ablation till is thus best avoided.

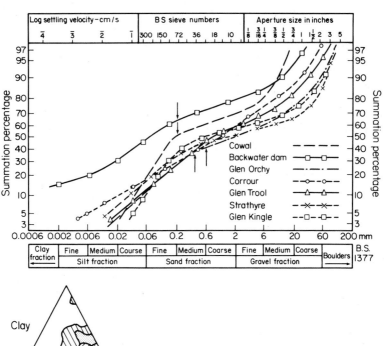

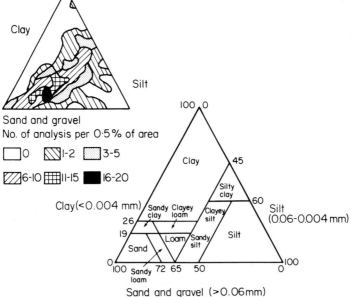

Figure 10.8 (a) Typical gradings of some Scottish morainic soils (after McGown (1971)). (b) Grain size distribution for about 500 tills (after Elson (1961))

The relative density of till deposits is often remarkably uniform, varying from 2.77–2.78. These values suggest the presence of fresh minerals in the fine fraction, that is, rock flour rather than clay minerals. Rock flour behaves more like granular material than cohesive and has a low plasticity. The consistency limits of tills are dependent upon water content, grain size distribution and the properties of the fine-grained fraction. Generally, however, the plasticity index is small and the liquid limit of tills decreases with increasing grain size. McKinley *et al.* (1974) in an investigation of lodgement till showed that plasticity tests on the matrix material placed this till in the CL group, just above the 'A' line on the plasticity chart (e.g. plastic limit = 16%, liquid limit = 28%). The till appeared to be fully saturated. The values of moisture content, bulk density and shear strength derived from unconfined compression tests for this till are given in Figure 10.9. This shows that the values of moisture content and bulk density are relatively constant and that the shear strength generally increases somewhat with depth.

The compressibility and consolidation of tills are principally determined by the clay content. For example, the value of compressibility index tends to increase linearly with increasing clay content whilst for moraines of very

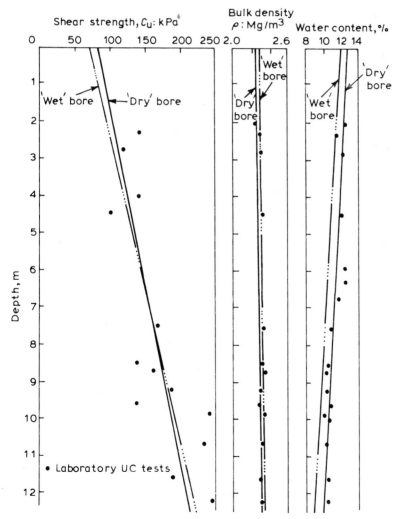

Figure 10.9 'Same day' tests on unconfined compression specimens (samples from 'dry' boreholes) (after McKinley *et al.* (1974))

low clay content, less than 2%, this index remains about constant ($C_c = 0.01$).

Klohn (1965) noted that dense, heavily overconsolidated till is relatively incompressible and when loaded undergoes very little settlement, most of which is elastic. For the average structure such elastic compressions are too small to be of concern and therefore can be ignored. But for certain structures they can be critical, so that their magnitude has to be estimated prior to construction. He carried out a series of plate loading tests which indicated that the modulus of elasticity of the till deposit concerned was very high, being of the order 1500 MPa. Observations subsequently, taken on all major structures, indicated that settlement occurred almost instantaneously on application of load.

In another survey of dense till Radhakrishna and Klym (1974) found the undrained shear strength, as obtained by pressuremeter and plate loading tests, to average around 1.6 MPa, while the values from triaxial tests ranged between 0.75 and 1.3 MPa. The average values of the initial modulus of deformation were around 215 MPa which was approximately twice the laboratory value. These differences between field and laboratory results were attributed to stress relief of material on sampling and sampling disturbance. Much lower values of shear strength were found for the Cromer Till by Kazi and Knill (1969). These ranged from 170–220 kPa.

10.6.4 Stratified drift and varved clays

Deposits of stratified drift are often subdivided into two categories, namely, those which develop in contact with the ice—the ice contact deposits—and those which accumulate beyond the limits of the ice, forming in streams, lakes or seas—the proglacial deposits.

Outwash fans are deposited by streams which emerge from the snout of a glacier and are composed of sediments ranging in size from coarse sands to boulders. When they are first deposited their porosity may be anything from 25–50% and they tend to be very pervious. The finer silt-clay fraction is transported further downstream. Several outwash fans may initially extend from the terminus of a glacier but often they gradually merge to form one deposit. On retreat they may bury the terminal moraine. Kames, kame terraces and eskers are also deposited by melt waters and usually consist of sands and gravels.

The most familiar proglacial deposits are varved clays. These deposits accumulated in proglacial lakes and are generally characterised by alternating laminae of finer and coarser grain size, each couplet being termed a varve. The thickness of the individual varve is frequently less than 2 mm although much thicker layers have been noted in a few deposits. Generally the coarser layer is of silt size and the finer of clay size.

Taylor et al. (1976) showed that in Devensian clays from Gale Common in Yorkshire, the clay minerals were well orientated around silt grains such that at boundaries between silty partings and matrix the clay minerals tended to show a high degree of orientation parallel to the laminae.

Usually very finely comminuted quartz, feldspar and mica form the major part of varved clays rather than clay minerals. For example, the clay mineral content may be as low as 10%, although instances where it has been as high as 70% have been recorded. What is more montmorillonitic clay has also been found in varved clays.

Wu (1958) noted that the principal constituents of the glacial lake clays found around the shores of the Great Lakes in North America are quartz, kaolinite and illite with small amounts of chlorite and vermiculite. The fabric of these clays varies from well orientated to almost random. He suggested that the latter must involve a flocculent or honeycomb structure. The clays possessed a low strength and high compressibility. Varved clays tend to be normally consolidated or lightly overconsolidated, although it is usually difficult to make the distinction. In many cases the precompression may have been due to ice loading. However, Saxena et al. (1978) reported that the upper part of the varved clay of Hackensack Valley, New Jersey was highly overconsolidated. This they attributed to the effects of fluctuating water levels or to desiccation.

The two normally discrete layers formed during the deposition of the varve present an unusual problem in that it may invalidate the normal soil mechanics analyses, based on homogeneous soils, from being used. As far as the Atterberg limits are concerned assessment of the liquid and plastic limits of a bulk sample may not yield a representative result. However, Metcalf and Townsend (1961) suggested that the maximum possible liquid limit obtained for any particular varved deposit must be that of the clayey portion, whereas the minimum value must be that of the silty portion. Hence they assumed that the maximum and minimum values recorded for any one deposit approximate to the properties of the individual layers. The range of liquid limits for varved clays tends to vary between 30 and 80% whilst that of plastic limit often varies between 15 and 30%. These limits, obtained from varved clays in Ontario, allow the material to be classified as inorganic silty clay of medium to high plasticity or compressibility. In some varved clays in Ontario the natural moisture content would appear to be near the liquid limit. They are consequently soft and have sensitivities generally of the order of 4. Since triaxial and unconfined compression tests tend to give very low strains at failure, around 3%, Metcalf and Townsend presumed that this indicates a structural effect in the varved clays. The average strength reported was about 5.9 MPa, with a range of 3.4–7 MPa. The effective stress parameters of apparent cohesion and angle of shearing resistance range from 0.7–2.8 MPa, and 22–25° respectively.

10.6.5 Quick clays

The material of which quick clays are composed is predominantly smaller than 0.002 mm but many deposits seem to be very poor in clay minerals, containing a high proportion of ground-down, fine quartz. For instance, it has been shown that quick clay from St Jean Vienney consists of very fine quartz and plagioclase. Indeed examination of quick clays with the scanning electron microscope has revealed that they do not possess clay based structures, although such work has not lent unequivocal support to the view that non-clay particles govern the physical properties.

Quick clays generally exhibit little plasticity, their plasticity index generally varying between 8 and 12%. Their liquidity index normally exceeds 1, and their liquid limit is often less than 40%. Quick clays are usually inactive, their activity frequently being less than 0.5. The most extraordinary property possessed by quick clays is their very high sensitivity. In other words a large proportion of their undisturbed strength is permanently lost following shear. The small fraction of the original strength gained after remoulding may be attributable to the development of some different form of interparticle bonding. The reason why only a small fraction of the original strength can ever be recovered is because the rate at which it develops is so slow. As an example the Leda Clay is characterised by exceptionally high sensitivity, commonly between 20 and 50, and a high natural moisture content and void ratio, the latter is commonly about 2. It has a low permeability, being around 10^{-10} m/s. The plastic limit is around 25%, with a liquid limit about 60%, and undrained shear strength of 700 kPa. When subjected to sustained load an undrained triaxial specimen of Leda Clay exhibits a steady time-dependent increase in both pore pressure and axial strain. Continuing undrained creep may often result in a collapse of the sample after long periods of time have elapsed.

Gillott (1979) has shown that the fabric and mineralogical composition of sensitive soils from Canada, Alaska and Norway are qualitatively similar. He pointed out that they possess an open fabric, high moisture content and similar index properties (Table 10.8).

An examination of the fabric of these soils revealed the presence of aggregations. Granular particles, whether aggregations or primary minerals, are rarely in direct contact, being linked generally by bridges of fine particles. Clay minerals usually are non-oriented and clay coatings on primary minerals tend to be uncommon, as are cemented junctions. Networks of platelets occur in some soils. Primary minerals, particularly quartz and feldspar, form a higher than normal proportion of the clay size fraction and illite and chlorite are the dominant phyllosilicate minerals. Gillot noted that the presence of swelling clay minerals varies from almost zero to significant amounts.

The open fabric which is characteristic of quick clays has been attributed to their initial deposition, during which time colloidal particles interacted to form loose aggregations by gelation and flocculation. Clay minerals exhibit strongly marked colloidal properties and other inorganic materials such as silica behave as colloids when sufficiently fine grained. Gillott suggested that the open fabric may have been retained during very early consolidation because it remained a near equilibrium arrangement. Its subsequent retention to the present day may be due to mutual interference between particles and buttressing of junctions between granules by clay and other fine constituents, precipitation of cement at particle contacts, low rates of loading, and low load increment ratio.

It has been argued that the open fabric of quick clays has become metastable because of environmental changes since their deposition. For example, Rosenqvist (1966) suggested that this was brought about as a result of isostatic uplift of the land after the disappearance of the ice sheets so that marine deposits emerged above sea level. This led to the salt content of their pore solutions being reduced by diffusion or leaching in the following fresh water regime. The flocculated fabric, favoured by high salt content, became metastable because fresh water favours dispersion of clay minerals. This concept

Table 10.8. Engineering properties of sensitive soils (after Gillott (1979))

Location*	Depth (m)	Natural moisture content (%)	Preconsolidation pressure (kPa)	Undrained strength (kPa)	Sensitivity	Liquid limit (%)	Plastic limit (%)	Liquidity index	Activity
O	13.7	60	450	160	—	49	23	1.4	0.35
Q	5.2	75	150	50	—	70	26	1.1	0.64
Q	14.3	81	150	50	—	65	28	1.4	0.45
O	2.6	65	60	20	100	55	22	1.3	0.73
Q	12.2	28	590	230	—	23	16	1.7	0.18
O	5.2	78	320	120	—	65	28	1.3	0.44
BC	20.1	38	—	20	30	28	22	2.7	0.22
BC	14.0	29	—	—	4	23	16	1.9	0.33
BC	35.4	37	—	60	5	28	23	2.8	0.17
A	61.3	17	—	—	—	26	21	0.8	—
A	60.7	—	—	—	—	23	20	—	—

* O = Ontario, Q = Quebec, BC = British Columbia, A = Alaska

has not found universal acceptance and perhaps only applies to Norwegian quick clays. Indeed sensitive soils of fresh water origin do occur. What is more Pusch and Arnold (1969) disputed the Rosenqvist theory, they having attempted to produce quick behaviour in a specially prepared soil composed largely of illite, following the stages outlined by Rosenqvist, and failed.

The fact that it appears that clay minerals are not quantitatively important in quick clays has led to the development of other theories to explain their peculiar properties. For example, Cabrera and Smalley (1973) suggested that these deposits owe their distinctive properties to the predominance of short range interparticle bonding forces which they maintained were characteristic of deposits in which there was an abundance of glacially produced, fine non-clay minerals. In other words they contended that the ice sheets supplied abundant ground quartz in the form of rock flour for the formation of quick clays. Certainly quick clays have a restricted geographical distribution, occurring in certain parts of the northern hemisphere which were subjected to glaciation during Pleistocene times.

Quick clays are associated with several serious engineering problems. Their bearing capacity is low, settlement is high and prediction of consolidation of quick clays by the standard methods is unsatisfactory. Slides in quick clays sometimes have proved disastrous but unfortunately the results of slope stability analyses are often unreliable. Dewatering leads to irreversible shrinkage.

Quick clays can liquefy on sudden shock. This has been explained by the fact that if quartz particles are small enough, having a very low settling velocity, and if the soil has a high water content, then the solid-liquid transition can be achieved.

10.7 Organic soils: peat

Peat is an accumulation of partially decomposed and disintegrated plant remains which have been fossilised under conditions of incomplete aeration and high water content. Physico-chemical and biochemical processes cause this organic material to remain in a state of preservation over a long period of time.

All present day surface deposits of peat in Britain have accumulated since the last ice age and therefore have formed during the last 20 000 years. On the other hand, some buried peats may have been developed during inter-glacial periods. Peats also have accumulated in post-glacial lakes and marshes where they are interbedded with silts and muds. Similarly they may be as-sociated with salt marshes. Fen deposits are thought to have developed in relation to the eustatic changes in sea level which occurred after the retreat of the last ice sheets from Britain. The most notable fen deposits in Britain are found south of the Wash. Similar deposits also are found in Suffolk and Somerset. These are areas where layers of peat interdigitate with wedges of estuarine silt and clay. However, the most familiar type of peat deposit in Britain is probably the blanket bog. These deposits are found on the cool wet uplands.

Approximately 95% of all deposits of peat have been formed from plants growing under aerobic conditions. The high water-holding capacity of peat maintains a surplus of water, which ensures continued plant growth and consequent peat accumulation. The rate of decomposition of plant detritus is relatively rapid under aerobic conditions but is slowed down several thousand-fold under anaerobic conditions. Drying out, groundwater fluctuations and snow loading bring about compression in the upper layers of a peat deposit. Indeed these mechanisms are often more important as far as near surface compression is concerned than effective overburden pressure. This is because the unit weight of peat may be less than that of water. As the water table in peat generally is near the surface, the effective overburden pressure is negligible.

10.7.1 Classification of peat

Macroscopically peaty material can be divided into three basic groups, namely, amorphous granular,* coarse fibrous and fine fibrous peat (see Radforth (1952)). The amorphous granular peats have a high colloidal fraction, holding most of their water in an adsorbed rather than a free state, the adsorption occurring around the grain structure. In the other two types the peat is composed of fibres, these usually being woody. In the coarse variety a mesh of second order size exists within the interstices of the first order network, whilst in fine fibrous peat the interstices are very small and contain colloidal matter.

Landva and Pheeney (1980) asserted that the proper identification of peat should include a description of the constituents, since the fibres of some plant remains are much stronger than others and are consequently of interest to the engineer. The degree of humification and water content should also be taken into account. They proposed a modified form of the Von Post classification of peat (Table 10.9).

10.7.2 Properties and behaviour of peat

The ash percentage of peat consists of the mineral residue remaining after its ignition, which is expressed as a fraction of the total dry weight. Ash contents may be as low as 2% in some highly organic peats, or may be as high as 50% as in some peats found on the Yorkshire moors (see Bell (1978)). The mineral material is usually quartz sand and silt. In many deposits the mineral content increases with depth. The mineral content does influence the engineering properties of peat.

The void ratio of peat ranges between 9, for dense amorphous granular peat, and 25, for fibrous types with a high content of sphagnum. It usually tends to decrease with depth within a peat deposit. Such high void ratios give rise to phenomenally high water contents. The latter is the most distinctive characteristic of peat. Indeed most of the peculiarities in the physical characteristics of peat are attributable to the amount of moisture present. This varies according to the type of peat, it may be as low as 500% in some

---

* Landva and Pheeney (1980) maintained that the term *amorphous granular* should be reserved for non-fibrous, truly amorphous peats only. They contended that most such material referred to in engineering literature is in fact moss peat, actual amorphous granular peat being rare. They suggested, therefore, that the term should be used with caution.

Table 10.9. Classification of peat (from Landva and Pheeney (1980))

(1) Genera	(2) Designation	(3) The degree of humification (H)	
		H Decomposition	Plant structure
Bryales (moss) = B	With few exceptions peats consist of a mixture of two or more genera. These are listed in decreased order of content, i.e. the principal component first, e.g. ErCS.	H_1 None	Easily identified
Carex (sedge) = C		H_2 Insignificant	Easily identified
Equisetum (horsetails) = Eq		H_3 Very slight	Still identifiable
Eriophorum (cotton grass) = Er		H_4 Slight	Not easily identified
Hypnum (moss) = H		H_5 Moderate	Recognisable but vague
Lignidi (wood) = W		H_6 Moderately strong	Indistinct
Nanoligmidi (shrubs) = N		H_7 Strong	Faintly recognisable
Phragmites = Ph		H_8 Very strong	Very indistinct
Scheuchzeria (acquatic herbs) = Sch		H_9 Nearly complete	Almost unrecognisable
Sphagnum (moss) = S		H_{10} Complete	Not discernible

(4) Water content (B)	(5) Fine fibres (F)	(6) Coarse fibres (R)	(7) Wood (W) and shrub remnants
Estimated from a scale of 1 (dry) to 5 (very high) and designated B_1, B_2, etc. Landva and Pheeney suggested the following ranges	These are fibres and stems less than 1 mm in diameter, F_0 = nil; F_1 = low content; F_2 = moderate content; F_3 = high content.	These fibres have a diameter exceeding 1 mm; R_0 = nil; R_1 = low content; R_2 = moderate content; R_3 = high content.	Wood and shrub is similarly graded: W_0 = nil; W_1 = low content; W_2 = moderate content; W_3 = high content.
B_2 less than 500%			
B_3 500–1000%			
B_4 1000–2000%			
B_5 Over 2000%			

The classification is affected by combining the letters from the groups, in the order 1, 2 ... 7, as appropriate.

amorphous granular varieties whilst, by contrast, values exceeding 3000% have been recorded from coarse fibrous varieties.

According to Wilson (1978) the water content of peat is held in the cells of plant remains, as well as in the voids. Water is also absorbed by the cell walls of the plant detritus. Landva and Pheeney (1980) estimated that about one-third of the water content of sphagnum peat is located in the voids, the remainder being in the cellular plant material. They further estimated that about half the water in sedge peat was contained in the plant remains. These three types of held water have different drainage characteristics. Water is forced out of the voids when peat undergoes stress. With continuing stress the particles are brought into contact and the cell structure begins to be distorted. Hence the water in the plant cells is pressurised. Some of this water moves through openings in the cell walls, but with increasing stress these begin to rupture. Water is thereby expelled, giving rise to increasing pore pressure in the voids. Wilson indicated that at this point the peat behaves as a material which has become rapidly softened. Further straining and rupture of the cell walls means that shear failure is imminent. Generally peat deposits are acidic in character, the pH values often varying between 5.5 and 6.5. However, due to admixture with chalky material some fen peats in Norfolk are neutral or even alkaline.

It has been found that the volumetric shrinkage of peat increases up to a maximum and then remains constant, the volume being reduced almost to the point of complete dehydration. The amount of shrinkage which can occur generally ranges between 10 and 75% of the original volume of the peat and it can involve reductions in void ratio from over 12 down to about 2.

As would be expected amorphous granular peat has a higher bulk density than the fibrous types. For instance, in the former it can range up to 1.2 t/m^3 whilst in woody fibrous peats it may be half this figure. However, the dry density is a more important engineering property of peat, influencing its behaviour under load. Dry densities of drained peat fall within the range 65–120 kg/m^3. The dry density is influenced by the mineral content and higher values than those quoted can be obtained when peats possess high mineral residues. The relative density of peat has been found to range from as low as 1.1 up to about 1.8, again being influenced by the content of mineral matter.

Because of the variability of peat in the field the value of its permeability as tested in the laboratory can be misleading. Nevertheless Hanrahan (1954) showed that the permeability of peat, as determined during consolidation testing, varied according to the loading and length of time involved as follows

(1) before test: void ratio = 12; permeability = 4×10^{-6} m/s
(2) after seven months loading at 55 kPa; void ratio = 4.5; permeability = 8×10^{-11} m/s.

Thus after seven months of loading the permeability of the peat was 50 000 times less than it was originally. Adams (1965) also has shown that there is a marked change in the permeability of peat as it volume is reduced under compression. The magnitude of construction pore water pressure is particularly significant in determining the stability of peat. Adams showed that the development of pore pressures in peat beneath embankments was appreciable, in one instance it approached the vertical unit weight of the embankment.

When loaded, peat deposits undergo high deformations but their modulus of deformation tends to increase with increasing load. If peat is very fibrous it appears to suffer indefinite deformation without planes of failure developing. On the other hand failure planes nearly always form in dense amorphous granular peats. Hanrahan and Walsh (1965) found that the strain characteristics of peat were independent of the rate of strain and that flow deformation, in their tests, was negligible. Strain often takes place in an erratic fashion in a fibrous peat. This may be due to the different fibres reaching their ultimate strengths at different strain values. The more brittle woody fibres fail at low strain, whereas the non-woody types maintain the overall cohesion of the mass up to much higher strains. The viscous behaviour of peat is generally recognised as being non-Newtonian and the relationship between stress and strain rate is a function of the void ratio. As the void ratio decreases so the effective viscosity increases and hence a certain value of stress produces a correspondingly smaller value of strain rate.

Apart from its moisture content and dry density the shear strength of a peat deposit appears to be influenced, first, by its degree of humification and, secondly, by its mineral content. As both these factors increase so does the shear strength. Conversely the higher the moisture content of peat, the lower is its shear strength. The dry density is influenced by the effective load to which a deposit of peat has been subjected. As the effective weight of 1 m^3 of drained peat is approximately 45 times that of 1 m^3 of undrained peat the reason for the negligible strength of the latter becomes apparent. Due to its extremely low submerged density, which may be between 15 and 35 kg/m^3, peat is especially prone to rotational failure or failure by spreading, particularly under the action of horizontal seepage forces.

In an undrained bog the unconfined compressive strength is negligible, the peat possessing a consistency approximating to that of a liquid. The strength is increased by drainage to values between 20 and 30 kPa and the modulus of elasticity to between 100 and 140 kPa.

Settlement in organic soils includes consolidation, secondary compression, shear strain and loss of organic solids due to decomposition. If the organic content of a soil exceeds 20%, by weight, consolidation becomes increasingly dominated by the behaviour of the organic material. For example, on loading, peat undergoes a decrease in permeability of several orders of magnitude which invalidates the Terzaghi theory of consolidation. Moreover residual pore pressure affects primary consolidation, and considerable secondary consolidation further complicates settlement prediction. Accordingly a modified method of settlement prediction for organic soils has been developed by Andersland and Al Khafaji (1980) which is related to the amount of organic matter in soil. The latter has to be accurately determined in order to use their method.

Differential and excessive settlement is the principal problem confronting the engineer working on a peaty soil. When a load is applied to peat, settlement occurs because of the low lateral resistance offered by the adjacent unloaded peat. Serious shearing stresses are induced even by moderate loads. Worse still, should the loads exceed a given minimum then settlement may be accompanied by creep, lateral spread, or in extreme cases by rotational slip and upheaval of adjacent ground. At any given time the total settlement in peat due to loading involves settlement with and without volume change. Settlement without volume change is the more serious for it can give rise to the types of failure mentioned. What is more it does not enhance the strength of peat.

When peat is compressed the free pore water is expelled under excess hydrostatic pressure. Since the peat is initially quite pervious and the percentage of pore water is high, the magnitude of settlement is large and this period of initial settlement is short (a matter of days in the field). Adams (1965) showed that the magnitude of initial settlement was directly related to peat thickness and applied load. The original void ratio of a peat soil also influences the rate of initial settlement. Excess pore pressure is almost entirely dissipated during this period. Settlement subsequently continues at a much slower rate which is approximately linear with the logarithm of time. This is because the permeability of the peat is significantly reduced due to the large decrease in volume. During this period the effective consolidating pressure is transferred from the pore water to the solid peat fabric. The latter is compressible and will only sustain a certain proportion of the total effective stress, depending on the thickness and permeability of the peat mass.

Adams (1963) maintained that the macro- and micro-structure of fibrous peat influences its consolidation. He considered that primary consolidation of such peats took place due to drainage of water from the macro-structure whilst secondary consolidation was due to the extremely slow drainage of water from the micropores into the macro-structure. Because the permeability of fine fibrous peat is higher than that of an amorphous granular peat its rate of primary consolidation is higher.

Due to the highly viscous water adsorbed around soil particles, amorphous granular peat exhibits a plastic structural resistance to compression and hence has a similar rheological behaviour to that of clay (see Adams (1963)). In this case secondary consolidation is believed to occur as a result of the gradual readjustment of the soil structure to a more stable configuration following the breakdown which occurs during the primary phase, due to dissipation of pore pressure. The rate at which this process takes place is controlled by the highly viscous adsorbed water surrounding each soil particle. The colloidal material which the former contains tends to plug the interstices, thereby reducing permeability. Wilson et al. (1965) suggested that amorphous granular peats exhibit considerable secondary consolidation and therefore settlement. Because of the highly complex structure of such peat they suggested that it may also exhibit phases of tertiary and quaternary consolidation.

One-dimensional consolidation theories have been developed by Berry and Poskitt (1972) for both amorphous granular and fibrous peat. These consider finite strain, decreasing permeability, compressibility and the influence of secondary compression with time. The different mechanisms involved in secondary compression of these two types of peat were found to give similar non-linear rheological models but their effective creep equations were fundamentally different. That for amorphous granular peat predicts an exponential increase in strain with incremental loading whilst that for fibrous peat predicts a linear increase.

With few exceptions improved drainage has no beneficial effect on the

rate of consolidation of peat. This is because efficient drainage only accelerates the completion of primary consolidation which is always completed
rapidly anyhow.

There is an extremely small increase in the void ratio which follows the
reduction of the load on a peat deposit. Hence the compressibility of preconsolidated peat is greatly reduced (see Bell (1978)). This can be illustrated
from the following figures.

Coefficient of compressibility of peat for a range of loading from 13.4–
26.8 kPa.

Normally loaded, $m_v = 12.214 \text{ m}^2/\text{MN}$
Preconsolidated, $m_v = 0.559 \text{ m}^2/\text{MN}$.

10.8 Fills

Because suitable building sites are becoming scarcer in urban areas the construction of buildings on fill or made-up ground has assumed a greater importance. A wide variety of materials are used for fills including domestic
refuse, ashes, slag, clinker, building waste, chemical waste, quarry waste and
all types of soils. The extent to which an existing fill will be suitable as a
foundation depends largely on its composition and uniformity. In the past
the control exercised in placing fill has frequently been insufficient to ensure
an adequate and uniform support for structures immediately after placement.
Consequently a time interval had to be allowed prior to building so that
the material would consolidate under its own weight. Although such a practice
may be suitable for small lightly loaded buildings it is unsatisfactory for more
heavily loaded structures. The latter can produce substantial amounts of
settlement.

The stability and potential settlement of foundation structures on fill are
largely governed by its density. Therefore random end tipping in thick layers,
giving low densities, produces an unsatisfactory condition since excessive and
non-uniform settlements are likely to occur under the load of the structure
erected. The nature and thickness of fill and the site condition (especially
the groundwater conditions) also affect the amount and rate of settlement.
Obviously the greater the thickness of a fill, the larger is its deformation
likely to be, as is the length of time over which deformation is likely to take
place.

The best materials with respect to minimum settlement are well graded,
hard and granular. Fills containing a large proportion of fine material, by
contrast, may take a long while to settle. Generally rock fills will settle 2.5%
of their thickness, sandy fills about 5% and cohesive material around 10%.
The rate of settlement decreases with time but in some cases it may take
10–20 years before movements are reduced within tolerable limits for building
foundations. In coarse-grained soils the larger part of movement generally
occurs within the first two years after the construction of the fill and after
five years settlements are usually very small. Indeed Meyerhof (1951) showed
that most settlement in a fill occurs within the first year of placement and
after two years is relatively small (Figure 10.10). Therefore two years generally
may be regarded as the lower limit before buildings are constructed on fills

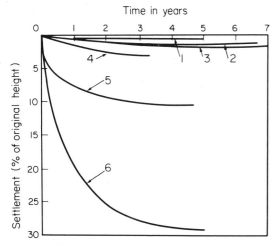

Figure 10.10 Observations of the settlement of various types of fill due to consolidation under its own weight (after Meyerhof (1951)). Description of curves: 1 = Well-graded sand, well compacted; 2 = Rockfill, medium state of compaction; 3 = Clay and chalk, lightly compacted; 4 = Sand, uncompacted; 5 = Clay, uncompacted; 6 = Mixed refuse, well compacted

consolidating under their own weight. Fills placed over low-lying areas of compressible or weak strata should be considered unsuitable unless tests demonstrate otherwise or the structure can be designed for low bearing capacity and irregular settlement. For example, clay soils beneath fills may undergo a prolonged period of consolidation. Frequently poorly compacted old fills continue to settle for years due to secondary consolidation. Mixed fills which contain materials liable to decay, which may leave voids or involve a risk of spontaneous combustion, afford very variable support and such sites should, in general, be avoided.

The material of new fills should be spread uniformly in thin layers and compacted at the optimum moisture content required to produce the maximum density. Properly compacted fills on sound ground can be as good as, or better than, virgin soil. According to Meyerhof (1951) the support afforded foundations is generally inadequate if the density of a cohesionless fill is less than 90% of the maximum value given by the standard compaction test. For cohesive material a somewhat higher degree of compaction may be necessary (say around 95%). It has been suggested that if rockfill is compacted to greater than 85% of the solid dry density of the rock then building can commence immediately. For instance, Kilkenny (1968) quoted the Expo '67 site in Montreal where a compacted density of 2.3 t/m^3 was achieved for the shale fill, as compared with a solid dry density of 2.7 t/m^3, and proved satisfactory. Well compacted rock, gravel, sand, shale and clay fills have shown settlements of only 0.5% of their thickness. Quarry wastes frequently have proved satisfactory fills for foundation purposes when properly compacted. Meyerhof indicated that hydraulic sand fills underwent very small settlements if placed above groundwater level, due to the consolidating influence of the downward percolation of the water. By contrast settlements are large and continue over a long time when clay fill is hydraulically placed.

Because the settlement distribution over the loaded area of a fill, even under uniform loading conditions, may be very irregular it is necessary to design a structure so that the total as well as the differential movements are restricted or can be withstood without damage. Hence the structure should either be sufficiently rigid to redistribute the loads and thereby reduce relative settlement, or should be relatively flexible to accommodate them without cracks appearing in the structure. Settlement observations on structures have revealed that a bearing pressure of 55 kPa is very conservative except on poorly compacted, fine-grained soils or industrial and domestic wastes consolidating under their own weight.

Waste disposal or sanitary land fills are usually very mixed in composition and suffer from continuing organic decomposition and physico-chemical breakdown. Methane and hydrogen sulphide are often produced in the process and accumulations of these gases in pockets in fills have led to explosions. The production of leachate is another problem. Some materials such as ashes and industrial wastes may contain sulphate and other products which are potentially injurious as far as concrete is concerned. The density of waste disposal fills varies from about 120–300 kg/m^3 when tipped. After compaction the density may exceed 600 kg/m^3. Moisture contents range from 10–50% and the average relative density of the solids from 1.7–2.5. Settlements are likely to be large and irregular. According to Sowers (1973) the mechanisms responsible for the settlement which occurs in waste disposal fills include mechanical distortion, bending, crushing and reorientation of materials which cause a reduction in the void ratio; ravelling, that is, the transfer of fines into the voids; physico-chemical and biochemical changes such as corrosion, combustion and fermentation and the interactions of these various mechanisms. The initial mechanical settlement of waste disposal fills is rapid and is due to a reduction in the initial void ratio. It takes place with no build-up of pore pressure. Settlement continues due to a combination of secondary compression (material disturbance) and physico-chemical and biochemical action, and Sowers has shown that the settlement-log time relationship is more or less linear. However, the rate of settlement produced by ravelling and combustion is erratic.

After studying the settlement records of three sanitary landfills for periods up to nine years, Yen and Scanlon (1975) concluded that the rate of settlement increases with the thickness of fill, although there seems to be a critical depth beyond which the rate does not increase. This critical depth was around 30 m in the fills examined. They suggested that such a critical depth was due to the anaerobic conditions that exist in deeper fills which inhibit decomposition. It was estimated that after completion of a fill settlement would range between 4.5 and 6% of the total depth of the fill, and that it may take place over 250 months. Yen and Scanlon found that the rate of settlement appeared to decrease linearly in proportion to the logarithm of median fill age.

Where urban renewal schemes are undertaken it may be necessary to construct buildings on areas covered by rubble fill. In most cases such fills have not been compacted to any appreciable extent and where the rubble has collapsed into old cellars large voids may be present. However, demolition rubble fill is usually comparatively shallow and the most economical method of constructing foundations is either to cut a trench through the fill and backfill it with lean concrete or to clear all the fill beneath the structure and replace it with compacted layers. Deep vibration techniques may prove economical in areas where old cellars make it difficult to operate backactor excavators.

Penman and Godwin (1975) noted that maximum rates of settlement occurred immediately after the construction of houses on an old opencast site which had been backfilled at Corby. These settlements decreased to small rates after about four years. They suggested that two of the causes of settlement in this fill were creep, which is proportional to log time, and partial inundation. The houses at Corby were constructed 12 years after the fill was placed. The amount of damage which they have suffered is relatively small and is attributable to differential settlement.

Where opencast fills exceed 30 m in depths, because greater settlements may occur, Kilkenny (1968) recommended that the minimum time which should elapse before development takes place should be 12 years after restoration is complete. He noted that settlement of opencast backfill appeared to be complete within 5–10 years after the operation. For example, comprehensive observations of the opencast restored area at Chibburn, Northumberland, 23–38 m in depth, revealed that the ultimate settlement amounted to approximately 1.2% of the fill thickness. Some 50% of the settlement was complete after two years and 75% within five years. In shallow opencast fills, up to 20 m deep, settlements of up to 75 mm have been observed.

10.9 Coarse colliery discard

There are two types of colliery discard, namely, coarse and fine. Coarse discard consists of run-of-mine material and reflects the various rock types which are extracted during mining operations. It contains varying amounts of coal which have not been separated by the preparation process. Some tips, particularly those with relatively high coal contents, may be partly burnt or burning and this affects their composition and therefore their engineering behaviour. Fine discard consists of either slurry or tailings from the washery, which is pumped into lagoons.

The moisture content of coarse discard would appear to increase with increasing content of fines, and generally falls within the range 5–15% (see Taylor (1978)). The range of relative density depends on the relative proportions of coal, shale, mudstone and sandstone (Table 10.1). Of particular importance is the proportion of coal, the higher the content, the lower the relative density. Tip material also shows a wide variation in bulk density and may in fact vary within a tip. Low densities are mainly a function of low relative density. The majority of tip material is essentially granular. Often most of it falls within the sand range, but significant proportions of gravel and cobble range also may be present. Due to breakdown, older and surface materials tend to contain a higher proportion of fines than that which occurs within a tip. In coarse discard the liquid and plastic limits are only representative of that fraction passing the 425 BS sieve, which frequently is less than 40% of the sample concerned. Nevertheless, the results of these consistency tests suggest a low to medium plasticity whilst in certain instances spoil has proved virtually non-plastic.

As far as effective shear strength of coarse discard is concerned, the angle of effective shearing resistance (ϕ') usually varies from 25–45°. It, and therefore the strength, increases in spoil which has been burnt. With increasing content of fine coal, on the other hand, the angle of shearing resistance is reduced. The shear strength of colliery spoil, and therefore its stability, is dependent upon the pore pressures developed within it. These are likely to be developed where there is a high proportion of fine material which reduces the permeability below 5×10^{-7}m/s.

Oxidation of pyrite within tip waste is governed by the access of air. However, the highly acidic oxidation products which result may be neutralised by alkaline materials in the waste. When this does not happen these chemical changes may give rise to pollution of drainage water. The sulphate content of weathered, unburnt colliery waste is usually high enough to warrant special precautions in the design of concrete structures which may be in contact with the discard or water issuing from it.

Spontaneous combustion of carbonaceous material, frequently aggravated by the oxidation of pyrite, is the most common cause of burning spoil. The problem of combustion has sometimes to be faced when reclaiming old tips (see Bell (1977)). The NCB (1973) recommend digging out, trenching, blanketing, injection with non-combustible material and water, and water spraying as methods by which spontaneous combustion in spoil material can be controlled. Moreover, spontaneous combustion may give rise to subsurface cavities in spoil heaps. Burnt ashes may also cover, to appreciable depths zones which are red hot. When steam comes in contact with red hot carbonaceous material, water gas is formed and when the latter is mixed with air it becomes potentially explosive. Explosions may also occur when burning spoil heaps are being reworked and a cloud of coal dust is formed near the heat surface. If the mixture of coal dust and air is ignited it may explode violently.

Noxious gases are emitted from burning spoil. These include carbon monoxide, carbon dioxide, sulphur dioxide and less frequently hydrogen sulphide. Carbon monoxide is the most dangerous since it cannot be detected by taste, smell or irritation and may be present in potentially lethal concentrations. By contrast sulphur gases are readily detectable in the aforementioned ways and are not usually present in high concentrations.

References

ADAMS, J. I. (1963). 'A comparison of field and laboratory measurement of peat', *Proc. 9th Muskeg Res. Conf. NRC-ACSSM Tech. Memo 81*, 117–35.

ADAMS, J. I. (1965). 'The engineering behaviour of Canadian muskeg', *Proc. 6th Int. Conf. Soil Mech. Foundation Engg, Montreal*, **1**, 3–7.

ALLAM, M. M. & SRIDHARAN, A. (1981). 'Effect of wetting and drying on shear strength', *Proc. ASCE, J. Geot. Engg Div.*, **107**, GT4, Paper 16178, 421–37.

ANDERSLAND, O. B. & AL KHAFAJI, A-A. W. N. (1980). 'Organic material and soil compressibility', *Proc. ASCE, J. Geot. Engg Div.*, **106**, GT7, Paper 15546, 749–58.

ARNOLD, M. & MITCHELL, P. W. (1973). 'Sand deformation in three-dimensional stress state', *Proc. 8th Int. Conf. Soil Mech. Foundation Engg, Moscow*, **1**, 11–18.

AUDRIC, T. & BOUQUIER, L. (1976). 'Collapsing behaviour of some loess soils from Normandy', *Q. J. Engg Geol.*, **9**, 265–78.

BARDEN, L. (1972). 'The relation of soil structure to the engineering geology of clay soil', *Q. J. Engg Geol.*, **5**, 85–102.

BELL, F. G. (1977). 'Coarse discard from mines', *Civ. Engg*, 37–9, Mar.

BELL, F. G. (1978). 'Peat: a note and its geotechnical properties', *Civ. Engg*, 45–9, 49–53, Jan., Feb.

BERRE, T. & BJERRUM, L. (1973). 'The shear strength of normally consolidated clays', *Proc. 8th Int. Conf. Soil Mech. Foundation Engg, Moscow*, **1**, 39–49.

BERRY, P. L. & POSKITT, T. J. (1972). 'The consolidation of peat', *Geotechnique*, **22**, 27–52.

BJERRUM, L. (1967). 'Progressive failure in slopes of overconsolidated plastic clay and clay shales', *Proc. ASCE Soil Mech. Foundation Engg Div.*, **93**, SM5, 2–49.

BJERRUM, L. (1972). 'Embankments on soft ground', (in *Performance of Earth and Earth-Supported Structures*), Vol. II, *ASCE Proc. Speciality Conf.*, Purdue University, Lafayette, Indiana, 32–3.

BOSWELL, P. G. H. (1961). *Muddy Sediments*, Heffer, Cambridge.

BOULTON, G. S. & PAUL, M. A. (1976). 'The influence of genetic processes on some geotechnical properties of glacial tills', *Q. J. Engg Geol.*, **9**, 159–94.

CABRERA, J. G. & SMALLEY, I. J. (1973). 'Quick clays as products of glacial action: a new approach to their nature, geology, distribution and geotechnical properties', *Engg Geol.*, **7**, 115–33.

CLARE, K. E. (1957a). 'Airfield construction on overseas soils. Part 2—Tropical black clays', *Proc. Inst. Civil Engrs*, **36**, Paper No. 6243, 223–31.

CLARE, K. E. (1957b). 'Airfield construction on overseas soils. Part 1—The formation, classification and characteristics of tropical soils', *Proc. Inst. Civil Engrs*, **36**, Paper No. 6243, 211–22.

CLEMENCE, S. P. & FINBARR, A. O. (1981). 'Design considerations for collapsible soils', *Proc. ASCE J. Geot. Engg Div.*, **107**, GT3, Paper 16106, 305–17.

CLEVENGER, W. A. (1958). 'Experiences with loess as a foundation material', *Trans Am. Soc. Civil Engrs*, **123**, (Paper No. 2961) 151–80.

CORNFORTH, D. H. (1964). 'Some experiments on the influence of strain conditions on the strength of sand', *Geotechnique*, **14**, 143–67.

CROOKS, J. H. A. & GRAHAM, J. (1976). 'Geotechnical properties of the Belfast estuarine deposits', *Geotechnique*, **26**, 293–315.

DIXON, H. H. & ROBERTSON, R. H. S. (1970). 'Some engineering experiences in tropical soils', *Q. J. Engg Geol.*, **3**, 137–50.

DUMBLETON, M. J. (1967). 'Origin and mineralogy of African red clays and Keuper Marl', *Q. J. Engg Geol.*, **1**, 39–46.

DUSSEAULT, M. B. & MORGENSTERN, N. R. (1979). 'Locked sands', *Q. J. Engg Geol.*, **12**, 117–32.

ELSON, J. A. (1961). 'Geology of tills', *Proc. 14th Canadian Conf. Soil Mech. Section 3*, 5–17.

EPPS, R. J. (July 1980). 'Geotechnical practice and ground conditions in coastal regions of the United Arab Emirates', *Ground Engg*, **13**, 19–25.

ESU, F. & GRISOLIA, M. (1977). 'Creep characteristics of overconsolidated jointed clay', *Proc. 9th Int. Conf. Soil Mech. Foundation Engg, Tokyo*, **1**, 93–100.

FOOKES, P. G. (1978). 'Engineering problems associated with ground conditions in the Middle East: inherent ground problems', *Q. J. Engg Geol.*, **11**, 33–50.

FOOKES, P. G., GORDON, D. L. & HIGGINBOTTOM, I. E. (1975). 'Glacial landforms, their deposits and engineering characteristics, in *The Engineering Behaviour of Glacial Materials*), *Proc. Symp. Midland Soil Mech. Foundation Engg Soc., Birmingham University*, 18–51.

GIDIGASU, M. D. (1976). *Laterite Soil Engineering*, Elsevier, Amsterdam.

GILLOTT, J. E. (1979). 'Fabric, composition and properties of sensitive soils from Canada, Alaska and Norway', *Engg Geol.*, **14**, 149–72.

GRATON, L. C. & FRAZER, H. J. (1935). 'Systematic packing of spheres with particular relation to porosity and permeability', *J. Geol.*, **43**, 785–909.

GRIM, R. E. (1962). *Applied Clay Mineralogy*, McGraw-Hill, New York.

HANRAHAN, E. T. (1954). 'An investigation of some physical properties of peat', *Geotechnique*, **4**, 108–23.

HANRAHAN, E. T. & WALSH, J. A. (1965). 'Investigation of the behaviour of peat under varying conditions of stress and strain', *Proc. 6th Int. Conf. Soil Mech. Foundation Engg, Montreal*, **1**, 226–30.

HARDY, R. M. (1966). 'Identification and performance of swelling soil types', *Canadian Geol. J.*, **2**, 141–53.

HOLTZ. W. G. & GIBBS, H. J. (1956a). 'Shear strength of pervious gravelly soils', *Proc. ASCE, J. Soil Mech. Foundation Engg Div.*, **82**, Paper 867.

HOLTZ, W. G. & GIBBS, H. J. (1956b). 'Engineering properties of expansive clays', *Trans. Am. Soc. Civil Engrs*, **121**, 641–63.

HORTA, J. C. DE S. O. (1980). 'Calcrete, gypcrete and soil classification', *Engg Geol.*, **15**, 15–52.

KAZI, A. & KNILL, J. L. (1969). 'The sedimentation and geotechnical properties of the Cromer Till between Happisburgh and Cromer, Norfolk, *Q. J. Engg Geol.*, **2**, 63–86.

KILKENNY, W. M. (1968). 'A study of the settlement of restored opencast coal sites and their suitability for building development', *Dept. Civil Engg Newcastle University*, Bull. No. 38.

KLOHN, E. J. (1965). 'The elastic properties of a dense glacial till deposit', *Canadian Geot. J.*, **2**, 116–28.

LANDVA, A. O. & PHEENEY, P. E. (1980). 'Peat fabric and structure', *Canadian Geot. J.*, **17**, 416–35.

LARIONOV, A. K. (1965). 'Structural characteristics of loess soils for evaluating their constructional properties', *Proc. 6th Int. Conf. Soil Mech. Foundation Engg Montreal*, **1**, 64–8.

McGOWN, A. (1971). 'The classification for engineering purposes of tills from moraines and associated landforms', *Q. J. Engg Geol.*, **4**, 115–30.

McGOWN, A. & DERBYSHIRE, E. (1977). 'Genetic influences on the properties of tills', *Q. J. Engg Geol.*, **10**, 389–410.

McKINLEY, D. G., TOMLINSON, M. J. & ANDERSON, W. F. (1974). 'Observations on the undrained strength of glacial till', *Geotechnique*, **24**, 503–16.

MAY, R. W. & THOMSON, S. (1978). 'The geology and geotechnical properties of till and related deposits in the Edmonton, Alberta, area', *Canadian Geot. J.*, **15**, 362–70.

METCALF, J. B. & TOWNSEND, D. L. (1961). 'A preliminary study of the geotechnical properties of varved clays as reported in Canadian engineering case records', *Proc. 14th Canadian Conf. Soil Mech.*, Section 13, 203–25.

MEYERHOF, G. G. (1951). 'Building on fill with special reference to the settlement of a large factory', *Struct. Engr*, **29**, No. 11, 297–305.

MIELENZ, R. C. & KING, M. E. (1955). 'Physical-chemical properties and engineering performance of clays', (in *Clays and Clay Technology*), Pask, J. A. & Turner, M. D. (eds), *California Division of Mines*, Bulletin 169, 196–254.

MINKOV, M., EVSTATIEV, D., ALEXIEV, A. P. & DOUCHEV, P. (1977). 'Deformation properties of Bulgarian loess soils', *Proc. 9th Int. Conf. Soil Mech. Foundation Engg Tokyo*, **1**, 215–18.

NATIONAL COAL BOARD. (1973). *Spoil Heaps and Lagoons Technical Handbook*, National Coal Board, London.

NIXON, I. K. & SKIPP, B. O. (1957a). 'Airfield construction on overseas soils. Part 5—Laterite', *Proc. Inst. Civil Engrs*, **36**, Paper No. 6258, 253–75.

NIXON, I. K. & SKIPP, B. O. (1957b). 'Airfield construction on overseas soils. Part 6—Tropical red clays', *Proc. Inst. Civil Engrs*, **36**, Paper No. 6258, 275–92.

OBERMEIER, S. F. (1974). 'Evaluation of laboratory techniques for measurement of swell potential of clays', *Bull. Ass. Engg Geologists*, **11**, 293–314.

OLA, S. A. (1978a). 'Geotechnical properties and behaviour of stabilized lateritic soils', *Q. J. Engg Geol.*, **11**, 145–60.

OLA, S. A. (1978b). 'The geology and engineering properties of the black cotton soils of north eastern Nigeria', *Engg Geol.*, **12**, 375–91.

OLA, S. A. (1980). 'Mineralogical properties of some Nigerian residual soils in relation with building problems', *Engg Geol.*, **15**, 1–13.

OLSON, R. E. (1974). 'Shearing strength of kaolinite, illite and montmorillonite', *Proc. ASCE, J. Geot. Engg Div.*, **100**, Paper No. 10947, 1215–29.

ORDEMIR, I., SOYDEMIR, C. & BIRAND, A. (1977). 'Swelling problems of Ankara Clay, Turkey', *Proc. 9th Int. Conf. Soil Mech. Foundation Engg, Tokyo*, **1**, 243–6.

PENMAN, A. D. M. (1953). 'Shear characteristics of a saturated silt in triaxial compression', *Geotechnique*, **3**, 312–15.

PENMAN, A. D. M. & GODWIN, E. W. (1975). 'Settlement of experimental houses on land left by opencast mining at Corby', (in *Settlement of Structures*), British Geotechnical Society, Pentech Press, London, 53–61.

POPESCU, M. E. (1979). 'Engineering problems associated with expansive clays from Romania', *Engg Geol.*, **14**, 43–53.

POULOS, S. J. (1981). 'The steady state of deformation', *Proc. ASCE, J. Geot. Engg Div.*, **107**, GT5, Paper 16241, 553–62.

PUSCH, R. & ARNOLD, M. (1969). 'The sensitivity of artificially sedimented organic free illite clay', *Engg Geol.*, **3**, 135–48.

RADFORTH, N. W. (1952). 'Suggested classification of muskeg for the engineer', *Engg J. (Canada)* **35**, 1194–210.

RADHAKRISHNA, H. S. & KLYM, T. W. (1974). 'Geotechnical properties of a very dense glacial till', *Canadian Geot. J.*, **11**, 396–408.

RIVARD, P. J. & LU, Y. (1978). 'Strength of soft fissured clays', *Canadian Geot. J.*, **15**, 382–90.

ROSENQVIST, I. TH. (1966). 'Norwegian research into the properties of quick clay—a review', *Engg Geol.*, **1**, 445–50.

SAXENA, S. K., HEDBERG, J. & LADD, C. C. (Sept. 1978). 'Geotechnical properties of Hackensack valley clays of New Jersey', *Geot. Test J.*, **1**, No. 3, 148–61.

SCHMERTMANN, J. H. (1969). 'Swell sensitivity', *Geotechnique*, **19**, 530–33.

SCHULTZE, E. & HORN, A. (1961). 'The shear strength of silt', *Proc. 5th Int. Conf. Soil Mech. Foundation Engg, Paris*, **1**, 350–53.

SCHULTZE, E. & KOTZIAS, A. B. (1961). 'Geotechnical properties of lower Rhine silt', *Proc. 5th Int. Conf. Soil Mech. Foundation Engg, Paris*, **1**, 329–33.

SHERWOOD, P. T. (1967). 'Classification tests on African red clays and Keuper Marl', *Q. J. Engg Geol.*, **1**, 47–56.

SKEMPTON, A. W. (1951). 'The bearing capacity of clays', *Building Research Congress*, Div. 1, 180–89.

SKEMPTON, A. W. & HUTCHINSON, J. N. (1969). 'Stability of natural slopes and embankment foundations', *Proc. 7th Int. Conf. Soil Mech. Foundation Engg*, State-of-the-Art Volume, Mexico, 221–42.

SOWERS, G. E. (1973). 'Settlement of waste disposal fills', *Proc. 8th Int. Conf. Soil Mech. Foundation Engg, Moscow*, **2**, 207–12.

TAYLOR, R. K. (1978). 'Properties of mining wastes with respect to foundations', (in *Foundation Engineering in Difficult Ground*), Bell, F. G. (ed.), Butterworths, London, 175–203.

TAYLOR, R. K., BARTON, R., MITCHELL, J. E. & COBB, A. E. (1976). 'The engineering geology of Devensian deposits underlying PFA lagoons at Gale Common, Yorkshire', *Q. J. Engg Geol.*, **9**, 195–218.

TERZAGHI, K. (1955). 'The influence of geological factors in the engineering properties of sediments', *Econ. Geol.*, 50th Ann. Vol., 557–618.

TOMLINSON, M. J. (1957). 'Airfield construction on overseas soils. Part 3—Saline calcareous soils', *Proc. Inst. Civil Engrs*, **36**, Paper No. 6239, 232–46.

VAN OLPHEN, H. (1963). *An Introduction to Clay Colloid Chemistry*, Wiley, New York.

WAHLS, H. E. (1962). 'Analysis of primary and secondary consolidation', *Proc. ASCE, J. Soil Mech. Foundation Engg Div.*, **88**, SM6, 207–31.

WEST, G. & DUMBLETON, M. J. (1970). 'The mineralogy of tropical weathering illustrated by some West Malaysian soils', *Q. J. Engg Geol.*, **3**, 25–40.

WILLIAMS, A. B. & JENNINGS, J. E. (1977). 'The *in situ* shear behaviour of fissured soils', *Proc. 9th Int. Conf. Soil Mech. Foundation Engg, Tokyo*, **2**, 169–76.

WILSON, N. E. (1978). 'The contribution of fibrous interlock to the strength of peat', *Proc. 17th Muskeg Res. Conf.*, Williams, G. P. (ed.), Nat. Res. Coun. Canada, Ass. Comm. Geotech. Res., Tech. Mem., No. 122, Ottawa, 5–10.

WILSON, N. E., RADFORTH, N. W., MacFARLANE, I. C. & LO, M. B. (1965). 'The rates of consolidation for peat', *Proc. 6th Int. Conf. Soil Mech. Foundation Engg, Montreal*, **1**, 407–12.

WU, T. H. (1958). 'Geotechnical properties of glacial lake clays', *Proc. ASCE, J. Soil Mech. Foundation Engg Div.*, **84**, SM3, Paper 1732, 1–36.

YEN, B. C. & SCANLON, B. (1975). 'Sanitary landfill settlement rates', *Proc. ASCE, J. Geot. Engg Div.*, **101**, GT5, Paper No. 11335, 475–87.

Chapter 11

Mechanical properties of rocks

11.1 Stages of deformation

If the perimeter of a particle is exposed to a force, then internal stresses are developed, and if these are strong enough, they bring about changes in the shape and/or size of the particle. The particle is then said to be strained, the term distortion is used to describe changes in shape, whilst dilatation refers to changes in size. Deformation refers to changes in shape accompanied by changes in size.

If any plane is taken within a solid body, then the internal components of stress may be resolved into normal stress, which acts at right angles to the plane, and tangential or shearing stress which acts parallel to the plane. Normal stress can be further resolved into tensile stress, if it tends to pull the material apart, and compressive stress, when it pushes the material together. At any point within a mass the stress to which it is subjected may be divided into three components, orientated at right angles to each other. These are the principal axes of stress, and when stresses are unequal, they are defined as the greatest, mean and least stress axes. The stress difference is the difference in value between the greatest and least stress.

Two types of strain can be distinguished, namely, homogeneous and heterogeneous. Homogeneous strain occurs when every part of a body is subjected to a strain of the same type and magnitude in any direction of displacement. Although the geometrical form is changed, similar parts possess the same orientation. If the strain is not the same throughout a body then it is described as heterogeneous. The changes which take place at any point along the principal axis of stress afford a measure of the amount of distortion.

Four stages of deformation have been recognised, elastic, elastico-viscous, plastic and rupture. The stages are dependent upon the elasticity, viscosity and rigidity of a rock, as well as on its stress history, temperature, time, pore water and anisotropy. An elastic deformation is defined as one which disappears when the stress responsible for it ceases. Ideal elasticity would exist if the deformation on loading and its disappearance on unloading were both instantaneous. This is never the case since there is always some retardation, known as hysteresis, in the unloading process. With purely elastic deformation the strain is a linear function of stress, that is, the material obeys Hooke's law. Therefore the relationship between stress and strain is constant and is referred to as Young's modulus. The latter is derived from the stress–strain curve. Rocks, however, only approximate to an ideal Hookean solid, the stress–strain relationships are not generally linear. Consequently Young's modulus is not a simple constant but is related to the level of stress applied.

The change at the elastic limit from elastic to plastic deformation is referred to as the yield point or yield strength. Yield strength has been defined by Robertson (1955) as the stress difference a body is able to withstand without yielding plastically and may be arbitrarily designated as the maximum stress difference at the elastic limit for constant experimental conditions.

If the stress on a material exceeds its elastic limit then it is permanently strained, the latter being brought about by plastic flow. Within the field of plastic flow there is a region where elastic stress is still important and this is referred to as the field of elastico-viscous flow. This term has been used to describe creep or continuous deformation which occurs in rocks when they are subjected to constant stress. Plasticity may be regarded as time-independent, non-elastic, non-recoverable, stress-dependent deformation under uniform sustained load. Solids are classified as brittle or ductile according to the amount of plastic deformation they exhibit. In brittle materials the amount of plastic deformation is zero or very little whilst it is large in ductile substances.

Rupture occurs when the stress exceeds the strength of the material involved. Rupture, or ultimate strength, was defined by Robertson (1955) as the maximum stress difference a body is able to withstand prior to loss of cohesion by fracturing for constant experimental conditions, fracturing being conceived as the breaking process leading to rupturing. Failure by rupture results from an excessively rapid application of stress which does not leave enough time for relief by plastic deformation. The initiation of rupture is marked by an increasing strain velocity.

Wawersik and Fairhurst (1970) suggested that the stress–strain characteristics of rock cores tested in uniaxial compression were sensibly the same during the rising part of the stress–strain curve and that in general terms the stress–strain paths between zero stress and the compressive strength could be divided into three regions (Figure 11.1). Region A was characterised by a steadily increasing slope and was usually more pronounced for more porous rocks and those containing numerous small fractures. In region B the stress–strain paths for brittle rocks were almost linear. The hysteresis (see Section 11.2) in both these regions is small. The slope of the curve decreases in region C and energy is dissipated which gives rise to large hysteresis loops upon unloading. Striking differences were noted in the behaviour of rocks in the post-failure region which allowed two classes to be recognised. In Class I, fracture propagation was stable in the sense that work must be done on the specimen for each incremental decrease in load carrying ability. Consequently rocks which exhibit such behaviour retain some strength after the compressive strength has been exceeded. Conversely in Class II the failure is unstable or self-sustaining. In other words, the elastic strain energy stored in the sample when applied stress equals compressive strength is

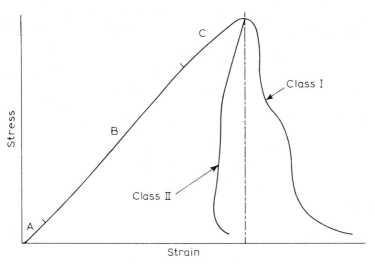

Figure 11.1 Three regions of the stress–strain curve (after Wawersik and Fairhurst (1970))

enough to maintain fracture propagation until the specimen has lost virtually all its strength. The fracture of such rocks cannot be controlled. No fabric changes occur in regions A and B and it is not before region C until fracture becomes evident.

Failure in region C is first of all marked by the formation of a large number of isolated fractures, both intergranular and intragranular. Such local fracturing characterises the relief of stress concentration produced by the mechanical inhomogeneities in the rock and most cracks are orientated parallel to the applied stress (Figure 11.2). This is quickly followed, however, by the development of two types of macroscopic shear failures, at the boundary and in the interior of the specimen, which suggests that most of the major sources of induced lateral tensile stresses are by now eliminated. The interior macroscopic shear failures are extended and become interconnected to form a conjugate set of open shear fractures. The central cones, which are now delineated, either abrade during the large shear displacement

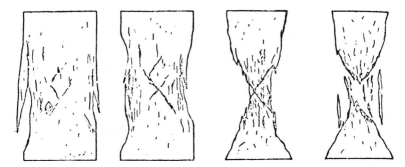

Figure 11.2 Stages in the development of fracturing with increasing unconfined compressive strength (after Wawersik and Fairhurst (1970))

or produce major fractures in the remnant cores. If one shear failure surface becomes dominant then cones are not developed and the sample ultimately splits in two.

Wawersik and Fairhurst point out that the compressive strength of rocks which show Class I behaviour is not governed by the formation of macroscopic shear failures but by local fracturing developed roughly parallel to the axis of loading. In rocks exhibiting Class II behaviour it seemed that the deformation band in which axial fracturing prevailed decreased rapidly until shear fracture took over as the principal manner of failure, determining the compressive strength of the rock. The apparent decrease of local fracture intensity leaves more energy for the propagation of macroscopic shear failures and provides an explanation for the self-sustaining nature of the failure process.

11.2 Elastic properties

As can be inferred from the above, elasticity is a property of an ideal material. Just how closely rocks approximate to the ideal depends upon three major factors, namely, homogeneity, isotropy and continuity. Homogeneity may be defined as a measure of the physical continuity of a material, that is, the constituent particles are evenly distributed throughout its volume. The elastic properties of the material are therefore the same at all points. Accordingly homogeneity is dependent largely upon scale. Isotropy in terms of a rock, can be defined as a measure of its directional properties. A rock is therefore only isotropic if it is monomineralic and its grains have a random orientation. Since most rocks are composed of two or more essential minerals, and many possess preferred orientation, they are strictly anisotropic. Hence they may react differently to forces in different directions, depending upon the degree of anisotropy. Continuity is a property specifically related to rocks and may be taken to refer to the amount of fracture and pore space in a particular mass of rock. The degree of continuity affects its cohesion and hence the transmission of an even stress distribution throughout the rock mass.

11.2.1 Young's modulus and Poisson's ratio

Young's modulus is the most important of the elastic constants and can be derived from the slope of the stress–strain curve, it being the ratio of stress to strain. Most crystalline rocks have S-shaped stress–strain curves (Figure 11.3). At low stresses the curve is non-linear and concave upwards, that is, Young's modulus increases as the stress increases. The initial tangent modulus is given by the slope of the stress–strain curve at the origin, as shown by line E_i. Gradually a level of stress is reached where the slope of the curve becomes approximately linear (that is, between a and b). In this region Young's modulus is defined as the tangent modulus, or the secant modulus, which is given by the slope of the line OP. At this stress level the secant has a lower value than the tangent modulus because it includes the initial 'plastic' stress history of the curve. Dhir and Sangha (1973) found that the 50% tangent modulus (the tangent modulus at half the failure load E_{t50})

provided the best value for summarising the stress–strain relationship.

Deere and Miller (1966), after testing 28 different rock types, classified the uniaxial stress–strain curves into six types (Figure 11.4). Types III, IV and V, however, are modifications of the representative S-curve. Type I represents the classical straight line behaviour of brittle materials which is typical of the more explosive failures of basalts, dolerites, quartzites, and strong dolostones and limestones. Softer limestones, siltstones and tuffs exhibit a more concave downwards curve as illustrated in Type II. These are usually somewhat more linear in the earlier and central portions, yielding 'plastically' as failure is approached. Type III is typical of sandstone, granite, some dolostones and dolerites, and schist cut parallel to the schistosity. Metamorphic rocks like marbles and gneiss are represented by Type IV. Schist cored along the schistosity has the long, sweeping S-shaped curve of Type V. Types III, IV and V are characterised by initial 'plastic' crack closing, followed by a steeper linear section. The upper part of such curves exhibits varying degrees of plastic yield as failure is approached. Type III rocks do not yield significantly, being more explosive with brittle-type failures (similar to Type I) than Types IV and V. The Type VI curve for rock salt has an initial small elastic straight line portion, followed by a combination of plastic deformation and continuous creep.

In addition to their non-elastic behaviour most rocks exhibit hysteresis. Under uniaxial stress the slope of the stress–strain curve during unloading is initially greater than during loading for all stress values (see Figure 11.3). As stress is decreased to zero a residual strain, OR, is often exhibited. On reloading the curve RS is produced, which in turn is somewhat steeper than OP. Further cycles of unloading and reloading to the same maximum stress give rise to hysteresis loops, which are shifted slightly to the right. These effects are associated with transient creep. The non-linear elastic behaviour and elastic hysteresis of brittle rocks under uniaxial compression have been explained as due to the presence of flaws or minute cracks in the rocks (see

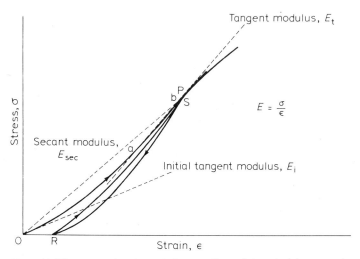

Figure 11.3 Representative stress–strain curve for rock in uniaxial compression

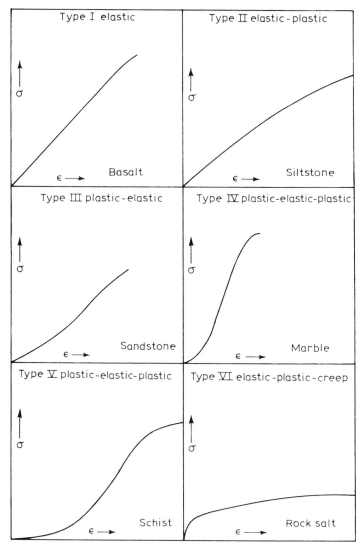

Figure 11.4 Typical stress–strain curves for rock in uniaxial compression loaded to failure (after Deere and Miller (1966))

Walsh (1965a)). At low stresses these cracks are open but they close as the stress is increased and the rock becomes elastically stiffer, that is, E increases with stress. Once the cracks are closed the stress–strain curve becomes linear. Nevertheless E is still lower in this portion of the curve than it would be for an uncracked solid and this has been attributed to sliding along crack surfaces. Since these cracks do not immediately slide in the opposite sense as the load is reduced, hysteresis loops are produced.

When hysteresis is large it is difficult to distinguish between elastic and plastic deformation, however, an elastic strain is related only to stress whereas a permanent strain is also related to the period over which the stress is applied.

As a consequence, in a strain experiment, with a constant load, the elastic deformation is characterised by a gradual decrease of the strain velocity which ultimately leads to a halt in the process. On the other hand, a permanent or plastic deformation continues indefinitely with a constant strain velocity. Under high temperature–pressure conditions permanent deformation may also take place by creep.

Farmer (1968) suggested that three types of elastic behaviour can be distinguished in rock testing. He regarded rocks as being quasi-elastic, semi-elastic or non-elastic, and maintained that they could be roughly delineated in terms of their apparent elastic moduli. A quasi-elastic rock possessed a value of E between 6.0 and 11.0×10^4 MPa, a semi-elastic rock between 4.0 and 7.0×10^4 MPa and a non-elastic rock less than 5.0×10^4 MPa. The values of E quoted are for the initial tangent modulus. The quasi-elastic rocks are fine-grained, massive and compact. In many ways such rocks approximate to a brittle elastic material having a near linear stress–strain relationship up to the point of failure (Figure 11.5a). The coarser grained igneous rocks and

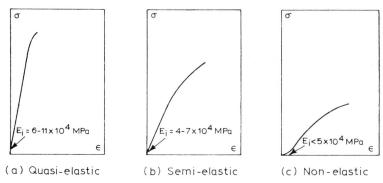

(a) Quasi-elastic   (b) Semi-elastic   (c) Non-elastic

Figure 11.5 Typical stress–strain relationships for rocks (after Farmer (1968))

fine-grained compacted sediments are less elastic, falling within the semi-elastic group. They exhibit a stress–strain relationship in which the slope of the curve decreases with increasing stress (Figure 11.5b). The third category, non-elastic rock, includes the less cohesive rocks with large pore space. It contains most of the weaker sedimentary rocks. The general curve shows an initial zone of increasing slope with increasing load, a feature which suggests compaction and crack closure before any near-linear deformation occurs (Figure 11.5c). These rocks tend to exhibit variable stress–strain characteristics.

When a specimen undergoes compression it is shortened and this is generally accompanied by an increase in its cross-sectional area. The ratio of lateral unit deformation to linear unit deformation, within the elastic limit, is known as Poisson's ratio. An idealised value for Poisson's ratio can be obtained by considering an idealised crystal structure, where contraction in one direction automatically leads to extension of the lattice in a perpendicular direction. In such a case, by considering the geometry of the structure, it can be shown that Poisson's ratio is 0.333. The work carried out by Deere and Miller (1966) showed that values of Poisson's ratio for rock must be regarded with some suspicion. These two authors gave the average initial tangent value of Poisson's ratio for all the rocks they studied as 0.125 and at a stress level of 50% ultimate stress as 0.341.

In a study of a number of rocks of low porosity, Walsh and Brace (1972) found that Poisson's ratio in uniaxial strain was up to 25% higher than the value obtained in direct static measurements. They assumed that movement along cracks had occurred during uniaxial strain and that this accounted for the discrepancy. In order to check this they undertook an elastic analysis based on a model of a crack-filled solid and they then found that they could predict the characteristics of Poisson's ratio observed experimentally. Poisson's ratio, when applied compressive stresses are high enough virtually to close all cracks, should accordingly be greater than the intrinsic value, because of the relative motion which may take place between crack faces when applied stresses are non-hydrostatic. At low stresses, when the cracks are open, Poisson's ratio should be less than the intrinsic value. This was shown to be the case for granite, gabbro and marble under uniaxial strain at high stress and at low stress for granite, gabbro and dolerite. Thus Walsh and Brace regarded the behaviour of low porosity rocks in uniaxial strain experiments as not being strictly elastic.

Rocks subjected to uniaxial compression tend to exhibit a common behaviour in that both Young's modulus and Poisson's ratio increase to more or less constant values as the stress is increased. As the compressive stress approaches the failure limit, Young's modulus falls, eventually reaching zero, while Poisson's ratio increases to values nearing or exceeding the theoretical maximum of 0.5 for an incompressible solid body. The opposite trend is observed when rocks are placed under uniaxial tension, namely, both Young's modulus and Poisson's ratio are initially high and they fall continuously as stress increases to failure point (see Hawkes et al. (1973)). Hawkes et al. found that the initial tangent moduli for the rocks they tested were similar in compression and tension. The value of Young's modulus in compression at half the load at failure (E_{t50}) is usually greater than the value in tension but there is considerable variation from one rock type to another. At very low stresses Poisson's ratio of a rock in tension can be greater than 0.5, indicating an initial decrease in volume, that is, a decrease in porosity. However, as stress is increased the ratio falls to comparatively low values (0.1).

As pointed out above, under uniaxial compression the cracks or flaws in rock tend initially to close and the rock becomes progressively stiffer as loading is increased (see Walsh (1965a); Morgenstern and Phukan (1966)). By contrast rocks become progressively less stiff under increasing tension and Young's modulus falls continuously until failure occurs. This was explained by Hawkes et al. (1973) who suggested that flaws underwent continuous opening when a rock specimen was subjected to tension.

11.2.2 Other elastic properties

Another elastic constant is compressibility, which is the ratio of change in volume of an elastic solid to change in hydrostatic pressure. A fourth elastic constant is rigidity, which refers to the resistance of a body to shear. The modulus of rigidity is the ratio of shear stress to shear strain, in a simple shear. The four elastic constants, Young's modulus (E), Poisson's ratio (v), compressibility (K) and rigidity (G) are not independent of each other and if any two are known it is possible to derive the other two from the following expressions

$$G = \frac{E}{2(1 + v)}$$

and

$$K = \frac{E}{3(1 - 2v)}$$

Young's modulus and Poisson's ratio are more readily determined experimentally.

Brace (1965) published the results of a number of linear compressibility tests which had been carried out on a group of rocks which were relatively isotropic, free from chemical alteration and had a simple mineralogy. Some of them, however, had a negligible porosity, whilst in others it was sufficiently high to cause pronounced initial changes in elastic properties with increasing pressure. The linear compressibilities were recorded up to 9 kbar and measurements were made in three mutually perpendicular directions. The rocks tended to reflect their relative isotropy in linear compressibility above 1 or 2 kbar but at lower pressures certain variations with direction became evident. For example, the two granites tested became increasingly anisotropic at lower pressures and at zero pressure, linear compressibility varied by almost a factor of 2. By contrast limestone and dolostone showed little variation with increasing pressure. The large initial decrease of compressibility can be related to porosity, grain size and lineation. Brace maintained that a certain amount of porosity in crystalline rocks occurs along grain boundaries and that the large initial decrease in compressibility with pressure is due to closure of these crack-like openings. He assumed that the differences in linear compressibility with direction must be the result of differences in the arrangement of cracks with direction. In micaceous rocks the grain boundary cracks are longer parallel with, than normal to bedding, hence larger changes of initial linear compressibility with pressure occur in a direction normal to the bedding.

Walsh (1965b) contended that the difference between the compressibilities of a porous rock body and a solid body of the same composition, is equal to the rate of change of porosity with externally applied pressure for the porous body. He maintained that this relationship was independent of the shape or concentration of the pores and depended on the assumption of linear elasticity and small strains. The influence of pore shape on compressibility was illustrated by comparing spherical pores with narrow cracks. The former were apparently two or three times more effective than the latter of the same radius in increasing compressibility. However, compressibility was found to be strongly affected by the presence of a few long cracks, indeed the increase in effective compressibility multiplied as the cube of the crack length. Walsh concluded that the influence of the longer cracks accounts for the apparent anisotropy at low pressures of seemingly isotropic rocks, as such are unlikely to have a random orientation.

The values of ultimate compressive (C_{ult}) and yield strength (C_{ys}) together with those of Young's modulus, allow the determination of the modulus ratio (M_E/C_{ult}) and the modulus of resilience (M_r). The modulus ratio is simply that between E_{t50} and the ultimate compressive strength. Three grades have been recognised by Deere and Miller (1966) namely: high, over 500; average, 200–500 and low, under 200. The modulus of resilience refers to the capacity of a material to absorb energy within the elastic range. It is equal to the area under the elastic portion of the stress–strain curve and has been defined as the strain energy absorbed per unit volume when the material is stressed to its elastic limit. Thus the yield strength and elasticity are used to derive the modulus of resilience as follows.

$$M_r = C_{ys}^2/2E_{t50}$$

The term resilience should not be confused with the modulus of resilience. Within the elastic limit the resilience is equal to the external work put into the material during deformation. Hence the total resilience of a material is the product of its volume and the modulus of resilience.

The modulus of toughness (M_t) represents the maximum amount of energy a unit volume of rock can absorb without fracture and it can be estimated as follows

$$M_t = \tfrac{2}{3}C_{ult}\varepsilon_f$$

where ε_f is the strain at failure. Toughness consequently reflects the ability of a rock to absorb energy during plastic deformation. In a static test this energy is measured by the area under the stress–strain curve which represents the work required to fracture the test specimen. Rocks with high toughness have high strength and ductility, whilst brittle materials usually have a low toughness since they show only small plastic deformation before fracture.

The constrained modulus of deformation (M_c) can be defined as the rate of change of vertical stress with respect to vertical strain under conditions of zero lateral strain. It is related to Young's modulus and Poisson's ratio by the following expression

$$M_c = E\left[\frac{(1-v)}{(1+v)(1-2v)}\right]$$

It can also be derived from rigidity (G) and compressibility (K) as follows

$$M_c = K + \tfrac{4}{3}G$$

11.3 Theories of failure

One of the most popular theories which was proposed to explain shear fractures was advanced by Coulomb (1773). The Coulomb criterion of brittle failure is based upon the idea that shear failure occurs along a surface if the shear stress acting in that plane is high enough to overcome the cohesive strength of the material and the resistance to movement. The latter is equal to the stress normal to the shear surface multiplied by the coefficient of internal friction of the material, whilst the cohesive strength is its inherent shear strength when the stress normal to the shear surface is zero. The relation between the failure criterion, the friction and the cohesion is then expressed by Coulomb's law

$$\tau = c + \sigma_n \tan \phi$$

where τ is the shearing stress, c is the apparent cohesion, σ_n is the normal stress and ϕ is the angle of internal friction or shearing resistance. The Coulomb criterion has been shown to agree with experimental data for rocks in which the relationship between the principal stresses at rupture is, to all intents, linear. If the relationship, however, is non-linear then it is assumed that either the angle of internal shearing resistance is dependent upon pressure or, more probably, the area of the grains in frictional contact increases as the normal pressure increases.

Coulomb's concept was subsequently modified by Mohr (1882). Mohr's hypothesis states that when a rock is subjected to compressive stress shear fracturing occurs parallel to those two equivalent planes for which shearing stress is as large as possible whilst the normal pressure is as small as possible. This statement assumes that a triaxial state of external stress is applied to a substance and that the maximum external stress is resolved into shear and normal components for any inclined potential shear planes existing in the stressed material. It has been suggested that shear fractures usually enclose an angle of less than 90° about the axis of maximum compression because the normal stresses which act across a shear plane are also involved in shear fracturing. According to Mohr the maximum or limiting tangential stress which a substance can withstand depends upon the normal stress across shear planes. Moreover, shearing across a potential shear plane can only occur when internal friction is overcome. As a result failure does not usually take place at 45° to the axis of maximum stress but at a point where the optimum ratio exists between the shearing components, the internal friction and the molecular cohesive force. The shearing angle is a constant which is governed by the ratio of the ultimate compressional to the ultimate tensile strength. It is therefore characteristic of a particular material at a given temperature and pressure. The more brittle a substance the smaller the angle. Conversely the angle is obtuse for ductile materials.

Griffith (1920) claimed that because of the presence of minute cracks or flaws, particularly in surface layers, the measured tensile strengths of most brittle materials are much less than those which would be inferred from the values of their molecular cohesive forces. Although the mean stress throughout a body may be relatively low, local stresses developed in the vicinity of the flaws were assumed to attain values equal to the theoretical strength. Under tensile stress the stress magnification around a flaw is concentrated where the radius of curvature is smallest, that is, at its ends. Hence the tensile stresses which develop around the flaw have most influence when the tensile stress zone coincides with the zone of minimum radius of curvature. The concentration of stress at the ends of flaws causes them to be enlarged and presumably with time they develop into fractures.

Griffith maintained that if a material is subjected to tensile stress, then the tensile strength, T, is given approximately by

$$T = \sqrt{(2E\gamma)/\pi a}$$

where E is Young's modulus, γ is the specific surface energy of the material and a is half the length of the Griffith crack. He concluded that the compressive strength, C is eight times the tensile strength. According to this concept, strength is inversely proportional to the square root of crack length so that the longest crack in a material determines its strength. It has been suggested from experimental studies that the length of a Griffith crack could be regarded as approximating to the grain size.

Brace (1964) showed that the fracture in hard rock was usually initiated in grain boundaries which could be regarded as the inherent flaws required by the Griffith theory. He supposed that as stress was increased prior to fracture, grain boundaries at numerous sites in a rock became loosened and that at the instant before fracture the rock was filled with loosened sections along grain boundaries which had various lengths and orientations. Cracks grew in such sections and ultimately gave rise to fractures.

Griffith's original theory was concerned with brittle fracture produced by applied tensile stress and he based his calculations upon the assumption that the inherent flaw, from which fracture is initiated, could be regarded as an elliptical opening. This simple assumption, however, is not valid in conditions of compressive stress. Accordingly McClintock and Walsh (1962) modified the Griffith theory to include the closing of flaws and the development of frictional forces across their surfaces as presumably occurs in compression. Brace (1964) noted that in compression tests grain boundaries, although loosened, were in frictional contact at the moment of fracture, and thereby lent support to McClintock and Walsh's modification of the Griffith theory (see also Dey and Wang (1981)).

Murrell (1963) extended the original Griffith theory and showed that the Griffith criterion of failure corresponds with a parabolic Mohr's envelope defined by the expression

$$\tau^2 = 4T^2 - 4T\sigma_n$$

where T is the universal tensile strength of the material and σ_n is the normal stress. This assumed that the Griffith cracks remained elliptical up to the point of failure, but in some rock materials this is not the case.

Although there is an encouraging agreement between experimental and theoretical results, the Griffith theory does not provide a complete description of the mechanism of rock failure for it is only strictly correct when applied to fracture initiation under conditions of static stress. Indeed Hoek (1966) contended that it was largely fortuitous that it could be so successfully applied to the prediction of fracture in rock since, once fracture is initiated, its propagation and ultimate failure is a relatively complex process. He further stated that the Griffith theory, when expressed in terms of the stresses at fracture, contributed little to the understanding of rock fracture under dynamic stress conditions, the energy changes associated with fracture or the deformation process of rock.

When the envelope fitted to a set of Mohr circles obtained from low pressure triaxial compression tests on brittle rocks is observed it is usually represented by a straight line as suggested by the Coulomb equation. As most engineering undertakings involve low confining pressures it is sufficient to assume that the coefficient of friction, which is derived from the slope of the envelope, is a constant. At high confining pressures, however, this assumption is usually erroneous. More importantly it is misleading to assume that the coefficient of friction of 'soft' rocks, like shales and mudstones, even at low confining pressures, is a constant. In fact their envelopes are curved. From experimental evidence it has been concluded that the coefficient of internal friction varies with normal compressive stress. The reason for this has been suggested by Hoek (1968) who supposed that it was connected with the interlocking nature of the asperities along the shear plane. He contended that the interlocking depended upon the intimacy of their contact which, in turn, depended upon the amount of normal stress.

One of the most recent empirical laws advanced to describe the shear strength of intact rock has been given by Barton (1976) and is as follows

$$\tau/\sigma_n = \tan\left(50 \log_{10}\left(\frac{\sigma_1 - \sigma_3}{\sigma_n}\right) + \phi_c\right)$$

where σ_n is the normal stress, σ_1 is the axial stress at failure, σ_3 is the effective confining pressure and ϕ_c is equal to 26.6°. Barton argued that the effective normal stress mobilised on conjugate shear surfaces was equal to the differential stress $(\sigma_1 - \sigma_3)$ at a critical state. This happens to be the limiting value of the dimensionless ratio $(\sigma_1 - \sigma_3)/\sigma_n$ used in formulating the empirical laws of friction and fracture strength. The critical state for intact rock was defined as the stress condition under which the Mohr envelope of peak shear strength reaches a point of zero gradient. This represents the maximum possible shear strength of the rock. There is a critical effective confining pressure for each rock above which strength cannot increase. The one-dimensional dilation associated with shearing is completely suppressed if the applied stress reaches the level of critical effective confining pressure.

11.4 Factors controlling the mechanical behaviour of rocks

The factors which influence the deformation characteristics and failure of rocks can perhaps be divided into internal and external categories. The internal factors include the inherent properties of the rock itself, whilst the external factors are those of its environment at a particular point in time. As far as the internal factors are concerned the mineralogical composition and texture are obviously important, but planes of weakness within a rock and the degree of mineral alteration are frequently more important (see Chapters 5 and 7 respectively). The temperature–pressure conditions of a rock's environment significantly affect its mechanical behaviour, as does its pore water content. In this respect the length of time which a rock suffers a changing stress and, to a lesser extent, temperature, and the rate at which these are imposed, also affect its deformation characteristics.

11.4.1 Composition and texture

The composition and texture of a rock are governed by its origin. For instance, the olivines, pyroxenes, amphiboles, micas, feldspars and silica minerals are the principal components in igneous rocks. These rocks have solidified from a magma. Solidification involves a varying degree of crystallisation, the greater the length of time involved, the greater the development of crystallisation. Hence glassy, microcrystalline, fine-, medium- and coarse-grained types of igneous rocks can be distinguished. In metamorphic rocks either partial or complete recrystallisation has been brought about by changing temperature-pressure conditions. Not only are new minerals formed in the solid state but the rocks may develop certain lineation structures. A varying amount of crystallisation is found within the sedimentary rocks, from almost complete, as in the case of certain chemical precipitates, to slight, as far as diagenetic crystallisation in the pores of, for example, certain sandstones.

Few rocks are composed of only one mineral species and even when they are the properties of that species vary slightly from mineral to mineral. Such variations within minerals may be due to cleavage, twinning, inclusions, cracking and alteration, as well as to slight differences in composition. This in turn is reflected in their physical behaviour. As a consequence few rocks can be regarded as homogeneous, isotropic substances. The size and shape relationships of the component minerals are also significant in this respect; generally the smaller the grain size, the stronger the rock.

One of the most important features of texture as far as physical behaviour, particularly strength, is concerned, is the degree of interlocking of the component grains. Fracture is more likely to take place along grain boundaries than through grains and therefore irregular boundaries make fracture more difficult. The bond between grains in sedimentary rocks is usually provided by the cement and/or matrix, rather than by grains interlocking. The amount and, to a lesser extent, the type of cement/matrix is important, influencing not only strength and elasticity, but also density, porosity and permeability.

Grain orientation in a particular direction facilitates breakage along that direction. This applies to all fissile rocks whether they are cleaved, schistose, foliated, laminated or thinly bedded. For example, Griggs (1951) performed tests on the Yule Marble at 1000 MPa confining pressure and ordinary temperatures, to observe the effects of anisotropy. All the specimens tested showed great plastic deformation. When subjected to compression, rock cylinders cut perpendicular to foliation were shown to be stronger than those parallel to the lineation.

Donath (1961) demonstrated that cores cut in Martinsburg Slate at 90° to the cleavage, possessed the highest breaking strength, whilst those cores cut at 30° exhibited the lowest. Similar tests were carried out by Brown *et al.* (1977) who showed that the compressive strength of the Delabole Slate is highly directional. Indeed it varies continuously with the angle made by the cleavage planes and the direction of loading (Figure 11.6). They found that even where the cleavage makes high or low angles with the major principal stress direction the mode of failure is mainly influenced by the cleavage. The water content and surface roughness are the principal factors governing shear strength along the cleavage planes. For instance, the average friction angle of smooth wet surfaces was determined as 20.5°, which was 9° less than that obtained for the same surfaces when dry. What is more it was found that surface roughness could add up to 40° to the basic friction angles. The degree of surface roughness was shown to vary appreciably according to direction along, and character of the cleavage plane concerned, which was reflected in the range of shear strength.

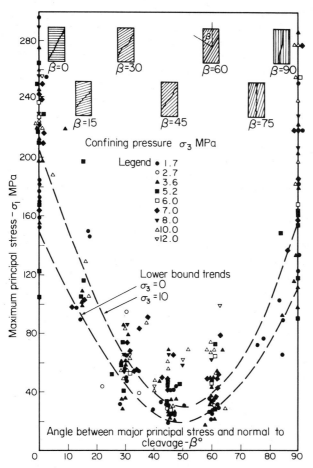

Figure 11.6 Compressive strength anisotropy in dark grey slate (after Brown *et al.* (1977))

11.4.2 Temperature–pressure conditions

Griggs (1936) noted the changes in rock behaviour with increasing pressure-temperature conditions, simulating increasing depth. He showed that the ultimate strength of the Solenhofen Limestone was increased by 360% under 10 000 atmospheres (1000 MPa). High temperatures tend to aid plastic deformation so that it becomes increasingly important with depth.

Although all rock types show an appreciable decrease in strength with increasing temperature and an increase in strength as the confining pressure is increased, the combined effect of these factors, as with increasing depth of burial, is notably different for different rock types. Tests on crystals of quartz, carried out by Griggs and Handin (1960) completely failed to achieve ductile deformation under any conditions. Subsequent experiments on quartzite and sandstone displayed cataclastic intergranular deformation. Experimental investigation has also shown that the effects of temperature changes on sedimentary rocks are of less consequence than those of pressure down to depths of 10 000 m. Nonetheless, with increasing temperatures there is a reduction in yield stress and strain hardening decreases. Heating particularly enhances the ductility, or the ability to deform permanently without loss of cohesion, of calcareous and evaporitic rocks, but it has little effect upon sandstones. Heard (1960) demonstrated that at lower confining pressures and higher temperatures, the Solenhofen Limestone changed from a brittle to a ductile material. For example, in tension the changeover pressure was reduced from 7300 atmospheres (730 MPa) at 25°C to 700 atmospheres (70 MPa) at 700°C, and from 1000 atmospheres (100 MPa) at 25°C to one atmosphere (100 kPa) at 480°C in compression. Granite, peridotite, pyroxenite and basalt become ductile between 300 and 500°C at five kilobars (500 MPa) confining pressure. They all exhibit a slight decrease in compressive strength, however, above 600°C basalt shows a sudden decrease.

The crushing strengths of the strongest rocks are in excess of 200 MPa but with high confining pressures they become effectively stronger and so more difficult to crush. This is particularly the case with calcareous rocks. At high pressures incipient fractures are closed and indeed the total flow of material without rupture may be indefinitely increased with increasing confining pressure.

Gowd and Rummel (1980) carried out a series of triaxial tests to examine the effect of high confining pressure on the behaviour of porous sandstone. They found that the transition from brittle to ductile deformation is characterised by an abrupt change from dilational behaviour at low pressures to compaction during inelastic axial strain at high pressures. This type of behaviour differs from that of rocks with low porosity. For instance, dilatancy persists well into the ductile field when Carrera Marble (porosity about 1%) is subjected to similar conditions (see Edmond and Paterson (1972)). Gowd and Rummel attributed the compaction which occurs during ductile deformation in porous sandstone at high confining pressure to the collapse of pore space and the rearrangement of quartz grains to give denser packing. At lower pressures, the dilation witnessed in porous sandstones was attributed to fracturing along grain boundaries as well as to fracturing of grains, and to the rearrangement of grains. During pre-peak dilation, fracturing is dominant over frictional sliding, which mainly controls post-peak deformation and leads to microscopic shear plane formation.

11.4.3 Pore solutions

Solutions increase the strain velocity and lower the fundamental strength of a substance. In experiments with alabaster, subjected to a load of 20 MPa, Griggs (1940) was able to demonstrate that a dry specimen soon reached its maximum strain of approximately 0.03%, whereas when a specimen had access to water the strain attained 1.75% in 36 days.

Colback and Wiid (1965) carried out a number of uniaxial and triaxial compression tests at eight different moisture contents, on quartzitic shale and quartzitic sandstone with porosities of 0.28 and 15% respectively. The moisture contents of the rock samples were controlled by keeping them in desiccators over saturated solutions of $CaCl_2$ at a constant temperature. The tests indicated that the compressive strengths of both rocks under saturated conditions were approximately half what they were under dry conditions

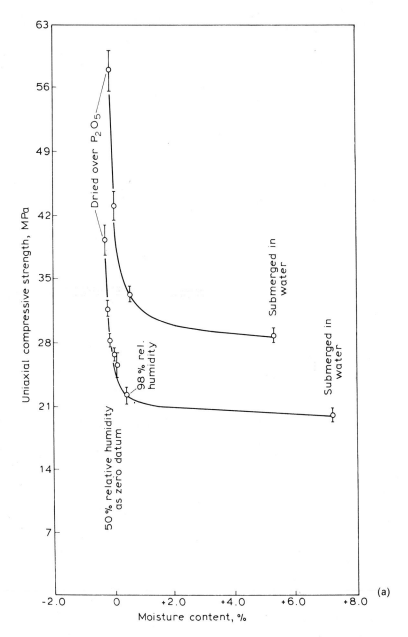

(Figure 11.7a). Similar tests carried out by Bernaix (1969) showed that the water content reduced the strength of the rocks concerned by 30–45%. From Figure 11.7b it will be noted that the slopes of the Mohr envelopes are not sensibly different, indicating that the coefficient of internal friction is not significantly affected by changes in moisture content. Colback and Wiid therefore tentatively concluded that the reduction in strength witnessed with increasing moisture content, was primarily due to a lowering of the tensile strength, which is a function of the molecular cohesive strength of the

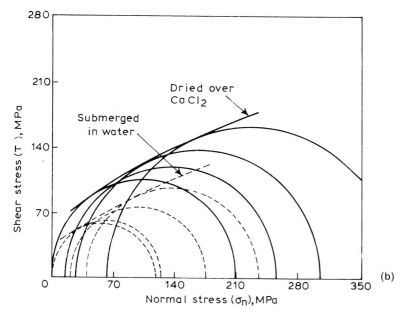

Figure 11.7 (a) Influence of pore water on the strength of quartzitic sandstone (after Colback and Wiid (1965)). (b) Mohr envelope for quartzitic shale at two moisture contents (after Colback and Wiid (1965)) (© Crown Copyright reserved. Queen's Printer, Canada, 1965)

material. Tests on specimens of quartzitic sandstone showed that their uniaxial compressive strength was inversely proportional to the surface tension of the different liquids into which they were placed. As the surface free energy of a solid submerged in a liquid is a function of the surface tension of the liquid, and since the uniaxial compressive strength is directly related to the uniaxial tensile strength, and this to the molecular cohesive strength, it was postulated that the influence of the immersion liquid was to reduce the surface free energy of the rock and hence its strength. The authors therefore concluded that the reduction in strength from the dry to the saturated condition of predominantly quartzitic rocks was a constant which was governed by the reduction of the surface free energy of quartz due to the presence of any given liquid.

Vutukuri (1974) investigated the effects of liquids on the tensile strength of limestone. He found that as the dialectic constant and surface tension of the liquid increased, the tensile strength of the limestone decreased.

It was observed by Heard (1960) that the Solenhofen Limestone changed from ductile to brittle as pore pressure was increased. He further noted that at temperatures below 480°C and confining pressures above 1000 atmospheres, the effects of interstitial fluids at any pressure upon this limestone was of no consequence when the magnitude of the pore pressure allowed the specimen to remain within the ductile field. Those samples which were deformed with access to fluids at pore pressures less than the confining pressure developed slickensides on the shear surfaces. When the ratio of pore pressure to confining pressure was decreased to the ductility transition point the slickensides gave place to friable mylonite.

11.4.4 Time-dependent behaviour

Most strong rocks, like granite, exhibit little time-dependent strain or creep, although creep in evaporitic rocks, notably salt, may greatly exceed the instantaneous elastic deformation. The time–strain pattern exhibited by a wide range of materials subjected to a constant uniaxial stress can, according to Price (1966), be represented diagrammatically as shown in Figure 11.8. The

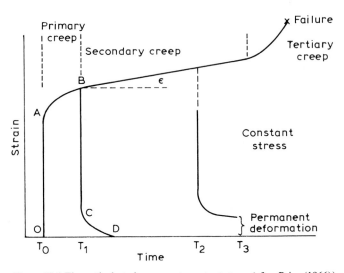

Figure 11.8 Theoretical strain curve at constant stress (after Price (1966))

instantaneous elastic strain, which takes place when a load is applied, is represented by OA. There follows a period of primary or transient creep (AB) in which the rate of deformation decreases with time. Primary creep is the elastic effect attributable to intragranular atomic and lattice displacements. If the stress is removed the specimen recovers. At first this is instantaneous (BC), but this is followed by a time elastic recovery, illustrated by curve CD. On the other hand, if the loading continues the sample begins to exhibit secondary or pseudo-viscous creep. This type of creep represents a phase of deformation in which the rate of strain is constant and is due principally to movements which occur on grain boundaries. The deformation is permanent and is proportional to the length of time over which the stress is applied. If the loading is further continued then the specimen suffers tertiary creep in which the strain rate accelerates with time and ultimately leads to failure. Creep deformation is limited at low temperatures and pressures but it may greatly exceed normal plastic flow when the pressures approach the limit of rupture. High temperatures also favour an increase in the rate and extent of creep.

From experiments carried out on Solenhofen Limestone under moderate hydrostatic pressure Robertson (1960) concluded that increased hydrostatic pressure (confining pressure being raised from 100 MPa to 200 MPa) caused a hundred-fold decrease in the primary creep rate per unit stress and that megascopic fracturing, chiefly intragranular, decreased rapidly as the confining pressure increased. This conclusion was based upon bulk density measurements and suggested that in limestone under a constant load, fracturing is an important process of primary creep. It also seems likely that confining pressure might decrease the size, number and propagation of fractures during creep and that it may facilitate their healing when complete unloading has taken place.

In experiments in which he applied stress to the Solenhofen Limestone Griggs (1936) introduced pauses in the rise of stress. He noted that during these pauses small non-elastic deformation occurred once the differential stress had reached a high enough threshold value (Figure 11.9). It was also

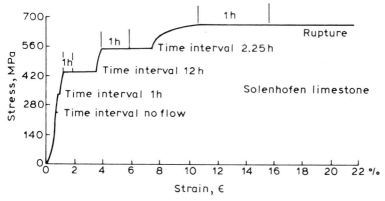

Figure 11.9 Stress–strain diagram of experiment on Solenhofen Limestone with pauses in increase of stress (after Griggs (1936))

observed that for the lowest stress application the strain did not increase with time as the threshold value was not attained. For each successive pause at a higher stress level the velocity of strain increased. It appeared that flowage in many experiments was masked by the rapid application of increased stress. Griggs also performed experiments which tested the change in ultimate strength with time. He found that time reduced the ultimate strength up to a certain point, beyond which there was no change. Moreover he showed that the amount of plastic deformation before rupture decreased with time.

Hobbs (1970) carried out a series of longitudinal strain–time measurements on cylindrical specimens of rock subjected to uniaxial compressive stresses ranging from 26.4–41.4 MPa, for periods ranging from a few minutes to more than a year. He found that after loading the creep rate at first decreased and then became approximately constant. Usually the measured longitudinal time strains were smaller than the instantaneous strains which occurred when the samples were loaded. An instantaneous increase in length occurred when the specimens were unloaded. This was followed by a time-dependent increase which ceased after 15 000 min. The instantaneous increase in length was less than the instantaneous decrease which took place on loading. Hobbs accordingly assumed that both instantaneous strain and primary creep under load were not completely recoverable and that the irrecoverable strain was possibly related to applied stress. During creep the volume of these specimens increased and their volume prior to rupture was larger than the initial unloaded volume.

11.5 Assessment of rock strength, hardness and elasticity

11.5.1 Compressive and shear strength

The uniaxial compressive strength of a rock is one of the simplest measures of strength to obtain. Although its application is limited, it does allow comparisons to be made between rocks and affords some indication of rock behaviour under more complex stress systems.

The behaviour of rock in uniaxial compression is influenced to some extent by the test conditions. The most important of these is the length-diameter or slenderness ratio of the specimen. Dhir and Sangha (1973) maintained that the most satisfactory slenderness ratio is 2.5. At lower ratios fractures take place in highly restrained specimen ends, while at higher ratios there is an undesirable release of elastic strain energy from the unfractured ends region to the fractured central zone during post-failure stressing. In other words such a ratio provides a reasonably good distribution of stress throughout. However, Obert and Duvall (1967) had previously suggested the use of the following empirical expression to relate the uniaxial strength to the length-diameter ratio

$$C_{\text{ult}} = \frac{C_{\text{act}}}{0.788 + 0.222D/L}$$

where C_{ult} is the compressive strength of a specimen of the same material having a $1:1$ length-diameter ratio and C_{act} is the compressive strength of a specimen for which $2 > (L/D) > 1/3$. Indeed Obert and Duvall reported that as far as the uniaxial compression of cylindrical specimens is concerned, the size of specimen has less effect than the natural variation in the values obtained from testing a given rock type when the specimen length–diameter ratio is kept constant.

The rate at which loading occurs is another test variable affecting the compressive strength (see Hawkes and Mellor (1970)). However, if a loading rate of between 7 and 70 kPa/s is used then this is unimportant (see Wuerker (1959)). The ends of core samples should be lapped so that they are exactly perpendicular to their long axes.

In a working party report to the Geological Society of London (Anon (1970)) the following scale of strength, based on uniaxial compressive tests, was recommended (see also ISRM and IAEG recommendations given in Chapter 12):

Strength (MPa)	Term
Less than 1.25	Very weak
1.25–5.00	Weak
5.00–12.50	Moderately weak
12.50–50	Moderately strong
50–100	Strong
100–200	Very strong
Over 200	Extremely strong

Where comparative assessment of rock strength has to be made it is some-times possible to dispense with the conventional uniaxial compressive test in favour of some other form of test. For example, a rock specimen with an irregular shape is used in the Protodyakonov test (see Protodyakonov (1963)). The weight and volume of each specimen is obtained prior to crushing and the mean crushing strength (P) of the specimen is related to its volume (V) as follows

$$\log P = \log f + 0.63 \log V$$

and

$$f = 0.19\,C$$

where C is the uniaxial compressive strength. It was suggested that 15–25 specimens should be tested to obtain the mean value.

Hobbs (1964) also carried out a series of tests on irregularly shaped specimens of rock. In this case the maximum height of each specimen was measured prior to testing. The contact area of the specimen ends on the loading platens was measured after the rock had failed. This was done by placing a piece of carbon paper and a piece of graph paper between the specimen and each of the loading platens. The carbon imprint on each of the graph papers is measured and so the average of the two areas is obtained. After an analysis of the results Hobbs found that the relationship between the compressive strength (C) and the average applied stress at fracture (P_{av}) was given by

$$C = 0.91\,P_{av} - 21.9\,\text{MPa}$$

Triaxial tests have to be carried out if the complete failure envelope of a rock material is required. As in triaxial testing of soils, a constant hydraulic pressure (the confining pressure) is applied to the cylindrical surface of the rock specimen, whilst applying an axial load to the ends of the sample. The axial load is increased up to the point where the specimen fails. A series of tests, each at a higher confining pressure, are carried out on specimens from the same rock. These enable Mohr circles and envelope to be drawn.

The procedure for determining the strength of rocks in triaxial com-pression has been outlined by Vogler and Kovari (1978) (see also ASTM (1967)). Testing of the rock sample is carried out within a specially con-structed high-pressure cell such as the Hoek cell (see Hoek and Franklin (1968); Figure 11.10). In routine triaxial testing of rock it is not usual to measure pore pressures. Indeed after a series of triaxial tests on rocks, Murrell (1963) showed that the pore pressure countered the effect of con-fining pressure. In other words, if the pore pressure is equal to the confining pressure, then fracture occurs at a constant value of the deviation stress, which is equal to the unconfined compressive strength.

A number of unconfined shear tests have been developed for rock testing (Figure 11.11). The shear strength in the single shear test is determined from the following expression

$$\tau = P/A$$

and from

$$\tau = P/2A$$

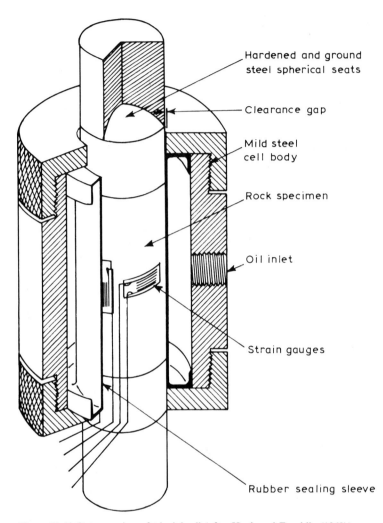

Hardened and ground
steel spherical seats

Clearance gap

Mild steel
cell body

Rock specimen

Oil inlet

Strain gauges

Rubber sealing sleeve

Figure 11.10 Cutaway view of triaxial cell (after Hoek and Franklin (1968))

for the double shear test, where A is the cross-sectional area of the specimen and P is the load required to bring about failure. In the punch test the specimen is placed within a cylindrical guide and a piston is forced through it (Figure 11.11c). The value of the shear strength is then calculated from

$$\tau = P/\pi DH$$

where D is the diameter of the punch and H is the thickness of the specimen. The torsional shear test is illustrated in Figure 11.11d, the shear strength this time being determined from the expression

$$\tau = 16M/\pi D^3$$

M being the applied torque at failure whilst D is the diameter of the sample.

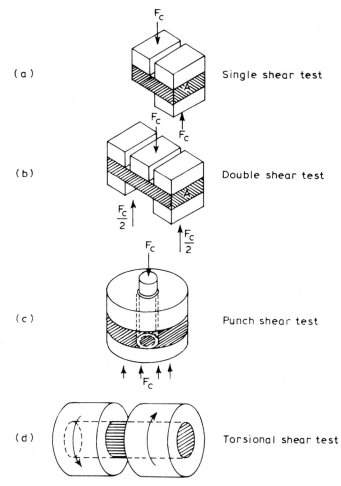

Figure 11.11 Unconfined shear strength tests

In all these shear tests the results are determined by the testing arrangement as well as by the rock material. As a consequence Everling (1964) suggested that the shear strength of a rock should only be derived by triaxial testing. He distinguished between the strength of a material in pure shear and the shear stress which is required to cause failure when the normal stress on the plane of fracture is zero. For most rocks there is a considerable difference between these two values.

11.5.2 Tensile strength

Rocks have a much lower tensile strength than compressive strength. Brittle failure theory predicts a ratio of compressive strength/tensile strength of about 8:1 but in practice it is generally between 15:1 and 25:1. The direct tensile strength of rocks has been obtained by attaching metal end caps with epoxy resins to specimens, which are then pulled into tension by wires. In direct tensile tests the slenderness ratio of cylindrical specimens should be 2.5–3.0 and the diameter preferably should not be less than NX core size (54 mm). The ratio of diameter of specimen to the largest grain in the rock should be at least 10:1 (see Bieniawski and Hawkes (1978)). Unfortunately the determination of the direct tensile strength has proved difficult since a satisfactory method has not been devised to grip the specimen without introducing bending stresses. Accordingly most tensile tests have been carried out by indirect methods.

In the flexural test a cylindrical specimen of rock is loaded between three points at a rate of 1.4 MPa/min until the sample fails. The flexural tensile strength (T_f) is then given by the expression

$$T_f = 8PL/\pi D^3$$

where P is the load at failure, L is the length between the supports and D is the diameter of the specimen. The flexural strength gives a higher value than that determined in direct tension.

In the Brazilian test a rock cylinder of length (L) and diameter (D) is loaded (with a load, P) in a diametrical plane along its axis. The sample usually fails by splitting along the line of diametrical loading and the tensile strength (T_b) can be obtained from

$$T_b = 2P/\pi LD$$

The use of the Brazilian test as an indirect method of assessing the tensile strength of rocks is based on the fact that most rocks in biaxial stress fields fail in tension when one principal stress is compressive. Failure, however, may be brought about by localised crushing along the axis of loading and not by diametral tension.

Disc-shaped specimens are also used in the Brazilian test. Curved jaw loading rigs are sometimes used when discs are tested, in an attempt to improve loading conditions. Uncertainties associated with the premature development of failure are sometimes removed by drilling a hole in the centre of a disc-shaped specimen. This has sometimes been referred to as the 'ring' test.

When a disc-shaped specimen is used the International Society for Rock Mechanics (ISRM) recommends that it is wrapped around its periphery with one layer of masking tape (see Bieniawski and Hawkes (1978)). In such cases the ISRM also recommends that the specimen should not be less than NX core size (54 mm in diameter) and that the thickness should be approximately equal to the radius of the specimen. A loading rate of 200 N/s was recommended. The tensile strength (T_b) of the specimen is obtained as follows

$$T_b = 0.636P/DH \text{ (MPa)}$$

where P is the load at failure (N), D is the diameter of the test specimen (mm) and H is the thickness of the test specimen measured at the centre (mm).

Mellor and Hawkes (1971) stated that the Brazilian test is useful for brittle materials but for other materials the test may give wholly erroneous results. Furthermore, Fairhurst (1964) concluded that the uniaxial tensile strengths of materials with low compression/tension ratios is underestimated by Brazilian tests in which radial loading is applied to disc-shaped specimens.

In the point load test the specimen is placed between opposing cone-shaped

Figure 11.12 Point load apparatus

platens and subjected to compression (Figure 11.12). This generates tensile stresses normal to the axis of loading. The tensile strength (T_p) is then derived by using the empirical expression (see Reichmuth (1963))

$$T_p = 0.96\,P/D^2 \text{ (or more simply } P/D^2)$$

where P is the load at failure and D is the diameter of the specimen or distance between the cones. The point load test has a number of variations such as the diametral test, the axial test and the irregular lump test. The latter is the least accurate (see Bieniawski (1975)). The diametral test is more convenient and simpler than the other two. Franklin and Broch (1972) suggested that the distance between the contact point and each end of the specimen should be at least $0.7D$, where D is the diameter of the core in the diametral test. In the axial test core specimens with a length to diameter ratio of 1.1 are suitable. They found that the point load strength tended to decrease when progressively larger specimens were tested and that saturated specimens underwent an average reduction in strength of approximately 33%. Rocks which are strongly anisotropic such as slates, schists, laminated sandstones, etc., should be tested both along and parallel to the lineation. Rock lumps with typical diameter of about 50 mm and with a ratio of longest to shortest diameter of between 1.0 and 1.4 are used. At least 20 lumps should be tested per sample.

The effect of size of specimens is greater in tensile than compression testing because in tension, cracks open and give rise to large strength reductions, whilst in compression the cracks close and so disturbances are appreciably reduced. In the axial and irregular lump tests the effects of size and shape

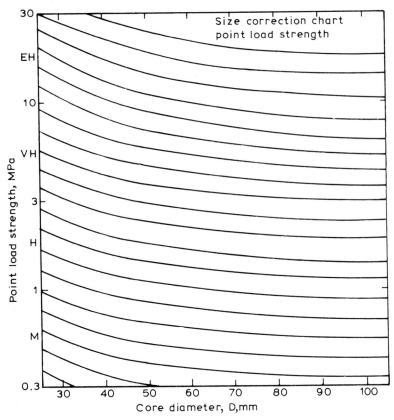

Figure 11.13 Size correction chart for point–load strength testing (after Broch and Franklin (1972))

are very pronounced. Accordingly Franklin and Broch (1972) recommended core with a diameter of 50 mm as a reference standard (T_{p50}) to which other sizes should be corrected by reference to a correction chart (Figure 11.13). Bieniawski (1975), however, considered that NX core (54 mm diameter) would be a better standard and that core specimens with diameter of less than BX size (42 mm) should not be used for point load testing. Brook (1977) discussed the influence of size on the results obtained from testing rocks with the point load apparatus. He suggested that both the shape effect and size effect of specimens can be accounted for by the following expression

$$T_p = \frac{\text{Load at fracture area 500 mm}^2}{500 \text{ mm}^2}$$

the load being determined either from a number of standard specimens of minimum cross-sectional area of 500 mm², 25 mm cores being the most convenient form, or from a log-log plot of load against area for a variety of shapes and sizes.

The standard deviation of the results of point load tests averages about 15% which Bieniawski (1975) maintained was acceptable for practical engineering purposes. The test is limited to rocks with uniaxial compressive strengths above 25 MPa (point load index above 1 MPa). In the case of rocks with apparently lower strengths, uniaxial compression testing is preferred.

Franklin and Broch (1972) suggested that the following point load strength scale could be used to classify rocks

	Point load strength index (MPa)	Equivalent uniaxial compressive strength (MPa)
Extremely high strength	Over 10	Over 160
Very high strength	3–10	50–160
High strength	1–3	15–60
Medium strength	0.3–1	5–16
Low strength	0.1–0.3	1.6–5
Very low strength	0.03–0.1	0.5–1.6
Extremely low strength	Less than 0.03	Less than 0.5

(Compare this classification of strength with that of the Geological Society of London, p. 508.)

They also suggested that uniaxial compressive strength (C) could be estimated from the point load index (T_{p50}) as follows

$$C = 24T_{p50}$$

11.5.3 Hardness

Hardness is one of the most investigated properties of materials, yet it is one of the most complex to understand. It does not lend itself to exact definition in terms of physical units. Indeed the numerical value of hardness is as much a function of the type of test used, as it is a material property. The concept of hardness has usually been associated with the surface of a material. Deere and Miller (1966) considered rock hardness as its resistance to the displacement of surface particles by tangential abrasive force, as well as its resistance to penetrating force, whether static or dynamic. They pointed out that rock hardness depended very much on the same factors as did toughness. The latter is controlled by the efficiency of the bond between the minerals or grains as well as the strength of these two components.

The hardness of a mineral is usually defined as its resistance to scratching and relative hardness has been used as a diagnostic property since the beginning of systematic mineralogy. As long ago as 1822 Mohs proposed a scale of hardness based upon ten minerals and this scale is still widely used today, although more sophisticated techniques are now available. Mohs' scale is as follows.

1. Talc
2. Gypsum
3. Calcite
4. Fluorspar
5. Apatite
6. Orthoclase
7. Quartz
8. Topaz
9. Corundum
10. Diamond

Each mineral in the scale is capable of scratching those of a lower order. The relative hardness of a given mineral can therefore be assessed by using a series of hardness pencils each of which are tipped by one of the minerals in Mohs' scale. The fingernail scratches minerals up to a hardness of about 2.5 and a penknife up to approximately 5.5. Attempts to assess the hardness of rock by summing the hardness values of its principal mineral constituents, according to their relative proportions, has not proved satisfactory.

A wide variety of penetrator and loading devices have been used for assessing the static indentation hardness. For example, the indenters used in the Brinell and Rockwell tests are spherical, and in the Knoop (see Winchell (1946)) and Vickers tests (see Das (1974)) they are pyramidal. The loading varies from 70 N in the Vickers test to 20 kN in that of Brinell. The Brinell and Rockwell hardness tests, however, are not generally applicable to rock because of its brittle nature. On the other hand, the Vickers test has been used to determine the microhardness of rock. A pyramidal-shaped diamond is applied to the surface of the material, and the surface area of the impression divided by the applied load provides a measure of

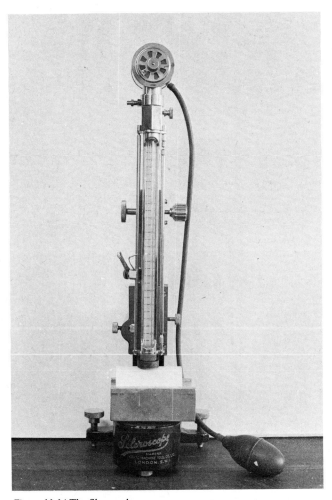

Figure 11.14 The Shore scleroscope

the hardness. Because a rock is not a homogeneous material, several hardness tests must be made over the surface of the specimen and the results averaged.

The Shore scleroscope is a non-destructive hardness-measuring device which indicates the relative values of hardness by the height of rebound of a small diamond pointed hammer which is dropped vertically onto a securely clamped test surface from a height of 250 mm (Figure 11.14). The ISRM recommends that the Shore scleroscope should be used to assess the hardness of rock surfaces ground smooth by using No. 1800 grade aluminium oxide abrasive powder. It also recommends that a specimen should have a minimum test surface of 10 cm^2 and a minimum thickness of 10 mm (see Atkinson *et al.* (1978)). At least 20 hardness determinations should be taken and averaged and each point of test should be at least 5 mm from any other.

Rabia and Brook (1979) examined the effects of specimen size on results and concluded that Shore hardness is dependent on the volume of the specimen tested and not simply on length or area. They suggested a minimum volume of 40 cm^3 for each test in order to obtain consistent values and that the mean of at least 50 readings on five specimens was required for a hardness value

Because they found a very good correlation between Shore hardness and uniaxial compressive strength Deere and Miller (1966) were able to devise the rock strength chart shown in Figure 11.15. They noted, however, that the chart appeared to be limited to rocks with strengths in excess of 35 MPa.

The Schmidt hammer was developed for measuring the strength of concrete and has since been adapted for assessing the hardness of rocks. It is a portable non-destructive device which expends a definite amount of stored energy from a spring and indicates the degree of rebound of a hammer mass within the instrument, following impact. Tests are made by placing the specimen in a rigid cradle and impacting the hammer at a series of points along its upper surface. The hammer is held vertically at right angles to the axis of the specimen (see Hucka (1965)). At least 20 readings should be taken from each sample and then averaged to give one value.

The ISRM recommends that specimens used for hardness testing with the Schmidt hammer should have a flat, smooth surface where tested and the rock material beneath this area should be free from cracks (see Atkinson *et al.* (1978)). Test locations should be separated by at least the diameter of the plunger. The ISRM suggests that a type L hammer with an impact energy of 0.74 Nm should be used. Atkinson *et al.* suggested that the lower 50% of the test values should be discarded and the average obtained from the upper 50%. This average is then multiplied by the correction factor to obtain the Schmidt hammer hardness

$$\text{Correction factor} = \frac{\text{Specified standard value of the anvil}}{\text{Average of 10 readings on the calibration anvil}}$$

The Schmidt hammer is not a satisfactory method for the determination of very soft or very hard rocks. However, Schmidt hardness shows a good correlation with compressive strength which allowed Deere and Miller (1966) to design another rock strength chart (Figure 11.16; see also Carter and Sneddon (1977)).

Abrasion tests measure the resistance of rocks to wear. Two abrasion tests

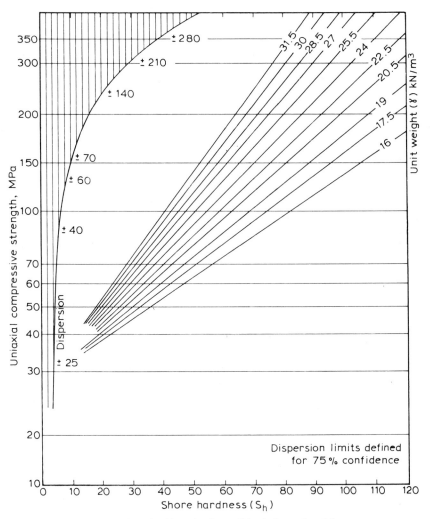

Figure 11.15 Correlation chart for Shore hardness (S_h) relating unit weight
of rock, compressive strength and hardness value (after Deere and Miller
(1966))

which have been used in the United States to measure hardness. They are
the Dorry and Los Angeles tests. The former test is carried out on a cylindrical
rock sample, 25 mm in both length and diameter, which is held against a
revolving disc and is under a pressure of 2.5 MPa (the total load equals 1.25
kg). Standard crushed quartz, sized between 30 and 40 ASTM mesh screens,
is fed onto the revolving disc (see ASTM, C-241 (1951)). The loss of weight,
obtained by subjecting both ends to a total of 1000 revolutions, gives the
hardness index (this is similar to the Dorry aggregate abrasion test (see
BS 812 (1967)). The Los Angeles abrasion test subjects a graded sample to
wear due to collision between rock pieces and also to impact forces produced
by an abrasive charge of steel spheres (see ASTM, C-131 (1969a); ASTM
C-535 (1969b)). In the Los Angeles test the rock aggregate and the abrasive

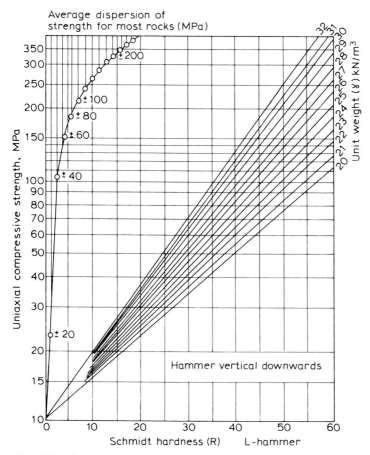

Figure 11.16 Correlation chart for Schmidt (L) hammer, relating unit weight of rock compressive strength and rebound number (after Deere and Miller (1966))

charge are placed in the machine and rotated at 30–33 rev/min. If the aggregate particles are smaller than 38 mm they are subjected to 500 revolutions and to 1000 revolutions if they are larger than 19 mm. After the test the sample is shaken through a No. 12 US sieve (approximately 1.7 mm aperture). The amount of wear is the loss in weight expressed as a percentage of the original weight.

Attrition has been defined as the resistance of one surface to the motion of another surface rubbing over it. The apparatus in the Deval test consists of a hollow cylindrical bucket, 340 mm long by 200 mm diameter, which is mounted in a frame which supports it at an angle of 30° to a horizontal axis of rotation (see ASTM, D-233 (1968)). The test sample comprises some 50 pieces of rock weighing about 5 kg. After revolving the cylinder 10 000 times at 30 rev/min the quantity of material finer than 0.06 mm is weighed and expressed as a percentage of the original weight, which gives a hardness value. This test is not often used today.

Toughness reflects the ability of a material to absorb energy during plastic deformation. In the impact toughness test a cylindrical sample of rock, 25 mm long by 25 mm diameter, is placed in the apparatus. A weight of 2 kg falls vertically between two guides, upon a spherical ended plunger weighing 1 kg, which rests in contact with the specimen. The height of the first blow is 10 mm and each successive blow is increased in height by that amount. The height of the blow which causes failure represents the toughness of the material. A minimum of six specimens of the same rock should be tested. If the rock material is laminated, cleaved or schistose, then three specimens should be prepared parallel to and three normal to these structural weaknesses. The average toughness in each of these directions is recorded.

11.5.4 Elastic properties

As noted in Section 11.1, Young's modulus and Poisson's ratio can be obtained by monitoring the vertical and lateral strain in rock specimens tested in uniaxial compression. Dhir and Sangha (1973) investigated the relationship between size, deformation and strength of cylindrical specimens loaded in uniaxial compression. They found that for a slenderness ratio of 2.5 the modulus determined from strain measurement over the central zone is 15% greater than that obtained from overall deflection measurements. A specimen diameter of 50 mm generally represents the transition in the size–strength relationship of a fine-grained material between the predominance of surface and internal flaws. Strain measurements on specimens less than this diameter are high and not representative of material behaviour.

According to Protodyakonov (1963) an approximate value of Young's modulus can be determined indirectly by using the Shore scleroscope hardness value (h) in the following equation

$$E = 1.07 \times \frac{h}{(154 - h)} \times 10^6 \text{ kg/cm}^2$$

A number of techniques have been used to determine the dynamic values of Young's modulus and Poisson's ratio, for example, Obert $et\ al.$ (1946) outlined a method of testing cylindrical specimens. These were supported at their centres and vibrated, to obtain the longitudinal frequency (f_1) and the torsional frequency (f_t). With these frequencies and the length (L) of the specimen, the longitudinal and torsional velocities of sound (v_1 and v_t respectively) can be obtained from

$$v_1 = 2f_1 L$$

$$v_t = 2f_t L$$

Young's modulus (E), Poisson's ratio (v) and the modulus of rigidity (G) can be obtained from the longitudinal and torsional velocities and the density (ρ) of the specimen

$$E = v_1{}^2 \rho$$

$$G = v_t{}^2 \rho$$

$$v = \frac{E}{2G} - 1$$

or

$$v = \frac{f_1^2}{2f_t^2} - 1$$

Hosking (1955) obtained the dynamic values of Young's modulus and Poisson's ratio by determining the velocities of propagation (v_p) in rock of an ultrasonic pulse and the sound at resonance (v_r) where L is the length of the specimen. With v_p and v_r it is possible to find Young's modulus and Poisson's ratio from the following expressions

$$E = \frac{v_r^2 \rho}{12g}$$

and

$$\frac{v_p}{v_r} = \frac{(1 - v)}{(1 + v)(1 - 2v)}$$

where ρ is the density and g is the acceleration due to gravity.

Deere and Miller (1966) used a similar method to derive the velocity of dilatational waves in rock samples while subjected to axial stress, the rock specimens being subjected to two complete loading and unloading cycles up to 34.5 MPa. They found that generally the denser rocks had higher dilatational velocities and that as axial pressures were increased the propagation velocities also increased, on average by some 10%. Deere and Miller also noticed that the velocities measured during unloading were usually higher than those measured during loading. The relationship between dilatational wave velocity (v_s), density, Young's modulus and Poisson's ratio is given by the following expression

$$v_s = \rho \frac{E(1 - v)}{(1 + v)(1 - 2v)}$$

This expression assumes that the static and dynamic properties are interchangeable. Indeed the static and dynamic values of Young's modulus obtained by these two authors were similar.

More recently the ISRM (see Rummel and van Heerden (1978)) have outlined laboratory methods of determining the velocity of propagation of elastic waves through rocks. Three methods were suggested, namely, the high frequency ultrasonic pulse technique, the low frequency ultrasonic pulse technique and the resonant method. The high frequency ultrasonic pulse method is used to determine the velocities of compressional and shear waves in rock specimens of effectively infinite extent compared to the wavelength of the pulse used. The condition of infinite extent is satisfied if the average grain size is less than wavelength of the pulse, which in turn is less than the minimum dimensions of the specimen. The low frequency ultrasonic pulse method is used to determine the velocity of dilatational and torsional waves in cylindrical or bar-shaped specimens of rock. The length to diameter ratio of specimens should be greater than 3 and the ratio of the wavelength of the pulse to the diameter should not be less than 5. By determination of the resonance frequency of both dilatational and torsional vibrations of cylindrical rock specimens (with a length to diameter ratio exceeding 3 and a wavelength to diameter ratio exceeding 6) the velocity of dilatational and torsional waves can be calculated.

11.6 Rock composition and texture in relation to physical properties

The micro-petrographic description of rocks for engineering purposes includes the determination of all parameters which cannot be obtained from a macroscopic examination of a rock sample, such as mineral content, grain size and texture, and which have a bearing on the mechanical behaviour of the rock or rock mass (see Hallbauer *et al.* (1978)). In particular a microscopic examination should include a modal analysis, determination of micro-fractures and secondary alteration, determination of grain size and, where necessary, fabric analysis. The ISRM recommends that the report of a petrographic examination should be confined to a short statement on the origin, classification and details relevant to the mechanical properties of the rock concerned. Wherever possible this should be combined with a report on the mechanical parameters (Figure 11.17).

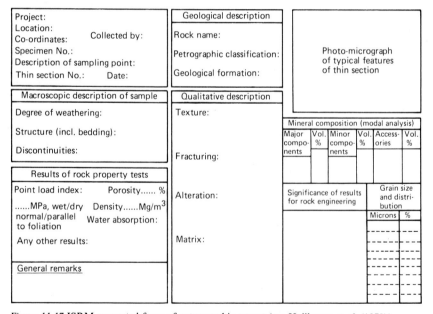

Figure 11.17 ISRM suggested form of petrographic report (see Hallbauer *et al.* (1978))

Mendes *et al.* (1966) proposed that quantitative micro-petrographic data could be used to formulate rock quality indices which were closely correlated with mechanical characteristics. A modal analysis was made of the mineralogical composition of the rock samples concerned, together with an analysis of their texture and micro-structure. As a result, the sample could be classified and its type and degree of alteration and the extent of micro-fracturing estimated. The percentage of sound minerals was determined. However, those sound minerals which had an adverse affect upon mechanical behaviour were grouped with the percentage of adverse minerals. Open micro-fissures were distinguished from those which were filled. In the latter type the nature of the cement is important, for example, silica provides a strong bond whilst other materials such as talc may lubricate movement along a micro-fissure. The quality index (K) was defined as

$$K = \frac{\sum_{i=1}^{n} p_i X_i}{\sum_{j=1}^{m} p_j Y_j}$$

in which n values of X_i are the percentages of sound minerals or minerals having a favourable influence upon mechanical behaviour, and the m values of Y_j are the percentages of altered minerals or sound minerals which have an adverse affect upon mechanical performance together with percentages of micro-fissures and voids. The coefficients p_i and p_j are weights which measure the influence on the mechanical characteristics of the rock sample of one or other mineral or peculiarity. The quality indices of granite and gneiss were correlated with their elasticity modulus (E) values. A good correlation was obtained for the granite material but that of gneiss was not so good.

A number of petrofabric techniques were developed by Willard and McWilliams (1969) in an attempt to gain a better understanding of the mechanical behaviour of rocks in relation to their micro-structure. They first subjected a number of rock samples to non-destructive (pulse velocity) tests and then examined their fabric. Next they took the same rock samples and this time subjected them to destructive (indirect tensile) tests and subsequently observed their fracture characteristics under the microscope. The five techniques they used were diametric mineralogical analysis, defect frequency orientation analysis, grain elongation analysis, macro-grid analysis and trans-granular–intergranular analysis. The first three of these techniques record micro-structural features of the rock which are not affected by changes brought about by destructive testing. As a consequence they help to explore the features which play a part in mechanical behaviour and that are not disturbed by cracking due to testing. The last two techniques record micro-structural features in the rock fabric which are brought about as a result of destructive testing. All five techniques have to be performed with reference to a three-dimensional coordinate system. With such a framework micro-structural features can be related to mechanical properties which vary in different directions.

Diametric mineralogical analysis simply consists of a modal analysis of the mineral components along given diameters of a circular thin section of rock, these sections being cut from the rock discs which are tested. In a test carried out on the Salisbury Granite Willard and McWilliams (1969) showed that the pulse velocity varied with direction and increased with increasing feldspar content. The object of defect frequency orientation (DFO) analysis is to describe and evaluate the frequency of defect occurrence. As far as Willard and McWilliams were concerned defects were either open or closed cracks or sites at which cracks would develop when the rock was subjected to critical tensile or shear stress. Therefore in addition to micro-fractures, defects included grain boundaries, mineral cleavages, twinning planes, inclusion trains and the elongation of shell fragments. Rocks are not uniformly coherent, homogeneous materials and defects occur as visible or microscopic linear or planar discontinuities associated with certain minerals. As is to be expected defects influence the ultimate strength of a rock and may act as surfaces of weakness which control the direction in which failure occurs. Between 500 and 1000 defect orientations were noted in each thin section of rock and each one was regarded as a vector of unit length with a range of 180°. From their investigations on the Barre Granite Willard and McWilliams noted that the frequency of defects tended to be inversely proportional to the breaking strength suggesting that the direction of weakest tensile strength was approximately normal to the direction of most defects. In rocks with a preferred orientation, grain elongation can be used as a method of correlating micro-structure with their mechanical properties. Higher failure strengths were found to be associated with line-loading at right angles to surfaces of lineation.

Macro-grid analysis is used to calculate the areal mineral percentages over the fracture surfaces. This is then compared with the volume percentages of minerals in the rock, which are obtained by modal analysis. The comparison reveals any tendency for a fracture surface to include unusually large or small amounts of particular minerals. When a fracture is produced in a granular rock it must either pass through grains or follow grain boundaries. In the first case it is described as transgranular (T) and in the second as intergranular (I). The relative lengths of the transgranular and intergranular parts along a fracture are summed and the failure expressed as a ratio T/I which can then be compared with the failure strength of the rock. When the T/I ratios were recorded from samples of Tennessee Marble it was found that they varied inversely with failure strength.

Subsequent work on the relationship between cracks in rocks and their elastic properties, by Simmons *et al.* (1975), emphasised the necessity to obtain quantitative petrographic data on crack dimensions, numbers of cracks per unit area or volume, and the distribution and orientation of cracks.

Onodera and Kumara (1980) found a linear relationship between Young's modulus and the grain boundary surface area per unit volume in granite. They also found a linear relationship between strength and grain size, that is, as the grain size of the granite decreased, the strength increased.

References

ANON. (1970). 'Working party report on the logging of cores for engineering purposes', *Q. J. Engg Geol.*, **3**, 1–24.

ASTM (1951). *Abrasion of Rock by use of the Dorry Machine*, C-241.

ASTM (1967). *Standard Method of Test for Triaxial Compressive Strength of Undrained Rock Core Specimens without Pore Pressure Measurement*, ASTM Designation D, 2664-67.

ASTM (1968). *Abrasion of Rock by use of the Deval Machine*, D-233.

ASTM (1969a). *Resistance to Abrasion of Small Size Coarse Aggregate by use of Los Angeles Machine*, C-131.

ASTM (1969b). *Resistance to Abrasion of Large Size Coarse Aggregate by use of Los Angeles Machine*, C-535.

ATKINSON, R. H., BAMFORD, W. H., BROCH, E., DEERE, D. U., FRANKLIN, J. A., NIEBLE, C., RUMMEL, F., TARKOY, P. S. & VAN DUYSE, H. (1978). 'Suggested methods for determining hardness and abrasiveness of rocks', ISRM Commission on Standardization of Laboratory and Field Tests. *Int. J. Rock Mech. Min. Sci. & Geomech. Abstr.*, **15**, 91–7.

BARTON, N. (1976). 'The shear strength of rock and rock joints', *Int. J. Roch Mech. Min. Sci. & Geomech. Abstr.*, **13**, 255–79.

BERNAIX, J. (1969). 'New laboratory methods of studying the mechanical properties of rocks', *Int. J. Roch Mech. Min. Sci.*, **6**, 43–90.

BIENIAWSKI, Z. T. (1975). 'The point load test in geotechnical practice', *Engg Geol.*, **9**, 1–11.

BIENIAWSKI, Z. T. & HAWKES, I. (1978). 'Suggested methods for determining tensile strength of rock materials', ISRM Commission on Standardization of Laboratory and Field Tests, *Int. J. Rock Mech. Min. Sci. & Geomech. Abstr.*, **15**, 101–103.

BRACE, W. F. (1964). 'Brittle fracture of rocks' (in *Symp. State of Stress in the Earth's Crust*), Judd, W. R. (ed.), Santa Monica, Elsevier, 111–80.

BRACE, W. F. (1965). 'Some new measurements of linear compressibility of rocks', *J. Geophys. Res.*, **70**, 391–8.

BRITISH STANDARDS INSTITUTION (1967). *Methods for Sampling and Testing of Mineral Aggregates, Sands and Fillers*, BS812, Br. Stand. Inst. London.

BROOK, W. F. (1977). 'A method of overcoming both shape and size effects in point load testing', *Proc. Conf. Rock Engg*, Newcastle University, **1**, 53–70.

BROWN, E. T., RICHARDS, L. R. & BARR, M. V. (1977). 'Shear strength characteristics of the Delabole Slates', *Proc. Conf. Rock Engg*, Newcastle University, **1**, 33–51.

CARTER, P. G. & SNEDDON, M. (1977). 'Comparison of Schmidt hammer, point load and unconfined compression tests in Carboniferous strata', *Proc. Conf. Rock Engg*, Newcastle University, **1**, 197–210.

COLBACK, P. S. B. & WIID, B. L. (1965). 'Influence of moisture content on the compressive strength of rock', *Symp. Canadian Dept. Min. Tech. Survey*, Ottawa, 65–83.

COULOMB, G. A. (1773). 'Sur une application des regles de maximus et minimus à quelques problèmes de statique relatifs à l'architecture', *Acad. Roy. des Sci., Mem de Math. et de Phys. par divers Sovans*, **7**, 343–82.

DAS, B. (1974). 'Vickers hardness concept in the light of Vickers impression', *Int. J. Rock Mech. Min. Sci. & Geomech. Abstr.*, **11**, 85–9.

DEERE, D. U. & MILLER, R. P. (1966). 'Engineering classification and index properties for intact rock', *Tech. Rep. No. AFWL-TR-65-115*, Air Force Weapons Lab., Kirtland Air Base, New Mexico.

DEY, T. N. & WANG, CHI-YUEN. (1981). 'Some mechanisms of microcrack growth and interaction in compressive rock failure', *Int. Rock Mech. Min. Sci. & Geomech. Abstr.*, **18**, 199–209.

DHIR, R. K. & SANGHA, C. M. (1973). 'Relationships between size, deformation and strength for cylindrical specimens loaded in uniaxial compression', *Int. J. Rock Mech. Min. Sci. & Geomech. Abstr.*, **10**, 699–712.

DONATH, F. A. (1961). 'Experimental study of shear failure in anisotropic rocks', *Bull. Geol. Soc. Am.*, **72**, 985–91.

EDMOND, J. M. & PATERSON, M. S. (1972). 'Volume changes during deformation of rocks at high pressure', *Int. J. Rock Mech. Min. Sci.*, **9**, 161–82.

EVERLING, G. (1964). 'Comments on the definition of shear strength', *Int. J. Rock Mech. Min. Sci.*, **1**, 145–54.

FAIRHURST, C. (1964). 'On the validity of the Brazilian test for brittle materials', *Int. J. Rock Mech. Min. Sci.*, **1**, 535–46.

FARMER, I. W. (1968). *Engineering Properties of Rocks*, Spon, London.

FRANKLIN, J. A. & BROCH, E. (1972). 'The point load strength test', *Int. J. Rock Mech. Min. Sci.*, **9**, 669–97.

GOWD, T. N. & RUMMEL, F. (1980). 'Effect of confining pressure on the fracture behaviour of a porous rock', *Int. J. Rock Mech. Min. Sci. & Geomech. Abstr.*, **17**, 225–9.

GRIFFITH, A. A. (1920). 'The theory of rupture', *Proc. 1st Conf. Appl. Mech.*, Delft, 55–70.

GRIGGS, D. T. (1936). 'Deformation of rocks under high confining pressures', *J. Geol.*, **44**, 541–77.

GRIGGS, D. T. (1940). 'Experimental flow of rocks under conditions favouring recrystallization', *Bull. Geol. Soc. Am.*, **51**, 1001–22.

GRIGGS, D. T. (1951). 'Deformation of Yule Marble', *Bull. Geol. Soc. Am.*, **62**, 853–62.

GRIGGS, D. T. & HANDIN, F. (1960). 'Observations on fracture' (in *Rock Deformation*), *Geol. Soc. Am. Mem.*, **79**, 347–65.

HALLBAUER, D. K., NIEBLE, C., BERARD, J., RUMMEL, F., HOUGHTON, A. BROCH, E. & SZLAVIN, J. (1978). 'Suggested methods for petrographic description', ISRM Commission on Standardization of Laboratory and Field Tests. *Int. J. Rock Mech. Min. Sci. & Geomech. Abstr.*, **15**, 41–5.

HAWKES, I. & MELLOR, M. (1970). 'Uniaxial testing in rock mechanics laboratories', *Engg Geol.*, **4**, 177–284.

HAWKES, I., MELLOR, M. & GARIEPY, S. (1973). 'Deformation of rocks under uniaxial tension', *Int. J. Rock Mech. Min. Sci. & Geomech. Abstr.*, **10**, 493–507.

HEARD, H. C. (1960). 'Transition from brittle to ductile flow in Solenhofen Limestone as a function of temperature, confining pressure and interstitial fluid pressure' (in *Rock Deformation*), *Geol. Soc. Am., Mem.*, **79**, 193–212.

HOBBS, D. W. (1964). 'A simple method for assessing the uniaxial compressive strength of rocks', *Int. J. Rock Mech. Min. Sci.*, **1**, 5–15.

HOBBS, D. W. (1970). 'Stress–strain–time behaviour in a number of Coal Measures rocks', *Int. J. Rock Mech. Min. Sci.*, **7**, 149–70.

HOEK, E. (1966). 'Rock mechanics—an introduction for the practical engineer', *Min. Mag.*, **114**, 236–55.

HOEK, E. (1968). 'Brittle fracture of rock' (in *Rock Mechanics in Engineering Practice*), Stagg, K. G. & Zienkiewicz, O. C. (eds), Wiley, London, 99–124.

HOEK, E. & FRANKLIN, J. A. (1968). 'Simple triaxial cell for field or laboratory testing of rock', *Trans. Inst. Min. Metall.*, **77**, Section A, A22–A26.

HOSKING, J. R. (1955). 'A comparison of tensile strength, crushing strength and elastic properties of roadmaking rocks', *Quarry Man. J.*, **39**, 200–212.

HUCKA, V. A. (1965). 'A rapid method for determining the strength of rock *in situ*', *Int. J. Rock Mech. Min. Sci.*, **2**, 127–34.

McCLINTOCK, F. A. & WALSH, J. B. (1962). 'Friction on Griffith cracks in rocks under pressure', *Proc. 4th Conf. Appl. Mech.*, 1015–21.

MELLOR, M. & HAWKES, I. (1971). 'Measurement of tensile strength by diametral compression of discs and annuli', *Engg Geol.*, **5**, 173–25.

MENDES, F. M., AIRES-BARROS, L. & RODRIGUES, F. P. (1966). 'The use of modal analysis in the mechanical characterization of rock masses', *Proc. 1st Int. Cong. Rock Mech.*, Lisbon, **1**, 217–23.

MOHR, O. (1882). *Abhandlungen aus dem Gebiete der Technische Meckanik*, Ernst und Sohn, Berlin.

MORGENSTERN, N. R. & PHUKAN, A. L. T. (1966). 'Non-linear deformation of a sandstone', *Proc. 1st Cong. Int. Soc. Rock Mech.*, Lisbon, **1**, 543–8.

MURRELL, S. A. F. (1963). 'A criterion for the brittle fracture of rocks and concrete under triaxial stress and the effect of pore pressure on the criterion', *Proc. 5th Symp. Rock Mech.*, University of Minnesota, Pergamon Press, New York, 563–77.

OBERT, L. & DUVALL, W. I. (1967). *Rock Mechanics and the Design of Structures*, Wiley, New York.

OBERT, L., WINDES, S. L. & DUVALL, W. I. (1946). 'Standardized tests for determining the physical properties of mine rock', *US Bur. Mines Rep. Invest.*, 3891.

ONODERA, T. F. & KUMARA, A. H. M. (1980). 'Relation between texture and mechanical properties of crystalline rocks', *Bull. Int. Soc. Engg Geol.*, No. 22, 173–7.

PRICE, N. J. (1966). *Fault and Joint Development in Brittle and Semi-Brittle Rock*, Pergamon Press, Oxford.

PROTODYAKONOV, M. M. (1963). 'Mechanical properties and drillability of rock', *Proc. 5th Symp. Rock Mech.*, University of Minnesota, Pergamon Press, New York, 103–118.

RABIA, H. & BROOK, N. (1979). 'The Shore hardness of rock', *Int. J. Rock Mech. Min. Sci. & Geomech. Abstr.*, **16**, 335–6.

REICHMUTH, D. R. (1963). 'Correlations of force-displacement data with physical properties of rock for percussive drilling', *Proc. 5th Symp. Rock Mech.*, University of Minnesota, Pergamon Press, New York.

ROBERTSON, E. C. (1955). 'Experimental study of the strength of rocks', *Bull. Geol. Soc. Am.*, **66**, 1275–1308.

ROBERTSON, E. C. (1960). 'Creep of Solenhofen Limestone under moderate hydrostatic pressure' (in *Rock Deformation*), *Geol. Soc. Am. Mem.*, **79**, 227–44.

RUMMEL, F. & VAN HEERDEN, W. L. (1978). 'Suggested methods for determining sound velocity', ISRM Commission on Standardization of Laboratory and Field Tests, *Int. J. Rock Mech. Min. Sci. & Geomech. Abstr.*, **15**, 55–8.

SIMMONS, G., TODD, T. & BALDRIDGE, W. S. (1975). 'Toward a quantitative relationship between elastic properties and cracks in low porosity rock', *Am. J. Sci.*, **275**, 318–45.

VOGLER, U. V. & KOVARI, K. (1978). 'Suggested methods for determining the strength of rock materials in triaxial compression', ISRM Commission on Standardization of Laboratory and Field Tests. *Int. J. Rock Mech. Min. Sci. & Geomech. Abstr.*, **15**, 47–51.

VUTUKURI, V. S. (1974). 'The effects of liquids on the tensile strength of limestone', *Int. J. Rock Mech. Min. Sci. & Geomech. Abstr.*, **11**, 27–9.

WALSH, J. B. (1965a). 'The effects of cracks on the uniaxial elastic compression of rocks', *J. Geophys. Res.*, **70**, 399–411.

WALSH, J. B. (1965b). 'The effect of cracks on the compressibility of rocks', *J. Geophys. Res.*, **70**, 381–9.

WALSH, J. B. & BRACE, W. F. (1972). 'Elasticity of rock in uniaxial strain', *Int. J. Rock Mech. Min. Sci.*, **9**, 7–15.

WAWERSIK, W. R. & FAIRHURST, C. (1970). 'A study of brittle rock fracture in laboratory compression experiments', *Int. J. Rock Mech. Min. Sci.*, **7**, 561–75.

WILLARD, R. J. & McWILLIAMS, J. R. (1969). 'Microstructural techniques in the study of physical properties of rocks', *Int. J. Rock Mech. Min. Sci.*, **6**, 1–12.

WINCHELL, H. (1946). 'Observations on orientation and hardness variations', *Am. Mineral.*, **31**, 149–52.

WUERKER, R. G. (1959). 'Influence of stress rate on the strength and elasticity of rocks', *Q. Colorado School Mines*, **54**, No. 3.

Chapter 12

Geotechnical description and classification of rocks and rock masses

12.1 Description of rocks and rock masses

Description is the initial step in an engineering assessment of rocks and rock masses. It should therefore be both uniform and consistent in order to gain acceptance.

The complete specification of a rock mass requires descriptive information on the nature and distribution in space of both the materials that constitute the mass (rock, soil, water and air-filled voids) and the discontinuities which divide it (see Anon (1977)). The intact rock may be considered as a continuum or polycrystalline solid consisting of an aggregate of minerals or grains whereas a rock mass may be looked upon as a discontinuum of rock material transected by discontinuities. The properties of the intact rock are governed by the physical properties of the materials of which it is composed and the manner in which they are bonded to each other. The parameters which may be used in a description of intact rock therefore include petrological name, mineral composition, colour, texture, minor lithological characteristics, degree of weathering or alteration, density, porosity, strength, hardness, intrinsic or primary permeability, seismic velocity and modulus of elasticity. Swelling and slake durability can be taken into account where appropriate, such as in the case of argillaceous rocks.

The behaviour of a rock mass is, to a large extent, determined by the type, spacing, orientation and characteristics of the discontinuities present (see Chapter 5). As a consequence, the parameters which ought to be used in a description of a rock mass include the nature and geometry of the discontinuities as well as its overall strength, deformation modulus, secondary permeability and seismic velocity. It is not necessary, however, to describe all the parameters for either a rock mass or intact rock.

The data collected should be recorded on data sheets for subsequent automatic processing. A data sheet for the description of rock masses and another for discontinuity surveys have been recommended by the Geological Society (see Anon (1977); Figures 12.1, 5.11). Because geological data often have strong spatial inter-relationships they are usually presented in cartographic, graphical or tabulated form which facilitates assessment.

528

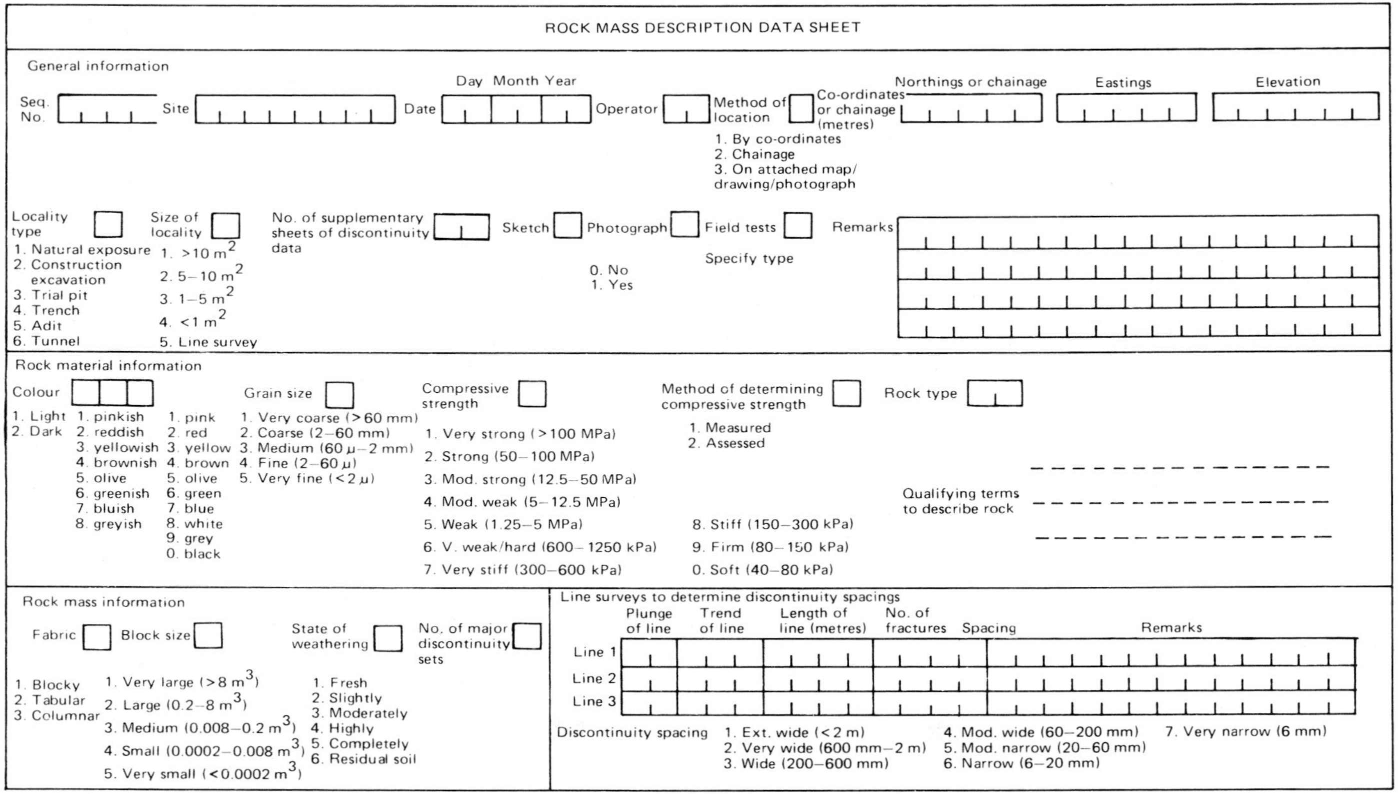

ROCK MASS DESCRIPTION DATA SHEET

General information

Seq. No. [ ][ ][ ][ ]   Date [ ][ ][ ] Day Month Year   Operator [ ]   Method of location [ ] Co-ordinates or chainage (metres)

1. By co-ordinates
2. Chainage
3. On attached map/drawing/photograph

Site [ ][ ][ ][ ][ ][ ][ ][ ][ ]   Northings or chainage   Eastings   Elevation

Sketch [ ]   Photograph [ ]   Field tests [ ]   Remarks

0. No
1. Yes

Specify type

Locality type [ ]
1. Natural exposure
2. Construction excavation
3. Trial pit
4. Trench
5. Adit
6. Tunnel

Size of locality [ ]
1. >10 m²
2. 5–10 m²
3. 1–5 m²
4. <1 m²
5. Line survey

No. of supplementary sheets of discontinuity data [ ]

Rock material information

Colour [ ][ ][ ]
1. Light
2. Dark

1. pink
2. red
3. yellowish
4. brown
5. olive
6. greenish
7. bluish
8. greyish

1. pink
2. red
3. yellow
4. brown
5. olive
6. green
7. blue
8. white
9. grey
0. black

Grain size [ ]
1. Very coarse (>60 mm)
2. Coarse (2–60 mm)
3. Medium (60 μ–2 mm)
4. Fine (2–60 μ)
5. Very fine (<2 μ)

Compressive strength [ ]
1. Very strong (>100 MPa)
2. Strong (50–100 MPa)
3. Mod. strong (12.5–50 MPa)
4. Mod. weak (5–12.5 MPa)
5. Weak (1.25–5 MPa)
6. V. weak/hard (600–1250 kPa)
7. Very stiff (300–600 kPa)
8. Stiff (150–300 kPa)
9. Firm (80–150 kPa)
0. Soft (40–80 kPa)

Method of determining compressive strength [ ]
1. Measured
2. Assessed

Rock type [ ]

Qualifying terms to describe rock

Rock mass information

Fabric [ ]
1. Blocky
2. Tabular
3. Columnar

Block size [ ]
1. Very large (>8 m³)
2. Large (0.2–8 m³)
3. Medium (0.008–0.2 m³)
4. Small (0.0002–0.008 m³)
5. Very small (<0.0002 m³)

State of weathering [ ]
1. Fresh
2. Slightly
3. Moderately
4. Highly
5. Completely
6. Residual soil

No. of major discontinuity sets [ ]

Line surveys to determine discontinuity spacings

	Plunge of line	Trend of line	Length of line (metres)	No. of fractures	Spacing	Remarks
Line 1						
Line 2						
Line 3						

Discontinuity spacing
1. Ext. wide (<2 m)
2. Very wide (600 mm–2 m)
3. Wide (200–600 mm)
4. Mod. wide (60–200 mm)
5. Mod. narrow (20–60 mm)
6. Narrow (6–20 mm)
7. Very narrow (6 mm)

Figure 12.1 **Rock mass data description sheet (after Anon (1977))**

12.2 Properties of rocks and rock masses

12.2.1 Geological properties

Intact rock may be described from a geological or engineering point of view. In the first case the origin and mineral content of a rock are of prime importance, as is its texture and any change which has occurred since its formation. In this respect the name of a rock provides an indication of its origin, mineralogical composition and texture (Figure 12.2). Only a basic petrographical description of the rock is required when describing a rock mass. A useful system of petrographical description has been provided by Dearman (1974) and was further developed by the IAEG (see Anon (1979a)). The report of the ISRM (International Association of Engineering Geology International Society for Rock Mechanics) (Anon (1978)) on petrographic description of rocks and rock masses has been referred to in Chapter 11. Obviously the engineering properties, mentioned in the previous section are included in the latter type of description.

The overall colour of a rock should be assessed by reference to a colour chart (e.g. the rock colour chart of the Geological Society of America). This is because it is difficult to make a quantitative assessment with the eye alone. The colour may be expressed in terms of hue (which is a basic colour or a mixture of basic colours), of chroma (which is the brilliance or intensity) and of value or lightness. A simple subjective scheme has been suggested by the Geological Society (see Anon (1977)) which involves the choice of a colour from column 3 below, supplemented if necessary by a term from column 2 and/or column 1

1	2	3
light	pinkish	pink
dark	reddish	red
	yellowish	yellow
	brownish	brown
	olive	olive
	greenish	green
	bluish	blue
		white
	greyish	grey
		black

The texture of a rock refers to its component grains and their mutual arrangement or fabric. It is dependent upon the relative sizes and shapes of the grains and their positions with respect to one another and the groundmass or matrix, when present. Grain size, in particular, is one of the most important aspects of texture, in that it exerts an influence on the physical properties of a rock. It is now generally accepted that the same descriptive terms for grain size ranges should be applicable to all rock types and should be the same as those used to describe soils (Table 12.1, page 533).

Table 12.1. Description of grain size

Term	Particle size	Equivalent soil grade
Very coarse grained	Over 60 mm	Boulders and cobbles
Coarse grained	2–60 mm	Gravel
Medium grained	0.06–2 mm	Sand
Fine grained	0.002–0.06 mm	Silt
Very fine grained	Less than 0.002 mm	Clay

GENETIC/GROUP	DETRITAL SEDIMENTARY		PYROCLASTIC	CHEMICAL/ORGANIC
Usual structure	BEDDED			
Composition	Grains of rock, quartz, feldspar and clay minerals	At least 50% of grains are of carbonate	At least 50% of grains are of fine-grained igneous rock	
GRAIN SIZE (mm)				
Very coarse-grained — BOULDERS COBBLES (RUDACEOUS)	Grains are of rock fragments. Rounded grains: CONGLOMERATE. Angular grains: BRECCIA	CARBONATE GRAVEL — CALCIRUDITE	Rounded grains AGGLOMERATE. Angular grains VOLCANIC BRECCIA LAPILLI TUFF	SALINE ROCKS Halite Anhydrite; Gypsum
60 — Coarse-grained — GRAVEL (RUDACEOUS)				
2 — Medium-grained — SAND (ARENACEOUS)	Grains are mainly mineral fragments. SANDSTONE: Grains are mainly mineral fragments. QUARTZ SANDSTONE: 95% quartz, voids empty or cemented. ARKOSE: 75% quartz, up to 25% feldspar: voids empty or cemented. GREYWACKE: 75% quartz, 15% fine detrital material: rock and feldspar fragments	CARBONATE SAND — CALCARENITE	TUFF	LIMESTONE and DOLOMITE (undifferentiated) / VOLCANIC ASH / LIMESTONE DOLOMITE CHERT FLINT
0.06 — Fine-grained — SILT (ARGILLACEOUS or LUTACEOUS)	SILTSTONE: 50% fine-grained particles	CARBONATE SILT — CALCISILTITE CHALK	Fine-grained TUFF	
0.002 — Very fine-grained — CLAY (ARGILLACEOUS or LUTACEOUS)	CLAYSTONE: 50% very fine grained particles	CARBONATE MUD — CALCILUTITE	Very fine-grained TUFF	PEAT LIGNITE COAL
GLASSY AMORPHOUS				

(Mudstone column: MUDSTONE, SHALE: fissile mudstone, MARLSTONE)

Figure 12.2 Rock type classification (after Anon (1979))

Figure 12.2—cont.

METAMORPHIC

FOLIATED — Quartz, feldspars, micas, acicular dark minerals

- GNEISS (ortho-, para-, Alternate layers of granular and flaky minerals)
- MARBLE
- GRANULITE
- MIGMATITE
- SCHIST
- QUARTZITE HORNFELS AMPHIBOLITE
- PHYLLITE
- SLATE MYLONITE

IGNEOUS — MASSIVE

Usual structure	Acid rocks (Light coloured minerals are quartz, feldspar, mica)	Intermediate	Basic rocks (Dark and light minerals)	Ultrabasic (Dark minerals)	GENETIC GROUP Composition	GRAIN SIZE (mm)
	PEGMATITE				Very coarse-grained	60
	GRANITE	DIORITE	GABBRO	PYROXENITE and PERIDOTITE	Coarse-grained	2
	MICROGRANITE	MICRODIORITE	DOLERITE	SERPENTINITE	Medium-grained	
	RHYOLITE	ANDESITE	BASALT		Fine-grained	0.06
					Very fine-grained	0.002
	OBSIDIAN and PITCHSTONE		TACHYLYTE		GLASSY AMORPHOUS	
	VOLCANIC GLASSES					

Other aspects of texture include the relative grain size and the grain shape. The IAEG (Anon (1979a)) suggested three types of relative grain size namely, uniform, non-uniform and porphyritic. Grain shape was described in terms of angularity (angular, subangular, subrounded and rounded); form (equi-dimensional flat, elongated, flat and elongated, and irregular) and surface texture (rough and smooth).

Rock material tends to deteriorate in quality as a result of weathering and/ or alteration. Weathering refers to those destructive processes, brought about by atmospheric agents at or near the Earth's surface, that produce a mantle of rock waste. Alteration refers to those changes which occur in the chemical or mineralogical composition of a rock brought about by per-meating hydrothermal fluids or by pneumatolytic action. Unlike weathering the effects of alteration may extend to considerable depths beneath the surface since the agents responsible may have originated from deeply emplaced igneous intrusions. Although weathering and alteration occur in the rock material, the processes are concentrated along the discontinuities in the rock mass. Qualitative classifications based on the estimation and descrip-tion of physical disintegration and chemical decomposition of originally sound rock frequently have been used to assess the degree of weathering (see Chapter 7). Such a classification of weathered rock masses as recommended by the IAEG (Anon (1979a)) and ISRM (Anon (1981)) is as follows.

Symbol	Degree of weathering (%)	Term	Description
W_0	0	Fresh	No visible sign of material weathering
W_1	Less than 25	Slightly	Discolouration indicates weathering of rock on major discontinuity surfaces
W_2	25–50	Moderately	Less than half the rock material is decomposed and/or disintegrated to a soil. Fresh or discoloured rock is present either as a discontinuous framework or as corestones
W_3	50–75	Highly	More than half the rock material is decomposed and/or disintegrated to a soil. Fresh or discoloured rock is present either as a discontinuous framework or as corestones
W_4	Over 75	Completely	Majority of rock material is decomposed and/or disintegrated to soil. The original structure of the rock mass is still intact
W_5	100	Residual soil	All material decomposed. No trace of rock structure preserved

Obviously all grades of weathering may not be present in a given rock mass. Furthermore the classification of weathering grade may have to be modified to suit certain types of rock masses and other classifications have been advanced for the Chalk (see Ward et al. (1968); and Chapter 13) and for Keuper Marl (see Chandler (1969); and Chapter 13).

12.2.2 Engineering properties

The IAEG (Anon (1979a)) grouped the dry density and porosity of rocks into five classes as shown below.

Class	Dry density (t/m³)	Description	Porosity (%)	Description
1	Less than 1.8	Very low	Over 30	Very high
2	1.8–2.2	Low	30–15	High
3	2.2–2.55	Moderate	15–5	Medium
4	2.55–2.75	High	5–1	Low
5	Over 2.75	Very high	Less than 1	Very low

Determination of the strength and deformability of intact rock is achieved with the aid of some type of laboratory test (see Chapter 11). If the strength of rock is not measured then it can be estimated as shown in Table 12.2.

Table 12.2. Estimation of the strength of intact rock (after Anon (1977))

Description	Approximate unconfined compressive strength (MPa)	Field estimation
Very strong	Over 100	Very hard rock—more than one blow of geological hammer required to break specimen
Strong	50–100	Hard rock—hand-held specimen can be broken with a single blow of hammer
Moderately strong	12.5–50	Soft rock—5 mm indentations with sharp end of pick
Moderately weak	5.0–12.5	Too hard to cut by hand
Weak	1.25–5.0	Very soft rock—material crumbles under firm blows with the sharp end of a geological hammer

Obviously such estimates can only be very approximate. As far as deformability is concerned the following five classes have been proposed by the IAEG (see Anon (1979a)).

Hardness can be defined as the mechanical competence of the intact rock.

Class	Deformability (MPa × 10³)	Description
1	Less than 5	Very high
2	5–15	High
3	15–30	Moderate
4	30–60	Low
5	Over 60	Very low

In the strict sense it is a surface property which is measured by using abrasion, indentation or rebound tests (see Chapter 11). These tests tend to reflect the hardnesses of individual grains rather than the intergranular bond or coherence of the rock. The point load strength provides a measure of coherence, in other words an indirect measure of tensile strength. Determination of hardness at natural moisture content can give misleading results from those materials, especially argillaceous rocks, that are water sensitive. Indeed most rocks show some hardness reduction when wetted, but mudstones and shales may actually disintegrate when subjected to stress relief combined with moisture content fluctuations. Hence Cottiss *et al.* (1971) recommended that in such instances the durability should be determined to supplement the values of hardness. Durability measures the susceptibility of rocks to weakening and disintegration in water (see Chapter 7, and Anon (1979b)).

As outlined in Chapter 5 discontinuities have a variety of origins. Where possible it is desirable to differentiate between origins since engineering properties may be strongly related to genesis (see Goodman and Duncan (1971)). The frequency and orientation of discontinuities within a rock mass influence its engineering performance (see Table 12.4). Hence enough data should be gathered to position each discontinuity in space. The shape (blocky, tabular, columnar) and size of blocks (see Chapter 5) formed by intersecting discontinuities should be recorded. The continuity of a discontinuity, although difficult to determine, is significant since persistent discontinuities may have an important influence on failure of a rock mass. Similarly the degree to which a discontinuity is open is important. If it is open then the nature of the infill material, if present, must be recorded and its unconfined compressive strength assessed. Lastly, the degree to which weathering or alteration has taken place along discontinuities should be described in relation to Table 7.3.

The permeability of intact rock (primary permeability) is usually several orders less than the *in situ* permeability (secondary permeability), as most

Table 12.3. Estimation of secondary permeability from discontinuity frequency (Anon (1977))

Rock mass description	Term	Permeability k (m/s)
Very closely to extremely closely spaced discontinuities	Highly permeable	10^{-2}–1
Closely to moderately widely spaced discontinuities	Moderately permeable	10^{-5}–10^{-2}
Widely to very widely spaced discontinuities	Slightly permeable	10^{-9}–10^{-5}
No discontinuities	Effectively impermeable	Less than 10^{-9}

water normally flows via discontinuities in rock masses (see Chapter 6). Although the secondary permeability is affected by the openness of discontinuities on the one hand and the amount of infilling on the other, a rough estimate of the permeability can be obtained from the frequency of discontinuities (Table 12.3). Admittedly such estimates must be treated with caution and cannot be applied to rocks which are susceptible to solution. The IAEG (Anon (1979a)) suggested the following grades of permeability which differ slightly from the class limits given in Table 12.3.

Class	Permeability k (m/s)	Description
1	Greater than 10^{-2}	Very highly
2	10^{-2}–10^{-4}	Highly
3	10^{-4}–10^{-5}	Moderately
4	10^{-5}–10^{-7}	Slightly
5	10^{-7}–10^{-9}	Very slightly
6	Less than 10^{-9}	Practically impermeable

The seismic velocity refers to the velocity of propagation of shock waves through a rock mass. Its value is governed by the mineral composition, density, porosity, elasticity and degree of fracturing within a rock mass. The IAEG (Anon (1979a)) recognised the following classes of sonic velocity for rocks.

Class	Sonic velocity (m/s)	Description
1	Less than 2500	Very low
2	2500–3500	Low
3	3500–4000	Moderate
4	4000–5000	High
5	Over 5000	Very high

Igneous rocks generally possess values of sonic velocity above 5000 m/s, those of metamorphic rocks range upwards from 3500 m/s and those of sedimentary rocks tend to vary between 1500 m/s and 4500 m/s. The latter range does not include unconsolidated deposits. The dynamic value of Young's modulus and Poisson's ratio can be derived from the seismic velocity (see Onodera (1963)) and both can be correlated with the degree of fracturing (see Grainger et al. (1973); and Table 5.4).

12.2.3 Basic geotechnical description of ISRM

The basic geotechnical description (BGD) of rock masses proposed by the ISRM (Anon (1981)) considered the following characteristics

(1) rock name with a simplified geological description,
(2) the layer thickness and fracture (discontinuity) intercept of the rock mass,
(3) the unconfined compressive strength of the rock material and the angle of friction of the fractures.

It was suggested that, where necessary, the rock mass should be divided into geotechnical units or zones. The division of the rock mass should be made in relation to the project concerned and the BGD should then be applied to each unit. The rock name is given in accordance with Figure 12.2. Although the simplified geological description depends upon the character of the rock masses involved, together with the requirements of the proposed scheme, it usually takes account of the mineralogical composition, texture and colour of the rock on the one hand and the degree of weathering, the nature of the discontinuities and the geological structure of the rock mass on the other. In addition the ISRM recommended that it would be advisable to provide a general geological description for the rock mass as well as one for each geotechnical unit.

The same class limits are used to describe the layer thickness of a geotechnical unit as are used for fracture intercept (Table 12.4; cf. Table 5.1). The ISRM defined the fracture intercept as the mean distance between successive fractures as measured along a straight line (Table 12.4). When the fracture spacing changes with direction, the value adopted in the BGD is that corresponding to the direction along the smallest mean intercept. Fractures or discontinuities can be grouped into sets. The average fracture intercept, measured perpendicular to the fractures, is recorded for each set and given as supplementary information.

Table 12.4. Classification of layer thickness and fracture intercept (after Anon (1981))*

Interval (m)	Layer thickness		Fracture intercept	
	Symbol†	Description	Symbol	Description
Over 2.0	L_1	Very large	F_1	Very large
0.6–2.0	L_2	Large	F_2	Wide
0.2–0.6	L_3	Moderate	F_3	Moderate
0.06–0.2	L_4	Small	F_4	Close
Less than 0.06	L_5	Very small	F_5	Very close

* The IAEG has proposed the same class limits
† If a unit is not layered then it is given the symbol L_0.

The unconfined compressive strength of the intact rock within a geo-technical unit of a rock mass represents the mean strength of rock samples taken from the zone. The BGD includes the following groupings; those suggested by the IAEG (Anon (1979a)) are given for comparison.

BGD strength (MPa)	Symbol	Description	IAEG strength (MPa)	Description
Over 200	S_1	Very high	Over 230	Extremely strong
60–200	S_2	High	120–230	Very strong
20–60	S_3	Moderate	50–120	Strong
6–20	S_4	Low	15–150	Moderately strong
Under 6	S_5	Very low	1.5–15	Weak

If the rock material is notably anisotropic, the mean strengths obtained in different directions should be recorded, and special note should be made of that direction along which the lowest mean strength occurs.

The angle of friction of fractures as defined by the ISRM refers to the slope of the tangent to the peak strength envelope at a normal stress of 1 MPa. The smallest mean value of the angle of friction is recorded when fracture sets differ in their shear strength. Table 12.5 shows the class limits for the angle of friction of fractures adopted by the BGD.

Table 12.5. Angle of friction of fractures

Interval	Symbol	Description
Over 45°	A_1	Very high
35–45°	A_2	High
25–35°	A_3	Moderate
15–25°	A_4	Low
Less than 15°	A_5	Very low

The data sheet given in Figure 12.3 is used for the BGD. Each zone is characterised by its rock name, followed by the class symbols corresponding to the parameter values, e.g. Granite L_0, F_3, S_2, A_3. Supplementary information is incorporated in the BGD when the rock mass concerned exhibits special features or if the requirements of the project so demand.

12.3 Principles of classification

Classifications of rocks devised by geologists usually have a genetic basis. Unfortunately, however, such classifications may provide little information relating to the engineering behaviour of the rocks concerned.

According to Coates (1964) classification is needed in geotechnical engineering in order to assist in making an initial assessment of a problem and to point to areas where additional information must be sought in order to obtain the required answer. Typical engineering problems that require rock classification include the assessment of slope stability, open excavation, subsurface excavation, foundation stability and the selection of rock for construction material.

Franklin (1970) contended that if a different system of rock classification is used for each engineering problem then much confusion and duplication of effort can arise. He therefore argued that it would be better if one system of classification could be used for a range of problems. However, there has been a trend towards the development of a multiplicity of classifications. Franklin admitted that a classification to some extent should be tailored to suit the application but he suggested that if classification criteria are carefully selected initially, then a change of emphasis from one application to the next, rather than a complete reorganisation, should suffice. Broad terms of reference therefore are necessary in designing such a classification system. Franklin went on to distinguish between basic and supplementary classification tests and observations. The former may be used to establish a universally applicable basis for the engineering classification of rocks. The latter are of less general relevance and may be used for added refinement in particular engineering problems.

Tests which are used for the engineering classification of rocks are termed index tests. If the right index tests are chosen then rocks having similar index properties, irrespective of their origin, will probably exhibit similar engineering performance. In order for an index test to be useful it must satisfy certain criteria. It should be simple to carry out, inexpensive and rapidly performed. The test results must be reproducible and the index property must be relevant to the engineering requirement. Generally these tests are carried out in large numbers so that a reliable picture of rock variation is obtained.

Classification systems of intact rock may be developed either by selecting individual index properties to represent others that are closely related, or by summing the values of closely related properties to derive a single score (see Cottiss et al. (1971)). Obviously if two types of test are closely related then the values of one may be used to predict the values of the other. Where economy of effort is important there is little to be gained by performing both tests. This has motivated the use of index tests in rock classification. For instance, D'Andrea et al. (1965) used nine index observations to predict the unconfined compressive strength of rock, employing multiple linear regression analysis for this purpose. Statistical methods may be used to decide which observations give the best prediction but the final selection of index properties must always take account of which properties are the easiest to evaluate.

Deere and Miller (1966) maintained that information concerning the physical properties of rocks and the nature of the discontinuities within rock masses is required in order to make rational predictions about their engineering behaviour under superimposed stresses. They further stated that the properties of intact rock should be investigated initially in an attempt to develop a meaningful system of evaluation of the *in situ* behaviour of rock. Appropriate reduction factors, attributable to the discontinuities, should then be determined for application to the intact rock data.

In fact two types of classification have been developed. First, there are those based upon some selected properties of the intact rock, as mentioned above. Secondly, and more importantly, there are those which take account of the properties of the rock mass, especially the nature of the discontinuities. The specific purpose for which a classification is developed obviously plays an important role in determining whether the emphasis is placed on the physical properties of the intact rock or on the continuity of the rock mass. The object of both types of classification is to provide a reliable basis for assessing rock quality.

12.4 Review of classifications

Since 1960 much effort has been devoted to the production of engineering classifications of rocks and rock masses. A review of the development of rock classification for engineering purposes has been provided by Dearman (1974).

Any classification of intact rock for engineering purposes should be relatively simple, being based on significant physical properties so that it has a wide application. For example, Deere and Miller (1966) based their engineering classification of intact rock on the unconfined compressive strength and the modulus ratio as follows.

(1) *Strength*

Class	Description	Unconfined compressive strength (MPa)
A	Very high strength	Over 224
B	High strength	112–224
C	Medium strength	56–112
D	Low strength	28–56
E	Very low strength	Less than 28

The strength categories follow a geometric progression and the dividing line between categories A and B was chosen at 224 MPa since it is about the upper limit of the strength of most rocks. A rock may be classified as CH, BH, DL, etc.

(2) *Modulus ratio*

Class	Description	Modulus ratio
H	High modulus ratio	Over 500
M	Medium modulus ratio	200–500
L	Low modulus ratio	Less than 200

Example of application of BGD

Type of work:[1] Concrete Dam

Investigation stage:[2] Preliminary   Exposure:[3] Outcrop

Location: Rocha da Galé Portugal   Observer:[4] Gomes Coelho   Date: June 77

[5]

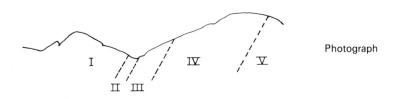

Photograph

Rock name and general geologic description:[6]
Isoclinal sequence of metasedimentary and meta-volcanic rocks composed by interbedded siliceous schists and greywackes (I), pyroclastic rocks like tuff and breccia (II), agglomerate with rhyolitic matrix (III), rhyolite (IV), and porphyritic quartz-diorite (V).

Supplementary geologic description:[7]
Zone I: Rock mass formed of grey to red siliceous schist and interbedded greywacke thinly bedded (2–20 cm), and very often thinly laminated (0.6–2 cm). Rock mass is crossed by widely to very widely spaced joints, open, without filling material. Rock is fresh (W_1) and strong.
Zone II: Interbedded zone 30 m thick, composed of pyroclastic tuff and breccia, moderately to highly weathered (W_3–W_4) and moderately weak.
Zone III: Rock mass formed by agglomerate with a matrix of rhyolitic composition; rock is fresh to slightly weathered (W_1–W_2) and strong to very strong.
Zone IV: Rock mass formed by rhyolite massive, fresh to slightly weathered (W_1–W_2) and strong to very strong.
Zone V: Rock mass formed by porphyritic quartz-diorite, with slight foliation and widely spaced bedding (0.6–2 m); rock is fresh to moderately weathered (W_1–W_2) and moderately strong.

Zones	Occurrence (%)[8]	Characterisation[9]	Zones	Occurrence (%)[8]	Characterisation[9]
I	20	Siliceous schist L_4; $F_{4,5}$; S_2; A_2	V	45	Quartz-diorite L_2; F_4; S_3; A_2
II	2	Breccia and Tuff L_4; F_4; S_4; A_3	VI		
III	8	Agglomerate L_0; F_4; S_2; A_2	VII		
IV	25	Rhyolite L_0; F_3; S_2; A_2	VIII		

Computation of parameters

Zone	Parameters	Samples				Average	Std. dev.	BGD symbols
		1	2	3	4			
I	Layer thickness (cm)	4	10	8	6	7		L_4
	Fracture interc. (cm)							$F_{4,5}$
	U. comp. strength (MPa)	66	65	150*	80	70		S_2
	Angle of friction (°)					35		A_2
II	Layer thickness (cm)	10	12	16	15	13		L_4
	Fracture interc. (cm)	15				15		F_4
	U. comp. strength (MPa)	15	20	12	15	15		S_4
	Angle of friction (°)					30		A_3
III	Layer thickness (cm)	—	—	—	—	—		L_0
	Fracture interc. (cm)	4	12	6	8	7		F_4
	U comp. strength (MPa)	236	250	150	170	200		S_2
	Angle of friction (°)					40		A_2
IV	Layer thickness (cm)	—	—	—	—	—		L_0
	Fracture interc. (cm)	22	25	50	45	36		F_3
	U comp. strength (MPa)	210	140	180	220	185		S_2
	Angle of friction (°)					40		A_2
V	Layer thickness (cm)	80	210	120	160	140		L_2
	Fracture interc. (cm)	4	7	12	18	10		F_4
	U comp. strength (MPa)	92*	55	60	50	55		S_4
	Angle of friction (°)					40		A_2

Remarks[10]

Layer thickness:	measured on outcrops
Fracture interc.:	measured and estimated
U comp. strength	lab. test and estimated
Angle of friction:	estimated

Supplementary information

* Normal to layering

(1) Main characteristics of the structure. (2) Preliminary, final, ... (3) Outcrop, trench, cores, ... (4) Name and quali-fication. (5) Stereo pair of photographs, with the zones outlined. Other stereo pairs may be added. Ordinary photographs and/or sketches can be resorted to. (6) Rock name, structure (folds, faults). Fracturing (fracture sets, fracture charac-teristics); weathering. (7) Specific aspects should be considered for each zone. (8) Estimated proportion, by volume, of the occurrence of each zone relative to the observed rock mass. (9) Rock name followed by the interval symbols of the parameter values. (10) Methods followed in the determination of the parameters and difficulties encountered.

Figure 12.3 Basic geotechnical description (after Anon. (1981))

Deere and Miller (1966) found that different rock types, when plotted on Figure 12.4 occupied different positions. For instance, the envelope enclosing sandstones and siltstones indicates that they have a unique position with respect to other rocks. It shows that they are more compressible in relation to their strength than most rock types. Granites also occupy a rather unique position in the centre of the zone of average modulus ratio. Deere and Miller suggested that specific rock types fall within certain areas on the classification chart because it is sensitive to mineralogy, fabric and direction of anisotropy.

Voight (1968), however, argued that the elastic properties of intact rock could be omitted from practical classifications since the elastic moduli as determined in the laboratory are seldom those which are required for engineering analysis.

Rocks possessing an interlocking fabric and little or no anisotropy generally fall into the medium modulus ratio category. Some limestones and dolostones, however, have a high modulus ratio which Deere (1968) attributed to their mineralogy, as well as to their interlocking fabric, they being composed of calcite and/or dolomite. The sandstone and shale envelopes on Figure 12.4 are open ended in their lower portions because several samples failed at strengths less than 7 MPa. Both sandstone and shale envelopes extend into the zone of low modulus ratio due to their anisotropy attributable to their bedding or lamination. In metamorphic rocks the gneiss envelope overlaps that of quartzite, as well as the two schist envelopes. This transition position indicates an increasing complexity in both mineralogy and fabric in going from quartzite to gneiss to schist.

Deere and Miller (1966) also considered that the Schmidt hammer and Shore sceleroscope hardnesses, sonic pulse velocity and unit weight could act as indices of the engineering behaviour of intact rock. In particular they found a correlation between Schmidt hammer hardness and Shore scleroscope hardness on the one hand and unconfined compressive strength and deformation on the other. Moreover they found that the correlation was improved when the unit weight was taken in conjunction with the indices of hardness and sonic velocity. These properties were therefore used to plot rock property charts (see Figures 11.15 and 11.16). The rock strength chart based on the Shore scleroscope hardness appears to be limited to rocks with unconfined compressive strengths in excess of 35 MPa.

A major advance in the development of the concepts governing engineering classifications of rock was made by Coates (1964). He considered the following five properties to be important.

(1) The principal reason for using the unconfined compressive strength of intact rock is that it indicates whether or not the strength is likely to be a source of trouble in itself. Furthermore there is a rough correlation between compressive strength and the modulus of deformation. Three categories were recognised
    (a) weak, less than 35 MPa,
    (b) strong, between 35 and 175 MPa,
    (c) very strong, greater than 175 MPa.
(2) The pre-failure deformation characteristics of the intact rock indicate whether creep could be expected in the material at stress levels less than those required to cause failure. In extreme cases it could also

E. tangent modulus at 50% ultimate strength

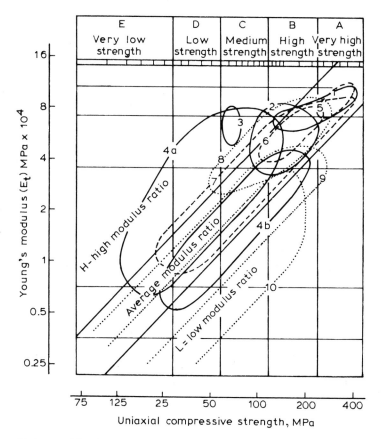

Metamorphic:—1—Quartzite,—2—Gneiss,—3—Marble,—4a—Schist,
steep foliation,—4b—Schist, flat foliation.
Igneous:--5--Diabase, --6--Granite,--7--Basalt and other flow rocks.
Sedimentary:····8····Limestone and dolomite,·····9··· Sandstone ,·····10··· Shale.

Figure 12.4 Engineering classification of intact rock based on uniaxial
compressive strength and modulus ratio. Fields are shown for igneous
sedimentary and metamorphic rocks (after Deere and Miller (1966))

indicate the possibility of ground heave. Two categories were recognised
(a) elastic,
(b) viscous (at a stress of 50% of the unconfined compressive strength
the strain rate is greater than two microstrain per hour).
(3) The failure characteristics of the intact rock influence the factor of
safety used in design, as well as the precautions to be taken during
construction. Two categories were distinguished
(a) brittle,
(b) plastic (more than 25% of the total strain before failure is
permanent).

(4) Gross homogeneity and isotropy of the formation
    (a) massive,
    (b) layered (i.e. generally including sedimentary and cleaved and schistose metamorphic rocks, as well as any other rocks with layering which produces parallel lines of weakness).
(5) The continuity of a rock mass, that is, whether it is divided into large or small blocks or is massive
    (a) solid (discontinuity spacing greater than 2 m),
    (b) blocky (discontinuity spacing between 75 mm and 2 m),
    (c) broken (in fragments that would pass through a 75 mm sieve).

12.5 The rating concept

As far as the classification of rock masses for engineering purposes is concerned most work has been done in relation to tunnelling and the construction of underground chambers. Engineers have been especially concerned with the determination of rock mass quality in relation to the time the rock mass can remain unsupported, and the type and amount of support necessary.

Terzaghi (1946) was one of the first workers to attempt an engineering classification of rock masses. In this he recognised the significance of discontinuities, their spacing and their filling materials as well as the influence of weathering (Table 12.6). However, Terzaghi's classification tended to overlook the properties of the rock. For instance, both chalk and granite could fall into the same class but their different character is obvious.

Table 12.6. Classification of in situ rock for predicting tunnel support requirements (after Terzaghi (1946))

Term	Description
Intact	Rock contains neither joints nor hair cracks
Stratified	Rock consists of individual strata with little or no resistance against separation along the boundaries between strata
Moderately jointed	Rock contains joints and hair cracks, but the blocks between joints are locally grown together or so intimately interlocked that vertical walls do not require lateral support
Blocky and seamy	Rock consists of chemically unweathered rock fragments which are entirely separated from each other and imperfectly interlocked. In such rock vertical walls may require support
Crushed	Chemically unweathered rock has the character of crusher run material
Squeezing	Rock that slowly advances into the tunnel without perceptible volume change
Swelling	Rock that advances into the tunnel chiefly on account of expansion caused by minerals with a high swelling capacity

John (1962) proposed a classification of rock masses which considered their compressive strength, degree of jointing and amount of alteration (Figure 12.5). Most subsequent classifications have followed and elaborated upon his ideas.

Wickham *et al.* (1972), Bieniawski (1973); (1974); (1975a); (1976) and Barton *et al.* (1975) proposed classifications of jointed rock masses which depended on various weighted aspects of both the rock material and the rock mass. Their objective was to obtain rock mass ratings which could be used for design purposes.

Wickham *et al.* (1972) introduced the concept of rock structure rating (RSR) which refers to the quality of rock structure in relation to ground support in tunnelling. Although their classification is specifically related to tunnels it did introduce the rating principle which has been adopted subsequently in other classifications. The rating principle allows several parameters to be taken into account and their influence on the rock mass is collectively assessed. The RSR system rates the relative effect on ground

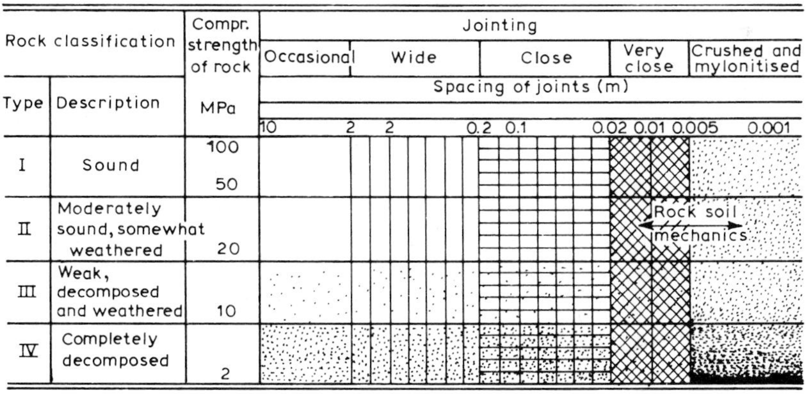

Figure 12.5 Representation of strength of a rock mass

support requirements of three parameters in relation to several geological factors and, where applicable, with respect to each other. Parameter A provides a general appraisal of the rock structure; parameter B is related to the joint pattern and the direction of drive and parameter C represents a general evaluation of the effect of groundwater flow on the type of support necessary (Table 12.7). The method allows for three conditions of joint surfaces, namely, tight or cemented, slightly weathered and severely weathered or open; and four types of water inflow. The RSR value of a particular rock mass is given by the numerical sum of the parameters A, B and C, and the values range from 25–100, reflecting the quality of the rock mass. For example, Wickham *et al.* concluded that rock masses with RSR values less than 27 would require heavy support, whilst those with ratings over 77 would probably stand unsupported.

Table 12.7. Rock structure rating (after Wickham et al. (1972))
(1) Parameter A—Geological structure

Basic rock type	Massive	Slightly faulted or folded	Moderately faulted or folded	Intensely faulted or folded
Igneous	30	26	15	10
Sedimentary	24	20	12	8
Metamorphic	27	22	14	9

(Maximum value of parameter A is 30)

(2) Parameters B—Joint pattern and direction of drive

Average joint spacing	Strike perpendicular to axis					Strike parallel to axis		
	Direction of drive							
	Both	With dip		Against dip			Both	
	Dip of prominent joints							
	<20°	20–50°	50–90°	20–50°	50–90°	<20°	20–50°	50–90°
<0.15 m Clearly jointed	14	17	20	16	18	14	15	12
0.15–0.3 m Moderately jointed	24	26	30	20	24	24	24	20
0.3–0.6 m Moderate to blocky	32	34	38	27	30	32	30	25
0.6–1.2 m Blocky to massive	40	42	44	36	39	40	37	30
>4.0 m Massive	45	48	50	42	45	45	42	36

(Maximum value of parameter B is 50)

(3) Parameter C—Groundwater, joint condition

Anticipated water inflow (l/s per 300 m)	Sum of parameters A + B					
	20–45			46–80		
	Joint condition*					
	1	2	3	1	2	3
None	18	15	10	20	18	14
Slight (<15)	17	12	7	19	15	10
Moderate (15–75)	12	9	6	18	12	8
Heavy (>75)	8	6	5	14	10	6

* 1 = tight or cemented; 2 = slightly weathered; 3 = severely weathered or open
(Maximum value of parameter C is 20)

The classification of rock masses advanced by Bieniawski (1973) initially incorporated the RQD (see Chapter 5); the unconfined compressive strength; the degree of weathering; the spacing, orientation, separation and continuity of the discontinuities; as well as the groundwater flow. Unfortunately, no account was taken of the roughness of joint surfaces or the character of the infill material. The unconfined compressive strength of the intact rock has an important bearing on the engineering performance of the rock mass when the discontinuities are widely spaced or the rock mass is weak. It is also important if the joints are not continuous. Because the unconfined compressive strength and the degree of weathering are two interdependent factors, Bieniawski (1974) subsequently revised his views and suggested that both factors should be regarded as one parameter, namely, the strength of the rock material. He chose a somewhat modified version of Deere's (1964) classification of intact strength for his classification (Table 12.8; *cf.* the classification of strength proposed by the Geological Society (Anon (1970)); see also Chapter 11; Anon (1970); (1979a); (1981) Section 12.2. Bieniawski argued that when assessing importance ratings the strength of rocks less than 25 MPa does not contribute to the overall mobility of the rock mass.

The point load test can be used to determine the intact strength on site as several authors (D'Andrea *et al.* (1965); Franklin and Broch (1972); and Bieniawski (1975b)) have found a close correlation between point load strength and the unconfined compressive strength.

Special caution should be exercised in using this classification in the case of shales and other swelling materials. For instance, Bieniawski (1973) suggested that their behaviour under conditions of alternate wetting and drying should be assessed by a slake-durability test (see Franklin and Chandra (1972)).

The presence of discontinuities reduces the overall strength of a rock mass and their spacing and orientation govern the degree of such reduction. Hence the spacing and orientation of the discontinuities are of paramount importance as far as the stability of structures in jointed rock masses is concerned. Bieniawski (1973) accepted the classification of discontinuity spacing which was proposed by Deere (1968).

Another revision of Bieniawski's ideas included the continuity and separation of discontinuities which he later (1974) grouped together under the heading, condition of discontinuities. This parameter also took account of the surface roughness of discontinuities and the quality of the wall rock. The condition of discontinuities is as important as the discontinuity spacing, for example, tight discontinuities with rough surfaces and no infill have a high strength. By contrast open discontinuities which are continuous mean a plane of weakness and furthermore facilitate unrestricted flow of groundwater. Obviously the condition of discontinuities influences the extent to which the rock material affects the behaviour of a rock mass.

Groundwater has an important effect on the behaviour of a jointed rock mass. However, as the pore water pressures are of greater significance in foundations than groundwater inflow, Pells (1974) suggested that the pore water pressure ratio (r_u), where r_u is defined as the ratio of the pore pressure to the major principal stress, should be included within the classification. This view was accepted by Bieniawski (1974) and the pore water pressure ratio was incorporated into his classification.

Bieniawski (1976) grouped each of the chosen rock mass parameters into five classes (Table 12.8). He considered five classes to be sufficient to provide a meaningful discrimination in all the chosen properties. Because these parameters vary in their relative importance from rock mass to rock mass and can contribute individually or collectively to its engineering performance, Bieniawski used a rating system. In other words a weighted numerical value was given to each class in each parameter. The total rock mass rating is the sum of the weighted values of the individual parameters, the higher the total rating, the better the rock mass condition (Table 12.8c). In describing a rock mass the class rating should be quoted with the class number, for example, class 3, rating 68.

Table 12.8 Engineering classification of jointed rock masses (after Bieniawski (1975a))

(a) Classification parameters and their ratings

1	Strength of intact rock material	Point load strength index (MPa)	> 8	4–8	2–4	1–2	Use of unaxial compressive test preferred
		Uniaxial compressive strength (MPa)	> 200 Very high strength	100–200 High strength	50–100 Medium strength	25–50 Low strength	< 25 Very low strength
	Rating		10	5	2	1	0
2	Drill core quality RQD		90–100%	75–90%	50–75%	25–50%	< 25%
	Rating		20	17	14	8	3
3	Spacing of discontinuities		> 3 m Very wide (solid)	1–3 m Wide (massive)	0.3–1 m Moderately close (blocky/seamy)	50–300 mm Close (fractured)	< 50 mm Very close (crushed)
	Rating		30	25	20	10	5
4	Orientations of discontinuities		Very favourable	Favourable	Fair	Unfavourable	Very unfavourable
	Rating		15	13	10	6	3
5	Condition of discontinuities		Extremely tight Very rough surfaces Not continuous No separation Hard joint wall rock	Very tight Slightly rough surfaces Separation < 0.1 mm Hard joint wall rock Not continuous	Tight Slightly rough surface Separation < 1 mm No gouge Soft joint wall rock	Open slickensided surfaces **or** Gouge < 5 mm thick **or** Joints open 1–5 mm Continuous joints	Very open Soft gouge > 5 mm thick **or** Joints open > 5 mm Continuous joints
	Rating		20	15	10	5	0

	None	> 25 l/min	25–125 l/min	> 125 l/min
Inflow per 10 mm tunnel length or				
joint water pressure / Ground water Ratio / major principal stress or	0	0.0–0.2	0.2–0.5	> 0.5
General conditions	Completely dry	Moist only (interstitial water)	Water under moderate pressure	Severe water problems
Rating	10	8	5	2

(b) *Rock mass classes and their ratings*

Class No.	I	II	III	IV	V
Description	Very good rock	Good rock	Fair rock	Poor rock	Very poor rock
Rating	100–90	90–70	70–50	50–25	< 25

(c) *Meaning of rock mass classes*

Class No.	I	II	III	IV	V
Average stand up time	10 years for 5 m span	6 months for 4 m span	1 week for 3 m span	5 hours for 1.5 m span	10 minutes for 0.5 m span
Cohesion of the rock mass	> 300 kPa	200–300 kPa	150–200 kPa	100–150 kPa	< 100 kPa
Friction angle of the rock mass	< 45°	40–45°	35–40°	30–35°	< 30°
Caveability of ore	Very poor	Will not cave readily. Large fragments.	Fair	Will cave readily. Good fragmentation	Very good

Laubscher (1977) developed a geomechanics classification for jointed rock masses which was based on Bieniawski's classification. This classification also was designed principally for use in relation to subsurface excavation. Laubscher's classification similarly involves the rating concept, the rating extending from 0–100, and is supposed to cover all variations of jointed rock masses from very poor to very good. There are five classes, each one being subdivided into an A and B group, and each group has a 10 point rating

Table 12.9. Classification of variations in jointed rock masses

Class	1		2		3		4		5	
Rating	100–81		80–61		60–41		40–21		20–0	
Description	Very good		Good		Fair		Poor		Very poor	
Sub-classes	A	B	A	B	A	B	A	B	A	B
Item										
(1) RQD, %	100–91	90–76	75–66	65–56	55–46	45–36	35–26	25–16	15–6	5–0
Rating	20	18	15	13	11	9	7	5	3	0
(2) IRS, MPa	141–136	135–126	125–111	110–96	95–81	80–66	65–51	50–36	35–21 20–6	5–0
Rating	10	9	8	7	6	5	4	3	2 1	0

(3) Joint spacing	Refer to Table 12.10	
Rating	30..0	

(4) Condition of joint	45° .... Static angle of friction .... 5°	
Rating	30.. Refer to Table 12.10 0	

(5) Groundwater	Inflow per 10 m length **or** Joint water pressure / Major principal stress **or** Description	0	25 l/min	25–125 l/min	125 l/min
		0	0.0–0.2	0.2–0.5	0.5
	Description	Completely dry	Moist only	Moderate pressure	Severe problems
Rating		10	7	4	0

(Table 12.9). The classification considers the rock quality designation (RQD), intact rock strength (IRS), joint spacing, joint condition and groundwater. The rating for the spacing of one, two or three joint sets is obtained by reference to Figure 12.6. Joint condition is based on the expression of the joint, its surface properties, the presence of alteration zones and the character of any fill material which is present. The rating of the joint condition is obtained by summing these various factors after they have been adjusted according to Table 12.10a, the total possible rating for joint condition being 30. Groundwater, particularly joint water pressure, was regarded as of paramount importance. The rating should be reduced if an excavation is orientated in an unfavourable direction with respect to geological structures, particularly the weakest joint sets. The magnitude of the adjustment depends on the attitude of the joints with respect to the vertical axis of the block (Table 12.10b). Laubscher's classification also involved adjustment of the class rating according to the influence of weathering, field and induced stresses, changes in stress and the influence of strike and dip orientations (Table 12.10c). For example, weathering affects the RQD, IRS and condition of the joints. As the RQD percentage can be decreased by weathering, Laubscher suggested that a reduction to 95% of the rating value was possible. Similarly the IRS is reduced if weathering occurs along microstructures. This time he suggested a decrease in rating to 96% for the bulk of the rock. An adjustment varying down to 82% of the rating was suggested for alteration along joints. Field and induced stresses may be responsible for compressing joints together and consequently the rating can be increased up to 120%. By contrast if the possibility of shear movement is increased, the rating would be decreased to 90% of its value. Similarly if joints open the rating is lowered by 24%.

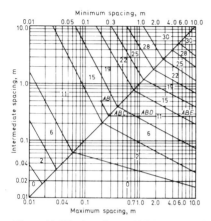

Figure 12.6 Ratings for multi-joint systems: example. Joint spacing: A = 0.2 m, B = 0.5 m, C = 0.6 m, D = 1.0 m, E = 7.0 m, AB = 15, ABC = 6, ABD = 11, ABE = 15

Barton *et al.* (1975) defined rock mass quality (*Q*) in terms of six parameters.

(1) The RQD or an equivalent system of joint density.
(2) The number of joint sets (J_n), which is an important indication of the degree of freedom of a rock mass. The RQD and the number of joint sets provide a crude measure of relative block size (RQD/J_n).
(3) The roughness of the most unfavourable joint set (J_r). The joint roughness and the number of joint sets determine the dilatancy of the rock mass.
(4) The degree of alteration or filling of the most unfavourable joint set (J_a). The roughness and degree of alteration of the joint walls or the filling materials provide an approximation of the shear strength of the rock mass (J_r/J_a).
(5) The degree of water seepage or the joint water reduction factor (J_w).
(6) The stress reduction factor (SRF). Squeezing and swelling are taken account of in the stress reduction factor. The active stress is defined as J_w/SRF.

The rock mass quality (*Q*) is then derived from

$$Q = (RQD/J_n) \times (J_r/J_n) \times (J_w/SRF)$$

The numerical value of *Q* ranges from 0.001 for exceptionally poor quality

Table 12.10. Adjustments to class ratings (after Laubscher (1977))

(1) Assessments of joint conditions (adjustments as combined percentages of total possible rating of 30)

Parameter	Description	Percentage adjustment
(a) Joint expression (large-scale)	Wavy unidirectional	99
		90
	Curved	89
		80
	Straight	79
		70
(b) Joint surface (small-scale)	Striated	99
		85
	Smooth	84
		60
	Polished	59
		50
(c) Alteration zone	Softer than wall rock	99
		70
(d) Joint filling	Coarse hard-sheared	99
		90
	Fine hard-sheared	89
		80
	Coarse soft-sheared	79
		70
	Fine soft-sheared	69
		50
	Gouge thickness < irregularities	49
		35
	Gouge thickness > irregularities	23
		12
	Flowing material > irregularities	11
		0

(2) Orientation adjustments for blocks with exposed bases

Number of joints defining block	Number of faces inclined away from vertical and adjustment percentage				
	70%	75%	80%	85%	90%
3	3		2		
4	4	3		2	
5	5	4	3	2	1
6	6		4	3	2,1

(3) Total possible reductions (%)

	RQD	IRS	Joint spacing	Condition of joints	Total
Weathering	95	96		82	75
Field and induced stresses				120–76	120–76
Changes in stress				120–60	120–60
Strike and dip orientation			70		70

squeezing ground, up to 1000 for exceptionally good quality rock which is practically unjointed.

In certain situations an assessment of rock mass quality can only be made from core material derived from drilling operations. This raises the question of the reliability of the value of rock mass quality determined solely from core material as compared with that obtained by *in situ* investigation. Unfortunately few attempts have been made to compare borehole predictions of rock mass quality ratings with related values obtained from *in situ* measurements. One of the few comparisons was made by Barton (1976) who determined the rock mass qualities of a massive biotite gneiss in which an underground power house was excavated. He found that the average values of rock mass quality (*Q*) as obtained from borehole cores were twice those determined *in situ*. By contrast Cameron-Clark and Budavari (1981), after investigations carried out on three tunnels, showed that values derived from core material tended to give a poorer picture of rock mass conditions than those obtained from *in situ* measurements. The opposing conclusions were explained by Cameron-Clark and Budavari in terms of the differing geological conditions examined. Barton had investigated massive rock whilst the rocks they were involved with were predominantly jointed. It was suggested that poorer values of rock mass quality would be obtained from core material obtained from jointed rock than those obtained by *in situ* investigation, whereas the converse situation applied in massive rocks. When comparisons were made in terms of the geomechanics classification, Cameron-Clark and Budavari showed that there was an approximate 80% probability of the results derived from core material falling within the same or within one class of those determined *in situ*. When they made comparisons using the *Q* classification system they found that the correlation between the rock class derived by core material and that derived from *in situ* investigation was somewhat better than that obtained by the geomechanics system.

References

ANON. (1970). 'Logging of cores for engineering purposes', Working Party Report, *Q. J. Engg Geol.*, **3**, 1–24.

ANON. (1972). 'The preparation of maps and plans in terms of engineering geology'. Working Party Report, *Q. J. Engg Geol.*, **5**, 293–382.

ANON. (1977). 'The description of rock masses for engineering purposes'. Working Party Report, *Q. J. Engg Geol.*, **10**, 355–88.

ANON. (1978). 'Suggested methods for petrographic description'. ISRM Commission on Standardization of Laboratory and Field Tests, *Int. J. Rock Mech. Min. Sci. & Geomech. Abstr.*, **15**, 41–5.

ANON. (1979a). 'Classification of rocks and soils for engineering geological mapping. Part 1—Rock and soil materials', *Bull. Int. Ass. Engg Geol.*, No. 19, 364–71.

ANON. (1979b). 'Suggested methods for determining water content, porosity, density, absorption and related properties and swelling and slake-durability index properties.' ISRM Standardization of Laboratory and Field Tests, *Int. J. Rock Mech. Min. Sci. & Geomech. Abstr.*, **16**, 325–41.

ANON. (1981). 'Basic geotechnical description of rock masses'. ISRM Commission on Classification of Rocks and Rock Masses, *Int. J. Rock Mech. Min. Sci. & Geomech. Abstr.*, **18**, 85–110.

BARTON, N. (1976). 'Recent experiences with the Q system in tunnel support design', *Proc. Symp. on Exploration for Rock Engineering*, Bieniawski, Z. T. (ed.), A. A. Balkema, Cape Town, **1**, 107–15.

BARTON, N. (1978). 'Suggested methods for quantitative description of discontinuities in rock masses', *Int. J. Rock Mech. Min. Sci. & Geomech. Abstr.*, **15**, 319–68.

BARTON, N., LIEN, R. & LUNDE, J. (1975). 'Engineering classification of rock masses for design of tunnel support', (*Rock. Mech.* **6**, 189–236, 1974) and *Norwegian Geot. Inst., Publ.*, **106**.

BIENIAWSKI, Z. T. (1973). 'Engineering classification of jointed rock masses', *Trans S. Afr. Inst. Civil Engrs*, **15**, 335–43.

BIENIAWSKI, Z. T. (1974). 'Geomechanics classification of rock masses and its application in tunnelling', *Proc. 3rd Int. Cong. Rock Mech., Denver*, **2**, 27–32.

BIENIAWSKI, Z. T. (ed.) (1975a). 'Engineering properties of rock with reference to tunnelling', (in *Tunnelling in Rock*), S. Afr. Inst. Civil Engrs/S. Afr. Nat. Gr. Rock Mech. *CSIR, Pretoria*, 105–23.

BIENIAWSKI, Z. T. (1975b). 'The point load test in geotechnical practice', *Engg Geol.*, **9**, 1–11.

BIENIAWSKI, Z. T. (ed.). (1976). 'Rock mass classification in rock engineering', *Proc. Symp. on Exploration for Rock Engineering*, A. A. Balkema, Cape Town, **1**, 97–106.

CAMERON-CLARK, I. S. & BUDAVARI, S. (1981). 'Correlation of rock mass classification parameters obtained from borecore and *in situ* observations', *Engg Geol.*, **17**, 19–53.

CHANDLER, R. J. (1969). 'The effects of weathering on the shear strength properties of Keuper Marl', *Geotechnique*, **19**, 321–34.

COATES, D. F. (1964). 'Classification of rocks for rock mechanics', *Int. J. Rock Mech. Min. Sci.*, **1**, 421–9.

COTTISS, G. I., DOWELL, R. W. & FRANKLIN, J. A. (1971). 'A rock classification system applied to civil engineering', *Civ. Engg Pub. Works Rev.*, **66**, Part 1–No. 777, 611–714; Part 2—No. 780, 736–43.

D'ANDREA, D. V., FISCHER, R. L. & FOGELSON, D. E. (1965). 'Prediction of the compressive strength of rock from other properties', *Rep. US Bur. Mines*, Invest. No. 6702.

DEARMAN, W. R. (1974). 'The characterisation of rock for civil engineering practice in Britain'. Centenaire de la Société Geologique de Bélgique, Colloque', *Géologie de L'Ingenieur*, Liège, 1–75.

DEERE, D. U. (1964). 'Technical description of cores for engineering purposes', *Rock Mech. Engg Geol.*, **1**, 17–22.

DEERE, D. U. (1968). 'Geologic considerations', (in *Rock Mechanics in Engineering Practice*), Stagg, K. G. & Zienkiewicz, O. C. (eds), Wiley, London, 1–19.

DEERE, D. U. & MILLER, R. P. (1966). *Engineering Classification and Index Properties for Intact Rock*, Tech. Rep. No. AFWL-TR-65-116, Air Force Weapons Lab., Kirtland Air Base, New Mexico.

FRANKLIN, J. A. (1970). 'Observations and tests for engineering description and mapping of rocks', *Proc. 2nd Int. Cong. Rock Mech, Belgrade*, **1**, Paper 1–3.

FRANKLIN, J. A. & BROCH, E. (1972). 'The point load strength test', *Int. J. Rock Mech. Min. Sci.*, **9**, 669–97.

FRANKLIN, J. A. & CHANDRA, A. (1972). 'The slake-durability test', *Int. J. Rock Mech. Min. Sci.*, **9**, 325–41.

GOODMAN, R. E. & DUNCAN, J. M. (1971). 'The role of structure and solid mechanics in the design of surface and underground excavations in rock', (in *Structure, Solid Mechanics and Engineering Design*), Te-eni, M. (ed.), Wiley, New York.

GRAINGER, P., McCANN, D. M. & GALLOIS, R. W. (1973). 'Application of seismic refraction techniques for the study of fracturing in the Middle Chalk at Mundford, Norfolk', *Geotechnique*, **23**, 219–32.

JOHN, K. W. (1962). 'An approach to rock mechanics', *Proc. ASCE J. Soil Mech. Foundation Engg Div.*, **88**, SM4, 1–30.

LAUBSCHER, D. H. (1977). 'Geomechanics classification of jointed rock masses—mining applications', *Trans. Inst. Min. Metall.*, Section A—Mining Industry, **86**, A1–A8.

ONODERA, T. F. (1963). 'Dynamic investigation of rocks *in situ*', *Proc. 5th Symp. Rock Mech*, Minnesota University, Rolla, Pergamon Press, New York, 517–33.

PELLS, P. J. H. (1974). 'Discussion: engineering classification of jointed rock masses', *Trans. S. Afr. Inst. Civil Engrs*, **16**, 242.

TERZAGHI, K. (1946). 'Introduction to tunnel geology', (in *Rock Tunneling with Steel Supports*), Proctor, R. & White, T., Commerical Shearing and Stamping Co., Youngstown, Ohio, 17–99.

VOIGHT, B. (1968). 'On the functional classification of rocks for engineering purposes', *Int. Symp. on Rock Mechanics, Madrid*, 131–5.

WARD, W. H., BURLAND, J. B. & GALLOIS, R. W. (1968). 'Geotechnical assessment of a site at Mundford, Norfolk, for a large proton accelerator', *Geotechnique*, **18**, 399–431.

WICKHAM, G. E., TIEDEMANN, H. R. & SKINNER, E. H. (1972). 'Support determination based on geologic predictions', *Proc. 1st N. Am. Tunneling Conf.*, AIME, New York, 43–64.

Chapter 13

Engineering performance of rocks

13.1 Igneous and metamorphic rocks

The plutonic igneous rocks are characterised by granular texture, massive structure and relatively homogeneous composition. In their unaltered state they are essentially sound and durable with adequate strength for any engineering requirement (Table 13.1). In some instances, however, intrusives may be highly altered, by weathering or hydrothermal attack. Furthermore fissure zones are by no means uncommon in granites. The rock mass may be very much fragmented along such zones, indeed it may be reduced to sand size material (see Terzaghi (1946)), and it may have undergone varying degrees of kaolinisation.

In humid regions valleys carved in granite may be covered with residual soils which extend to depths often in excess of 30 m. Fresh rock may only be exposed in valley bottoms which have actively degrading streams. At such sites it is necessary to determine the extent of weathering and the engineering

Table 13.1. Some physical properties of igneous and metamorphic rocks

	Relative density	Unconfined compressive strength (MPa)	Point load strength (MPa)	Shore scleroscope hardness	Schmidt hammer hardness	Youngs modulus ($\times 10^3$ MPa)
Mount Sorrel granite	2.68	176.4	11.3	77	54	60.6
Eskdale granite	2.65	198.3	12.0	80	50	56.6
Dalbeattie granite	2.67	147.8	10.3	74	69	41.1
Markfieldite	2.68	185.2	11.3	78	66	56.2
Granophyre (Cumbria)	2.65	204.7	14.0	85	52	84.3
Andesite (Somerset)	2.79	204.3	14.8	82	67	77.0
Basalt (Derbyshire)	2.91	321.0	16.9	86	61	93.6
Slate* (North Wales)	2.67	96.4	7.9	41	42	31.2
Slate† (North Wales)		72.3	4.2			
Schist* (Aberdeenshire)	2.66	82.7	7.2	47	31	35.5
Schist†		71.9	5.7			
Gneiss	2.66	162.0	12.7	68	49	46.0
Hornfels (Cumbria)	2.68	303.1	20.8	79	61	109.3

* Tested normal to cleavage or schistosity.
† Tested parallel to cleavage or schistosity.

555

properties of the weathered products. Generally the weathered product of plutonic rocks has a large clay content although that of granitic rocks is sometimes porous with a permeability comparable to that of medium-grained sand.

Joints in plutonic rocks are often quite regular, steeply dipping structures in two or more intersecting sets. Sheet joints tend to be approximately parallel to the topographic surface. Consequently they may introduce a dangerous element of weakness into valley slopes. For example, in a consideration of Mammoth Pool Dam foundations on sheeted granite, Terzaghi (1962) observed that the most objectionable feature was the sheet joints orientated parallel to the rock surface. In the case of dam foundations such joints, if they remain untreated, may allow the escape of large quantities of water from the reservoir. This, in turn, may lead to the development of hydrostatic pressures in the rock downstream which are high enough to dislodge sheets of granite.

Generally speaking the older volcanic deposits do not prove a problem in foundation engineering, ancient lavas having strengths frequently in excess of 200 MPa. But volcanic deposits of geologically recent age at times prove treacherous, particularly if they have to carry heavy loads such as concrete dams. This is because they often represent markedly anisotropic sequences in which lavas, pyroclasts and mudflows are interbedded. Hence foundation problems in volcanic sequences arise because weak beds of ash, tuff and mudstone occur within lava piles which give rise to problems of differential settlement and sliding. In addition weathering during periods of volcanic inactivity may have produced fossil soils, these being of much lower strength.

The individual laval flows may be thin and transected by a polygonal pattern of cooling joints. They also may be vesicular or contain pipes, cavities or even tunnels (see Chapter 1).

Pyroclasts usually give rise to extremely variable ground conditions due to wide variations in strength, durability and permeability. Their behaviour very much depends upon their degree of induration; for example, many agglomerates have a high enough strength to support heavy loads such as concrete dams and also have a low permeability. By contrast, ashes are invariably weak and often highly permeable. One particular hazard concerns ashes, not previously wetted, which are metastable and exhibit a significant decrease in their void ratio on saturation. Tuffs and ashes are frequently prone to sliding. Montmorillonite is not an uncommon constituent in the weathered products of basic ashes.

Slates, phyllites and schists are characterised by textures which have a marked preferred orientation. Platy minerals such as chlorite and mica tend to segregate into almost parallel or subparallel bands alternating with granular minerals such as quartz and feldspar. This preferred alignment of platy minerals accounts for the cleavage and schistosity which typify these metamorphic rocks and means that slate, in particular, is notably fissile. Obviously such rocks are appreciably stronger across, than along the lineation (Table 13.1; see also Chapter 11). Not only do cleavage and schistosity adversely affect the strength of metamorphic rocks, they also make them more susceptible to decay. Generally speaking, however, slates, phyllites and schists weather slowly but the areas of regional metamorphism in which they occur have suffered extensive folding so that in places rocks may be fractured and deformed. Some schists, slates and phyllites are variable in quality, some being excellent foundations for heavy structures; others, regardless of the degree of their deformation or weathering, are so poor as to be wholly undesirable. For instance, talc, chlorite and sericite schists are weak rocks containing planes of schistosity only a millimetre or so apart. Some schists become slippery upon weathering and therefore fail under a moderately light load.

The engineering performance of gneiss is usually similar to that of granite. However, some gneisses are strongly foliated which means that they possess a texture with a preferred orientation. Generally this will not significantly affect their engineering behaviour. They may, however, be fissured in places and this can mean trouble. For instance, it would appear that fissures opened in the gneiss under the heel of the Malpasset Dam, which eventually led to its failure (see Jaeger (1963)).

Fresh, thermally metamorphosed rocks such as quartzite and hornfels are very strong and afford good ground conditions. Marble has the same advantages and disadvantages as other carbonate rocks.

13.2 Arenaceous sedimentary rocks

Simply defined a sandstone is an indurated sand. In other words a sandstone is a clastic sediment in which mineral grains or rock fragments are bound together with cement and/or matrix. However, there are several fundamental types of sandstone depending on their composition, more particularly the proportions of feldspar, quartz and detrital matrix they contain. For example, quartz arenites contain over 95% quartz whereas greywackes contain 15–25% detrital matrix with little or no cement (see Pettijohn *et al.* (1975); and Chapter 3). Sandstones may vary from thinly laminated micaceous types to very thickly bedded varieties. Moreover they may be cross bedded and are invariably jointed. With the exception of shaly sandstone, sandstone is not subject to rapid surface deterioration on exposure.

The dry density and especially the porosity of a sandstone are influenced by the amount of cement and/or matrix material occupying the pores. Usually the density of a sandstone tends to increase with increasing depth below the surface (see Bell (1978a)).

The compressive strength of a sandstone is influenced by its porosity, amount and type of cement and/or matrix material, as well as the composition of the individual grains. Price (1960; 1963) showed that the strength of sandstone with a low porosity (less than 3.5%) was controlled by the quartz content and degree of compaction. In those sandstones with a porosity in excess of 6% he found that there was a reasonably linear relationship between dry compressive strength and porosity, that is, for every 1% increase in porosity the strength decreased by approximately 4%. If cement binds the grains together then a stronger rock is produced than one in which a similar amount of detrital matrix performs the same function. However, the amount of cementing material is more important than the type of cement, although if two sandstones are equally well cemented, one having a siliceous, the other a calcareous cement, then the former is the stronger. For example, ancient quartz arenites in which the voids are almost completely occupied with siliceous material are extremely strong with crushing strengths exceeding 240 MPa. By contrast poorly cemented sandstones may possess crushing strengths less than 3.5 MPa.

In a consideration of the Fell Sandstones, Bell (1978a) found that a highly significant relationship existed between unconfined compressive strength on the one hand and tensile strength and Shore scleroscope and Schmidt hammer hardness measures on the other and a significant relationship existed with Young's modulus. Both Shore scleroscope and Schmidt hammer hardness possessed highly significant relationships with Young's modulus, density and porosity. Bell was unable to demonstrate any statistically significant relationship between mineralogical composition on the one hand and density, hardness, strength and deformability on the other. However, both the packing density and packing proximity had highly significant or significant relationships with most of the index properties investigated.

Singh et al. (1978) examined the influence of anisotropy in Chunar Sandstone on its hardness and uniaxial compressive strength. They showed that the average Shore scleroscope hardness normal to the bedding was 46 whereas parallel to the bedding it was 44. The average uniaxial compressive strength normal and parallel to the bedding was respectively 97 and 96 MPa. Incidentally, they found a similar relationship in the Singrauli Coal.

The pore water plays a very significant role as far as the compressive strength and deformation characteristics of a sandstone are concerned. This is illustrated by the Fell Sandstone (Northumberland) and Bunter Sandstone (Nottinghamshire) which may suffer a reduction of uniaxial compressive strength, when a dry specimen is saturated, of nearly 30 and 60% respectively (Table 13.2). The difference in the reduction of strength on saturation probably is explained by the difference in cementation, the Bunter Sandstone being much more poorly cemented and so more porous (see Table 13.2).

West (1979) found that Bunter Sandstone obtained from Warrington exhibited a general tendency to decrease in strength with increasing moisture content (the average air-dry strength was 37 MPa compared with 23 MPa when saturated). Young's modulus also underwent a reduction from a mean value of 7.2×10^2 MPa for the sandstone when dry to 5.5×10^2 MPa for saturated sandstone.

Moore (1974) derived a value of Young's modulus of 1100 MPa for the

Table 13.2. Some physical properties of arenaceous sedimentary rocks

	Fell Sandstone (Rothbury)	Chatsworth Grit (Stanton-in-the-Peak)	Bunter Sandstone (Edwinstowe)	Keuper Waterstones (Edwinstowe)	Horton Flags Helwith Bridge)	Bronyllwyn Grit (Llanberis)
Relative density	2.69	2.69	2.68	2.73	2.70	2.71
Dry density (t/m³)	2.25	2.11	1.87	2.26	2.62	2.63
Porosity (%)	9.8	14.6	25.7	10.1	2.9	1.8
Dry unconfined compressive strength (MPa)	74.1	39.2	11.6	42.0	194.8	197.5
Saturated unconfined compressive strength (MPa)	52.8	24.3	4.8	28.6	179.6	190.7
Point load strength (MPa)	4.4	2.2	0.7	2.3	10.1	7.4
Scleroscope hardness	42	34	18	28	67	88
Schmidt hardness	37	28	10	21	62	54
Young's modulus ($\times 10^3$ MPa)	32.7	25.8	6.4	21.3	67.4	51.1
Permeability ($\times 10^{-9}$ m/s)	1.74	1.96	3.51	0.022	—	—

Bunter Sandstone from long term (up to 18 months) plate load testing. The modulus was found to increase with depth. At the highest loading, 5.6 MPa, settlement did not exceed 4 mm. Creep accounted for 20–30% of the total settlement at loads varying between 0.3 and 1.5 MPa, but at 3.0 and 5.6 MPa it was lower. Moore and Jones (1975) concluded that at fairly low stresses the Bunter Sandstone, even though weathered near the surface, provided a sound foundation. Moreover the rapid reduction in settlement with

Figure 13.1 Folding produced by valley bulging, as revealed during the excavation for the dam for Howden reservoir (courtesy of the Severn–Trent Water Authority)

depth presumably means that simple spread foundation structures may be suitable even for sensitive buildings.

Many sandstones in the valleys excavated in the Millstone Grit series (Namurian) have been fractured by valley bulging or cambering. Some of the most spectacular valley bulges in Britain were revealed in the excavations for the Howden, Derwent and Ladybower Dams (Figure 13.1). In the latter the folding was present to a depth of almost 60 m (see Hill (1949)). A further consequence of valley bulges is the opening up of tension fissures in sandstones forming the valley sides. These fissures run parallel to the valley and may be up to 250 mm wide close to the valley side, but they become progressively narrower and finally disappear when followed into the hillside. Valley bottom and valley side disturbances appear to be the result of a number of factors including stress relief, moisture up-take by underlying shales, artesian water pressures, valley notch concentration of stresses or frozen ground conditions.

Frequently thin beds of sandstone and shale are interbedded. Foundations on such sequences may give rise to problems of shear, settlement, and rebound, the magnitude of these factors depending upon the character of the shales. In some cases this even accentuates the undesirable properties of the shale by permitting access of water to the shale-sandstone contacts. Contact seepage may weaken shale surfaces and cause slides in dipping formations.

13.3 Argillaceous sedimentary rocks

13.3.1 Siltstones

Siltstones have a high quartz content with a predominantly siliceous cement. They therefore tend to be hard, tough rocks (Table 13.3). Like sandstones, their disintegration is governed by their fracture pattern. After several months

Table 13.3. Engineering properties of some Coal Measures rocks

	Mudstone	Siltstone	Shale	Barnsley Hards coal	Deep Duffryn coal
Relative density	2.69	2.67	2.71	1.5	1.2
Dry density (Mg/m³)	2.32	2.43	2.35	–	–
Dry unconfined compressive strength (MPa)	45.5	83.1	20.2	54.0	18.1
Saturated unconfined compressive strength (MPa)	21.3	64.8	–	–	–
Point load strength (MPa)	3.8	6.2	–	4.1	0.9
Scleroscope hardness	32	49	–	–	–
Schmidt hardness	27	39	–	–	–
Young's modulus ($\times 10^3$ MPa)	25	45	5.2	26.5	–

of weathering debris in excess of cobble size may be produced. Subsequent degradation down to component grain size takes place at a very slow rate.

13.3.2 Shales

Quartz usually accounts for approximately one-third of a normal shale, clay minerals, including micas and chlorite, for another third and other minerals such as feldspar, calcite, dolomite, pyrite, hematite and limonite, together with some carbonaceous matter, make up the remainder. The mineral content of shales influences their geotechnical properties, the most important factor in this respect being the quartz-clay minerals ratio. For example, the liquid limit of clay shales increases with increasing clay mineral content, the amount of montmorillonite, if present, being especially important. Mineralogy also affects the activity of an argillaceous material, again this increases with clay mineral content, particularly with increasing content of montmorillonite. Activity influences the slaking characteristics of a shale.

Consolidation with concomitant recrystallisation on the one hand and the parallel orientation of platy minerals, notably micas, on the other give rise to the fissility of shales. An increasing content of siliceous or calcareous material gives a less fissile shale whilst carbonaceous or organic shales are exceptionally fissile. Moderate weathering increases the fissility of shale by partially removing the cementing agents along the laminations or by expansion due to the hydration of clay particles. Intense weathering produces a soft clay-like soil. Ingram (1953) recognised massive, flaggy and flaky varieties of shale according to their degree of fissility. Some shales exhibit all degrees of fissility in the same bed. Spears (1980) related the fissility of shales to their quartz content (see Chapter 3).

Shale is frequently regarded as an undesirable material to work in. Certainly there have been many failures of structures founded on slopes in shales. Nevertheless many shales have proved satisfactory as foundation rocks. Hence it can be concluded that shales vary widely in their engineering behaviour and that it is therefore necessary to determine the problematic types. This variation in engineering behaviour to a large extent depends upon their degree of compaction and cementation. Indeed Mead (1936) divided shales into compaction and cementation types in his classification (Table 13.4). The cemented shales are invariably stronger and more durable. Marine shales may be impregnated with carbonate cement which makes them appreciably stronger and, of course, they may grade into impure limestones. Carbonaceous shales contain a significant proportion of organic matter and are therefore softer.

The natural moisture content of shales varies from less than 5%, increasing to as high as 35% for some clayey shales. When the natural moisture content of shales exceeds 20% they frequently are suspect as they tend to develop potentially high pore pressures. Usually the moisture content in the weathered zone is higher than in the unweathered shale beneath.

Depending upon the relative humidity, many shales slake almost immediately when exposed to air (see Kennard et al. (1967)). Desiccation of shale, following exposure, leads to the creation of negative pore pressures and consequent tensile failure of the weak intercrystalline bonds. This leads to

Table 13.4. Classification of shale (after Mead (1936)

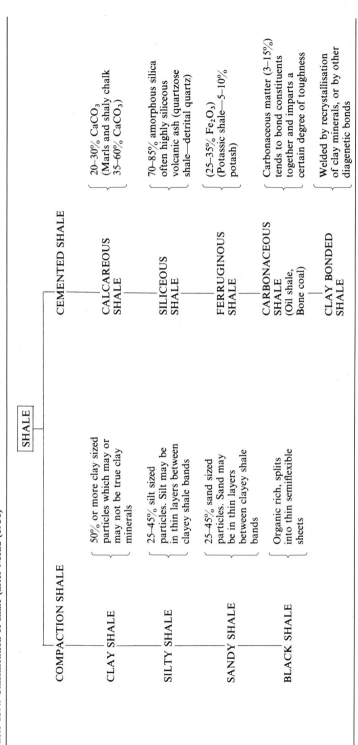

SHALE

COMPACTION SHALE

CLAY SHALE — 50% or more clay sized particles which may or may not be true clay minerals

SILTY SHALE — 25–45% silt sized particles. Silt may be in thin layers between clayey shale bands

SANDY SHALE — 25–45% sand sized particles. Sand may be in thin layers between clayey shale bands

BLACK SHALE — Organic rich, splits into thin semiflexible sheets

CEMENTED SHALE

CALCAREOUS SHALE — 20–30% $CaCO_3$ (Marls and shaly chalk 35–60% $CaCO_3$)

SILICEOUS SHALE — 70–85% amorphous silica often highly siliceous volcanic ash (quartzose shale—detrital quartz)

FERRUGINOUS SHALE — (25–35% Fe_2O_3) (Potassic shale—5–10% potash)

CARBONACEOUS SHALE (Oil shale, Bone coal) — Carbonaceous matter (3–15%) tends to bond constituents together and imparts a certain degree of toughness

CLAY BONDED SHALE — Welded by recrystallisation of clay minerals, or by other diagenetic bonds

the production of shale particles of coarse sand, fine gravel size. Alternate wetting and drying causes a rapid breakdown of compaction shales. Low-grade compaction shales, in particular, undergo complete disintegration after several cycles of drying and wetting. Indeed De Graft-Johnson *et al.* (1973) found that the compacted variety of Accra Shale could be distinguished from the cemented variety by wetting and drying tests. The compacted shales generally crumbled to fine material after 2 or 3 cycles, whilst the cemented samples withstood 6 cycles, none of the samples exceeding a loss of 8%. This indicates that well cemented shales are fairly resistant to slaking.

Morgenstern and Eigenbrod (1974) used a water absorption test to assess the amount of slaking undergone by argillaceous material. This test measures the increase of water content in relation to the number of drying and wetting cycles undergone. They found that the maximum slaking water content increased linearly with increasing liquid limit and that during slaking all materials eventually reached a final water content equal to their liquid limit. Materials with medium to high liquid limits exhibited very substantial volume changes during each wetting stage, which caused large differential strains, resulting in complete destruction of the original structure. Thus materials characterised by high liquid limits are more severely weakened during slaking than materials with low liquid limits.

The primary problem attributable to slaking of shale during construction is that upon exposure it becomes coated with mud when wetted, which prevents the development of bond between concrete and rock. This can be prevented by coating the surface with a protective material, or by pouring a protective concrete cover immediately after exposure. Slaking of shales after construction causes ravelling and spalling of cut slopes and is sometimes the cause of undermining and collapse of more competent beds, but this is rarely a serious problem. Mudstones tend to break down along irregular fracture patterns, which when well developed, can mean that those rocks disintegrate within one or two cycles of wetting and drying.

Lamination effects an important control on the breakdown of shales (see Taylor and Spears (1970)). Other controls on the breakdown of shaly materials include air breakage and dispersal of colloidal material (see Chapter 7 and Badger *et al.* (1956)). A feature of the breakdown of shales and mudstones is their disintegration to produce silty clays (see Grice (1969)).

The swelling properties of certain clay shales have proven extremely detrimental to the integrity of many civil engineering structures. Swelling is attributable to the absorption of free water by certain clay minerals, notably, montmorillonite, in the clay fraction of a shale. Casagrande (1949) noted that highly fissured overconsolidated shales have greater swelling tendencies than poorly fissured clayey shales, the fissures providing access for water.

The degree of packing, and hence the porosity, void ratio and density of shale, depends on its mineral composition and grain size distribution, its mode of sedimentation, its subsequent depth of burial and tectonic history, and the effects of diagenesis. The porosity of shale may range from slightly under 5 to just over 50%, with natural moisture contents of 3–35%. Argillaceous materials are capable of undergoing appreciable suction before pore water is removed, drainage commencing when the necessary air-entry suction is achieved (about pF 2). Under increasing suction pressure the incoming air drives out water from a shale and some shrinkage takes place in the fabric before air can offer support. Generally as the natural moisture content and liquid limit increase so the effectiveness of soil suction declines.

It would appear that the strength of compacted shales decreases exponentially with increasing void ratio and moisture content. However, in cemented shales the amount and strength of the cementing material are the important factors influencing intact strength.

Morgenstern and Eigenbrod (1974) carried out a series of compression softening tests on argillaceous materials. They found that the rate of softening of these materials when immersed in water largely depends upon their degree of induration. Poorly indurated materials soften very quickly and they may undergo a loss of up to 90% of their original strength within a few hours. Mudstones, at their natural water content, remain intact when immersed in water. However, they swell slowly, hence decreasing in bulk density and strength. This time-dependent loss in strength is a very significant engineering property of mudstones. A good correlation exists between their initial compressive strength and the amount of strength loss during softening.

The value of shearing resistance of a shale chosen for design purposes will lie somewhere between the peak and residual strengths. In weak compaction shales cohesion may be lower than 15 kPa and the angle of friction as low as 5°. By contrast, Underwood (1967) quoted values of cohesion and angle of friction of 750 kPa and 56° respectively for dolomitic shales of Ordovician age, and 8–23 MPa and 45–64° respectively for calcareous and quartzose shales from the Cambrian period. Generally shales with a cohesion less than 20 kPa and an apparent angle of friction of less than 20° are likely to present problems. The elastic moduli of compaction shales range from less than 140–1400 MPa, whilst those of well cemented shales have elastic moduli in excess of 14 000 MPa (see Table 13.5).

Unconfined compressive strength tests on Accra Shales carried out by De Graft-Johnson et al. (1973) indicated that the samples usually failed at strains between 1.5 and 3.5%. The compressive strengths varied from 200 kPa–20 MPa, with the values of the modulus of elasticity ranging from 6.5–1400 MPa. Those samples which exhibited the high strengths were generally cemented types.

According to Zaruba and Bukovansky (1965) the mechanical properties of the Ordovician shales in the Prague area are controlled by their lithological composition, degree of weathering and amount of tectonic disturbance. They quoted results of numerous loading tests which indicated that the moduli of elasticity were influenced more by the degree of weathering than by the lithological composition of the rock. However, when tests were performed in galleries on unweathered shales, the material which was predominantly pelitic gave lower values of the moduli of elasticity than that which was predominantly quartzose.

The higher the degree of fissility possessed by a shale, the greater is the anisotropy with regard to strength, deformation and permeability. For instance, the influence of fissility on Young's modulus can be illustrated by two values quoted by Chappell (1974), 6000 and 7250 MPa, for cemented shale tested parallel and normal to the lamination respectively. Zaruba and Bukovansky (1965) found that the values of Young's modulus were up to five times greater when they tested shale normal as opposed to parallel to the direction of lamination.

Table 13.5. An engineering evaluation of shales (after Underwood (1967))

Laboratory tests and in situ observations	Physical properties — Average range of values		Probable in situ behaviour*						
	Unfavourable	Favourable	High pore pressure	Low bearing capacity	Tendency to rebound	Slope stability problems	Rapid sinking	Rapid erosion	Tunnel support problems
Compressive strength (kPa)	350–2070	2070–3500	✓	✓✓					✓✓
Modulus of elasticity (MPa)	140–1400	1400–14 000			✓✓				✓✓
Cohesive strength (kPa)	35–700	700–>10 500			✓✓	✓			✓
Angle of internal friction (deg)	10–20	20–65			✓	✓			
Dry density (t/m³)	1.12–1.78	1.78–2.56	✓					✓(?)	✓
Potential swell (%)	3–15	1–3	✓			✓		✓✓	
Natural moisture content (%)	20–35	5–15	✓↓			✓↓			
Coefficient of permeability (m/s)	10^{-7}–10^{-12}	$>10^{-7}$	✓			✓	↓		
Predominant clay minerals	Montmorillonite or illite	Kaolinite and chlorite	✓			✓	✓		
Activity ratio	0.75–>2.0	0.35–0.75				✓		✓	
Wetting and drying cycles	Reduces to grain sizes	Reduces to flakes		✓✓		✓✓		✓(?)	✓✓
Spacing of rock defects	Closely spaced	Widely spaced							
Orientation of rock defects	Adversely oriented	Favourably oriented			✓	✓			✓
State of stress	>Existing over-burden load	≃Over-burden load			✓				✓

Note: According to S. Irmay (*Israel Journal of Technology*, 1968, **6**, No. 4, 165–72), the maximum possible $\phi = 47.5°$
*The ticks relate to the unfavourable range of values

According to Burwell (1950) well cemented shales, under structurally sound conditions, present few problems for large structures such as dams, though their strength limitations and elastic properties may be factors of importance in the design of concrete dams of appreciable height. They, however, have lower moduli of elasticity and generally lower shear values than concrete and therefore in general are unsatisfactory foundation materials for arch dams.

The problem of settlement in shales generally resolves itself into one of reducing the unit bearing load by widening the base of structures or using spread footings. In some cases appreciable differential settlements are provided for by designing articulated structures capable of accommodating differential movements of individual sections without damage to the structure. Severe settlements may take place in low-grade compaction shales. However, compaction shales contain fewer open joints or fissures which can be compressed beneath heavy structures than do cemented shales. Where concrete structures are to be founded on shale and it is suspected that the structural load will lead to closure of defects in the rock, in situ tests should be conducted to determine the elastic modulus of the foundation material.

Uplift frequently occurs in excavations in shales and is attributable to swelling and heave. Rebound on unloading of shales during excavation is attributed to heave due to the release of stored strain energy. The conserved strain energy tends to be released more slowly than it is in harder rocks. Shale relaxes towards a newly excavated face and sometimes this occurs as offsets at weaker seams in the shale. The greatest amount of rebound occurs in heavily overconsolidated compaction shales; for example, at Garrison Dam, North Dakota, just over 0.9 m of rebound was measured in the deepest excavation in the Fort Union Clay Shales. What is more high horizontal residual stresses caused saw cuts 75 mm in width to close within 24 h (see Smith and Redlinger (1953)). Some 280 mm of rebound was recorded in the underground workings in the Pierre Shale at Oake Dam, South Dakota (see Underwood et al. (1964)). Moreover differential rebound occurred in the stilling basin due to the presence of a fault. Differential rebound movements require special design provision.

The stability of slopes in excavations can be a major problem in shale both during and after construction. This problem becomes particularly acute in dipping formations and in formations containing expansive clay minerals. Sulphur compounds are frequently present in shales, clays, mudstones and marls. An expansion in volume large enough to cause structural damage can occur when sulphide minerals such as pyrite and marcasite suffer oxidation to give anhydrous and hydrous sulphates.

According to Fasiska et al. (1974) the pyrite structure may be regarded as a stacking of almost close packed hexagonal sheets of sulphide ions with iron ions in the interstices between the sulphide layers. The packing density is related to the radius of the sulphide ion which is 1.85 Å (volume = 26.1 Å^3). In the sulphate structure each atom of sulphur is surrounded by four atoms of oxygen in tetrahedral coordination. The packing density is related to the radius of the sulphate ion which is 2.8 Å, giving a volume of 92.4 Å. This represents an increase in volume per packing unit of approximately 350%. Hydration involves a further increase in volume. In fact such a reaction is electrolytic, that is, water is required and the sulphate ion exists in solution.

Any cation in the system may cause the precipitation of sulphate crystals. If calcium carbonate is present, gypsum may be formed, which may give rise to an eight-fold increase in volume over the original sulphide, exerting pressures of up to about 0.5 MPa. This leads to further disruption and weakening of the rocks involved.

Penner *et al.* (1973) quoted a case of heave in a black shale of Ordovician age in Ottawa which caused displacement of the basement floor of a three-storey building. The maximum movement totalled some 107 mm, the heave rate being almost 2 mm per month. When examined, the shale in the heaved zone was found to have been altered to a depth of between 0.7 and 1 m. Beneath, the unaltered shale contained numerous veins of pyrite, indeed the sulphur content was as high as 1.6%. The heave was attributable to the breakdown of pyrite to produce sulphur compounds which combine with calcium to form gypsum and jarosite. The latter minerals formed in the fissures and between the laminae of the shales in the altered zone. Measurements of the pH gave values ranging from 2.8–4.4. It was concluded that the alteration of the shales was the result of biochemical weathering brought about by autotrophic bacteria. The heaving was therefore arrested by creating conditions unfavourable for bacterial growth. This was done by neutralising the altered zone by introducing a potassium hydroxide solution into the examination pits. The water table in the altered zone was also kept artificially high so that the acids would be diffused and washed away, and to reduce air entry.

When a load is applied to an essentially saturated shale foundation the void ratio in the shale decreases and the pore water attempts to migrate to regions of lesser load. Because of its relative impermeability water becomes trapped in the voids in the shale and can only migrate slowly. As the load is increased there comes a point when it is in part transferred to the pore water, resulting in a build-up of pore pressure. Depending on the permeability of the shale and the rate of loading, the pore pressure can more or less increase in value so that it equals the pressure imposed by the load. This greatly reduces the shear strength of the shale and a serious failure can occur, especially in the weaker compaction shales. For instance, high pore pressure in the Pepper Shale was largely responsible for the foundation failure at Waco Dam, Texas (see Underwood (1967)). Pore pressure problems are not so important in cemented shales.

Clay shales usually have permeabilities of the order of 1×10^{-8} m/s to 10^{-12} m/s; whereas sandy and silty shales and closely jointed cemented shales may have permeabilities as high as 1×10^{-6} m/s. However, Jumikis (1965) noted that where the Brunswick Shale in New Jersey was highly fissured it could be used as an aquifer. He also noted that the build-up of groundwater pressure along joints could cause shale to lift bedding planes and lead to slabs of shale breaking from the surface. Hence in this shale formation subsurface water must be drained by an efficient system to keep excavations dry. The Brunswick Shale is generally covered with a mantle of weathered material which may be up to 1.2 m thick which grades upwards into a residual soil, which also may, in places, be a metre or so thick. These materials disintegrate rapidly on wetting and drying, and on freezing. In fact the silty residual soil is subject to frost heave and frost boils. The weathered material cannot support foundations for heavy structures.

13.3.3 Marls

The marls of the Keuper series in Britain consist of between 50 and 90% clay minerals. The marl may contain thin veins or beds of gypsum. In such cases the groundwaters contain sulphates. According to Dumbleton (1967) illite accounts for 28–56% of the clay minerals and chlorite may total some 39%. Usually more than half the chlorite is of the swelling type. Quartz tends to vary between 5 and 35%. The other minerals include calcite and dolomite (which usually comprise less than 20%), and hematite 1 or 2%. Occasionally other clay minerals such as sepiolite and palygorskite have been found.

The clay particles tend to be aggregated mainly into silt-size units, the aggregated structure being extremely variable (see Davis (1968)). Particles composing the aggregate are held together by cement. Sherwood (1967) suggested that silica might be the cementing agent whilst Lees (1965) assumed that the clay particles were bound together by physical forces. The former explanation seems the more likely. Aggregation leads to the lack of correlation between consistency limits and shear strength on the one hand and clay content on the other. Because engineering behaviour is controlled by the aggregates, rather than the individual clay minerals, the plasticity, according to Davis (1967) is lower than would be expected. He consequently proposed the aggregation ratio (A_r) as a means of assessing the degree of aggregation. The aggregation ratio was defined as the percentage weight of clay as determined by mineralogical analysis, expressed as a ratio of the percentage weight of clay particles determined by sedimentation techniques. On the other hand Sherwood (1967) maintained that aggregation did not give anomalous plasticity values for Keuper Marls and that they can generally be classified as materials of low to medium plasticity (Table 13.6). The activity of the Keuper Marls increases as the degree of aggregation.

Certain marls exhibit rapid softening when exposed to wet conditions. Such deterioration can be estimated by assessing their moisture adsorption potential (see Kolbuszewski et al. (1965)). These marls are fissured, weathering and water penetrating the fissures and thereby further reducing the strength of the material. The fissures close with increasing depth.

Skempton and Davis (see Chandler (1969)) proposed the following classification of weathered Keuper Marl.

Zone	Description
V Fully weathered	Matrix only. Plastic, slightly silty clay. May be fissured
IV Highly weathered	Matrix with occasional pellets <3 mm diameter. Little trace of original structure, although clay may be fissured. Permeability less than underlying layers
III Moderately weathered	Matrix with frequent lithorelicts up to 25 mm. As weathering progresses, lithorelics become less angular. Water content of matrix greater than that of lithorelics
II Slightly weathered	Angular blocks of unweathered marl. First indications of weathering; matrix starting to encroach along joints leading to 'spheroidal' weathering
I Unweathered	Marl (often fissured). Moisture content varies due to different lithology

Weathering first develops along fissures in Zone II material, the weathered product consisting of a thin veneer of silt. A significant proportion of the

Table 13.6. Soil classification tests on Keuper Marl (from Sherwood (1967))

	Clay content by sedimentation (%)	Clay content by X-ray analysis (%)	Liquid limit (%)	Plastic limit (%)	Plasticity index (%)
(1)	26	94	71	40	31
(2)	36	58	33	19	14
(3)	30	87	46	28	25
(4)	12	77	48	29	19

marl is weathered in Zone III, the unweathered material occurring as angular fragments set in a weathered matrix which is predominantly silty. The water content of the matrix exceeds that of the lithorelicts. In Zone IV the lithorelicts are mainly of coarse sand size, and the marl has lost much of its silty texture, indeed up to 50% may be composed of clay-sized particles. This indicates that particle aggregation is broken down upon weathering. Little or no trace of the original structure now remains and the material has a lower permeability than has Zone III, 5×10^{-9} to 5×10^{-10} m/s as compared with 1×10^{-8} to 1×10^{-9} m/s respectively. Finally the marl is completely weathered, becoming a plastic, slightly silty, clay. Chandler (1969) showed that highly and fully weathered marls could be distinguished from material from Zones I, II and III by their particle size distribution and plasticity index (Table 13.7).

An extensive review of the Keuper Marl as a foundation material has been provided by Meigh (1976).

Table 13.7. Some index properties of Keuper Marl (after Chandler (1969))

Index property	Weathering zone		
	I and II	III	IV
Bulk density (t/m^3)	2.5–2.3	2.3–2.1	2.2–1.8
Dry density (t/m^3)	2.4–1.9	2.1–1.8	1.8–1.4
Natural moisture content (%)	5–15	12–20	18–35
Liquid limit (%)	25–35	25–40	35–60
Plastic limit (%)	17–25	17–27	17–33
Plasticity index	10–15	10–18	17–35
% clay size (BS 1377)	10–35	10–35	30–50
Aggregation ratio (A_r)	10–2.5	10–2.5	2.5
c' (kPa)	≥ 27.6	≤ 17.2	≤ 17.2
ϕ' (deg)	40	42–32	32–25
ϕ'_r (deg)	32–23	29–22	24–18
*$\dfrac{\tau_{max} - \tau_{res}}{\tau_{max}}$ (%)	55	55–30	35–20

* Percentage reduction from peak to residual strength

13.3.4 Seatearths

Seatearths are almost invariably found beneath coal seams. Indeed they represent fossil soils and as such are characterised by the presence of fossil rootlets. These tend to destroy the lamination and bedding. The character of a seatearth depends on the type of deposit which was laid down immediately before the establishment of plant growth. If this was mud then a fireclay underlies the coal, whereas if silts and sands were deposited, then a gannister was subsequently formed. Many gannisters are pure siltstones and because they are usually well cemented they are hard and strong. Fire-clays with a low quartz content are typically highly slickensided and break easily along randomly orientated listric surfaces. The presence of listric surfaces may mean that a fireclay will disintegrate within a few cycles of wetting and drying (see Chapter 7).

13.4 Carbonate sedimentary rocks

Carbonate rocks contain more than 50% of carbonate minerals, amongst which calcite and/or dolomite predominate. Normally the term limestone is used for those rocks in which the carbonate fraction is composed principally of calcite and the term dolostone is reserved for those rocks in which dolomite accounts for more than half the carbonate fraction. Chalk is a rather peculiar type of soft, remarkably pure limestone which is characteristically developed in the Upper Cretaceous system.

The engineering properties of carbonate sediments are influenced by grain size and those post-depositional changes which bring about induration, and thereby increase strength. Consequently Fookes and Higginbottom (1975) chose these factors as the basis for their engineering classification of near-shore carbonate sediments (Figure 13.2). Induration of a carbonate sediment frequently starts during the early stages of deposition as a result of cementation which occurs where individual grains are in contact. Thus cementation is not solely dependent upon the influence of consolidation due to increasing overburden pressures. Because induration can take place at the same time as deposition is occurring, this means that carbonate sediments can sustain high overburden pressures, which, in turn, means that they can retain high void ratios to considerable depths, indeed a layer of cemented grains may overly one that is uncemented. Eventually, however, high overburden pressures, creep and recrystallisation produce crystalline limestone with very low porosity. A review of classifications of carbonate rocks in relation to engineering has been given by Burnett and Epps (1979).

Representative values of some physical properties of carbonate rocks are listed in Table 13.8. It can be seen that generally the density of these rocks

Table 13.8. Some physical properties of carbonate rocks

	Carbon-iferous Limestone (Buxton)	Magnesium Limestone (Anston)	Ancaster Freestone (Ancaster)	Bath Stone (Corsham)	Middle Chalk (Hillington)	Upper Chalk (Northfleet)
Relative density	2.71	2.83	3.70	2.71	2.70	2.69
Dry density (t/m³)	2.58	2.51	2.27	2.30	2.16	1.49
Porosity (%)	2.9	10.4	14.1	15.3	19.8	41.7
Dry unconfined compressive strength (MPa)	106.2	54.6	28.4	15.6	27.2	5.5
Saturated unconfined compressive strength (MPa)	83.9	36.6	16.8	9.3	12.3	1.7
Point load strength	3.5	2.7	1.9	0.9	0.4	—
Scleroscope hardness	53	43	38	23	17	6
Schmidt hardness	51	35	30	15	20	9
Youngs modulus (×10³ MPa)	66.9	41.3	19.5	16.1	30.0	4.4
Permeability (×10⁻⁹ m/s)	0.3	40.9	125.4	160.5	1.4	13.9

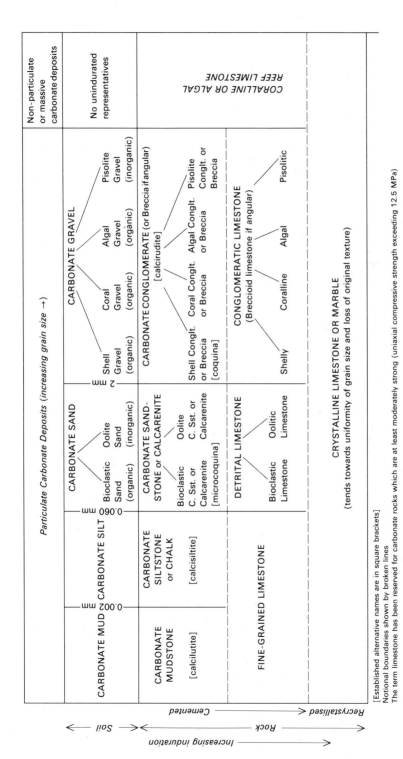

Figure 13.2 Proposed classification of pure carbonate sediments for engineering purposes (after Fookes and Higginbottom (1975))

increases with age, whilst the porosity is reduced. Diagenetic processes mainly account for the lower porosities of the Carboniferous and Magnesian Limestones quoted in the table. On the other hand the high porosity values of the Upper Chalk may be due to the presence of hollow tests and the complex shapes of the constituent particles. What is more the Upper Chalk is poorly cemented and has not suffered the same degree of pre-consolidation loading as the Middle and Lower Chalk.

Attewell (1971) noted that the saturation moisture content for the Great Limestone in the north of England, whether dolomitised or not, was less than 3%. This meant that when subjected to freeze–thaw testing the material exhibited no deleterious effects. The uniaxial compressive strength of the limestone ranged from around 110–210 MPa and it had an average porosity of some 4%. That limestone which had been dolomitised had a higher porosity, averaging approximately 7.5%, and it possessed a lower uniaxial compressive strength, varying from 70–165 MPa.

Age often has an influence on the strength and deformation characteristics of carbonate and other sedimentary rocks. From Table 13.8 it can be seen that Carboniferous Limestone is generally very strong (see Bell (1981)). Conversely the Bath Stone (Great Oolite, Jurassic) is only just moderately strong. Similarly the oldest limestones tend to have the highest values of Young's modulus.

A review of the engineering properties of the Miami Limestone in relation to its performance as a foundation has been provided by Kaderabek and Reynolds (1981).

13.4.1 Limestones and solution

Thick bedded, horizontally lying limestones relatively free from solution cavities afford excellent foundations. On the other hand thin bedded, highly folded or cavernous limestones are likely to present serious foundation problems. A possibility of sliding may exist in thinly bedded, folded sequences. Similarly if beds are separated by layers of clay or shale, especially when inclined, these may serve as sliding planes and result in failure.

Limestones are commonly transected by joints. These have generally been subjected to various degrees of dissolution so that some may gape (Figure 7.24a). Rainwater is generally weakly acidic and further acids may be taken into solution from organic or mineral matter. The degree of aggressiveness of water to limestone can be assessed on the basis of the relationship between the dissolved carbonate content, the pH value and the temperature of the water. At any given pH value, the cooler the water the more aggressive it is. If solution continues its rate slackens and it eventually ceases when saturation is reached. Hence solution is greatest when the bicarbonate saturation is low. This occurs when water is circulating so that fresh supplies with low lime saturation are continually made available. As James and Kirkpatrick (1980) noted, a material dissolves at a rate and in a manner which is influenced by its solubility (c_s) and specific solution rate constant (K) (Table 13.9). The form a particular mineral adopts does not influence its solution rate. Not only is the rate of flow significant but the area of material exposed to flowing water is also important. Non-saline water can dissolve up to 400 ppm of calcium carbonate.

Sinkholes may develop where joints intersect and these may lead to an integrated system of subterranean galleries and caverns. The latter are characteristic of thick massive limestones. The progressive opening of discontinuities by solutioning leads to an increase in mass permeability. Sometimes dissolution produces a highly irregular pinnacled surface on limestone pavements. The size, form, abundance and downward extent of the aforementioned features depends upon the geological structure and the presence of interbedded impervious layers. Individual cavities may be open; they may be partially or completely filled with clay, silt, sand or gravel mixtures, or they may be water-filled conduits. Solution cavities present numerous problems in the construction of large foundations such as for dams, among which bearing strength and watertightness are paramount. Few sites are so bad that it is impossible to construct safe and successful structures upon them, but the cost of the necessary remedial treatment may be prohibitive. Dam sites should be abandoned where the cavities are large and numerous and extend to considerable depths.

An important effect of solution in limestone is enlargement of the pores, which enhances water circulation thereby encouraging further solution. This brings about an increase in stress within the remaining rock framework which reduces the strength of the rock mass and leads to increasing stress corrosion. On loading the volume of the voids is reduced by fracture of the weakened cement between the particles and by the reorientation of the intact aggregations of rock that become separated by loss of bonding. Most of the resultant settlement takes place rapidly within a few days of the application of load. Sowers (1975) related a case of settlement of two- and three-storey reinforced concrete buildings founded on soft porous oolitic limestone in Florida, which had been subjected to dissolution. The limestone had been leached and was highly porous with little induration remaining 2.1 m below the rock surface. One of the buildings was surrounded by a crack in the more intact surface crust and settlement exceeding 100 mm was recorded. Standard penetration resistances obtained from the weakened oolite ranged between 1 and 2 blows per 0.3 m.

Rapid subsidence can take place due to the collapse of holes and cavities within limestone which has been subjected to prolonged solution, this occurring when the roof rocks are no longer thick enough to support themselves. It must be emphasised, however, that the solution of limestone is a very slow process; contemporary solution is therefore very rarely the cause of collapse. For instance, Kennard and Knill (1968) quoted mean rates of surface lowering of limestone areas in the British Isles which ranged from 0.041–0.099 mm annually. They also quoted experiments carried out in flowing non-saline water which produced an average solution rate of approximately 1 mm per year.

Nevertheless solution may be accelerated by man-made changes in the groundwater conditions or by a change in the character of the surface water that drains into limestone. For instance, James and Kirkpatrick (1980), in a consideration of the location of hydraulic structures on soluble rocks, wrote that if such dry fissured rocks are subjected to substantial hydraulic gradients then they will undergo dissolution along these fissures, hence leading to rapidly accelerating seepage rates. From experimental work which they carried out on the Portland Limestone they found that the values of solution

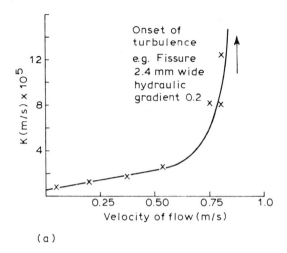

(a)

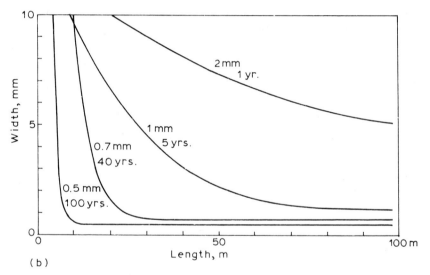

(b)

Figure 13.3 (a) Rate of solution (K) of Portland Stone and its dependence on flow velocity (after James and Kirkpatrick (1980)). (b) Enlargement of fissures in calcium carbonate rock by pure flowing water (after James and Kirkpatrick (1980))

rate constant (K) increased appreciably at flow velocities which corresponded to a transitional flow regime. They showed that such a flow regime occurred in fissures about 2.5 mm in width which experienced a hydraulic gradient of 0.2 (Figure 13.3a). According to these two authors solution takes place along a small fissure by retreat of the inlet face due to removal of soluble material (Table 13.9; Figure 13.3b). Dissolution of larger fissures gives rise to long tapered enlargements, that enable seepage rates to increase rapidly and runaway situations to develop.

A classic example of a failure of a hydraulic structure founded on limestone is the Hales Bar Dam and Reservoir on the Tennessee (see Frink (1946)). There solution rates varying between 1.5 and 3.1 mm/year were recorded. North (1951) described collapses of cavities in the Lias Limestone in Bridgend, Wales. These collapses were caused by ancient cavities being enlarged by excessive water discharging through them. In one case water was escaping from a water main, in another leakage was occurring from storm drains. North noted that one collapsed cavity which was filled, collapsed again the following year. The reason given was that the supporting fill was washed by flowing water from the cavity into a fissure. The obvious conclusion is that where possible such troublesome cavities should be lined prior to treatment with fill.

Differential settlement under loading has occurred in limestone which at ground level appeared competent. Unfortunately, immediately beneath the surface the limestone consisted of long narrow pinnacles of rock separated by solution channels occupied by broken limestone rubble in a clay matrix (see Early and Dyer (1964)).

According to Sowers (1975) ravelling failures are the most widespread and

Table 13.9. Solution of soluble rocks (after James and Kirkpatrick (1980))

Rock	(a) Solubility (c_s) in pure water c_s (kg/m³ at 10°C)*	(b) Solution rate constants (K) at 10°C (flow velocity—0.05 m/s)	
		m/s × 10⁵*	m^4 (kg/s × 10⁶)
		$\theta = 1$	$\theta = 2$
Halite	360.0	0.3	
Gypsum	2.5	0.2	
Anhydrite	2.0		0.8
Limestone	0.015	0.4	

* c_s is dependent upon temperature and the presence of other dissolved salts

* K is dependent on temperature, flow velocity and other dissolved salts
θ = order of dissolution reaction

(c) Limiting widths of fissures in massive rock*

Rock	(i) Upper limit for stable inlet face retreat	(ii) For a rate of retreat of 0.1 m/year (e.g. cavern formation)	(iii) Maximum safe lugeon value†
Halite	0.05	0.05	
Gypsum	0.2	0.3	50
Anhydrite	0.1	0.2	7
Limestone	0.5	1.5	700

* These values are for pure water, at a fissure spacing in massive rock of one per metre and a hydraulic gradient of 0.2. For water containing 300 mg/l of CO_2, the stable limit width of fissure becomes 0.4 mm in limestone.
† One lugeon unit is equal to a flow of 1 l/m/min at a pressure of 1 MPa (it is approximately equal to a coefficient of permeability or hydraulic conductivity of 10^{-7} m/s).

(d) Solution of particulate deposits*

Rock	(i) Limiting seepage velocity (m/s)	(ii) Width of solution zone (m)
Halite	6.0×10^{-9}	0.0002
Gypsum	1.4×10^{-6}	0.04
Anhydrite	1.6×10^{-6}	0.09
Limestone	3.0×10^{-4}	2.8

* Rate of movement of solution zone—0.1 m/year. Mineral particles of 50 mm diameter. Pure water.

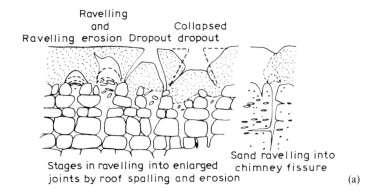

Ravelling
and   Collapsed
Ravelling erosion Dropout dropout

Sand ravelling into
chimney fissure

Stages in ravelling into enlarged
joints by roof spalling and erosion

(a)

(b)

Figure 13.4 (a) Mechanisms of ravelling collapse. (b) House being swallowed by ravelling collapse in Central Florida (courtesy Professor George F. Sowers and P. Kehoe

probably the most dangerous of all the subsidence phenomena associated with limestones. Ravelling occurs when solution enlarged openings extend upward to a rock surface overlain by soil (Figure 13.4). The openings should be interconnected and lead into channels through which the soil can be eroded by groundwater flow. Initially the soil arches over the openings but as they are enlarged a stage is reached when the soil above the roof can no longer support itself and so it collapses. A number of conditions accelerate the development of cavities in the soil and initiate collapse. Rapid changes in moisture content lead to aggravated slabbing or roofing in clays and flow in cohesionless sands. Lowering the water table increases the downward seepage gradient and accelerates downward erosion; reduces capillary attraction in sand and increases instability of flow through narrow openings and gives rise to shrinkage cracks in highly plastic clays which weakens the mass in dry weather and produces concentrated seepage during rains. Increased infiltration often initiates failure, particularly when it follows a period when the water table has been lowered.

A review of catastrophic collapses which have occurred since the 1930s in the karst terrain of Missouri has been provided by Williams and Vineyard (1976). Changes in soil moisture content and groundwater flow account for most failures in karst terrain and the type of failure (slow subsidence or rapid collapse) is partly related to the intensity and magnitude of the changes. The mechanical properties of the soil overlying bedrock and the hydrological characteristics of the region also influence failure. For example, collapses apparently are not common in poorly drained surface soils. However, some of the residual soils overlying limestone are well drained with macroscopic fractures and relict fabric preserved. Williams and Vineyard maintained that the downward weathering of bedrock was a more common cause of collapse and subsidence than failure by upwards migration of cavern roofs. They noted that most collapses in Missouri had occurred where there was a reasonably thick soil cover, few collapses having occurred where this was less than 12 m. Nonetheless, no collapses have been recorded in Missouri where soil thicknesses are greater than 65 m, but they frequently occur where soil thickness ranges from 12–30 m. It would appear that the pH value of soil is higher where soil is thicker, ranging from 4–5 in suspect areas investigated. It is around 7 in areas of thin soil cover.

Williams and Vineyard described vertical shafts developed in soil cover, which were revealed during excavation of a site at Fort Leonards Wood. These shafts had not reached the surface but were over 20 m in depth and several metres in diameter. They probably would have given rise to collapse and cratering at some future date. The soil material was being transported away by streams flowing through passages in the underlying limestone.

Solution features are also developed in dolostone. Jennings et al. (1965) described the collapse of a three-storey crusher plant into a sinkhole at Westdriefontein. They noted that some of the sinkholes in the Transvaal Dolomite were 100 m in diameter and 40 m in depth. This particular collapse took place suddenly and the authors suggested that it was influenced by the lowering of the water table, by up to 120 m, due to pumping from mines. This meant that fissures in the roofs of caverns in the dolostone were enlarged by dissolution due to percolating water. Avens (i.e. chimneys) were developed in the roofs and they spread into the overlying mantle of soil (Figure 13.5). Hence the arches above individual caverns were seriously weakened, this ultimately leading to collapse. Mine dewatering has also led to catastrophic collapses in limestone areas in the United States (see Williams and Vineyard (1976)).

13.4.2 Chalk

With the exception of certain horizons in the Lower Chalk of south-east England which possess an appreciable mud content, the Chalk is a remarkably pure soft limestone, usually containing over 95% calcium carbonate. Generally chalk can be divided into coarse and fine fractions. The coarse fraction, which may constitute 20–30%, falls within the 10–100 μm range.

Negligible ground movement, arch partially or fully developed

Ground movement appreciable, arch not formed

Air-filled void

1946 water table

(a)

(b)

Figure 13.5 (a) Section through ground showing conditions leading to a void liable to cause a sinkhole (left) and conditions leading to caving subsidence (right) (after Jennings *et al.* (1965)). (b) Collapse of crushing plant into a sinkhole. Westdriefontein Mine, December 1962 (courtesy Gold Fields of South Africa Ltd)

This contains material derived from the mechanical breakdown of large shelled organisms and, to a lesser extent, from foraminifera. The fine fraction, which takes the form of calcite particles about one micrometre or less in size, is almost entirely composed of coccoliths and may form up to, and sometimes over 80% of certain horizons.

Bell (1977) found a notable range in the dry density of chalk, which substantiated the work of Higginbottom (1965). For example, low values have been recorded from the Upper Chalk of Kent (1.35–1.64 t/m³) whilst those from the Middle Chalk of Norfolk and the Lower Chalk of Yorkshire frequently exceed 2.0 t/m³. In other words the density of the Chalk tends to increase with depth, presumably due to increasing overburden pressures, the void ratio decreasing. The porosity of chalk tends to range between 30 and 50%. Carter and Mallard (1974) found that chalk compressed elastically up to a critical pressure, the apparent preconsolidation pressure. Marked breakdown and substantial consolidation occurs at higher pressures. The apparent preconsolidation pressure is influenced by consolidation cementation and possibly creep. They obtained coefficients of consolidation (c_v) and compressibility (m_v) similar to those found by Meigh and Early (1957) and Wakeling (1965); $c_v = 1135$ m²/year, $m_v = 0.019$ m²/MN. The unconfined strength of the Chalk ranges from moderately weak (much of the Upper Chalk) to moderately strong (much of the Lower Chalk of Yorkshire and the Middle Chalk of Norfolk) according to the strength classification recommended by the Geological Society (1970). However, the unconfined compressive strength of chalk undergoes a marked reduction when it is saturated. For instance, according to Bell, some samples of Upper Chalk from Kent suffer a dramatic loss on saturation amounting to approximately 70%. Samples from the Lower and Middle Chalk may show a reduction in strength averaging some 50%. The pore water also has a critical influence on the triaxial strengths of chalk and obviously its modulus of deformation (see Meigh and Early (1957)).

Bell (1977) noted that the mode of failure of the Upper Chalk when tested in triaxial conditions was influenced by the confining pressure. Diagonal shear failure occurred at lower confining pressures but at 4.9 MPa confining pressure and above, plastic deformation took place, giving rise to barrel-shaped failures in which numerous small inclined shear planes were developed. Meigh and Early (1957) previously had explained such behaviour suggesting that at high cell pressures specimens of chalk underwent disaggregation. This they demonstrated by wetting and drying failed specimens which brought about their structural collapse.

Chalk is a non-elastic rock ($E_i = < 5 \times 10^4$ MPa), and the Upper Chalk from Kent is particularly deformable, a typical value of Young's modulus being 5×10^3 MPa. In fact the Upper Chalk exhibits elastic-plastic deformation, with perhaps incipient creep, prior to failure. The deformation properties of chalk in the field depend upon its hardness, and the spacing, tightness and orientation of its discontinuities. The values of Young's modulus are also influenced by the amount of weathering chalk has undergone. Using these factors, Ward et al. (1968) classified the Middle Chalk at Mundford, Norfolk, into five grades, and showed that the value of Young's modulus varies with grade (Table 13.10). They pointed out that Grades IV and V were largely the result of weathering and were therefore independent

Table 13.10. Correlation between grades and the mechanical properties of Middle Chalk at Mundford (after Ward et al. (1968))*

Grade	Description	Approx. range of E (MPa)	Approx. value of E_{dyn} (MPa) (after Abbiss (1979))	Range of compression wave velocities (km/s) (after Grainger et al. (1973))	Bearing pressure causing yield (kPa)	Creep properties	SPT N value (after Wakeling (1970))†	Rock mass factor (after Burland and Lord (1970))
V	Structureless melange. Unweathered and partly weathered angular chalk blocks and fragments set in a matrix of deeply weathered remoulded chalk. Bedding and jointing are absent.	Below 500	Below 500	0.65–0.75	Below 200	Exhibits significant creep	Below 15	0.1
IV	Friable to rubbly chalk. Unweathered or partially weathered chalk with bedding and jointing present. Joints and small fractures closely spaced, ranging from 10–60 mm apart.	500–1000	800	1.0–1.2	200–400	Exhibits significant creep	15–20	0.1–0.2
III	Rubbly to blocky chalk. Unweathered medium to hard chalk with joints 60–200 mm apart. Joints open up to 8 mm, sometimes with secondary staining and fragmentary infillings.	1000–2000	4000	1.6–1.8	400–600	For pressures not exceeding 400 kPa creep is small and terminates in a few months	20–25	0.2–0.4
II	Medium hard chalk with widely spaced, closed joints. Joints more than 200 mm apart. Fractures irregularly when excavated, does not break along joints. Unweathered.	2000–5000	7000	2.2–2.3	Over 1000	Negligible creep for pressure of at least 400 kPa	25–35	0.6–0.8
I	Hard, brittle chalk with widely spaced, closed joints. Unweathered.	Over 5000	Over 10 000	Over 2.3	Over 1000	Negligible creep for pressure of at least 400 kPa	Over 35	Over 0.8

* Ward et al. emphasised that their classification was specifically developed for the site at Mundford and hence its application elsewhere should be made with caution
† The correlation between SPT N value and grade may be different in the Upper Chalk (see Dennehy (1976))

of lithology whereas Grades I and II were completely unweathered so that the difference between them was governed by their lithological character. Grades V, IV and III occur in succession from the surface down whilst Grade I may overlie Grade II or vice versa. Burland and Lord (1970) observed that both the full-scale tank test and plate load tests at Mundford indicated that at low applied pressures even Grade IV chalk behaves elastically. At higher pressures chalk exhibits yielding behaviour. They pointed out that for stresses up to 1.0 MPa the plate load tests showed that Grades IV and V exhibit significant creep and in the long term, creep deflections may be considerably larger than immediate deflections. Creep in Grade III is smaller and terminates more rapidly whilst Grades II and I undergo negligible creep.

The dynamically determined value of Young's modulus in the Middle Chalk at Mundford has been shown to increase with increasing depth from 1.8×10^3 MPa at 5 m to 10.2×10^3 MPa at 19.3 m (see Abbiss (1979)). These values are more than double those of Young's modulus determined by static methods (Table 13.10). The corresponding values of seismic velocities for these five grades of chalk were determined by Grainger et al. (1973) and are given in Table 13.10.

Burland et al. (1974) found that settlements of a five-storey building founded in soft low-grade chalk at Reading were very small. Their findings agreed favourably with those previously obtained at Mundford, as did the results of an investigation carried out at Basingstoke by Kee et al. (1974).

As in limestone, the discontinuities are the fundamental factors governing the mass permeability of chalk (see Ineson (1962)). Chalk is also subject to dissolution along discontinuities. However, subterranean solution features tend not to develop in chalk since it is usually softer than limestone and so collapses as solution occurs. Nevertheless, solution pipes and swallow holes are present in the Chalk, being commonly found near the contact of the Chalk with the overlying Tertiary and drift deposits. West and Dumbleton (1972) suggested that high concentrations of water, such as run-off from roads, can lead to the re-activation of swallow holes and the formation of small pipes within a few years. They also recorded that new swallow holes often appear at the surface without warning after a period of heavy rain or following the passage of plant across a site. Moreover they found that voids can gradually migrate upwards through chalk due to material collapsing. Lowering of the chalk surface beneath overlying deposits due to solution can occur, disturbing the latter deposits and lowering their degree of packing. Hence the chalk surface may be extremely irregular in places.

Chalk during cold weather may suffer frost heave (see Lewis and Croney (1965)), ice lenses up to 25 mm in thickness being developed along bedding planes. Higginbottom (1965) suggested that a probable volume increase of some 20–30% of the original thickness of the ground may ultimately result.

Solifluction deposits known as head or combe deposits, and which were formed under periglacial conditions during Pleistocene times, are commonly found along valley bottoms carved in the Chalk of southern England. Head is a poorly stratified, poorly sorted deposit of angular chalk fragments, frequently set in a matrix of remoulded, pasty, fine chalk detritus. Combe deposits are, to a varying degree, cemented with secondary carbonate. Frost shattering also took place during Pleistocene times and its effects in the Chalk of southern England may extend to depths of several metres.

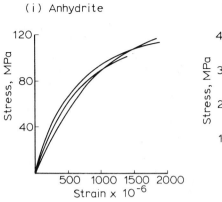

(i) Anhydrite

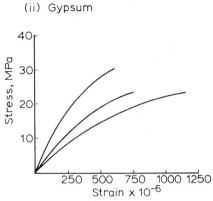

(ii) Gypsum

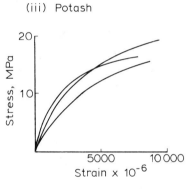

(iii) Potash

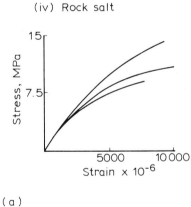

(iv) Rock salt

(a)

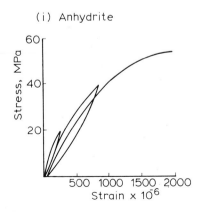

(i) Anhydrite

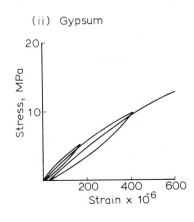

(ii) Gypsum

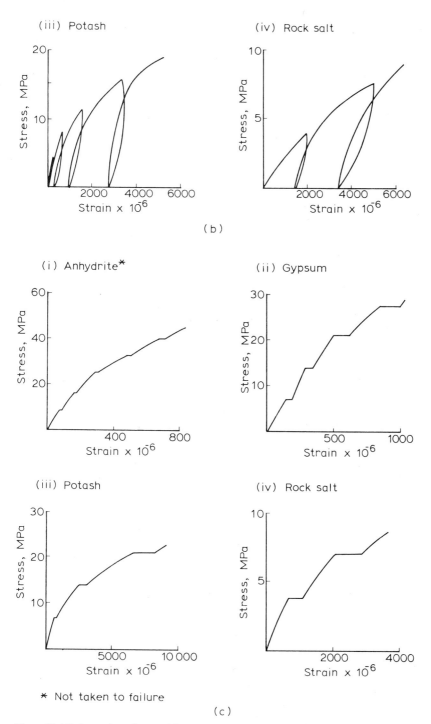

Figure 13.6 Deformation of evaporitic rocks. (a) Some typical stress–strain curves. (b) Hysteresis curves. (c) Incremental creep curves

13.5 Evaporitic sedimentary rocks

Representative relative densities and dry densities of gypsum, anhydrite, rock salt and potash are given in Table 13.11, as are porosity values. Anhydrite according to the classification of unconfined compressive strength (Geological Society (1970)) is a strong rock, gypsum and potash are moderately strong, whilst rock salt is moderately weak (Table 13.11). Values of Young's modulus are also given in Table 13.11, from which it can be ascertained that gypsum and anhydrite have high values of modulus ratio whilst potash and rock salt have medium values. Evaporitic rocks exhibit varying degrees of plastic deformation before failing. For example, in rock salt the yield strength may be as little as one-tenth the ultimate compressive strength, whereas anhydrite undergoes comparatively little plastic deformation prior to rupture (Figure 13.6). Creep may account for anything between 20 and 60% of the strain at failure when these evaporitic rocks are subjected to incremental creep tests (see Bell (1981b)). Rock salt is most prone to creep.

Table 13.11. Some physical properties of evaporitic rocks

	Gypsum (Sherburn-in-Elmet)	Anhydrite (Sandwith)	Rock salt (Winsford)	Potash (Loftus)
Relative density	2.36	2.93	2.2	2.05
Dry density (t/m³)	2.19	2.82	2.09	1.98
Porosity (%)	4.6	2.9	4.8	5.1
Unconfined compressive strength (MPa)	27.5	97.5	11.7	25.8
Point load strength (MPa)	2.1	3.7	0.3	0.6
Scleroscope hardness	27	38	12	9
Schmidt hardness	25	40	8	11
Young's modulus ($\times 10^3$ MPa)	24.8	63.9	3.8	7.9
Permeability ($\times 10^{-9}$ m/s)	0.62	0.03	—	—

Justo and Zapico (1975) recorded that the amount of settlement which occurred when gypsum was subjected to plate load testing the maximum load being 1.2 MPa, was negligible.

Gypsum is more readily soluble than limestone, 2100 ppm can be dissolved in non-saline waters as compared with 400 ppm. Sinkholes and caverns can therefore develop in thick beds of gypsum (see Eck and Redfield (1965)) more rapidly than they can in limestone. Indeed in the United States such features have been known to form within a few years where beds of gypsum are located beneath dams. Extensive surface cracking and subsidence has occurred in certain areas of Oklahoma and New Mexico due to the collapse of cavernous gypsum (see Brune (1965)). The problem is accentuated by the fact that gypsum is weaker than limestone and therefore collapses more readily. Kendal and Wroot (1924) quoted vivid, but highly exaggerated, accounts of subsidences which occurred in the Ripon district, Yorkshire, in the eighteenth and nineteenth centuries due to the solution of gypsum. They wrote that wherever beds of gypsum approach the surface their presence can be traced by broad funnel-shaped craters formed by the collapse of overlying marl into areas from which gypsum has been removed by solution. Apparently these craters only took a matter of minutes to appear at the surface. However, where gypsum is effectively sealed from the ingress of water by overlying impermeable strata such as marl, dissolution does not occur (see Redfield (1968)).

The solution rate of gypsum or anhydrite is principally controlled by the area of their surface in contact with water and the flow velocity of water associated with a unit area of the material. Hence the amount of fissuring in a rock mass, and whether it is enclosed by permeable or impermeable beds, is most important. Solution also depends on the sub-saturation concentration of calcium sulphate in solution. According to James and Lupton (1978) the concentration dependence for gypsum is linear whilst that for anhydrite is a square law. The salinity of the water also is influential. For example, the rates of solution of gypsum and anhydrite are increased by the presence of sodium chloride, carbonate and carbon dioxide in solution. It is therefore important to know the chemical composition of the groundwater.

Massive deposits of gypsum are usually less dangerous than those of anhydrite because gypsum tends to dissolve in a steady manner forming caverns or causing progressive settlements. For instance, if small fissures occur at less than 1 m intervals solution usually takes place by removal of gypsum as a front moving 'downstream' at less than 0.01 m/year. However, James and Lupton (1978) showed that if the following conditions were met:

(1) the rock temperature was 10°C,
(2) the water involved contained no dissolved salts, and
(3) a hydraulic gradient of 0.2 was imposed,

then a fissure, 0.2 mm in width and 100 m in length, in massive gypsum, would, in 100 years, have widened by solution so that a block 1 m^3 in size could be accommodated in the entrance to the fissure. In other words a cavern would be formed. If the initial width of the fissure exceeds 0.6 mm, large caverns would form and a runaway situation would develop in a very short time. In long fissures the hydraulic gradient is low and the rate of flow is reduced so that solutions become saturated and little or no material is removed. Indeed James and Lupton implied that a flow rate of 10^{-3} m/s was rather critical in that if it was exceeded, extensive solution of gypsum could take place. Solution of massive gypsum is not likely to give rise to an accelerating deterioration in a foundation if precautions such as grouting are taken to keep seepage velocities low.

Massive anhydrite can be dissolved to produce uncontrollable runaway situations in which seepage flow rates increase in a rapidly accelerating manner. Even small fissures in massive anhydrite can prove dangerous. If anhydrite is taken in the above example not only is a cavern formed, but the fissure is enlarged as a long tapering section. Within about 13 years the flow rate increases to a runaway situation. However, if the fissure is 0.1 mm in width then the solution becomes supersaturated with calcium sulphate and gypsum is precipitated. This seals the outlet from the fissure and from that moment any anhydrite in contact with the water is hydrated to form gypsum. Accordingly 0.1 mm width seems to be a critical fissure size in anhydrite.

If soluble minerals occur in particulate form in the ground then their removal by solution can give rise to significant settlements. In such situations the width of the solution zone and its rate of progress are obviously important as far as the location of hydraulic structures are concerned (see James and Kirkpatrick (1980); Table 13.9). James and Lupton (1978) have provided equations for the determination of these two factors in gypsiferous deposits. Again the solubility of the mineral and its solution rate constant are the most important parameters. Of lesser consequence is information relating to the volumetric proportion of soluble minerals in the ground and details relating to their size, distribution and shape. This does not mean to say that such data should not be obtained, for they obviously aid the determination of the amount of settlement which might take place.

Another point which should be borne in mind, and this particularly applies to conglomerates cemented with soluble material, is that when this is removed by solution the rock is reduced greatly in strength. A classic example of this is associated with the failure of the St Francis Dam in California in 1928. One of its abutments was founded in conglomerate cemented with gypsum which was gradually dissolved, the rock losing strength, and ultimately the abutment failed.

Hawkins (1979) noted the presence of voids in rocks of Keuper age in the area around the Severn estuary. These voids were formed as a result of gypsum being removed in solution. He also noted the effects of dissolution in dolomitic rocks which contain gypsum. This action may lead to an enrichment in calcium, as magnesium sulphate is lost in solution. Anhydrite is less likely to undergo catastrophic solution in a fragmented or particulate form than gypsum. Conversion to gypsum is more characteristic of extensive deposits of permeable granular anhydrite.

Uplift is a problem associated with anhydrite. This takes place when anhydrite is hydrated to form gypsum, in so doing there is a volume increase of between 30 and 58% which exerts pressures that have been variously estimated between 2 and 69 MPa. It is thought that no great length of time is required to bring about such hydration. When it occurs at shallow depths it causes expansion but the process is gradual and is usually accompanied by the removal of gypsum in solution. At greater depths anhydrite is effectively confined during the process. This results in a gradual build-up of pressure and the stress is finally liberated in an explosive manner. According to Brune (1965) such uplifts in the United States have taken place beneath reservoirs, these bodies of water providing a constant supply for the hydration process, percolation taking place via cracks and fissures. Examples are known where the ground surface has been elevated by about 6 m. The rapid explosive movement causes strata to fold, buckle and shear which further facilitates access of water into the ground.

Sulphuric acid and sulphate are produced when gypsum is subjected to weathering. Aqueous solutions of sulphate and sulphuric acid react with tricalcium aluminate in Portland cement to form calcium sulphoaluminate or ettringite (see Anon (1975)). This reaction is accompanied by expansion. The rate of attack is very much influenced by the permeability of the concrete or mortar and the position of the water table. This is because sulphates can only continue to reach cement by movement of their solutions in water. Thus if a structure is permanently above the water table it is unlikely to be attacked. By contrast, below the water table, movement of water may replenish the sulphates removed by reaction with cement, thereby allowing the reaction to continue.

Salt is even more soluble than gypsum and the evidence of slumping, brecciation and collapse structures in rocks which overlie saliferous strata bear witness to the fact that salt has gone into solution in past geological times.

Considerable natural solution of salt in Cheshire means that it does not outcrop at the surface, it usually terminates in a solution surface at a depth ranging from 70–150 m. The regions beneath such surfaces, which are subjected to circulating groundwater and solution, are referred to as wet rock head. Where beds of rock salt are beyond the reach of circulating groundwater their original thickness is preserved and these regions are referred to as dry rock head. Natural subsidence in wet rock head areas is believed to have been responsible for the formation of the Cheshire Meres but generally speaking this type of subsidence, although it operates over large areas, takes place extremely slowly, it is perhaps imperceptible within the context of historic time. Some of the meres are floored by peat deposits and it has been suggested that they may be pre-Pleistocene in age. The Keuper Marl which generally overlies the saliferous beds is incompetent which means that large solution cavities have not been able to form since the collapse of the marl was probably contemporaneous with salt removal. These collapses produced linear subsidence depressions at the surface. Where a brine run occurs beneath glacial deposits cavities have formed in the latter. These then collapse to produce rather circular shaped, small hollows at the surface. They make their appearance almost instantaneously but fortunately they only occur infrequently. It would seem that as the sand collapses it chokes the brine run, thereby preventing further subsidence.

It is generally believed, however, that in areas underlain by saliferous beds, measurable surface subsidence is unlikely to occur except where salt is being extracted (see Bell (1975)). Perhaps this is because equilibrium has been attained between the supply of unsaturated groundwater and the salt available for solution. Exceptionally cases have been recorded of rapid subsidence, such as the 'Meade salt sink' in Kansas. Johnson (1901) explained its formation as due to solution of beds of salt located at depth. This area of water, about 60 m in diameter, formed as a result of rapid subsidence in March 1879. At the same time, 64 km to the south west, the railway station at Rosel and several buildings disappeared due to the sudden appearance of a sinkhole.

13.6 Carbonaceous sedimentary rocks

Most coal seams are composite in character. At the base the coal is often softer and is sometimes simply referred to as 'bottom coal'. Bright coal is often of most importance in the centre of a seam whilst dull coal may predominate in the upper part of a seam.

Coal generally breaks into blocks which have three faces approximately perpendicular to each other. These surfaces are referred to as cleat. The cleat direction is usually fairly constant and is best developed in bright coal. Cleat perhaps may be coated with films of mineral matter, commonly calcite, ankerite and pyrite. Coal seams may split or be replaced, totally or partially, by washouts.

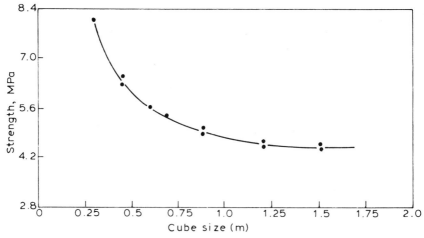

Figure 13.7 Experimental relationship between the strength and cube side length of coal specimens tested underground (courtesy Elsevier Scientific Publishing Company)

Coal is more suspect to mechanical than chemical weathering. Its crushing strength varies but generally it is less than 20 MPa (Table 13.3).

It was noted by Bieniawski (1975) that when testing coal to failure in the laboratory there are two disadvantages. First, when coal is removed from a mine it deteriorates as a result of the changes in temperature and humidity. Secondly, the size of the specimen has a marked influence on the results obtained. This is largely because of the presence of cleat. However, after a series of large-scale *in situ* loading tests, Bieniawski found that the strength of cubic specimens of coal decreased with increasing size, tending to an asymptotic value at about 1.5 m (Figure 13.7). The modulus of elasticity also showed a general tendency to decrease with increasing size of specimen. Failure of *in situ* specimens was found to be associated with the gradual opening of vertical cleats and spalling from one or more faces. Ultimately a double pyramid type of failure was produced, although a central core of fairly solid coal sometimes remained. For details relating to the strength of pillars of coal, see Hustrulid (1976) or Bell (1978b).

References

ABBISS, C. P. (1979). 'A comparison of the stiffness of the Chalk at Mundford from a seismic survey and large-scale tank test', *Geotechnique*, **29**, 461–8.

ANON. (1975). 'Concrete in sulphate bearing soils and ground waters', *Digest*, **174**, Building Research Establishment, Watford, HMSO, London.

ATTEWELL, P. B. (1971). 'Geotechnical properties of the Great Limestone in northern England', *Engg Geol.*, **5**, 89–112.

BADGER, C. W., CUMMINGS, A. D. & WHITMORE, R. L. (1956). 'The disintegration of shale', *J. Inst. Fuel*, **29**, 417–23.

BELL, F. G. (1975). 'Salt and subsidence in Cheshire, England', *Engg Geol.*, **9**, 237–47.

BELL, F. G. (1977). 'A note on the geotechnical properties of chalk', *Engg Geol.*, **11**, 221–2.

BELL, F. G. (1978a). 'The physical and mechanical properties of the Fell Sandstone', *Engg Geol.*, **12**, 1–29.

BELL, F. G. (1978b). 'Subsidence due to mining operations' (in *Foundation Engineering in Difficult Ground*), Bell, F. G. (ed.), Butterworths, London, 322–62.

BELL, F. G. (1981a). 'A survey of the physical properties of some carbonate rocks', *Bull. Int. Ass. Engg Geol.*, No. 24, 105–110.

BELL, F. G. (1981b). 'Geotechnical properties of evaporites', *Bull. Int. Ass. Engg Geol.*, No. 24, 137–144.

BIENIAWSKI, Z. T. (1975). '*In situ* strength and deformation characteristics of coal', *Engg Geol.*, **5**, 325–40.

BRUNE, G. (1965). 'Anhydrite and gypsum problems in engineering geology', *Bull. Ass. Engg Geologists*, **3**, 26–38.

BURLAND, J. B., KEE, R. & BURFORD, D. (1974). 'Short-term settlement of a five storey building on soft chalk', (in *Settlement of Structures*), British Geotechnical Society, Pentech Press, London, 259–65.

BURLAND, J. B. & LORD, J. A. (1970). 'The load deformation behaviour of Middle Chalk at Mundford, Norfolk: a comparison between full-scale performance and *in situ* and laboratory measurements' (in *In Situ Investigations in Soils and Rocks*), British Geotechnical Society, London, 3–16.

BURNETT, A. D. & EPPS, R. J. (Mar. 1979). 'The engineering geological description of the carbonate suite rocks and soils', *Ground Engg*, **12**, No. 2, 41–8.

BURWELL, E. B. (1950). 'Geology in dam construction, Part I' (in *Applications of Geology to Engineering Practice*), Berkey Volume, Geol. Soc. Am., 11–31.

CARTER, P. G. & MALLARD, D. J. (1974). 'A study of the strength, compressibility and density trends within the Chalk of south-east England', *Q. J. Engg Geol.*, **7**, 43–56.

CASAGRANDE, A. (1949). *Notes on Swelling Characteristics of Clay-Shales*, Harvard Soil Mechanics Series, Harvard University, Cambridge, Mass.

CHANDLER, R. J. (1969). 'The effect of weathering on the shear strength properties of Keuper Marl', *Geotechnique*, **19**, 321–34.

CHAPPELL, B. A. (1974). 'Deformational response of differently shaped and sized test pieces of shale rock', *Int. J. Rock Mech. Min. Sci.*, **11**, 21–8.

DAVIS, A. G. (1967). 'On the mineralogy and phase equilibrium of Keuper Marl', *Q. J. Engg Geol.*, **1**, 25–46.

DAVIS, A. G. (1968). 'The structure of Keuper Marl', *Q.J. Engg Geol.*, **1**, 145–53.

DE GRAFT-JOHNSON, J. W. S., BHATIA, H. S. & YEBOA, S. L. (1973). 'Geotechnical properties of Accra Shales', *Proc 8th Int. Conf. Soil Mech. Foundation Engg*, Moscow, **2**, 97–104.

DENNEHY, J. P. (1976). 'Correlating the SPT value with chalk grade for some zones of the Upper Chalk', *Geotechnique*, **26**, 610–14.

DUMBLETON, M. J. (1967). 'Origin and mineralogy of African red clays and Keuper Marl', *Q. J. Engg Geol.*, **1**, 39–46.

EARLY, K. R. & DYER, K. R. (1964). 'The use of a resistivity survey on a foundation site underlain by karst dolomite', *Geotechnique*, **14**, 341–8.

ECK, W. & REDFIELD, R. C. (1965). 'Engineering geology problems at Sanford Dam, Texas', *Bull. Ass. Engg Geologists*, **3**, 15–25.

FASISKA, E., WAGENBLAST, H. & DOUGHERTY, M. T. (1974). 'The oxidation mechanism of sulphide minerals', *Bull. Ass. Engg Geologists*, **11**, No. 1, 75–82.

FOOKES, P. G. & HIGGINBOTTOM, I. E. (1975). 'The classification and description of near-shore carbonate sediments for engineering purposes', *Geotechnique*, **25**, 406–11.

FRINK, J. W. (1946). 'The foundations of Hales Bar Dam', *Econ. Geol.*, **41**, 576–92.

GEOLOGICAL SOCIETY ENGINEERING GROUP (1970). 'Working Party Report on the Logging of Cores for Engineering Purposes', *Q. J. Engg Geol.*, **3**, 1–24.

GRAINGER, P., McCANN, D. M. & GALLOIS, R. W. (1973). 'The application of seismic refraction to the study of fracturing in the Middle Chalk at Mundford, Norfolk', *Geotechnique*, **23**, 219–32.

GRICE, R. H. (1969). 'Test procedures for the susceptibility of shale to weathering', *Proc. 7th Int. Conf. Soil Mech. Foundation Engg*, Mexico City, **3**, 884–9.

HAWKINS, A. B. (1979). 'Case histories of some effects of solution/dissolution in the Keuper rocks at the Severn estuary', *Q. J. Engg Geol.*, **12**, 31–40.

HIGGINBOTTOM, I. E. (1965). 'The engineering geology of the Chalk', *Proc. Symp. on Chalk in Earthworks*, Inst. Civil Engrs, London, 1–14.

HILL, H. P. (1949). 'The Ladybower Reservoir', *J. Inst. Water Engrs*, **3**, 414–33.

HUSTRULID, W. A. (1976). 'A review of coal pillar strength formulae', *Rock Mech.*, **8**, 115–145.

INESON, J. (1962). 'A hydrogeological study of the permeability of chalk', *J. Inst. Water Engrs*, **16**, 255–86.

INGRAM, R. L. (1953). 'Fissility of mudrocks', *Bull. Geol. Soc. Am.*, **64**, 869–78.

JAEGER, C. (1963). 'The Malpasset Report', *Water Power*, **15**, 55–61.

JAMES, A. N. & KIRKPATRICK, I. M. (1980). 'Design of foundations of dams containing soluble rocks and soils', *Q.J. Engg Geol.*, **13**, 189–98.

JAMES, A. N. & LUPTON, A. R. R. (1978). 'Gypsum and anhydrite in foundations of hydraulic structures', *Geotechnique*, **28**, 249–72.

JENNINGS, J. E., BRINTO, A. B. A., LOUW, A. & GOWAN, G. D. (1965). 'Sinkholes and subsidences in the Transvaal Dolomite of South Africa', *Proc. 6th Conf. Soil Mech. Foundation Engg*, Montreal, **1**, 51–4.

JOHNSON, W. D. (1901). 'The high plains and their utilization', *U.S. Geol. Surv., 21st Annual Rep.*, Part 4, 601–741.

JUMIKIS, A. R. (1965). 'Some engineering aspects of the Brunswick Shale', *Proc. 6th Int. Conf. Soil Mech. Foundation Engg*, Montreal, **2**, 99–102.

JUSTO, J. L. & ZAPICO, L. (1975). 'Compression between measured and estimated settlements at two Spanish aqueducts on gypsum rock' (in *Settlement of Structures*), British Geotechnical Society, Pentech Press, London, 266–74.

KADERABEK, A. M. & REYNOLDS, R. T. (1981). 'Miami Limestone foundation design and construction', *Proc. ASCE, J. Geot. Engg Div.*, **107**, GT7, 859–72.

KEE, R., PARKER, A. S. & WEHALE, J. E. C. (1974). 'Settlement of a twelve storey building on piled foundations in chalk at Basingstoke' (in *Settlement of Structures*), British Geotechnical Society, Pentech Press, London, 275–82.

KENDAL, P. F. & WROOT, H. E. (1924). *The Geology of Yorkshire*, Printed Privately.

KENNARD, M. F. & KNILL, J. L. (1968). 'Reservoirs on limestone, with particular reference to the Cow Green Scheme', *J. Inst. Water Engrs*, **23**, 87–113.

KENNARD, M. F., KNILL, J. L. & VAUGHAN, P. R. (1967). 'The geotechnical properties and behaviour of Carboniferous shale at Balderhead Dam', *Q. J. Engg Geol.*, **1**, 3–24.

KOLBUSZEWSKI, J., BIRCH, N. & SHOJOBI, J. O. (1965). 'Keuper Marl research', *Proc. 6th Int. Conf. Soil Mech. Foundation Engg*, Montreal, **1**, 59–63.

LEES, G. (1965). 'Geology of the Keuper Marl', *Proc. Geol. Soc. London*, No. 1621, 46.

LEWIS, W. A. & CRONEY, D. (1965). 'The properties of chalk in relation to road foundations and pavements', *Proc. Symp. on Chalk in Earthworks*, Inst. Civil Engrs, London, 27–42.

MEAD, W. J. (1936). 'Engineering geology of damsites', *Trans. 2nd Int. Cong. Large Dams*, Washington, D.C., **4**, 183–98.

MEIGH, A. C. (1976). 'The Triassic rocks, with particular reference to predicted and observed performance of some major foundations', *Geotechnique*, **26**, 391–452.

MEIGH, A. C. & EARLY, K. R. (1957). 'Some physical and engineering properties of chalk', *Proc. 4th Int. Conf. Soil Mech. Foundation Engg*, London, **1**, 68–73.

MOORE, J. F. A. (1974). 'A long-term plate test on Bunter Sandstone', *Proc. 3rd Int. Cong. Rock Mech.*, Denver, **2**, 724–32.

MOORE, J. F. A. & JONES, C. W. (1975). '*In situ* deformation of Bunter Sandstone' (in *Settlement of Structures*), British Geotechnical Society, Pentech Press, London, 311–19.

MORGENSTERN, N. R. & EIGENBROD, K. D. (1974). 'Classification of argillaceous soils and rocks', *Proc. ASCE, J. Geot. Engg Div.*, GT10, **100**, 1137–56.

NORTH, F. J. (1951) 'Some geological aspects of subsidence not due to mining', *Proc. S. Wales Inst. Engrs*, **52**, 127–58.

PENNER, E., EDEN, W. J. & GILLOTT, J. E. (1973). 'Floor heave due to biochemical weathering of shale', *Proc. 8th Int. Conf. Soil Mech. Foundation Engg*, Moscow, **2**, 151–8.

PETTIJOHN, P. J., POTTER, P. E. & SIEVER, R. (1975). *Sands and Sandstones*, Springer-Verlag, Berlin.

PRICE, N. J. (1960). 'The compressive strength of Coal Measure rocks', *Colliery Guardian*, **199**, 283–92.

PRICE, N. J. (1963). 'The influence of geological factors on the strength of Coal Measure rocks', *Geol. Mag.*, **100**, 428–43.

REDFIELD, R. C. (1968). 'Brantley reservoir site—an investigation of evaporite and carbonate facies', *Bull. Ass. Engg Geologists.*, **6**, 14–30.

SHERWOOD, P. T. (1967). 'Classification tests on African red clays and Keuper Marl', *Q. J. Engg Geol.*, **1**, 47–56.

SINGH, D. P., NATH, R. & SINGH, J. B. (Aug. 1978). 'A comparative study of anisotropy of Singrauli Coal and Chunar Sandstone, *J. Mines Metals Fuels*, **26**, No. 8, 283–90.

SMITH, C. K. & REDLINGER, J. F. (1953). 'Soil properties of the Fort Union clay shale', *Proc. 3rd Int. Conf. Soil Mech. Foundation Engg*, Zurich, **1**, 56–61.

SOWERS, G. F. (1975). 'Failures in limestones in the humid subtropics', *Proc. ASCE, J. Geot. Engg Div.*, **101**, GT8, 771–87.

SPEARS, D. A. (1980). 'Towards a classification of shales', *J. Geol. Soc.*, **137**, 125–30.

TAYLOR, R. K. & SPEARS, D. A. (1970). 'The breakdown of British Coal Measures rocks', *Int. J. Rock Mech. Min. Sci.*, **7**, 481–501.

TERZAGHI, K. (1946). 'Introduction to tunnel geology', (in *Rock Tunnelling with Steel Supports*), Procter, R. & White, T. (eds), Commercial Shearing and Stamping Co., Youngstown, Ohio, 17–99.

TERZAGHI, K. (1962). 'Dam foundations on sheeted rock', *Geotechnique*, **12**, 199–208.

UNDERWOOD, L. B. (1967). 'Classification and identification of shales', *Proc. ASCE, Soil Mech. Foundation Engg Div.*, **93**, SM6, 97–116.

UNDERWOOD, L. B., THORFINNSON, S. T. & BLACK, W. T. (1964) 'Rebound in redesign of the Oake Dam hydraulic structures', *Proc. ASCE, J. Soil Mech. Foundation Engg Div.*, **90**, SM2, Paper 3830, 859–68.

WAKELING, T. R. M. (1965). 'Foundations on chalk', *Proc. Symp. on Chalk in Earthworks*, Inst. Civil Engrs, London, 15–23.

WAKELING, T. R. M. (1970). 'A comparison of the results of standard site investigation methods against the results of a detailed geotechnical investigation in Middle Chalk at Mundford, Norfolk', (in *In Situ Investigations in Soils and Rocks*), British Geotechnical Society, London, 17–22.

WARD, W. H., BURLAND, J. B. & GALLOIS, R. W. (1968). 'Geotechnical assessment of a site at Mundford, Norfolk, for a large proton accelerator', *Geotechnique*, **18**, 399–431.

WEST, G. (Sept. 1979). 'Strength properties of Bunter Sandstone', *Tunnels and Tunnelling*, **7**, No. 7, 27–9.

WEST, G. & DUMBLETON, M. J. (1972). 'Some observations on swallow holes and mines in the Chalk', *Q. J. Engg Geol.*, **5**, 171–8.

WILLIAMS, J. H. & VINEYARD, J. D. (1976). 'Geologic indicators of catastropic collapse in Karst terrain in Missouri', *Transportation Res. Rec.*, **612**, 31–7.

ZARUBA, Q. & BUKOVANSKY, M. (1965). 'Mechanical properties of Ordivician shales of Central Bohemia', *Proc. 6th Int. Conf. Soil Mech. Foundation Engg*, Montreal, **3**, 421–4.

Chapter 14

Photogeology and maps for engineering purposes

The last decade has seen much progress in the development of engineering geological maps. Engineering geological maps may be simply geological maps to which engineering geological data have been added, or alternatively the rocks and soils in the area concerned may be presented as mapped units defined in terms of engineering properties or behaviour. The boundaries of such units may bear no relationship to the lithostratigraphical units or geological structure generally represented on conventional geological maps. This type of map is often referred to as a geotechnical map.

Aerial photographs, and more recently, remote imagery, have proved important aids in the production of engineering geological maps, particularly in those underdeveloped regions of the world where good topographic maps are not available. However, the amount of useful information obtainable from aerial photographs and imagery depends upon their characteristics as well as the nature of the terrain they portray. Remote imagery and aerial photographs prove of most value during the planning and reconnaissance stages of a project. The information they provide can be transposed to the base map and this is checked during fieldwork. This information not only allows the fieldwork programme to be planned much more readily, but it also should help to shorten the period spent in the field.

The final map is produced from the information collected from various sources (literature survey, aerial photographs and imagery, and fieldwork). Large engineering projects may require that a series of derivative maps are produced. The latter sometimes take the form of transparent overlays. Once the data have been gathered there are then the problems of how they should be represented on the map and at which scale should the map be drawn. The latter is very much influenced by the requirement, in other words the more detailed a map needs to be, the larger its scale. As far as presentation is concerned, this may involve not only the choice of colours and symbols, but also the use of overprinting. It is suggested that the symbols recommended by Anon (1972) should be used for engineering geological maps. Overprinting frequently takes the form of striped or stippled shading, both of which can be varied, for instance, according to frequency, pattern, dimension or colour.

14.1 Remote sensing

Remote sensing commonly represents one of the first stages of land assessment in underdeveloped areas. It involves the identification and analysis of phenomena on the Earth's surface by using devices borne by aircraft or spacecraft (see Beaumont (1979a)). Most techniques used in remote sensing depend upon recording energy from part of the electromagnetic spectrum, ranging from gamma rays, through the visible spectrum to radar (Figure 14.1). The two principal systems of remote sensing are infrared linescan (IRLS) and side-looking airborne radar (SLAR). The scanning equipment used measures both emitted and reflected radiation and the employment of suitable detectors and filters permits the measurement of certain spectral bands. Signals from several bands of the spectrum can be recorded simultaneously by multi-spectral scanners. Lasers are being developed for use in remote sensing.

14.1.1 Infrared linescanning

Infrared linescanning is dependent upon the fact that all objects emit electromagnetic radiation generated by the thermal activity of their component atoms. Emission is greatest in the infrared region of the electromagnetic spectrum for most materials at ambient temperature. The reflected infrared region ranges in wavelength from 0.7–3.0 μm and includes the photographic infrared band. This can be detected by certain infrared-sensitive film. The thermal infrared region ranges in wavelength from 3.0–14.0 μm (the most effective waveband used for thermal infrared linescanning for geological purposes is 8–14 μm, according to Warwick et al. (1979)). Photoelectrical detectors and optical mechanical scanners are used to record images in the thermal infrared spectral region, not cameras, as thermal infrared radiation is absorbed by glass.

Infrared linescanning involves scanning a succession of parallel lines across the track of an aircraft with a scanning spot. The spot travels forwards and backwards in such a manner that nothing is missed between consecutive

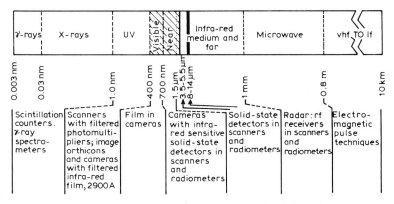

Figure 14.1 Electromagnetic spectrum showing wavelengths of thermal infrared radiation to other wavelengths (after Warwick et al. (1979))

passes. Since only an average radiation is recorded the limits of resolution depend on the size of the spot. The diameter of the spot is usually around 2–3 milliradians, which means that if the aircraft is flying at a height of 1000 m then the spot measures 2–3 m across. The radiation is picked up by a detector which converts it to electrical signals, which in turn are transformed into visible light via a cathode ray tube thereby enabling a record to be made on film or magnetic tape. The data can be processed in colour as well as black and white, colours substituting for grey tones. Relatively cold areas are depicted as purple and relatively hot areas as red on the positive print, whereas hot areas are shown as black and cold as white on the black and white negative film base (see Mathur (1979)). Unfortunately prints are increasingly distorted with increasing distance from the line of flight which limits the total useful angle of scan to about 60° on either side. In order to reduce the distortion along the edges of the imagery flight lines have a 50–60% overlap. According to Warwick et al. (1979) a temperature difference of 0.15°C between objects of 500 mm diameter can be dectected by an aircraft at an altitude of 300 m. The spatial resolution is, however, much lower than aerial photographs, in which the resolution at this height would be 80 mm. At higher altitudes the difference becomes more marked.

The use of infrared linescan depends on clear, calm weather. What is more some thought must be given to the fact that thermal emissions vary significantly throughout the day. The time of the flight is therefore important. From the point of view of engineering geology pre-dawn flying proves most suitable for thermal infrared linescan. This is because radiant temperatures are fairly constant and reflected energy is not important whereas during a sunny day radiant and reflected energy are roughly equal so that the latter may obscure the former. Also because sun-facing slopes are warm and shade slopes cool rough topography tends to obliterate the geology in post-dawn imagery.

Although temperature differences of 0.1°C can be recorded by infrared linescan, these do not represent differences in the absolute temperature of the ground but in emission of radiation. Careful calibration is therefore needed in order to obtain absolute values. Emitted radiation is determined by the temperature of the object and its emissivity, which can vary with surface roughness, soil type, moisture content and vegetative cover.

A grey scale can be used to interpret the imagery, it being produced by computer methods from linescan data which have been digitised. This enables maps of isoradiation contours to be produced. Colour enhancement also has been used to produce isotherm contour maps, with colours depicting each contour interval. This method has been used in the preparation of maps of engineering soils (see Reeves et al. (1975)).

Identification of grey tones is the most important aspect as far as the interpretation of thermal imagery is concerned, since these provide an indication of the radiant temperatures of a surface. Warm areas give rise to light, and cool areas to dark tones. Thermal inertia is important in this respect since rocks with high thermal inertia, such as dolostone or quartzite, are relatively cool during the day and warm at night. Rocks and soils with low thermal inertia, for example, shale, gravel or sand, are warm during the day and cool at night. In other words the variation in temperature of materials with high thermal inertia during the daily cycle is much less than those with low thermal inertia. Because clay soils possess relatively high thermal inertia they appear warm in pre-dawn imagery whereas sandy soils, because of their relatively low thermal inertia, appear cool (Table 14.1). The moisture content of a soil influences the image produced, that is, soils which possess high moisture content appear cool irrespective of their type. Consequently high moisture content may mask differences in soil types. Fault zones are often picked out because of their higher moisture content. Similarly the presence of old landslides can frequently be discerned due to their moisture content differing from that of their surroundings.

Texture can also help interpretation. For instance, outcrops of rock may have a rough texture due to the presence of bedding or jointing, whereas soils usually give rise to a relatively smooth texture. However, where soil cover is less than 0.5 m, the rock structure is usually observable on the imagery since deeper, more moist soil occupying discontinuities gives a darker signature. Free-standing bodies of water are usually readily visible on thermal imagery, however, the high thermal inertia of highly saturated organic deposits may approach that of water masses, the two may therefore prove at times difficult to distinguish.

Table 14.1. Thermal properties of geological materials and water at 20°C (after Warwick et al. (1979))

Geological materials	Thermal conductivity k. (cal cm^{-1} s^{-1} °C^{-1})	Density ρ (gm cm^{-3})	Thermal capacity c (cal gm^{-1} °C^{-1})	Thermal diffusity k (cm^2 s^{-1})	Thermal inertia P (cal cm^{-2} s$^{-\frac{1}{2}}$ °C^{-1})	$1/P$ (often used as thermal inertia value)
(1) Basalt	0.0050	2.8	0.20	0.009	0.053	19
(2) Clay soil (moist)	0.0030	1.7	0.35	0.005	0.042	24
(3) Dolomite	0.0120	2.6	0.18	0.026	0.075	13
(4) Gabbro	0.0060	3.0	0.17	0.012	0.055	18
(5) Granite	$\left\{\begin{array}{l}0.0075 \\ 0.0065\end{array}\right.$	2.6	0.16	0.016	0.052	19
(6) Gravel	0.0048	2.5	0.17	0.011	0.045	22
(7) Limestone	0.0048	2.5	0.17	0.011	0.045	22
(8) Marble	0.0055	2.7	0.21	0.010	0.056	18
(9) Obsidian	0.0030	2.4	0.17	0.007	0.035	29
(10) Peridotite	0.0110	3.2	0.20	0.017	0.084	12
(11) Pumice, loose	0.0006	1.0	0.16	0.004	0.009	111
(12) Quartzite	0.0120	2.7	0.17	0.026	0.074	14
(13) Rhyolite	0.0055	2.5	0.16	0.014	0.047	21
(14) Sandy gravel	0.0060	2.1	0.20	0.014	0.050	20
(15) Sandy soil	0.0014	1.8	0.24	0.003	0.024	42
(16) Sandstone quartz	$\left\{\begin{array}{l}0.0120 \\ 0.0062\end{array}\right.$	2.5	0.19	0.013	0.054	19
(17) Serpentine	$\left\{\begin{array}{l}0.0063 \\ 0.0072\end{array}\right.$	2.4	0.23	0.013	0.063	16
(18) Shale	$\left\{\begin{array}{l}0.0042 \\ 0.0030\end{array}\right.$	2.3	0.17	0.008	0.034	29
(19) Slate	0.0050	2.8	0.17	0.011	0.049	20
(20) Syenite	$\left\{\begin{array}{l}0.0077 \\ 0.0044\end{array}\right.$	2.2	0.23	0.009	0.047	21
(21) Tuff, welded	0.0028	1.8	0.20	0.008	0.032	31
(22) Water	0.0013	1.0	1.01	0.001	0.037	27

14.1.2 Side-looking airborne radar short pulses

In side-looking airborne radar, short pulses of energy, in a selected part of the radar waveband are transmitted sideways to the ground from antennae on both sides of an aircraft (see Sabins (1978)). The pulses of energy strike the ground along successive range lines and are reflected back at time intervals related to the height of the aircraft above the ground. The reflected pulses are transformed into black and white photographs with the aid of a cathode ray tube. Returning pulses cannot be accepted from any point within 45° from the vertical so that there is a blank space under the aircraft along its line of flight. Also the image becomes increasingly distorted towards the track of the aircraft. The belt covered by normal SLAR imagery varies from 2–50 km and, although the scanning is oblique, the system converts it to an image which is more or less planimetric.

There are some notable differences between SLAR images and aerial photographs. For instance, although variations in vegetation produce slightly different radar responses, a SLAR image depicts the ground more or less as it would appear on aerial photographs devoid of vegetation. As noted above displacements of relief are to the side towards the imaging aircraft and not radial about the centre as in aerial photographs. Furthermore radar shadows fall away from the flight line and are normal to it. The shadows on SLAR images form black areas which yield no information, whereas most areas of shadow on aerial photographs are partially illuminated by diffused lighting. The subtle changes of tone and texture which occur on aerial photographs are not observable on SLAR images.

Because the wavelengths used in SLAR are not affected by cloud cover; imagery can be obtained at any time. This is particularly important in equatorial areas, which are rarely free of cloud. Consequently this technique provides an ideal means of reconnaissance survey in underdeveloped areas.

Typical scales for radar imagery available commercially are 1:100000–1:250000, with a resolution of between 10 and 30 m. Smaller objects than this can appear on the image if they are strong reflectors, and the original material can be enlarged. Mosaics are suitable for the identification of regional geological features and for preliminary identification of terrain units. Lateral overlap of radar cover can give a stereoscopic image, which offers a more reliable assessment of the terrain. Furthermore imagery recorded by radar systems can provide appreciable detail of landforms as they are revealed due to the low angle of incident illumination.

14.1.3 Satellite imagery

Beaumont and Beaven (1977) noted that imagery of the Earth's surface obtained from space gives a broad view of an area, illustrating conditions as they exist at a particular time, and indicates the interrelationships between geology, landform, climate, vegetation and land-use. Small-scale space imagery provides a means of initial reconnaissance which allows areas to be selected for further, more detailed investigation, either by aerial and/or ground survey methods. Indeed in many parts of the world a Landsat image may provide the only form of base map available.

The large areas of the ground surface which satellite images cover give a regional physiographic setting and permit the distinction of various landforms according to their characteristic photo-patterns. Accordingly such imagery can provide a geomorphological framework from which a study of the component landforms is possible. The character of the landforms may afford some indication of the type of material of which they are composed and geomorphological data aid the selection of favourable sites for field investigation on larger-scale aerial surveys. Small-scale imagery may enable regional geological relationships and structures to be identified which are not noticeable on larger-scale imagery or mosaics (see Norman (1980)).

The capacity to detect surface features and landforms from imagery obtained by multispectral scanners on Landsat satellites is facilitated by energy reflected from the ground surface being recorded within four specific wavelength bands (see Beaumont (1977)). These are visible green (0.5–0.6 μm), visible red (0.6–0.7 μm) and two invisible infrared bands (0.7–0.8 μm and 0.9–1.0 μm). The images are reproduced on photographic paper and are available for the four spectral bands plus two false colour composites. The infrared band between 0.7 and 0.8 μm is probably the best for geological purposes. Because separate images within different wavelengths are recorded at the same time the likelihood of recognising different phenomena is significantly enhanced. Since the energy emitted and reflected from objects commonly varies according to wavelength, its characteristic spectral pattern or signature in an image is determined by the amount of energy transmitted to the sensor within the wavelength range in which that sensor operates. As a consequence a unique tonal signature frequently may be identified for a feature if the energy which is being emitted and/or reflected from it is broken into specially selected wavelength bands.

In Figure 14.2 the reflectance curves for four different rock types illustrate the higher reflectance of brown sandstone at longer (orange–red) wavelengths and the lower reflectance of the siltstone in the shorter (blue) wavelengths of the visible spectrum (see Beaumont (1979b)). This indicates that if reflected energy from the shorter and longer ends of the visible spectrum

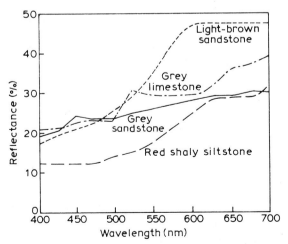

Figure 14.2 Spectral reflectance curves for four different rock types (from Beaumont (1979a))

are recorded separately, differentiation between rock types can be achieved. The ability to distinguish between different materials increases when imagery is recorded by different sensors outside the visible spectrum, the spectral characteristics then being influenced by the atomic composition and molecular structure of the materials concerned.

In addition to the standard photographs at a scale of 1 : 1 000 000, both transparencies (positive and negative) and enlargements, at scales of 1 : 250 000 and 1 : 500 000, are available, as are false colour composites. The latter often show up features not easily observable on black and white images.

Landsat images may be interpreted in a similar manner to aerial photographs, although the images do not come in stereopairs. Nevertheless a pseudostereoscopic effect may be obtained by viewing two different spectral bands (band-lap stereo) of the same image or by examining images of the same view taken at different times (time-lap stereo). There is also a certain amount of side-lap, which improves with latitude. This provides a true stereographic image across a restricted strip of a print, however, significant effects are only produced by large relief features. Interpretation of Landsat data may also be accomplished by automated methods using digital data directly or by using interactive computer facilities with visual display devices (see Beaumont (1977)). An example of the interpretation of suitably enlarged and enhanced Landsat imagery, together with airborne multispectral and thermal imagery has been given by Cole (1977), who showed that geological structures and lithological units could be distinguished.

The value of space imagery is important where existing map coverage is inadequate. For example, it can be of use for the preparation of maps of terrain classification, for regional engineering soil maps, for maps used for route selection, for regional inventories of construction materials, and for inventories of drainage networks and catchment areas (see Beaumont (1978)). A major construction project is governed by the terrain, optimum location requiring a minimum disturbance of the environment. In order to assess the ground conditions it is necessary to make a detailed study of all the photopattern elements that comprise the landforms on the satellite imagery. Important evidence relating to soil types, or surface or subsurface conditions may be provided by erosion patterns, drainage characteristics or vegetative cover. Engineering soil maps are frequently prepared on a regional basis for both planning and location purposes in order to minimise construction costs, the soils being delineated for the landforms within the regional physiographic setting.

14.2 Aerial photographs and photogeology

Aerial photographs are generally taken from an aeroplane which is flying at an altitude of between 800 and 9000 m, the height being governed by the amount of detail that is required. Photographs may be taken at different angles ranging from vertical to low oblique (excluding horizon) to high oblique (including horizon). Vertical photographs, however, are the most relevant for photogeological purposes (see Norman (1968a)). Oblique photographs have occasionally been used for survey purposes but, because their scale of distortion from foreground to background is appreciable, they are not really suitable. Nevertheless because they offer a graphic visual image of the ground they constitute good illustrative material.

Normally vertical aerial photographs have 60% overlap on consecutive prints on the same run, and adjacent runs have a 20% overlap or side-lap. As a result of tilt (the angular divergence of the aircraft from a horizontal flight path) no photograph is ever exactly vertical but the deviation is almost invariably less than 1°. Scale distortion away from the centre of the photograph represents another source of error.

Not only does a study of aerial photographs allow the area concerned to be divided into topographical and geological units, but it also enables the engineering geologist to plan fieldwork and to select locations for sampling. This should result in a shorter, more profitable period in the field. It has been suggested by Rengers and Soeters (1980) that when a detailed interpretation of aerial photographs is required, the photographs can be enlarged up to approximately twice the scale of the final map to be produced.

Examination of consecutive pairs of aerial photographs with a stereoscope allows observation of a three-dimensional image of the ground surface (see Dumbleton and West (1970)). This is due to parallax differences brought about by photographing the same object from two different positions. The three-dimensional image means that heights can be determined and contours can be drawn, thereby producing a topographic map. However, the relief presented in this image is exaggerated so that slopes appear steeper than they actually are. Nonetheless this helps the detection of minor changes in slope and elevation. Unfortunately exaggeration proves a definite disadvantage in mountainous areas as it becomes difficult to distinguish between steep and very steep slopes. A camera with a longer focal lens reduces the amount of exaggeration and therefore its use may prove preferable in such areas.

Aerial photographs may be combined in order to cover larger regions. The simplest type of combination is the uncontrolled print laydown which consists of photographs, laid along side each other, which have not been accurately fitted into a surveyed grid. Photomosaics represent a more elaborate type of print laydown, requiring more care in their production and controlled photomosaics are based on a number of geodetically surveyed points. They can be regarded as having the same accuracy as topographic maps.

14.2.1 Types of aerial photographs

There are four main types of film used in normal aerial photography, namely, black and white, infrared monochrome, true colour and false colour. Black and white film is used for topographic survey work and for normal interpretation purposes. The other types of film are used for special purposes. For example, infrared monochrome film makes use of the fact that near-infrared radiation is strongly absorbed by water. Accordingly, it is of particular value when mapping shorelines, the depth of shallow underwater features and the presence of water on land, as for instance, in channels, at shallow depths underground or beneath vegetation. Furthermore, it is more able to penetrate haze than conventional photography. True colour photography displays variations of hue, value and chroma, rather than tone only and generally offers much more refined imagery. As a consequence colour photographs have an advantage over black and white ones as far as photogeological inter- pretation is concerned, in that there are more subtle changes in colour in the former than in grey tones in the latter, hence they record more geological information. However, colour photographs are more expensive and it is difficult to reproduce slight variations in shade consistently in processing. Another disadvantage is the attenuation of colour in the atmosphere, with the blue end of the spectrum suffering a greater loss than the red end. Even so at the altitudes at which photographs are normally taken the colour dif- ferentiation is reduced significantly. Obviously true colour is only of value if it is closely related to the geology of the area shown on the photograph. False colour is the term frequently used for infrared colour photography since on reversed positive film, green, red and infrared light are recorded respec- tively as blue, green and red. False colour provides a more sensitive means of identifying exposures of bare grey rocks than any other type of film. Lineaments, variations in water content in soils and rocks, and changes in vegetation which may not be readily apparent on black and white photo- graphs are often clearly depicted by false colour. The choice of the type of photographs for a project is governed by the uses they will have to serve during the project (see Allum (1970)). For example, Norman *et al.* (1975) found that in the detection of unstable slopes, the best results were obtained from false colour photographs.

14.2.2 Photogeology

Allum (1966) pointed out that when stereopairs of aerial photographs are observed the image perceived represents a combination of variations in both relief and tone. However, relief and tone on aerial photographs are not absolute quantities for particular rocks. For instance, relief represents the relative resistance of rocks to erosion as well as the amount of erosion which has occurred. Tone is important since small variations may be indicative of different types of rock. Unfortunately tone is affected by light conditions, which vary with weather, time of day, season, etc., and processing. Never- theless basic intrusions normally produce darker tones than acid intrusions. Quartzite, quartz schist, limestone, chalk and sandstone tend to give light tones; whilst slates, micaceous schists, mudstones and shales give medium tones; and amphibolites give dark tones.

According to Allum (1966) the factors which affect the photographic appearance of a rock mass include climate, vegetative cover, absolute rate of erosion, relative rate of erosion of a particular rock mass compared with that of the country rock, colour and reflectivity, composition, texture, structure, depth of weathering, physical characteristics, and factors inherent in the type of photography and the conditions under which the photograph was obtained. Many of these factors are interrelated. Regional geological structures are frequently easier to recognise on aerial photographs, which provide a broad synoptic view, than they are in the field.

Lineaments have been defined by Allum (1966) as any alignment of features on an aerial photograph. The various types recognised include topographic, drainage, vegetative and colour alignments. Bedding is portrayed by lineaments which usually are few in number and occur in parallel groups. If a certain bed is more resistant than those flanking it, then it forms a clear topographic lineament (see Norman (1968)). Even if bedding lineaments are interrupted by streams they usually are persistent and can be traced across the disruptive feature.

Foliation may be indicated by lineaments. It can often be distinguished from bedding since parallel lineaments which represent foliation tend to be both numerous and impersistent.

Care must be exercised in the interpretation of the dip of strata from stereopairs of aerial photographs. For example, dips of 50 or 60° may appear almost vertical, and dips between 15 and 20° may look more like 45° because of vertical exaggeration. However, Allum (1966) maintained that with practice dips can be estimated reliably in the ranges, less than 10°, 10–25°, 25–45°, and over 45° (Figure 14.3). Furthermore displacement of relief makes all vertical structures appear to dip towards the central or principal point of a photograph. Because relief displacement is much less in the central areas of photographs than at their edges, it is obviously wiser to use the central areas when estimating dips. It must also be borne in mind that the topographic slope need bear no relation to the dip of the strata composing the slope. However, scarp slopes do reflect the dip of rocks (Figure 14.4) and as dipping rocks cross interfluves and river valleys they produced crescent and V-shaped traces respectively. The pointed end of the V always indicates the direction of dip and the sharper the angle of the V, the shallower the dip. If there are no dip slopes it may be possible to estimate the dip from bedding traces. Vertical beds are independent of relief.

The axial trace of a fold can be plotted and the direction and amount of its plunge can be assessed when the direction and amount of dip of the strata concerned can be estimated from aerial photographs. Steeply plunging folds have well rounded noses and the bedding can be traced in a continuous curve. On the other hand gently plunging folds occur as two bedding lineaments meeting at an acute angle (the nose) to form a single lineament. Also the presence of repeated folding may sometimes be recognised by plotting bedding plane traces on aerial photographs.

Straight lineaments which appear as slight negative features on aerial photographs usually represent faults (Figure 14.5) or master joints. In order to identify the presence of a fault there should be some evidence of movement. Usually this evidence consists of the termination or displacement of other structures. In areas of thick soil or vegetation cover, faults may be less obvious. Nevertheless Norman (1976) has shown that in some situations traces of old faulting and folding can be detected from aerial photographs even though these structures are covered by superficial deposits or unconformable rocks. Faults running parallel to the strike of strata also may be difficult to recognise. Joints, of course, show no evidence of displacement. Jointing patterns may assist the recognition of certain rock types, as for example, in limestone or granite terrains.

Dykes and veins also gave rise to straight lineaments, which are at times indistinguishable from those produced by faults or joints. If, however, dykes or veins are wide enough they may give a relief or tonal contrast with the country rock. They are then distinctive. Acid dykes and quartz veins often are responsible for light coloured lineaments and basic dykes for dark

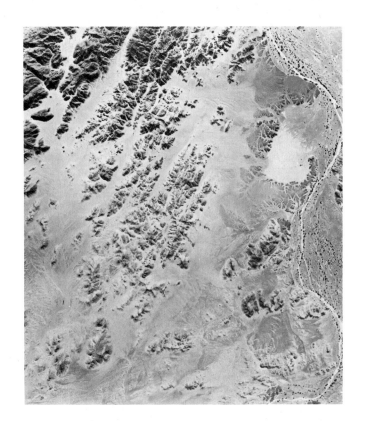

(a)

(b)

Figure 14.3 (a) The Tibesti Mountains of Northern Chad. (b) Eroded syncline in well-bedded sedimentary rocks, with lake in central core. Murnau er Mulde (western) West Germany (courtesy of Aerofilms Ltd)

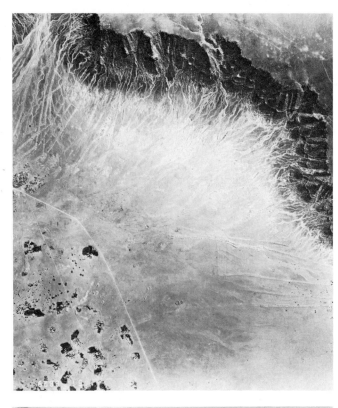

Figure 14.4 Escarpment and cuesta in tilted well-bedded sedimentary rocks (courtesy of Aerofilms Ltd)

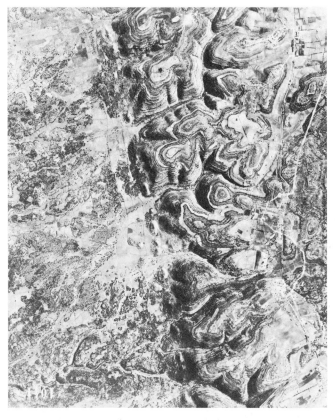

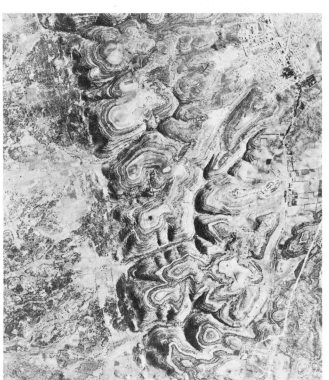

Figure 14.5 Faulted escarpment, with deeply incised drainage in horizontal well-bedded sedimentary rocks, Jebel Akhdar, Libya (courtesy of Aerofilms Ltd)

lineaments. Even so because relative tone depends very much on the nature of the country rock positive identification cannot be made from aerial photographs alone.

If the area portrayed by the aerial photographs is subject to active erosion then it is frequently possible to differentiate between different rock masses, although it is not possible to identify the rock types. Normally only general rather than specific rock types are recognisable from aerial photographs, for example, superficial deposits, sedimentary rocks, metamorphic rocks, intrusive rocks and extrusive rocks. Allum (1966) suggested that superficial deposits could be grouped into transported and residual categories. Transported superficial deposits can be recognised by their blanketing effect on the geology beneath; by their association with their mode of transport and with diagnostic landforms such as meander belts, river terraces, drumlins, eskers, sand-dunes, etc. and their relatively sharp boundaries. Residual deposits generally do not blanket the underlying geology completely and in places there are gradational boundaries with rock outcrops. Obviously no mode of transport can be recognised. It is usually possible to distinguish between metamorphosed and unmetamorphosed sediments as metamorphism tends to make rocks more similar as far as resistance to erosion is concerned. Metamoprhism should also be suspected when rocks are tightly folded and associated with multiple intrusions. By contrast rocks which are horizontally bedded or gently folded, and are unaffected by igneous intrusions are unlikely to be metamorphic. As noted above acid igneous rocks give rise to light tones on aerial photographs and they may display evidence of jointing. The recognition of volcanic cones indicates the presence of extrusive rocks (Figure 14.6).

The suggestions put forward by Norman and Huntington (1974) relating to the use of photogeological interpretation during the early stages of an engineering geological survey can be taken as a summary of this section. These include

(1) mapping and analysis of folding,
(2) mapping of regional fault systems and recording any evidence of relatively recent fault movements and shear surfaces,
(3) determination of the number and geometry of joint systems,
(4) a study of the lithology and resistance to erosion of surface rocks in relation to relief and landforms,
(5) recording the distribution of superficial deposits,
(6) drainage studies which would take into account surface run-off characteristics; boundaries of catchment areas, stream divides, etc.; areas liable to periodic flooding; areas of subsurface drainage, especially of caverous limestone as illustrated by surface solution features; relative permeability of the principal rock and soil types at the surface,
(7) an assessment of the stability of slopes (aerial photographs are particularly useful in detecting old failures which are difficult to appreciate on the ground),
(8) aerial photographs can also be used as an aid in the detection of old mine workings (see Anon (1976a)).

Figure 14.6 Agua Volcano, Guatemala. Flying height 4000 m above ground. Negative scale 1 : 26000 (by kind permission of Wild Heerbrugg Ltd, Switzerland)

14.3 Morphological maps

The classical method of landform mapping is through surveyed contours. Waters (1958), however, devised a technique, that was further refined by Savigear (1965), which defined the geometry of the ground surface in greater detail than is normally found on contour maps. They proposed that the ground surface consisted of planes which intersected in convex and concave, angular or curved 'discontinuities'. An angular discontinuity was defined as a break of slope and a curved discontinuity as a change of slope. A morphological map is therefore divided into slope units which are delineated by breaks of slope so defining the pattern of the ground (distinction between breaks and changes of slope provides a more precise appreciation of landform than is possible from reading of contours).

Morphometry has been defined by Kertesz (1979) as the measurement and mathematical analysis of the configuration of the Earth's surface and the shape and dimensions of its landforms. Morphometric data such as the inclination of slopes, amplitude of relief, valley density and depth, can be obtained from detailed topographical maps (see Demek (1972); Miller (1953)).

When available, aerial photographs should be used for preliminary morphological mapping since they furnish an idea of the terrain and may be used to locate discontinuities between the morphological units. These are recorded on the photographs and then transferred, by plotter, to the base map. The best scale for the field map depends on the objectives of the survey. Whatever the scale, however, some units will be recognised which have boundaries which are too close together to be represented separately. If small features, which are regarded as important, cannot be incorporated on the base map, then they should be mapped on a larger scale. For example, Savigear (1965) maintained that certain features, such as cliff units, should always be represented in morphological mapping. Most standard geomorphological features can be represented on a base map which has a scale of 1:10000. However, not only does clear representation of all morphological information on one map provide difficult cartographic problems, but it also gives rise to difficulties in interpretation and use, thus limiting the value of the map. This problem can, to some extent, be overcome by using overlays to show some special aspect of the land surface.

Savigear (1965) distinguished three types of morphological survey. The reconnaissance survey was carried out at scales between 1:75000 and 1:25000, only first-order discontinuities being mapped. Indeed it is rarely possible to record meaningful slope discontinuities at scales less than 1:75000. A detailed survey is carried out with scales between 1:25000 and 1:7000, when first-, second- and third-order discontinuities are mapped. The scales for a precise survey are larger than 1:7000.

Convex and concave boundaries are distinguished in morphological mapping and measurements can be made of slope steepness and, if present, slope curvature. Knowledge of slope angles is needed for the study of present day processes and to understand the development of relief. Steepness can be shown by an arrow lying normal to the slope, pointing downhill, and the angle of the slope is marked in degrees on the arrow (Figure 14.7a). Special symbols are used for very steep slopes such as cliffs. Differences in slope

Table 14.2a. Slope angle classification for detailed geomorphological maps (after Demek (1972))

Slope category	Slope terminology	Gradient (%)	Slope length to slope height ratio
0–0°30′	Plain		
0–30′–2°	Slightly sloping	0–3.5	∞–28.6
2–5°	Gently inclined	3.5–8.7	28.6–11.4
5–15°	Strongly inclined	8.7–26.8	11.4–3.7
15–35°	Steep	26.8–70	3.7–1.4
(25–35°)	(Very steep)		
35–55°	Precipitous	70–143	1.4–0.7
55–90°	Vertical	143–∞	0.7–0
Over 90°	Overhanging		

Table 14.2b Critical slope steepness for certain activities*

Steepness (%)	Critical for
1	International airport runways
2	Main-line passenger and freight rail transport
	Maximum for loaded commercial vehicles without speed reduction
	Local aerodrome runways
	Free ploughing and cultivation
	Below 2%—flooding and drainage problems in site development
4	Major roads
5	Agricultural machinery for weeding, seeding
	Soil erosion begins to become a problem
	Land development (constructional) difficult above 5%
8	Housing, roads
	Excessive slope for general development
	Intensive camp and picnic areas
9	Absolute maximum for railways
10	Heavy agricultural machinery
	Large-scale industrial site development
15	Site development
	Standard wheeled tractor
20	Two-way ploughing
	Combine harvesting
	Housing-site development
25	Crop rotations
	Loading trailers
	Recreational paths and trails

* From Cooke, R. U. and Doornkamp, J. C. (1974). *Geomorphology in Environmental Management*, Clarendon Press, Oxford.

steepness can be emphasised by shading or colours according to defined slope classes (Figure 14.7b). Demek (1972) suggested several categories of slope (see Table 14.2a). Slope category maps depict average inclination over an area and make it easier to perceive the distribution of steep slopes, planation surfaces, valley asymmetry, than is possible from contour maps (see Kertesz (1979)). Slope steepness is of considerable importance in land management, for example, it frequently poses the restricting factor in route location, urban development and agriculture (see Table 14.2b).

Mapping the surface form is the first step in geomorphological mapping. The next is to make interpretations regarding the forms and to ascribe an

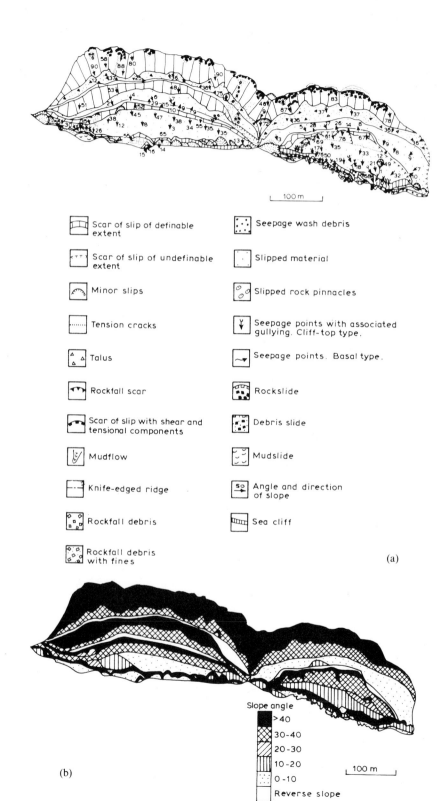

Scar of slip of definable extent

Scar of slip of undefinable extent

Minor slips

Tension cracks

Talus

Rockfall scar

Scar of slip with shear and tensional components

Mudflow

Knife-edged ridge

Rockfall debris

Rockfall debris with fines

Seepage wash debris

Slipped material

Slipped rock pinnacles

Seepage points with associated gullying. Cliff-top type.

Seepage points. Basal type.

Rockslide

Debris slide

Mudslide

Angle and direction of slope

Sea cliff

(a)

Slope angle

>40

30-40

20-30

10-20

0-10

Reverse slope

100 m

(b)

origin to them. This must be done in relation to the geological materials which compose each feature and in relation to the past and present processes operating in the area concerned. A morphogenetic map indicates the origin of the landforms recognised in the field, while a process map identifies the spatial pattern of geomorphological processes which are currently active (Figure 14.8).

Pitts (1979) used morphological mapping techniques to illustrate the distribution of a complex array of slopes which occur in the cliffs between Exmouth in Devon and Lyme Regis, Dorset. He maintained that simple techniques of surveying, involving the use of a tape, a prismatic compass and an inclinometer, in conjunction with an ability to 'read' ground, may yield good results. In this way he recorded the breaks and changes of slope, the direction of slope and the maximum slope angle, and so was able to define slope units. Pitts was able to discern significant interrelationships between the morphological units he mapped and the local geology.

Knott *et al.* (1980) mapped the surface morphology of an area in Northamptonshire by drawing the boundaries between slopes of different steepness, as well as recording slope angles (see Figure 14.9a). Their interpretation of how these slopes had originated provided the basis for all the other geomorphological assessments which were made (Figure 14.9b and c). Morphological mapping may prove useful as a quick reconnaissance exercise prior to a site investigation or as a more extensive undertaking where difficult or inaccessible terrain is concerned and therefore restricts the use of some site investigation techniques.

14.4 Engineering geomorphological maps

The purpose of geomorphological maps is to portray the forms of the surface, the nature and properties of the materials of which these surfaces are composed and to indicate the type and magnitude of the processes in operation. As such they provide a comprehensive, integrated statement of landform and drainage. Consequently they contain much information of potential value as far as construction projects and land-use planning are concerned.

Brunsden *et al.* (1975a) contended that the recording of geomorphological data on maps for engineering purposes was of value since surface form and aerial pattern of geomorphological processes often influence the choice of a site. In other words geomorphological mapping gives a rapid appreciation of the nature of the ground and thereby helps the design of more detailed investigations, as well as focusing attention on problem areas. It involves the recognition of landforms along with their delimitation in terms of size and shape (Figure 14.10). The principal object during a reconnaissance survey is the classification of every component of the land surface in relation to its origin, present evolution and likely material properties, based on techniques of landform interpretation. A site investigation provides additional data which enable the preliminary views on causative processes to be revised, when

Figure 14.7 (a) Morphological map of the Haven and Culverhole cliffs landslips (after Pitts (1979)). (b) Slope categories of the Haven and Culverhole cliffs landslips (after Pitts (1979))

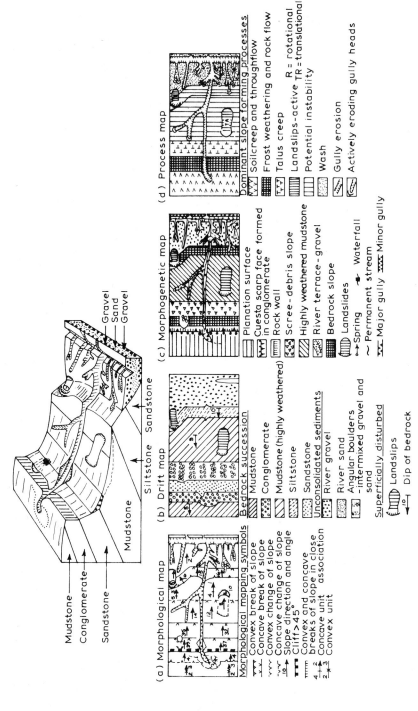

Mudstone

Conglomerate

Sandstone

Mudstone — Conglomerate — Sandstone — Siltstone Sandstone

Gravel
Sand
Gravel

(a) Morphological map

Morphological mapping symbols
ᐯᐯᐯ Convex break of slope
ᐱᐱ Concave break of slope
ᐯᐯᐯ Convex change of slope
ᐱᐱᐱ Concave change of slope
↑₁₀ Slope direction and angle
⊤⊤⊤ Cliff >45°
ᵢₗₗₗ Convex and concave
⁴⁺² breaks of slope in close
²⁺³ association
4 ↑ 2 Concave unit
2 ↑ 3 Convex unit

(b) Drift map

Bedrock succession
▨ Mudstone
◦◦◦ Conglomerate
▨ Mudstone (highly weathered)
∷∷ Siltstone
⋅⋅⋅ Sandstone

Unconsolidated sediments
∷∷∷ River gravel
⋅⋅⋅ River sand
▨ Angular boulders
  — intermixed gravel and
  sand

Superficially disturbed
◠◠ Landslips
↓₁₀ Dip of bedrock

(c) Morphogenetic map

▥ Planation surface
ᐯᐯᐯ Cuesta scarp face formed
  in conglomerate
▥ Rock wall
▨ Scree-debris slope
▨ Highly weathered mudstone
▨ River terrace-gravel
▦ Bedrock slope
◠◠ Landslides
↝ Spring   ⤵ Waterfall
∿ Permanent stream
⨯⨯ Major gully ⨯⨯ Minor gully

(d) Process map

Dominant slope forming processes
ᐯᐯ Soilcreep and throughflow
▨ Frost weathering and rock flow
⊤⊤ Talus creep
▥ Landslips-active R = rotational
      TR =translational
▥ Potential instability
▥ Wash
▨ Gully erosion
◌ Actively eroding gully heads

Figure 14.8 Schematic diagram to illustrate the principles and products of geomorphological mapping. (a) The morphological map emphasises slope form and steepness. (b) The materials or drift map indicates the nature of surface materials. (c) The morphogenetic map indicates the origin of the landforms. (d) The process map identifies the distribution of currently active geomorphological processes (after Savigear (1965))

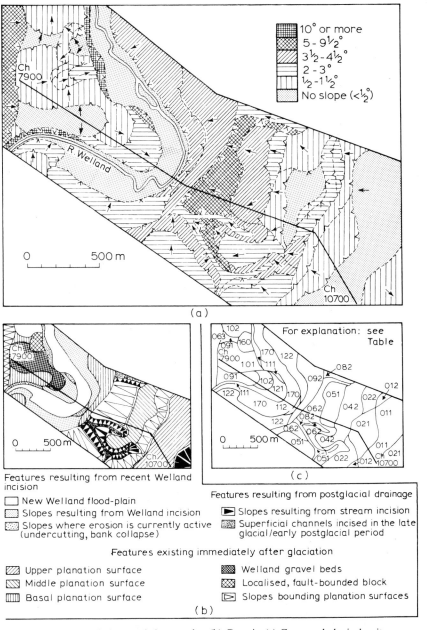

Figure 14.9 (a) Morphology and slope-angles. (b) Genesis. (c) Geomorphological units
(see p. 614)

1	2	3	4	5	6	7	8	9
Unit No.	Slope (degrees)	Morphogenetic characteristics	Topsoil	Water movement	Creep class	Soil transport	Drainage class	Drainage control
011	0–1	Upper surface	SCq	I	1	N	P	MG
012	0–1	Upper surface	SC	I	1	N	P	MG
021	1½–4	Upper surface upper bounding slope	SC	It	2	Ll	I	G
022	1½–4	Upper surface upper bounding slope	SC	T	2	S	I	G
042	0–1	Middle surface	CSb	It	1	Ll	W	G
051	2–4	Middle surface upper bounding slope	CSg	T	2	S	M	MG
062	4½–9	Middle surface middle bounding slope	CSg	Tw	3	S	W	M
082	1–3	Incised superficial channel	CSg	It	2	Ll	M	M
091	½–4	Basal surface	CSb	I	1	N	W	G
092	½–4	Basal surface	CSg	T	2	S	W	M
101	0–2	Welland gravel terraces	SGc	I	1	N	W	G
102	0–2	Welland gravel terraces	SGc	It	2	Ll	W	G
111	5–20	Meander bluff	CSg	W	5	L	M	M
112	5–20	Meander bluff	SC	W	3	L	M	M
122	1½–4	Meander slip-off slope	SC	T	2	Ll	M	M
160	0–1½	Localised fault-bounded block	SCsh	I	1	N	M	MG
170	0–½	Welland flood-plain	MC	Iw	1	Ar	P	MG

NOTES: Classes

Topsoil	Creep	Water movement	Soil transport	Drainage class
C Clay	1 Slope-sine 0–0.1	I Infiltration	N No loss	P Poor
S Sand	2 Slope-sine 0.1–0.2	T Throughflow	Ll Little loss	I Imperfect
M Silt	3 Slope-sine 0.2–0.3	W Wash	L Loss	M Moderate
G Gravel	4 Slope-sine 0.3–0.4		S Steady state	W Well-drained
b Blocks	5 Slope-sine 0.4+		Ar Accumulation with	
q chalk erratics			removal in flood	Drainage control
sh shale fragments				
				M Morphology
				G Geology

Figure 14.9 (c) Specifications of geomorphological units

necessary. Further precision can be afforded geomorphological interpretations by obtaining details from climatic, hydrological or other records and by analysis of the stability of landforms. What is more an understanding of the past and present development of an area is likely to aid prediction of its behaviour during and after construction operations. Geomorphological maps should therefore show how surface expression will influence an engineering project and should provide an indication of the general environmental relationship of the site concerned.

The aims of such a geomorphological survey are to guide and complement a site investigation and have been summarised by Brunsden *et al.* (1975b) as follows.

(1) Identification of the general characteristics of the terrain of an area, thereby providing a basis for evaluation of alternative locations and avoidance of the worst hazard areas.

(2) Identification of factors outside the site which may influence it, such as mass movement.

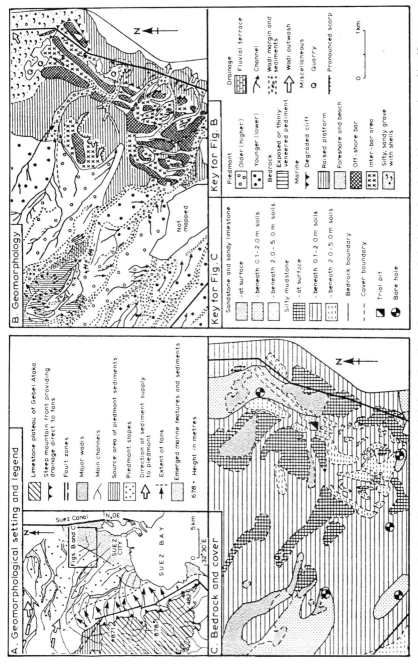

Figure 14.10 Rationalisation of borehole and trial pit information through a geomorphological and geological interpretation of landforms, Suez, Egypt (after Doornkamp *et al.* (1979))

Table 14.3. Summary outline of working practice (after Doornkamp et al. (1979))

Phase	Liaison with engineers	Desk studies (home based)	Field studies (based on site)
I	Brief received from client-discussions with senior engineers and engineering geologist involved Brief re-examined	Familiarisation with project Examination of available literature and maps Air-photo interpretation	
II	Continuing discussions with engineer's field staff		Field mapping —investigation of landforms, materials and processes —review of trial pit and borehole information (if available) Geomorphological map compilation
III	Report with maps passed to client	Derivative maps compiled Data additional to initial brief compiled Site investigation suggestions defined	

(3) Provision of a synopsis of geomorphological development of the area which includes:
  (a) a description of the extent and degree of weathering,
  (b) a classification of slopes based on their steepness, material composition, mode of development and stability,
  (c) a description of the location, pattern and magnitude of the surface and subsurface drainage features (including karst development),
  (d) definition of the shape and extent of geomorphological units such as fans, scree slopes, terraces, etc.
  (e) recognition of specific hazards such as flooding and landslides.
(4) Location of suitable supplies of construction materials.

Obtaining such information should facilitate the planning of a subsequent site investigation. For instance, it should aid the location of boreholes, and these hopefully will confirm what has been discovered by the geomorphological survey. Geomorphological mapping may therefore help to reduce the cost of a site investigation.

Doornkamp et al. (1979) pointed out that the recognition of both the interrelationships between landforms on site and their individual or combined relationships to landforms beyond the site is fundamental. This is necessary in order to appreciate not only how the site conditions will affect the engineering but, just as importantly, how the engineering will affect the site and the surrounding environment. In particular, knowledge of ground conditions, the sensitivity of the landscape to change, and environmental hazards, are required for good design.

Doornkamp *et al.* (1979) suggested that if engineers are to obtain maximum advantage from a geomorphological survey, then derivative maps should be compiled from the geomorphological sheets. Such derivative maps generally are concerned with some aspect of ground conditions, such as land-slip areas or areas prone to flooding or over which sand-dunes migrate (see Figure 8.15). An example of such an approach has been provided by Knott *et al.* (1980), who produced several maps for an area which straddled the river Welland in Northamptonshire, in an effort to demonstrate the relation-ship between geomorphological mapping units and soils (see Figure 14.9).

The general procedure in an engineering geomorphological investigation is summarised in Table 14.3. Phase I, which is carried out prior to the field work, involves familiarisation with the project and the landscape. The amount of information which can be obtained from a literature survey varies with location. In some underdeveloped countries little or nothing may be available, even worthwhile topographical maps, which are normally a prerequisite of a geomorphological mapping programme, may not exist. Base maps can then be made from aerial photographs which can be specially commissioned.

A study of aerial photographs enables many of the significant landforms and their boundaries to be defined prior to the commencement of fieldwork. The scale of the photographs is usually 1 : 10 000. Field mapping permits the correct identification of landforms, recognised on aerial photographs, as well as geomorphological processes, and indicates how they will affect the engineering project. Mapping of the site can provide data on the nature of the surface materials.

The scale of a geomorphological map is influenced by the engineering requirement and the map should focus attention on the information rele-vant to the particular project. Maps produced for extended areas, such as needed for route selection, are drawn on a small scale. Small-scale maps have also been used for planning purposes, land-use evaluation, land reclamation, flood plain management, coastal conservation, etc. These general geo-morphological maps concentrate on portraying the form, origin, age and distribution of landforms, along with their formative processes, rock type and surface materials (see Figure 14.10). In addition, if information is avail-able, details of the actual frequency and magnitude of the processes can be shown by symbols, annotation, accompanying notes or successive maps of temporal change.

On the other hand, large-scale maps and plans of local investigations provide an accurate portrayal of surface form, drainage characteristics and the properties of surface materials, as well as an evaluation of currently active processes. They may be classified as reconnaissance, site investigation and construction maps and plans (see Brunsden *et al.* (1975b)). In practice, however, a map or plan produced for one stage of a project may be used at other stages. This is especially true of reconnaissance maps which may prove adequate for the site investigation.

14.5 Terrain evaluation

Terrain evaluation is only concerned with the uppermost part of the land sur-face of the Earth, that is, with that which lies at a depth of less than 6 m, excluding permanent masses of water. Mitchell (1973) described terrain evaluation as involving analysis (the simplification of the complex phenomena which make up the natural environment), classification (the organisation of data in order to distinguish and characterise individual areas), and appraisal (the manipulation, interpretation and assessment of data for practical ends) of an area of the Earth's surface which is of interest to engineers. There are two different approaches to this in terrain evaluation, namely, parametric evaluation and landscape classification. Parametric land evaluation refers to the classification of land on a basis of selected attribute values appropriate to the particular study, such as class of slope or the extent of a certain kind of rock. The simplest form of parametric map is one which divides a single factor into classes. Landscape classification is based on the principal geomorphological features of the terrain.

In terrain evaluation the initial interpretation of landscape can be made from large-scale maps and aerial photographs (see Webster and Beckett (1970)). Observation of relief should give particular attention to direction (aspect) and angle of maximum gradient, maximum relief amplitude and the proportion of the total area occupied by bare rock or slopes. In addition an attempt should be made to interpret the basic geology and the evolution of the landscape. An assessment of the risk of erosion (especially the location of slopes which appear potentially unstable) and the risk of excess deposition of water-borne or wind-blown debris should also be made (see Dowling and Beaven (1969)).

Terrain evaluation provides a method whereby the efficiency and accuracy of preliminary surveys can be improved. In other words it allows a subsequent site investigation to be directed towards the relevant problems. It also offers a rational means of correlating known and unknown areas, that is, of applying information and experience gained on one project to a subsequent project. This is based on the fact that landscape systems of terrain evaluation have indicated that landscapes in different parts of the world are sufficiently alike to make predictions from the known to the unknown.

The following units of classification of land have been recognised for purposes of terrain evaluation, in order of decreasing size, namely, land zone, land division, land province, land region, land system, land facet and land element (see Brink *et al.* (1966)). The land system, land facet and land element are the principal units used in terrain evaluation (see Lawrance (1972); (1978); Anon (1978)).

A land systems map shows the subdivision of a region into areas with common physical attributes which differ from those of adjacent areas. Land systems are usually recognised from aerial photographs, the boundaries between different land systems being drawn where there are distinctive differences between landform assemblages. For example, the character of land units can be largely determined from good stereo-pairs of photographs with an optimum scale of about 1 : 20 000, depending on the complexity of the terrain. Field work is necessary to confirm the landforms and to identify soils and bedrock.

In order to establish the pattern identified on the aerial photographs as a land system, it is necessary to define the geology and range of small topographic units referred to as land facets. A land system extends to the limits of a geological formation over which it is developed or until the pre-

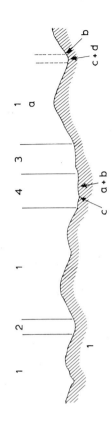

Land region:	Lowland sediments
Climate (rainfall):	1750–2500 mm p.a.
Geology:	Shales and mudstones. A very small area of granite exists south-east of Masjid Tanah in Malacca (Alor Gajah Variant)
Landscape:	Low hills with uneven slopes and small gullies; frequent broad river valleys with well-developed terraces. All slopes are gentle, and the terraces are particularly advantageous for road location
Soils:	Clays and silty clays, often with laterite horizons (sometimes massive). BSCS: GPF (laterite gravels); I, E (clays and silty clays)
Vegetation:	Rubber plantations; padi in major valleys
Relief:	20–50 m
Altitude:	Few–90 m
Area:	847 km²

Figure 14.11 Alor Gajah land system (after Lawrence (1978))

Table 14.4. Alor Gajah land system (see Figure 14.7)

Land facet	Land form	Soils, materials, hydrology and vegetation	Engineering properties
1	**Hill**		Widespread occurrence of laterite, characteristically lacking sand-sized particles (GPF). Suitable for sub-base when unstabilised (CBR about 20–40 at B.S. optimum). Suitable for base when stabilised with cement (CBR about 150 after 7 days cure with 4% cement added). Vertical bedding liable to very rapid weathering on exposure in cutting; silty horizons (particularly the pallid zone) very prone to erosion and slipping. Suggested safe angle of cut 35–40 (−45)°. Excavation possible mostly by scraper, but freshly-exposed unweathered rock may require ripping. Poor material for fill, especially when very silty. Upper, less silty, soil layers should be stored and laid on top of siltier soils in lower parts of embankment.
	(a) Slope Moderate slopes (5−) 7–12°, usually bumpy and irregular in detail. Overall slope is straight or gently convex, but often slightly concave in upper portion to give a more prominent hill top. Slope 70–100 m long.	Up to 1 m of red, red-brown or yellowish brown clay over massive or nodular laterite up to 2 m thick, over a silty clay mottled zone. Weathered rock occurs below about 4 m. The irregular development of laterite on the slopes tends to form benches on the bumpy slopes, and it may outcrop on the steeper parts. Rubber plantations.	
			Soil test data.

	(b) Gully side Up to 5 m deep, 20°. Fades into hillside towards gully head. Active soil creep on gully sides.		
	(c) Gully floor 6–10 m wide, flat or gently sloping towards stream. 100–300 (−750) m long; fairly straight.		
	(d) Stream Up to 3 m wide and may be incised 2 m.		
2	**Minor valley** 30–100 m wide 500–1000 m long; flat in section, very gently sloping upstream. Narrow stream a few m wide.	Variable soils, the weathering products of local rocks, mostly clays and silty clays. Poor to well drained, depending on materials; water table within 2 m. Mostly minor cultivation and rubber, but some padi in wetter valleys.	Soils weak and poorly drained. Embankment and small culvert required.

Soil test data (under Engineering properties, facet 1):

	LL	PI	LS
n	22	22	22
x̄	69	36	14
s	23	14	4

Table 14.4—cont.

3	*Terrace* Level or very gently undulating; up to 120 m across, but varies considerably along its length. Not continuous for more than about 2 km; usually much less. Merges with valley floor, or may steepen slightly in a short bluff.	Variable soils of MSS Local Alluvium Association, usually light brown or clayey sand with poor profile development. Usually freely draining; depth to water table at least 3–4 m. Minor cultivation.	Variable subgrade strength, but generally freely draining and dry. Favourable for road location. Few earthworks.
4	*Main valley* (a) Valley floor Flat, 100–130 (–600) m wide.	Similar to 3(a) but wetter and more clayey with organic material. Periodically flooded; water table within 1–2 m of surface. Padi cultivation.	Very weak and poorly drained (FS). Continuous embankment at least 1.3 m high required. Culverts may be necessary to carry small streams, supported in the weakest areas on wooden piles. Soil compressible, but fairly stable after initial compaction.
	(b) Abandoned channel About 20 m wide, sinuous, up to several hundred m long.	As 3(b) but wetter. Padi cultivation.	
	(c) River Few –20 m wide, winding.	Permanent flow. Tree lined banks.	Small bridge required. Foundation on piles.
5	*Coastal swamp* Restricted to coast, occasional. Narrow strips about 1 km long and 200 m wide extend up to 500 m (very rarely 1500 m) inland in a strip 200–400 m wide.	Organic marine and estuarine clays. Waterlogged to surface and tidally flooded. Mangrove and other swamp vegetation.	BSCS: (FO), (FSO). Very weak, compressible and poorly drained soils. Should be avoided altogether if possible. Continuous embankment 1–1½ m high if necessary. Culverts may need to be supported on wooden piles.

N.B. Alor Gajah variant. alor Gajah land system is developed on GRANITE over about 30 km³ on the coast SE of Masjid Tanah in Malacca State. Soils on facet 1 are sandy clays of MSS Regam Series, and in places are only a few m deep over good solid granite, which has been quarried locally. Soils on facets 2 and 3 are local alluvial soils of granite origin, and are therefore more likely to be sandy.
(After Lawrance (1978))

Fine fraction of plastic soils has a very high plasticity but not all soils are plastic. Freely-draining profile, stable in cut (safe angle 50–60°) and not prone to erosion or slip. Stable in embankments.

Table 14.5. Aspects of engineering geological mapping

Geological aspects	Geomorphological aspects	Hydrogeological aspects	Mines and quarries	Site investigations
The character of rocks and soils should include their distribution, stratigraphical and structural arrangements, age, genesis, lithology, physical state or their physical and mechanical properties.	Geomorphological features should be shown on a map whenever possible.	Hydrological conditions affect land-use, planning, site selection and the cost, durability and even safety of structures.	All mines and quarries, whether active or abandoned, should be recorded, plus any details relating to working.	Indication should be made on maps where and when site investigations have been carried out.
Division of rocks and soils into mappable units should be based on engineering geology.	Evaluation of geomorphological conditions in engineering geological mapping should include an explanation of the relationship between surface conditions and geological setting; the origin, development and age of individual geomorphological elements; the influence of geomorphological conditions on hydrology and geodynamic processes. The prediction of such factors as lateral abrasion of riverbanks, movement of dunes, collapse of karstic features, etc., is	There should be a note about the general hydrogeological conditions. Conditions should be quantified wherever possible. Hydrogeological conditions include the distribution of surface water and water-bearing soils and rocks, infiltration conditions, zones of saturated open discontinuities, depth to water table and its range of fluctuations, regions of confined water and piezometric levels, storage	The nature of any infill or made-ground should be noted. Notes on subsidence should be given whenever possible.	A summary of available information should be given. All boreholes, pits, trenches, etc., should be marked. Details of test results should be given.
Soils and rocks should be described as set out in Chapters 9 and 12 respectively.				
Distribution and grade of weathering should be recorded (see Chapter 7).				
Shear zones and details of faulting and fault zones should be recorded.				

Table 14.5—cont.

As much detail as possible should be given about the character of discontinuities (see Chapter 5). Isopachytes of selected horizons may be given.	important. Geodynamic phenomena, include erosion and deposition, aeolian phenomena, permafrost, slope movements, formation of karstic conditions, suffusion, subsidence, volume changes in soil, data on seismic phenomena including active faults, current regional tectonic movements and volcanic activity. All these features are important in planning and construction. It is important to show not only the features but also the condition favouring their development, frequency of occurrence, the rate at which each process is going on, and their intensity.	coefficients, direction and velocity of flow; springs, rivers, lakes and the limits and occurrence interval of flooding; pH, salinity, corrosiveness and the presence of bacterial or other pollutants. Boreholes and wells provide data on the position of the water table, piezometric levels, hydraulic conductivity, storage coefficient and groundwater chemistry. Aquifers, aquitards or aquicludes should be distinguished and the position of springs should be indicated. isopachytes may be used to show the position of the water table and isopiestic lines to show the groundwater pressure conditions.

vailing land forming process gives way and another land system is developed. Land systems maps are usually prepared at scales of 1 : 500 000 or 1 : 1 000 000. More detailed maps may be required in complex terrain. They provide the engineer with background information which can be used in a preliminary assessment of the ground conditions in the area with which he is concerned and permit locations to be identified where detailed investigations may prove necessary.

A land system comprises a number of land facets. Each land facet possesses a simple form, generally being developed on a single rock type or superficial deposit. The soils, if not the same throughout the facet, at least vary in a consistent manner. An alluvial fan, a levee, a group of sand-dunes or a cliff are examples of a land facet. Indeed geomorphology frequently provides the basis for the identification of land facets. Land facets occur in a given pattern within a land system. They may be mapped from aerial photographs at scales between 1 : 10 000 and 1 : 60 000.

A land facet may, in turn, be composed of a small number of land elements, some of which deviate somewhat in a particular property, such as soils, from the general character. They represent the smallest unit of landscape that is normally significant. For example, a hill slope may consist of two land elements, an upper steep slope and a gentle lower slope. Other examples of land elements include small river terraces, gully slopes and small outcrops of rock.

Although nearly all terrain evaluation mapping is carried out at the land system level, the land region may be used in a large feasibility study for some engineering operation. A land region consists of land systems which possess the same basic geological composition and have an overall similarity of landforms. Land regions are usually mapped at a scale between 1 : 1 000 000 and 1 : 5 000 000.

Most land system maps are accompanied by a report which gives the basic information used to establish the classification of landforms within the area surveyed. The occurrence of land facets is normally shown on a block diagram (Figure 14.11), cross section or a map; maps are more often used in areas where the relative relief is very small such as alluvial plains. The descriptions of land facets include data on slope and soil profile, with vegetation and water regime referred to where appropriate (Table 14.4).

The scale of aerial photography used in terrain evaluation varies widely but can be roughly divided into three ranges (see Mitchell (1973)). Large-scale photographs have a scale greater than about 1 : 20 000 and are used for detailed interpretation of small features such as beach ridges, river terraces, periglacial deposits, etc. Medium-scale photographs range from around 1 : 20 000–1 : 50 000. These are most frequently used in terrain study. Small-scale photographs range from approximately 1 : 50 000–1 : 80 000. They provide a method of cheap, rapid reconnaissance of low-value terrain such as deserts.

14.6 Engineering geological maps and plans

Geological maps are usually constructed by superimposing the geological information collected in the field on a topographic base sheet. If topographic maps are not available then they must be made either by conventional surveying methods or from aerial photographs. The geological field work requires that exposures of rock (or soil) in the area represented on the map are examined and that their lithology, age and significant structures are recorded (see Low (1957); Lahee (1961); Himus and Sweeting (1968)). The position of each exposure must be determined so that field data can be plotted on the base map.

A geological map can be described as a general purpose map in the sense that it places a similar emphasis on all the geological facts which it represents. In the United Kingdom the Institute of Geological Sciences usually produces two maps for each sheet area, namely, the solid and drift editions, the superficial deposits being marked in outline on the former map whilst they are shown in full on the latter.

Both topographical and geological maps (together with accompanying memoirs) contain much basic data which can be used during the planning stage of any construction operation, when a decision on the choice of sites has to be made (see Eckel (1951); Varnes (1974)). Nevertheless, from the engineer's point of view one of the shortcomings of conventional geological maps is that the boundaries are stratigraphical and more than one type of rock may be included in a single mappable unit. What is more the geological map is lacking in quantitative information which the engineer requires, concerning such facts as the physical properties of the rocks and soils, the nature of the discontinuities, the amount and degree of weathering, the hydrogeological conditions, etc. However, special maps can be prepared from a geological map. For example, such special maps could interpret the geology in terms of potential supplies of construction materials. They would show the spatial distribution of each kind of material, its topographical position and accessibility to existing means of transportation. Cross-sections, symbols and abbreviated logs could show the thickness of each unit and whether it is at or near the surface. A geological map also can be used to indicate those rocks which should be investigated as potential producers of water. Some idea of the thickness of potential aquifers and the depth at which they occur can be obtained from sections. Perhaps some idea of the quality of the water may be gleaned from the type of rock forming the aquifer.

Geological maps can be made more relevant to the engineer by the addition of engineering data; for example, this can be incorporated into an enlarged key. If the engineering information is extensive, it can then take the form of notes accompanying the map. These notes should describe superficial deposits and bedrock in detail. Where possible rocks and soils should be classified according to their engineering behaviour. Details of geological structures should be recorded, especially fault and shear zones, as should the nature of the discontinuities and grade of weathering, where appropriate.

14.6.1 Purpose of engineering geological maps

An engineering geological map provides an impression of the geological environment, surveying the range and type of engineering geological conditions, their individual components and their interrelationships (Table 14.5). A map, however, represents a simplified model of the facts and the complexity of various dynamic geological factors can never be portrayed in its entirety. The amount of simplification required is governed principally by the purpose and scale of the map, the relative importance of particular engineering geological factors or relationships, the accuracy of the data and on the techniques of representation employed (see Anon (1976b)). Engineering geological maps should be accompanied by cross-sections and an explanatory text and legend. More than one map of the area may be required to record all the information which has been collected during the survey.

Engineering geological maps and plans are used mainly for planning purposes. They provide planners and engineers with information which will assist them in the planning of land-use and the location, construction and maintenance of engineering structures of all types (see Clark and Johnson (1975)). An appreciation of which aspects of the engineering geological environment might affect the successful completion of a project is of special importance, as is the critical assessment of the influence that the construction operations, and the finished structure, might have on the environment. A harmonious relationship between the two should be maintained as far as possible. Accordingly Dearman and Matula (1977) and Anon (1976b) sug-

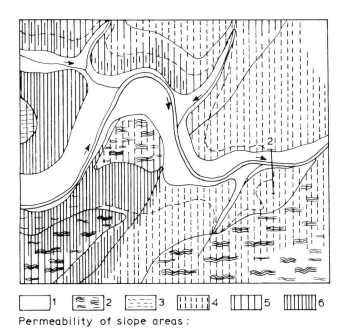

Permeability of slope areas :

1 - low,   2 - medium,   3 - high.

Estimated stability of the ground related to the conditions imposed by water storage reservoirs :

4 - high,   5 - moderate,   6 - low

Figure 14.12 Special-purpose map of engineering geological zoning for planning of hydrotechnical developments (after Dearman and Matula (1976))

gested that an engineering geological map should fulfil the following requirements

(1) it should portray the information required to evaluate the engineering geological aspects of the environment, especially in relation to regional planning or civil or mining engineering purposes,

(2) it should facilitate the prediction of changes in the engineering geological environment likely to occur due to the proposed undertaking and, when necessary, to suggest preventive works.

(3) it should present information in a manner which can be readily understood by professional users.

Engineering geological maps usually are produced on the scale of 1:10 000 or smaller whereas engineering geological plans, being produced for a particular engineering purpose, have a larger scale (see Anon (1972)). Dearman *et al.* (1977) maintained that the main principles of engineering geological mapping are applicable to all maps and plans, irrespective of the scale. The major differences between maps of different scales are, first, the amount of data they show and, second, the manner in which it is presented.

There are two basic types of engineering geological plans, namely, the site investigation plan, and the construction or foundation stage plan (see Dearman and Fookes (1974)). In the case of the former type the scale varies from 1:5000 to as large as 1:500 or even 1:100, depending on the size and nature of the site and the engineering requirement. The foundation plan records the ground conditions exposed during construction operations. It may be drawn to the same scale as the site investigation plan or the construction drawings. Plans may be based on large-scale topographic maps or large-scale base maps produced by surveying or photogrammetric methods.

14.6.2 Types of engineering geological maps

Engineering geological maps may serve a special purpose or a multipurpose (see Dearman and Matula (1972); Anon (1976b)). Special-purpose maps provide information on one specific aspect of engineering geology such as grade of weathering, jointing patterns or mass permeability (Figure 14.12). On the other hand, special-purpose maps may serve one particular purpose as, for example, the engineering geological conditions at a dam site or along a routeway, or for zoning for land-use in urban development. Multipurpose maps cover various aspects of engineering geology and provide information for planning or engineering purposes.

In addition, engineering geological maps may be analytical or comprehensive. Analytical maps provide details, or evaluate individual components, of the geological environment. Examples of such maps include those showing degree of weathering or seismic hazard. Comprehensive maps either depict all the principal components of the engineering geological environment or are maps of engineering geological zoning, delineating individual territorial units on a basis of uniformity of the most significant attributes of their engineering geological character.

Engineering geological (i.e. geotechnical) maps and plans indicate the distribution of units, defined in terms of engineering properties rather than by age as do conventional geological maps. For instance, engineering geological maps can be produced in terms of index properties, rock quality or grade of weathering. A plan for a foundation could be made in terms of design parameters. The unit boundaries are then drawn for changes in the particular property. Frequently the boundaries of such units coincide with stratigraphical boundaries. In other instances, as for example, where rocks are deeply weathered, they may bear no relation to geological boundaries. Unfortunately one of the fundamental difficulties in preparing geo-technical maps arises from the fact that changes in physical properties of rocks and soils are frequently gradational. As a consequence regular checking of visual observations by *in situ* testing or sampling is essential to produce a map based on engineering properties (see Fookes (1969)).

Classification of rocks and soils on engineering geological maps should be based on the principle that the physical or engineering geological properties of the material in its present state are governed by the combined effects of mode of origin, subsequent diagenetic, metamorphic or tectonic history, and by weathering processes (see Anon (1976b); (1979)). This principle of classification makes it possible to determine the reasons for the lithological and physical characteristics of rocks and soils as well as their spatial distribution.

The selection of an appropriate method of plotting the boundaries of mapping units in the field depends on the purpose for which the mapping is being carried out. This in turn governs the scale and the latter dictates the basic taxonomic or mapping unit. Anon (1976b) suggested the following classification of rock and soil units, based on origin and lithology, for use on engineering geological maps.

(1) The engineering geological type (ET) has the highest degree of homo-geneity and should be uniform in lithological character and physical state. These units can be characterised by statistically deter-mined values derived from determinations of mechanical properties. They are generally portrayed only on large-scale maps, that is, with scales larger than 1:5000.

(2) A lithological type (LT) is homogeneous throughout in com-position, texture and structure but generally is not uniform in physical state. Reliable values of average mechanical properties cannot

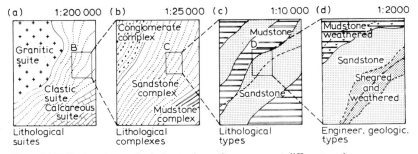

Figure 14.13 Engineering geological mapping units on maps at different scales (after Anon (1976b))

be given for the entire unit, normally only a general idea of engineering properties, with a range of values, can be presented. These units are used on large-scale, and where possible, medium-scale maps (from 1:5000–1:10 000).

(3) A lithological complex (LC) is formed of a set of genetically related lithological types developed under specific palaeogeographical and tectonic conditions. The spatial arrangement of lithological types in a complex is uniform and distinctive. However, a lithological complex is not necessarily uniform in either lithological character or physical state. As a result it is impossible to define the mechanical properties of an entire complex, only its general behaviour being indicated. Data on individual lithological types can be given. The lithological complex is used as a mapping unit on medium and some small-scale maps (from 1:10 000–1:200 000).

(4) The lithological suite (LS) consists of many lithological complexes which were developed under generally similar palaeogeographical and tectonic conditions. A general uniformity is imparted to the suite by the possession of certain common lithological characteristics in the various lithological complexes which comprise the suite. Only very general engineering geological properties can be defined and such units are used only on small-scale maps.

An illustration of the mapping units outlined above is given in Figure 14.13 which shows their use on progressively larger scales.

14.6.3 Production of engineering geological maps

The preparation of engineering geological maps of urban areas frequently involves systematic searches of archives (see Dearman *et al.* (1977)). Information from site investigation reports, records of past and present mining activity, successive editions of Ordnance Survey plans, etc. may prove extremely useful (see Chaplow (1975); Dumbleton and West (1971)). The information thereby obtained is plotted in plan and indexed on a documentation map. Preparation in this way of a series of engineering geological maps can reduce the amount of effort involved in the preliminary stages of a site investigation, and indeed may allow site investigations to be designed for the most economical confirmation of the ground conditions.

In a review of engineering geological mapping of urban areas in Belgium, De Beer *et al.* (1980) described the production of a geotechnical 'atlas' of each area which was mapped. This 'atlas' comprised a documentation map (this is a topographical map which shows the location of boreholes and the data derived therefrom); an individual map showing the isopachytes of each formation and another giving the isohypses of the upper surface of the formation in question; a hydrogeological map; a map of engineering geological zones; a number of engineering geological cross sections; and an explanatory key.

An example of a multivariate approach to rock characterisation for engineering geological mapping, carried out at the site investigation stage of an engineering project, has been provided by Cottiss *et al.* (1971). The project concerned was the scheme for a ring road system around Edinburgh.

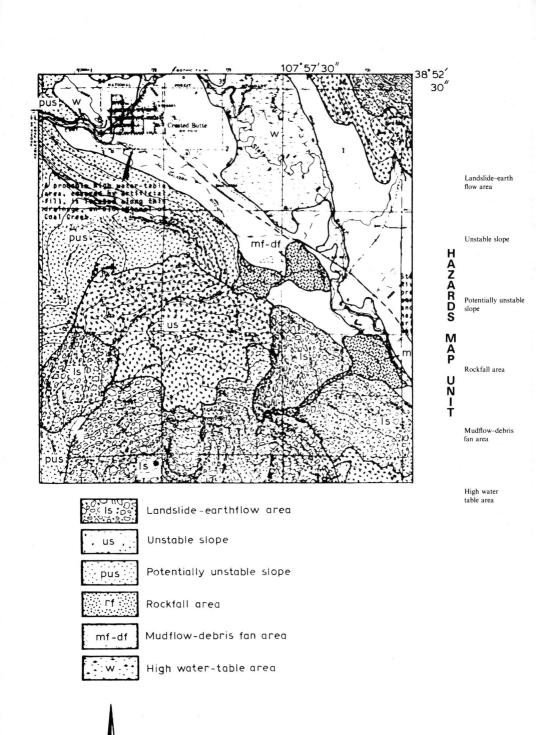

Landslide–earth
flow area

Unstable slope

Potentially unstable
slope

Rockfall area

Mudflow–debris
fan area

High water
table area

HAZARDS MAP UNIT

ls	Landslide-earthflow area
us	Unstable slope
pus	Potentially unstable slope
rf	Rockfall area
mf-df	Mudflow-debris fan area
w	High water-table area

- N -

0   0.5   1 km

(a)

Residential Development		Roads	Utilities	Open space recreational complexes including ski areas, but not associated structures	Commercial and industrial development, including larger residential buildings such as condominiums and apt buildings	Low-value lightweight agricultural buildings	Agriculture uses grazing and similar
High density	Low density						
3+ ABCDEFH	3 ABCDFH	3+ ABCDEFH	2 ABCEFH	1+ AB	3+ ABCFH	2+ AFH	1+ ABE
Remedial engineering typically is prohibitively expensive.	May be possible with careful engineering.	Typically not feasible without careful engineering.	Compatible with open-space land use.	Commonly feasible. Maintenance costs may be high.	Maintenance costs probably will be high.	Maintenance costs may be high.	Usually minor problems except for irrigation ditches and fences.
3 ACDEFH	2+ ACDEFH	3 ACDEFH	1+ ACFH	1 AF	3 ACDEFH	2 AFH	1 ADE
Remedial engineering usually necessary.	Remedial engineering may be necessary.	Remedial engineering usually necessary.	Remedial engineering may be necessary.	Commonly feasible.	Remedial engineering necessary.	Remedial engineering necessary.	Usually minor problems except where ditch leakage causes earthflows.
3 CDEF	2 CDEF	2+ ABCDEFH	1 CDE	0 F	2 ACFH	2 ACFH	0 E
Remedial engineering may be necessary.	Remedial engineering may be necessary.	May experience difficulties without careful planning/engineering.	Careful planning can minimise hazard.	Typically no difficulties.	Remedial engineering necessary.	Remedial engineering necessary	Usually minor problems except in areas of intense cultivation of hillslopes.
3+ ABCFH	3 ABCFH	3 ABCFH	2 ABCH	1+ B	2 ABCH	2+ ABCFH	1 B
Rarely compatible without elaborate and expensive mitigation.	Careful siting typically necessary to minimise hazard.	Remedial engineering can minimise hazard.	Careful planning can minimise hazard.	Careful siting can minimise hazard.	Maintenance costs probably will be high.	Careful siting can minimise hazard.	Usually few or minor problems.
3+ BCDFH	3 BCDFH	3 BDH	2 BFH	2 B	2 BCDFH	2 BDFH	0 B
Rarely compatible without elaborate and expensive mitigation.	Rarely compatible without elaborate and expensive mitigation.	Compatible only with elaborate and expensive mitigation.	Possibly excessive maintenance necessary.	Commonly feasible if rise is acceptable.	Occasional very high maintenance costs can be expected.	Occasional very high maintenance costs can be expected.	Usually few or minor problems.
3 GH	2+ GH	3 GH	0 G	1 G	2+ GH	2 GH	0 G
Basements and septic tank sewage disposal usually not feasible.	Basements and septic tank sewage disposal usually not feasible.	Usually difficult—depends on type of development. Flood plain determination may be necessary.	Some remedial engineering may be necessary in unusual cases.	Usually little difficulty. Possibility of flood damage.	May require special construction techniques remedial engineering.	May require special construction techniques remedial engineering.	Desirable for many kinds of agriculture.

EXPLANATION OF CHART SYMBOLS

3 High hazard
2 Moderate hazard
1 Low hazard
0 Very low, if any hazard

Typical potential hazard for indicated land use — Conditions affecting actual degree of hazard — Comments applicable to most cases

Example

(b)

MEANING OF LETTER SYMBOLS

A Especially severe on slopes greater than 30%.

B Slope movement intermittent dependent on variation in weather or other factors.

C Oversteepening or cutting of slopes can increase hazard greatly.

D Artificial or natural increase in ground moisture can increase hazard greatly.

E Removal of natural vegetation can increase hazard greatly.

F Hazard may decrease considerably as slope decreases.

G Varies seasonally.

H Detailed engineering geology studies necessary during pre-planning stages of development.

Figure 14.14 (a) An example of engineering geological mapping in the Crested Butte–Gunnison area (after Soule (1980)). (b) The matrix is formatted so as to indicate to the map user that several geological and geology-related factors should be considered when contemplating the indicated land use in a given type of mapped area. This matrix can also serve to recommend additional types of engineering geological studies that may be needed for a site. Thus the map can be used to model or anticipate the kinds of problems that a land-use planner or land developer may have to overcome before a particular activity is permitted or undertaken

The geology of the area is quite variable, including basalts and welded tuffs, and a variety of sandstones and siltstones. The superficial cover includes glacial and fluvio-glacial materials; recent river, lake and beach deposits and fill. Because of the variety of rock types and associated geotechnical problems it was decided that a broad coverage of the site, using classification testing and mapping rather than sophisticated testing at a few locations, would be more suitable. To simplify the data for engineering application a rock quality score was computed by combining the assessed results of tests by means of a standard formula. The site was divided into five zones on the basis of the test results. Each zone differed significantly from the next in mechanical characteristics.

The first map produced by the Institute of Geological Sciences which involved engineering geology was that of Belfast (see Wilson (1972)). Use was made of the plentiful information from boreholes which was available, particularly that referring to glacial and post-glacial deposits. This enabled isopachytes of the estuarine clays (the principal engineering problem) and contours indicating depth to rockhead, to be superimposed on the outlines of the solid geology. A large-scale cross-section is also provided. Detailed engineering geological information is given, in tabular form, on the reverse side of the map. For example, the table of rock and soil characteristics summarises the various rock and soil groups, listing their mode of occurrence, their thickness, their structure, and their hydrogeological and geotechnical properties (ranges are given for density, void ratio, permeability, undrained shear strength, compressibility and aggregate properties). Thirty selected borehole logs are also recorded on the reverse side and an explanation of the local rock terms is included.

Cratchley and Denness (1972) have described how an engineering geological map of Milton Keynes was produced for the Institute of Geological Sciences. The field staff first prepared a map in terms of stratigraphic units. As mentioned above, stratigraphic units often contain a range of lithologies. This occurred at Milton Keynes and unfortunately it proved impossible to represent every lithological unit within the broader stratigraphical horizons, since there were too many of them and they were often too thin or inconsistent to be mappable. Accordingly it was decided that, first, those lithological units which were measurably thick and spatially consistent, and possessed similar geotechnical characteristics, should be grouped into one engineering geological unit irrespective of their stratigraphical description. Secondly, irregularly layered and closely variable beds of very limited spatial consistency were also regarded as one engineering geological unit, the component members having a range of engineering properties within the behavioural characteristics of the mass in relation to a particular engineering application. Matters were further complicated by the fact that the findings of the field survey indicated that there may be more variation in the engineering properties of till within a few centimetres than occurs in the Oxford Clay of the Milton Keynes area over several kilometres.

Ultimately the information from the investigation was presented on base maps of engineering geology which were accompanied by transparent overlays, for example, showing the thickness of drift. A table incorporating the range of values of the geotechnical properties of the materials concerned and their interpreted behavioural characteristics, together with a detailed report, were also produced. The scale of the base maps was 1:25 000. A map detailing the superficial deposits of the Ouzel Valley was drawn on a scale of 1:10 000.

Soule (1980) outlined a method of mapping areas prone to geological hazards, by using map units based primarily on the nature of the potential hazards associated with them. The resultant map, together with its explanation, are combined with a land-use/geological hazard area matrix which provides some idea of the engineering problems which may arise in the area represented by the map. For instance, the matrix would indicate the effects of any changes in slope or the mechanical properties of rocks or soils, and attempts to evaluate the severity of hazard for various land-uses. As an illustration of this method, Soule used a landslide hazard map of the Crested Butte-Gunniston area, Colorado (Figure 14.14a). This map attempts to show which factors within individual map units have the most significance as far as potential hazard is concerned. The accompanying matrix outlines the problems likely to be encountered as a result of human activity (Figure 14.14b).

A review of the procedures used to map soil profiles during a site investigation programme for earthworks or foundations has been provided by Marsland *et al.* (1980). They noted the value of various types of samplers in sensitive soils, of *in situ* testing, and the use of large-diameter boreholes for inspection purposes. They also emphasised the importance of describing soil properties, including their fabric, adequately (see also McGown *et al.* (1980)). The objective of such mapping of soils is to produce a three-dimensional picture of the soil conditions at the site in question, to a depth where the behaviour of the soil has little influence on the proposed project.

References

ALLUM, J. A. E. (1966). *Photogeology and Regional Mapping*, Pergamon, Oxford.

ALLUM, J. A. E. (1970). 'Consideration of the relative values of true and infrared red colour photography for geological purposes', *Trans. Inst. Min. Metall.*, **79**, Section B, Applied Earth Science, B76–87.

ANON. (1972). The preparation of maps and plans in terms of engineering geology—Working Party Report', *Q. J. Engg Geol.*, **5**, 293–381.

ANON. (1976a). *Reclamatiom of Derelict Land: Procedure for Locating Abandoned Mine Shafts*, Dept. of Environment, London.

ANON. (1976b). *Engineering Geological Maps. A Guide to their Preparation*, Unesco Press, Paris.

ANON. (1978). 'Terrain evaluation for highway engineering and transport planning', *Transp. Road Res. Lab.*, Rep. SR448, DOE, Crowthorne.

ANON. (1979). 'Classification of rocks and soils for engineering geological mapping. Part I—Rock and soil materials', Report of the Commission of Engineering Geological Mapping. *Int. Ass. Engg Geol.*, No. 19, 364–71, Krefeld.

BEAUMONT, T. E. (1977). 'Techniques for the interpretation of remote sensing imagery for highway engineering purposes', *Transp. Road Res. Lab.*, Rep. LR753, DOE, Crowthorne.

BEAUMONT, T. E. (1978). Remote sensing for transport planning and highway engineering in developing countries', *Transp. Road Res. Lab.*, Sup. Rep. 433, DOE, Crowthorne.

BEAUMONT, T. E. (1979a). 'Remote sensing survey techniques', *J. Inst. Highway Engrs*, **26**, No. 4, 2–12.

BEAUMONT, T. E. (1979b). 'Remote sensing for location and mapping of engineering construction materials', *Q. J. Engg Geol.*, **12**, 147–58.

BEAUMONT, T. E. & BEAVEN, P. J. (1977). 'The use of satellite imagery for highway engineering in overseas countries', *Transp. Road Res. Lab.*, Rep. SR279, DOE, Crowthorne.

BRINK, A. B. A., MABBUTT, J. A., WEBSTER, R. & BECKETT, P. H. T. (1966). 'Report of the working group on land classification and data storage', *Military Engg Exp. Estab.*, Rep. No. 940, Christchurch.

BRUNSDEN, D., DOORNKAMP, J. C., FOOKES, P. G., JONES, D. K. C. & KELLY, J. H. M. (1975a). 'The use of geomorphological mapping techniques in highway engineering', *J. Inst. Highway Engrs*, **22**, 35–41.

BRUNSDEN, D., DOORNKAMP, J. C., FOOKES, P. G., JONES, D. K. C. & KELLY, J. H. M. (1975b). 'Large-scale geomorphological mapping and highway engineering design', *Q. J. Engg Geol.*, **8**, 227–53.

CHAPLOW, R. C. (1975). 'Engineering geology and site investigation', *Ground Engg*, **8**, 34–8.

CLARK, A. R. & JOHNSON, D. K. 'Geotechnical mapping as an integral part of site investigation—two case histories', *Q. J. Engg Geol.*, **8**, 211–24.

COLE, M. M. (1977). 'Landsat and airborne multispectral and thermal imagery used for geological mapping and identification of ore horizons in the Lady Annie-Lady Loretta and Dugald river areas, Queensland, Australia', *Trans. Inst. Min. Metall.*, **86**, Section B, Applied Earth Science, B195–215.

COTTISS, G. I., DOWELL, R. W. & FRANKLIN, J. A. (1971). 'A rock classification system applied to civil engineering', *Civ. Engg Publ. Works Rev.*, 611–14, 736–8. (1971).

CRATCHLEY, C. R. & DENNESS, B. (1972). 'Engineering geology in urban planning with an example from the new city of Milton Keynes', *Proc. Int. Geol. Congr.*, 24th Session, Montreal, Section 13, 13–22.

DEARMAN, W. R. & FOOKES, P. G. (1974). 'Engineering geological mapping for civil engineering practice in the United Kingdom,' *Q. J. Engg Geol.*, **7**, 223–56. (1974).

DEARMAN, W. R. & MATULA, M. (1977). 'Environmental aspects of engineering geological mapping', *Bull. Int. Ass. Engg Geol.*, No. 14, 141–6.

DEARMAN, W. R., MONEY, M. S., COFFEY, R. J., SCOTT, P. & WHEELER, M. (1977). 'Engineering geological mapping of the Tyne and Wear conurbation, north-east England', *Q. J. Engg Geol.*, **10**, 145–68.

DE BEER, E., FAGNOUL, A., LOUSBERG, E., NUYENS, J. & MAETENS, J. (1980). 'A review on engineering geological mapping in Belgium', *Bull. Int. Ass. Engg Geol.*, No. 21, 91–8.

DEMEK, J. (1972). *Manual of Detailed Geomorphological Mapping*, Academia, Prague.

DOORNKAMP, J. C., BRUNSDEN, D., JONES, D. K. C., COOKE, R. U. & BUSH, P. R. (1979). 'Rapid geomorphological assessments for engineering', *Q. J. Engg Geol.*, **12**, 189–204.

DOWLING, J. W. P. & BEAVEN, P. J. (1969). 'Terrain evaluation for road engineers in developing countries', *J. Inst. Highway Engrs*, **16**, 5–15.

DUMBLETON, M. J. & WEST, G. (1970). 'Air-photograph interpretation for road engineers in Britain', *Transp. Road Res. Lab.*, Rep. LR369, DOE, Crowthorne.

DUMBLETON, M. J. & WEST, G. (1971). 'Preliminary sources of information for site investigations in Britain', *Trans. Road Res. Lab.*, Rept. LR403, DOE, Crowthorne.

ECKEL, E. B. (1951). 'Interpreting geologic maps for engineers', *Symp. on Surface and Subsurface Reconnaissance, ASTM, Spec. Tech. Publ.*, No. 122, 5–15.

FOOKES, P. G. (1969). 'Geotechnical mapping of soils and sedimentary rock for engineering purposes with examples of practice from the Mangla Dam project', *Geotechnique*, **19**, 52–74.

HIMUS, G. W. & SWEETING, G. S. (1968). *Elements of Field Geology*, University Tutorial Press, London.

KERTESZ, A. (1979). 'Representing the morphology of slopes on engineering geomorphological maps with special reference to slope morphometry; *Q. J. Engg Geol.*, **12**, 235–41.

KNOTT, P. A., DOORNKAMP, J. C. & JONES, R. H. (1980). 'The relationship between soils and geomorphological mapping units—a case study from Northamptonshire', *Bull. Int. Ass. Engg Geol.*, No. 21, 186–93.

LAHEE, F. H. (1961). *Field Geology*, McGraw Hill, New York.

LAWRANCE, C. J. (1972). 'Terrain evaluation in West Malaysia, Part 7—Terrain Classification and survey methods', *Transp. Road Res. Lab.*, Rep. LR506, DOE, Crowthorne.

LAWRANCE, C. J. (1978). 'Terrain evaluation in West Malaysia, part 2—Land systems of south west Malaysia', *Transp. Road Res. Lab.*, Rep. SR378, DOE, Crowthorne.

LOW, J. W. (1957). *Geologic Field Methods*, Harper and Row, New York.

McGOWN, A., MARSLAND, A., RADMAN, A. M. & GABR, A. W. A. (1980). 'Recording and interpreting soil microfabric data', *Geotechnique*, **30**, 417–47.

MARSLAND, A., McGOWN, A. & DERBYSHIRE, E. (1980). 'Soil profile mapping in relation to site evaluation for foundations and earthworks', *Bull. Int. Ass. Engg Geol.*, No. 21, 139–55.

MATHUR, B. S. (1979). 'Remote sensing sensors for environmental studies', *Proc. ASCE, Trans. Engg J.*, **105**, TE2, Paper 14707, 439–55.

MILLER, A. A. (1953). *The Skin of the Earth*, Methuen, London.

MITCHELL, C. W. (1973). *Terrain Evaluation*, Longmans, London.

NORMAN, J. W. (1968a). 'The air photograph requirements of geologists', *Photogrammetric Record*, **6**, 133–49.

NORMAN, J. W. (1968b). 'Photogeology of linear features in areas covered by superficial deposits', *Trans. Inst. Min. Metall.*, **78**, Section B, Applied Earth Science, B60–77.

NORMAN, J. W. (1976). 'Photogeological fracture trace analysis as a subsurface exploration technique', *Trans. Inst. Min. Metall.*, **85**, Section B, Applied Earth Science, B52–62.

NORMAN, J. W. (1980). 'Causes of some crustal failure zones interpreted from Landsat images and their significance in regional mineral exploration', *Trans. Inst. Min. Metall.*, **89**, Section B, Applied Earth Science, B63–73.

NORMAN, J. W. & HUNTINGTON, J. F. (1974). 'Possible applications of photogeology to the study of rock mechanics', *Q. J. Engg Geol.*, **7**, 107–19.

NORMAN, J. W., LIEBOWITZ, T. H. & FOOKES, P. G. (1975). 'Factors affecting the detection of slope instability with aerial photographs in an area near Sevenoaks, Kent,' *Q. J. Engg Geol.*, **8**, 159–76.

PITTS, J. (1979). 'Morphological mapping in the Axmouth–Lyme Regis undercliffs, Devon', *Q. J. Engg Geol.*, **12**, 205–18.

REEVES, R. G. ANSEN, A. & LINDEN, D. (eds). (1975). *Manual of Remote Sensing*, American Society of Photogrammetry.

RANGERS, N. & SOETERS, R. (1980). 'Regional engineering geological mapping from aerial photographs', *Bull. Int. Ass. Engg Geol.*, No. 21, 103–11.

SABINS, F. F. (1978). *Remote Sensing—Principles and Interpretation*, W. Freeman and Co., San Francisco.

SAVIGEAR, R. A. G. (1965). 'A technique of morphological mapping', *Ann. Ass. Am. Geog.*, **55**, 514–38.

SOULE, J. M. (1980). 'Engineering geologic mapping and potential geologic hazards in Colorado', *Bull. Int. Ass. Engg Geol.*, No. 21, 121–31.

VARNES, D. J. (1974). 'The logic of geological maps with reference to their interpretation and use for engineering purposes', *US Geol. Surv.*, Prof. Paper 873.

WARWICK, D., HARTOPP, P. G. & VILJOEN, R. P. (1979). 'Application of thermal infrared linescanning technique to engineering geological mapping in South Africa,' *Q. J. Engg Geol.*, **12**, 159–80.

WATERS, R. S. (1958). 'Morphological mapping', *Geography*, **43**, 10–17.

WEBSTER, R. & BECKETT, P. H. T. (1970). 'Terrain classification and evaluation using air photography', *Photogrammetria*, **26**, 51–75.

WILSON, H. E. (1972). 'The geological map and the civil engineer', *Proc. Int. Geol. Congr.*, 24th Session, Montreal, Section 13, 83–6.

Appendix 1

Classification of silicate minerals

The silicate minerals constitute about a quarter of all known minerals and account for more than 90% of the Earth's crust. The fundamental structural unit of the silicate minerals is the SiO_4 tetrahedron. The manner in which such tetrahedra are linked together provides a basis for the classification of these minerals.

The simplest structural arrangement is that in which single SiO_4 tetrahedra are linked together by cations. This arrangement occurs in the orthosilicates. In the sorosilicates an oxygen ion is shared by two tetrahedra. These double tetrahedral units are joined by cations. If two of the oxygens of the SiO_4 groups are shared with adjacent tetrahedra, then they may be linked in such a way as to form ring structures. The ring silicates are generally referred to as cyclosilicates. When two oxygen ions of an SiO_4 group are shared with two other tetrahedra, one on either side, the structure may be extended 'indefinitely' to form a chain. The individual tetrahedra are so arranged within the chains that their vertices point alternatively in opposite directions. These chain silicates are termed inosilicates. Both single and double chain structures occur. In the latter type an oxygen ion is shared by the alternate tetrahedra of two parallel chains, so cross linking the chains. In the sheet silicates or phyllosilicates three of the oxygen ions of the SiO_4 tetrahedra are shared with those of neighbouring groups to form 'indefinitely' extended flat sheets. The final class of silicate structures is that which is developed when the SiO_4 groups are so arranged that the ratio of silicon to oxygen is $1:2$. In other words, in these framework or tectosilicates every ion of oxygen in a tetrahedron is shared with adjacent tetrahedra.

A note on the physical properties of minerals

(1) Crystal system—six crystal systems are distinguished by the relative lengths and angular relationships of their crystallographic axes. These are the cubic system ($a_1 = a_2 = a_3$; $\alpha = \beta = \gamma = 90°$); the hexagonal system ($a_1 = a_2 = a_3 \neq c$; $\beta_1 = \beta_2 = \beta_3 = 90°$, $\gamma = 120°$); the tetragonal

system $(a_1 = a_2 \neq c; \quad \alpha = \beta = \gamma = 90°)$; the orthorhombic system $(a \neq b \neq c; \quad \alpha = \beta = \gamma = 90°)$; the monoclinic system $(a \neq b \neq c; \quad \alpha = \gamma = 90° \neq \beta)$; and the triclinic system $(a \neq b \neq c; \alpha \neq \beta \neq \gamma)$.

(2) Cleavage is the tendency of a mineral to split in certain definite directions thereby producing a more or less smooth surface. It indicates a direction of minimum cohesion, reflecting weakness in the atomic structure. Cleavage planes have a definite relationship to the crystal faces.

(3) Colour is the most obvious property of a mineral, however, as a diagnostic feature it is of little importance since the same mineral can occur in different colours in different specimens.

(4) The streak of a mineral is the colour of its powder, which may be obtained by crushing, filing or scratching it. The most satisfactory method of obtaining a powder is to rub the mineral on a streak plate. The latter is made of unglazed porcelain. The streak plate cannot be used on minerals with a hardness of 7 or above. Streak is a more constant property than colour.

(5) The lustre of a mineral can be defined as the appearance of its surface in reflected light. Two principal types of lustre are distinguished, namely, metallic and non-metallic. A metallic lustre is exhibited by metals and minerals with a metallic appearance. There are several types of non-metallic lustre, for example, vitreous (of broken glass), resinous (of resin) pearly (of pearl) and silky (of silk).

(6) The fracture of a mineral refers to the surface along which it breaks, independent of the cleavage parting. The following types of fracture are recognised: even (surface more or less flat), uneven (rough and irregular surface), conchoidal (concave curved surface displaying percussion rings), hackly (sharp serrated surface) and earthy (similar to the surface developed on broken chalk).

(7) The hardness of a mineral is usually defined as its resistance to scratching. The Mohr scale of hardness is based upon 10 minerals and is referred to in Chapter 11.

(8) The relative density (specific gravity) indicates the number of times a mineral is heavier than an equal volume of water, that is, it is the ratio of the density of a substance to the density of water.

Appendix 2

Silicate minerals

	Crystal system	Cleavage	Colour	Streak	Lustre	Fracture	Hardness	Relative density
(1) ORTHOSILICATES								
OLIVINES Fosterite Mg_2SiO_4 Fayalite Fe_2SiO_4	Orthorhombic	Poor	Olive green	White	Vitreous	Conchoidal	6.5–7.0	3.2–3.4
GARNETS								
Pyrope $Mg_3Al_2Si_3O_{12}$	Cubic	None	Red to black	White	Vitreous	Conchoidal	6.5–7.5	3.7–3.8
Almandine $Fe_3Al_2Si_3O_{12}$	Cubic	None	Brown, red, black	White	Vitreous	Subconchoidal	6.5–7.5	4.32
Grossularite $Ca_3Al_2Si_3O_{12}$	Cubic	None	Yellow, red, brown	White	Vitreous	Subconchoidal	6.5–7	3.4–3.6
Andradite $Ca_3Fe_2Si_3O_{12}$	Cubic	None	Red, black	White	Vitreous	Subconchoidal	7	3.75–4.1
IDOCRASE $Ca_{10}Al_4(MgFe)_2(Si_2O_7)_2(SiO_4)_5(OH)_4$	Tetragonal	Indistinct	Brown	White	Vitreous	Subconchoidal	6.5	3.35–3.45
SILLIMANITE Al_2SiO_5	Orthorhombic	Perfect	Greyish yellowish brown	White	Vitreous	Uneven	6–7	3.23
ANDALUSITE Al_2SiO_5	Orthorhombic	Poor	Grey, green, yellow	White	Vitreous	Uneven	7.5	3.1–3.2
KYANITE Al_2SiO_5	Triclinic	Good	Light blue, white	White	Vitreous to pearly	Fibrous	4.5–7	3.5–3.7
STAUROLITE $(FeMg)_2(AlFe)_9O_6(SiO_4)_4(OOH)_2$	Orthorhombic	Indistinct	Reddish brown	White, grey	Subvitreous to resinous	Conchoidal	7.5	3.7–3.8
(2) SOROSILICATES								
EPIDOTE $CA_2Fe^{III}Al_2O_3.OH.Si_2O_7.SiO_4$	Monoclinic	Perfect	Dark green	White	Vitreous	Uneven	6–7	3.2–3.4
ZOISITE $Ca_2Al.Al_2O_2.OH.Si_2O_7.SiO_4$	Orthorhombic	Perfect	White, grey	White	Vitreous	Uneven	6–6.5	3.1
LAWSONITE $CaAl_2(OH)_2(Si_2O_7)H_2O$	Orthorhombic	Perfect	Blue	White	Vitreous	Conchoidal	8	3.2–3.3
PUMPELLYITE $Ca_4(MgFe^{III}Mn)(AlFe^{III}Ti)_5O(OH)_3 (Si_2O_7)_2(SiO_4)_2.2H_2O$	Monoclinic	Distinct	Bluish green, brown	White	Vitreous	Conchoidal	5.5	3.2–3.3

Silicate minerals—cont.

	Crystal system	Cleavage	Colour	Streak	Lustre	Fracture	Hardness	Relative density
(3) CYCLOSILICATES								
CORDIERITE $(MgFe)_2Al_3(Si_5AlO_{18})$	Orthorhombic	Imperfect	Blue	White	Vitreous	Conchoidal	7–7.5	2.53–2.78
(4) INOSILICATES								
PYROXENES								
Hypersthene $(MgFe)Si_2O_6$	Orthorhombic	Good	Brownish green, black	Greyish green	Submetallic	Uneven	5.6	3.4–3.5
Diopside $(CaMg)Si_2O_6$	Monoclinic	Good	White, green	White	Vitreous	Uneven	5.6	3.2–3.38
Augite $(CaMgFeAl)_2(SiAl)_2O_6$	Monoclinic	Good	Brown, black	Greyish green	Vitreous to resinous	Conchoidal or uneven	5–6	3.2–3.5
WOLLASTONITE $CaSiO_3$	Triclinic	Perfect	White, grey	White	Vitreous	Hackly	4.5–5	2.8–2.9
AMPHIBOLES								
Anthophyllite $(MgFe)_7Si_8O_{22}(OHF)_2$	Orthorhombic	Perfect	White, green brown	White	Vitreous	Uneven	6	2.9–3.4
Cummingtonite $(MgFe)_7Si_8O_{22}(OH)_2$	Monoclinic	Perfect	Brown green	Greyish white	Vitreous	Uneven	5–6	3.1–3.4
Tremolite $Ca_2Mg_5Si_8O_{22}(OHF)_2$	Monoclinic	Perfect	Colourless to grey	White	Vitreous	Uneven	5–6	2.9–3.1
Actinolite $Ca_2(MgFe)_5Si_8O_{22}(OHF)_2$	Monoclinic	Perfect	Green	White	Vitreous	Uneven	5–6	3.0–3.2
Hornblende $(CaNaK)_{2-3}(MgFe^{II}Fe^{III}Al)_5$ $(SiAl)_2Si_6O_{22}(OHF)_2$	Monoclinic	Good	Greenish black	Brownish green	Vitreous	Uneven	5–6	3.0–3.45
Glaucophane $Na_2Mg_3Al_2Si_8O_{22}(OHF)_2$	Monoclinic	Perfect	Colourless to blue	Bluish	Vitreous	Uneven	6	3.1–3.3
(5) PHYLLOSILICATES								
MICAS								
Muscovite $K_2Al_4(Al_2Si_6)O_{20}(OH)_4$	Monoclinic	Perfect	Colourless white	White	Pearly	Flexible	2.5–3	2.77–2.88
Biotite $K_2(MgFe)_{4-0}(Fe^{III}AlTi)_{0-2}$ $(Al_{2-3}Si_{5-6})O_{20}(OHF)_{2-4}$	Monoclinic	Perfect	Green, black	White	Pearly	Flexible	2.5–3	2.7–3.1
CHLORITE $(MgFeAl)_6(SiAl)_4O_{10}(OH)_8$	Monoclinic	Perfect	Green, brown	Greenish white	Pearly	Hackly	1.5–2.5	2.6–2.94
TALC $Mg_3Si_4O_{10}(OH)_2$	Monoclinic	Perfect	White, green	White	Pearly	Uneven	1	2.7–2.8
SERPENTINE $Mg_5Si_2O_5(OH)_4$	Monoclinic	Indistinct	Green, black, brown	White	Resinous	Conchoidal	3–4	2.5–2.6
PREHNITE $Ca_2Al_2Si_3O_{10}(OH)_2 \cdot nH_2O$	Orthorhombic	Poor	Colourless pale green	Colourless white	Vitreous	Conchoidal	8	3.1

Silicate minerals—cont.

	Crystal system	Cleavage	Colour	Streak	Lustre	Fracture	Hardness	SG
(6) TECTOSILICATES								
QUARTZ SiO_2	Hexagonal	None	Colourless, white	White	Vitreous	Conchoidal	7	2.65
FELDSPARS								
Sanidine $KAlSi_3O_8$	Monoclinic	Perfect	Colourless, white	White	Vitreous	Conchoidal	6	2.57
Orthoclase $KAlSi_3O_8$	Monoclinic	Perfect	White, pinkish	White	Vitreous	Conchoidal	6	2.57
Microcline $KAlSi_3O_8$	Triclinic	Perfect	White	White	Vitreous	Conchoidal	6–6.5	2.56
Anorthoclase	Triclinic	Perfect	Colourless, white	White	Vitreous	Conchoidal	6	2.56
Albite $NaAlSi_3O_8$	Triclinic	Perfect	White	White	Vitreous	Uneven	6–6.5	2.6–2.62
Oligoclase	Triclinic	Perfect	White	White	Vitreous	Uneven	6–6.5	2.64
Andesine	Triclinic	Perfect	White	White	Subvitreous	Uneven	6	2.67
Labradorite	Triclinic	Perfect	Grey bluish greenish	White	Vitreous	Uneven	6	2.67–2.7
Bytownite	Triclinic	Perfect	Grey bluish	Grey	Subvitreous	Uneven	6	2.72–2.73
Anorthite $CaAl_2Si_2O_8$	Triclinic	Perfect	Colourless, white	White	Vitreous	Conchoidal	6–6.5	2.74–2.76
FELDSPATHOIDS								
Nepheline $Na_3KAl_4Si_4O_{16}$	Hexagonal	Distinct	Colourless, white	White	Vitreous	Subconchoidal	5.5–6	2.5–2.6
Leucite $KAlSi_2O_6$	Tetragonal	Poor	White, grey	Colourless	Vitreous	Conchoidal	5.5–6	2.5
ZEOLITE								
$CaNa_2K_2(Al_2Si_7O_{18}).7H_2O$	Orthorhombic /Monoclinic	Perfect	White	White	Vitreous	Conchoidal	3.5–5.5	2.1–2.4

Non-silicate minerals

	Crystal system	Cleavage	Colour	Streak	Lustre	Fracture	Hardness	Relative density
(1) OXIDES								
SPINEL $MgAl_2O_4$	Cubic	Poor	Red brown black	White	Vitreous	Conchoidal	9	3.5–4.1
CORUNDUM Al_2O_3	Hexagonal (trigonal)	None	Grey, greenish, reddish	White	Vitreous	Conchoidal	9	3.9–4.1
HEMATITE Fe_2O_3	Hexagonal (trigonal)	None	Black, red	Red	Metallic	Uneven	5.5–6.5	4.9–5.3
MAGNETITE Fe_3O_4	Cubic	Poor	Black	Black	Metallic	Subconchoidal	5.5–6.5	5.18
(2) HYDROXIDES								
LIMONITE $Fe.O.OH.nH_2O$	Limonite is not a true mineral. It is composed of a cryptocrystalline mixture of goethite (FeO.OH) and lepidocrosite (FeO.OH), along with some water. Some hematite may also be present.							
GIBBSITE $Al(OH)_3$	Monoclinic	Perfect	White grey	White	Pearly	Hackly	2.5–3	2.4
BRUCITE $Mg(OH)_2$	Hexagonal	Perfect	White, greyish, bluish	White	Pearly	Hackly	2.5	2.39
(3) CARBONATES								
CALCITE $CaCO_3$	Hexagonal (trigonal)	Perfect	Colourless, white	White	Vitreous	Conchoidal	3	2.71
DOLOMITE $CaMg(CO_3)_2$	Hexagonal (trigonal)	Perfect	Colourless, white	Colourless	Vitreous	Conchoidal	3.5–4	2.8–2.9
SIDERITE $FeCO_3$	Hexagonal (trigonal)	Perfect	Yellowish brown	White	Vitreous	Uneven	3.5–4.5	3.7–3.9
(4) SULPHATES								
GYPSUM $CaSO_4.nH_2O$	Monoclinic	Perfect	Colourless, white, grey, pinkish	White	Pearly	Conchoidal	2	2.3–2.37
(5) SULPHIDES								
PYRITE FeS_2	Cubic	Poor	Brassy yellow	Brownish black	Metallic	Uneven	6–6.5	4.9–5.1

Index